# Studies i

Mathematical Logic

Volume 58

# Handbook of Mathematical Fuzzy Logic

Volume 3

Volume 47
Logic Across the University: Foundations and Applications. Proceedings of the Tsinghua Logic Conference, Beijing, 2013
Johan van Benthem and Fenrong Liu, eds.

Volume 48
Trends in Belief Revision and Argumentation Dynamics
Eduardo L. Fermé, Dov M. Gabbay, and Guillermo R. Simari

Volume 49
Introduction to Propositional Satisfiability
Victor Marek

Volume 50
Intuitionistic Set Theory
John L. Bell

Volume 51
Metalogical Contributions to the Nonmonotonic Theory of Abstract Argumentation
Ringo Baumann

Volume 52
Inconsistency Robustness
Carl Hewitt and John Woods, eds.

Volume 53
Aristotle's Earlier Logic
John Woods

Volume 54
Proof Theory of N4-related Paraconsistent Logics
Norihiro Kamide and Heinrich Wansing

Volume 55
All about Proofs, Proofs for All
Bruno Woltzenlogel Paleo and David Delahaye, eds

Volume 56
Dualities for Structures of Applied Logics
Ewa Orłowska, Anna Maria Radzikowska and Ingrid Rewitzky

Volume 57
Proof-theoretic Semantics
Nissim Francez

Volume 58
Handbook of Mathematical Fuzzy Logic. Volume 3
Petr Cintula, Christian G. Fermüller and Carles Noguera, eds.

Studies in Logic Series Editor
Dov Gabbay dov.gabbay@kcl.ac.uk

# Handbook of Mathematical Fuzzy Logic

Volume 3

Edited by

Petr Cintula,

Christian G. Fermüller

and

Carles Noguera

ISBN 978-1-84890-193-3

College Publications
Scientific Director: Dov Gabbay
Managing Director: Jane Spurr
Department of Informatics
King's College London, Strand, London WC2R 2LS, UK

http://www.collegepublications.co.uk

Original cover design by Orchid Creative www.orchidcreative.co.uk
This cover produced by Laraine Welch
Printed by Lightning Source, Milton Keynes, UK

---

# Preface to the third volume

Mathematical fuzzy logic (MFL) is a subdiscipline of mathematical logic. Comprising the mathematical study of a certain family of formal logical systems whose algebraic semantics involve some notion of *truth degree*. The main motivations for MFL, its historical origins, and its later evolution were briefly explained in the preface of the first volume of this handbook series.

Fuzzy logics can be seen as a particular kind of many-valued logical system where the intended semantics is typically based on algebras of linearly ordered truth values. As a subdiscipline of mathematical logic, MFL has acquired the typical core agenda of this field and is studied by many mathematically-minded researchers regardless of its original motivations, as witnessed by the plethora of works on a diversity of topics that have been accumulating over the years.

This handbook series aims to be an up-to-date, systematic presentation of the best developed areas of MFL. The first two volumes were released in 2011, also in *Studies in Logic, Mathematical Logic and Foundations* (vols. 37 and 38), College Publications. The first volume starts with a gentle introduction to MFL assuming only some basic knowledge of classical logic. The second chapter presents and develops a general and uniform framework for MFL based on the notions and methods of abstract algebraic logic. The third chapter is a presentation of the deeply developed proof theory of fuzzy logics. The fourth chapter presents the standard algebraic semantics for fuzzy logics based on classes of semilinear residuated lattices. The fifth chapter closes the first volume with the study of Hájek's BL logic and its algebraic counterpart.

The second volume of the series starts with the sixth chapter and is devoted to another widely studied fuzzy logic, Łukasiewicz logic Ł, and MV-algebras as its algebraic semantics. The seventh chapter deals with a third distinguished fuzzy logic, Gödel–Dummett logic, and its variants. The eighth chapter studies fuzzy logics in expanded languages providing greater expressive power. The ninth chapter collects results on functional representations of fuzzy logics and their free algebras. The last two chapters are dedicated to complexity issues: Chapter X studies the computational complexity of propositional fuzzy logics, while Chapter XI is devoted to the arithmetical hierarchy of first-order predicate fuzzy logics.

Intense research in MFL has continued in recent years, bringing to a higher level of maturity some of the topics that were omitted from the previous volumes of the handbook. This third volume collects chapters on seven such areas. It starts with Chapter XII, a systematic study of the most prominent part of the algebraic semantics for fuzzy logics: integral residuated chains. The thirteenth chapter presents a different type of semantics for some prominent fuzzy logics obtained by extending Hintikka's and Giles' semantic games. In a similar fashion, the fourteenth chapter is devoted to another game-theoretic interpretation of t-norm-based fuzzy logics, namely, that given by Ulam–Rényi games arising from the theory of error correcting codes. The fifteenth chapter surveys a series of works on fuzzy logics with evaluated syntax in which intermediate truth-values are incorporated as syntactical devices that accompany each formula as a lower bound for its truth-degree. Chapter XVI surveys another area that has recently been intensively developed: fuzzy description logics, understood as tractable fragments of first-order fuzzy logics amenable to knowledge representation. The seventeenth chapter focuses on another application of MFL by presenting the theory of MV-algebras endowed with functions, called *states*, that allow one to model finitely additive probability measures. The volume is concluded by Chapter XVIII offering a philosophical discussion of the role of MFL in the study of vagueness.

It should be emphasized that this is not a handbook written by a single team of authors, but a collection of chapters prepared by distinguished experts in each area. Nevertheless, the editors have encouraged a reasonable level of homogeneity between the chapters, as regards their structure and notation. The series is conceived as a unit with consecutive page and chapters enumeration. The majority of chapters in this volume contain a purely theoretical (mathematical) presentation, making it mostly a book on mathematical logic, focusing on the study of a particular family of many-valued non-classical logics. However, as briefly outlined before, some chapters feature connections to other mathematical fields, to applications in computer science and, in the case of the last chapter, to developments in analytical philosophy. The intended audience of the book is quite wide, comprising at least the following groups of readers: (1) students of logic looking for a systematic presentation of MFL where they can study the discipline from scratch, (2) experts on MFL that may use it as a reference book for consultation, (3) readers interested in fuzzy set theory and its applications looking for the logical foundations of (some parts) of the area, and (4) readers interested in philosophical and linguistic issues related to reasoning in presence of vagueness, looking for a mathematical apparatus that can be applied to some aspects of those issues.

**Acknowledgements** The editors wish to express their gratitude to

- all authors for their great effort in writing the chapters,
- the readers assigned to chapters for their invaluable help in improving them (Jesse Alama, Brunella Gerla, Giangiacomo Gerla, Siegfried Gottwald, Gejza Jenca, Ondrej Majer, Vincenzo Marra, George Metcalfe, Franco Montagna, Daniele Mundici, Ulrike Sattler, Stewart Shapiro, and Esko Turunen),
- Petra Ivaničová, Christoph Roschger, and Jane Spurr for technical and administrative support, and
- the joint project of Austrian Science Fund (FWF) I1897-N25 and Czech Science Foundation (GAČR) 15-34650L for supporting the preparation of the book.

**Dedication** This volume is dedicated to two great researchers who passed away during the preparation of the book: Siegfried Gottwald and Franco Montagna. Siegfried was one of the pioneers in the area of MFL, which he deeply influenced with his monograph *A Treatise on Many-Valued Logics*, and remained a constant contributor to the field and a close collaborator of this handbook series. Franco joined the MFL community later, but he soon became one of the field's major figures, as witnessed by the fact that he is the only author who contributed to each of the three volumes of this series.

Petr Cintula, Christian G. Fermüller, and Carles Noguera
*Editors*

# *Contents of Volume 1*



# *Contents of Volume 2*

# *Volume 3*

## LIST OF AUTHORS

FERNANDO BOBILLO
Department of Computer Science and Systems Engineering
University of Zaragoza
C. Maria de Luna 1
50018 Zaragoza, Spain
Email: fbobillo@unizar.es

MARCO CERAMI
Department of Computer Science
Palacký University Olomouc
17. listopadu 2
771 46 Olomouc, Czech Republic
Email: marco.cerami@upol.cz

FERDINANDO CICALESE
Department of Computer Science
University of Verona
Strada Le Grazie 15
37134 Verona, Italy
Email: ferdinando.cicalese@univr.it

FRANCESC ESTEVA
Artificial Intelligence Research Institute (IIIA)
Spanish National Research Council (CSIC)
Campus de la Universitat Autònoma de Barcelona s/n
08193 Bellaterra, Catalonia, Spain
Email: esteva@iiia.csic.es

CHRISTIAN G. FERMÜLLER
Theory and Logic Group 185.2
Vienna University of Technology
Favoritenstraße 9-11
1040 Vienna, Austria
Email: chrisf@logic.at

TOMMASO FLAMINIO
Dipartimento di Scienze Teoriche e Applicate
Universitá dell'Insubria
Via Mazzini 5
21100 Varese, Italy
Email: tommaso.flaminio@uninsubria.it

ÀNGEL GARCÍA-CERDAÑA
Artificial Intelligence Research Institute (IIIA)
Spanish National Research Council (CSIC)
Campus de la Universitat Autònoma de Barcelona s/n
08193 Bellaterra, Catalonia, Spain
and
Information and Communication Technologies Department
Universitat Pompeu Fabra
Tànger 122-140
08018 Barcelona, Catalonia, Spain
Email: angel@iiia.csic.es

TOMÁŠ KROUPA
Institute of Information Theory and Automation
Czech Academy of Sciences
Pod Vodárenskou věží 4
182 08 Prague, Czech Republic
Email: kroupa@utia.cas.cz

FRANCO MONTAGNA
Department of Information Engineering and Mathematics
University of Siena
Via Roma 56
53100 Siena, Italy

VILÉM NOVÁK
Institute for Research and Applications of Fuzzy Modeling
University of Ostrava & NSC IT4Innovations
30. dubna 22
701 03 Ostrava 1, Czech Republic
Email: Vilem.Novak@osu.cz

RAFAEL PEÑALOZA
KRDB Research Centre
Free University of Bozen-Bolzano
Piazza Domenicani 3
I-39100 Bozen-Bolzano BZ, Italy
Email: rafael.penaloza@unibz.it

NICHOLAS J.J. SMITH
Department of Philosophy
Main Quadrangle A14
University of Sydney
NSW 2006, Australia
Email: njjsmith@sydney.edu.au

UMBERTO STRACCIA
Istituto di Scienze e Tecnologie dell'Informazione (ISTI)
Consiglio Nazionalle delle Ricerche (CNR)
Via G. Moruzzi 1
56124 Pisa, Italy
Email: straccia@isti.cnr.it

THOMAS VETTERLEIN
Department of Knowledge-Based Mathematical Systems
Johannes Kepler University Linz
Altenberger Straße 69
4040 Linz, Austria
Email: Thomas.Vetterlein@jku.at

# CONTENTS

## XVIII Fuzzy Logics in Theories of Vagueness
Nicholas J.J. Smith 1237

# Chapter XII: Algebraic Semantics: The Structure of Residuated Chains

THOMAS VETTERLEIN

## 1 Introduction

Chapter IV of this Handbook is devoted to the algebraic semantics of substructural logics. Its central topic are FL-algebras, also called pointed residuated lattices, which represent an algebraic counterpart of FL, the Full Lambek logic.

In the present chapter, we resume this topic, adopting however a somewhat narrower point of view. We are interested in the generalised semantics of fuzzy logics and we focus to this end on semilinear, integral residuated lattices. Semilinearity means that the subdirect irreducible algebras are totally ordered. In fact we will restrict the discussion to residuated chains, in accordance with the fact that the most basic property characterising standard semantics in fuzzy logic is linearity. Integrality means that the monoidal identity is the top element. We generally require also this condition to hold, in accordance with the fact that in fuzzy logic it is usually assumed that a fully true proposition behaves in conjunctions neutrally.

We are thus interested in a classification of integral residuated chains. We seek ways to reduce these algebras to simpler ones, or ways to construct these algebras from structures that are better understood. Our point of view is in this sense "constructive" and we will not much develop universal-algebraic aspects. However, our analysis is based on established algebraic procedures; constructions methods that are incompatible with our algebraic framework will not be taken into account.

The problem that we address is, in the general case, far from a solution. No single approach is known at present that is expected to have the potential to cover eventually all structures in which we are interested. Here, we select two approaches, which are totally different in nature, and demonstrate their capabilities and limits. We shall see that each approach leads to an insight into, certainly not all but, quite a range of residuated structures.

The first approach takes up the apparently central role played by lattice-ordered groups within the variety of residuated lattices. Indeed, residuated lattices can be built from lattice-ordered groups in many ways, the best-known examples being MV-algebras or the more comprehensive BL-algebras.

A natural candidate for representations by means of groups are cancellative residuated lattices. For this topic we refer to [23]. Here, we will exploit the property of divisibility, a condition fulfilled, e.g. by BL-algebras. We will, however, not assume commutativity. As the main result, we will prove a representation theorem for divisible, integral residuated chains, also known as totally ordered pseudohoops [9]. This result generalises the well-known structure theorem for BL-algebras [1]. What makes our exposition special is the employed method, which differs, e.g. from the one employed in [9]. The idea is to represent residuated structures by means of partial algebras, as has been proposed in the context of quantum structures [10].

Our second approach deals with the commutative case, but this time we will not assume a property that is as particular as divisibility. Our focus will thus be on integral, commutative residuated chains. The example that we have in mind is the standard semantics of the fuzzy logic MTL. Recall that MTL is the logic of left-continuous t-norms together with their residua; a standard MTL-algebra is the real unit interval endowed with these two operations. In accordance with this example, we will focus on almost complete chains, where almost completeness is, roughly speaking, the same as completeness except for the possible absence of a bottom element.

The residuated chains considered in this second part possess a totally ordered set of quotients induced by filters. Our main concern is the question how to construct coextensions, that is, structures whose quotient is a given one. Under the condition that the congruence classes are isomorphic to real intervals, we can describe all those coextensions that are, in a natural sense, indivisible. In particular, all standard MTL-algebras that possess a quotient induced by an Archimedean filter are fully specifiable in terms of this quotient and the order type of the congruence classes.

We note that for the specification as well as for the visual representation of integral, commutative residuated chains, their quotients, and their coextensions we have a simple, yet efficient tool to our disposal: the regular representation of monoids [8], adapted to our context in the straightforward way. On this basis, we can, for instance, specify a coextension in a "modular" way. We will in fact do so when formulating our main result on real Archimedean coextensions.

We will proceed as follows. The following Section 2 puts up our favourite algebraic framework, providing the basic definitions around totally ordered monoids. The subsequent Section 3 is devoted to the partial-algebra method for the representation of residuated chains and Section 4 discusses quotients and coextension of residuated chains. We conclude with some more background information as well as hints for further reading in Section 5.

## 2 Residuated totally ordered monoids

We begin by recalling the algebraic notions relevant for this chapter; cf. Chapter IV of this Handbook. A *residuated lattice* is an algebra $\langle L, \wedge, \vee, \odot, /, \backslash, 1\rangle$ such that (i) $\langle L, \wedge, \vee\rangle$ is a lattice, (ii) $\langle L, \odot, 1\rangle$ is a monoid, and (iii) $/, \backslash$ are the left and right residuals of $\odot$, respectively. The latter condition means that, for $a, b, c \in L$ we have

$$a \odot b \leq c \quad \text{if and only if} \quad b \leq a\backslash c \quad \text{if and only if} \quad a \leq c/b.$$

In other words, for any $a \in L$, the right translation $\cdot \odot a$ and the mapping $\cdot/a$ form a Galois connection; similarly for the left translation $a \odot \cdot$ and the mapping $a\backslash\cdot$.

Moreover, we call a residuated lattice *commutative* if so is the monoidal operation $\odot$, and we call it *integral* if the monoidal identity 1 is the top element of the lattice. Finally, a *chain* is meant to be a totally ordered set, and we refer to a residuated lattice whose underlying order is a chain as a *residuated chain*.

In fuzzy logic, the implications are usually considered as the most basic connectives. Accordingly, we could consider the residuals $/$ and $\backslash$ as the primary operations of a residuated lattice. We will, however, not do so. The residuals on the one hand and the monoidal operation on the other hand are, given the lattice order, mutually uniquely determined. To explore the structure of residuated lattices it is thus not necessary to deal with all three operations. The monoidal operation may be viewed as a product, the residuals may be viewed as divisions; we will describe the structure of residuated lattices on the basis of the former, the product-like operation alone.

Moreover, we focus in this chapter exclusively on residuated chains. Accordingly, we will include in our signature the total order relation instead of the infimum and supremum operations.

DEFINITION 2.0.1. *A structure* $\boldsymbol{L} = \langle L, \leq, \odot, 1\rangle$ *is a* totally ordered monoid, *or* tomonoid *for short, if* (i) $\langle L, \odot, 1\rangle$ *is a monoid,* (ii) $\langle L, \leq\rangle$ *is a chain, and* (iii) $\leq$ *is compatible with* $\odot$*, that is,* $a \leq b$ *implies* $a \odot c \leq b \odot c$ *and* $c \odot a \leq c \odot b$.

*A tomonoid* $\boldsymbol{L}$ *is called* residuated *if, for any* $a, b \in L$*, there is a largest element* $c$ *such that* $a \odot c \leq b$ *and there is a largest element* $d$ *such that* $d \odot a \leq b$*. Moreover,* $\boldsymbol{L}$ *is called* commutative *if so is* $\odot$*, and* $\boldsymbol{L}$ *is called* negative *if* $1$ *is the top element.*

It is clear that residuated tomonoids are in a one-to-one correspondence with residuated chains. Under this correspondence, the properties of commutativity coincide, and so do the properties of integrality and negativity, respectively.

In semigroup theory, there are two competing ways of denoting the associative operation. Instead of the multiplicative notation $\langle L, \odot, 1\rangle$ we may equally well use the additive one $\langle L, \oplus, 0\rangle$. In logics, the former notation is clearly preferred because $\odot$ typically models the conjunction, 1 stands for the full truth, and the set of propositions is traditionally ordered such that stronger statements are modelled by smaller elements. However, the additive notation is often more practical, and authors indeed regularly switch to the additive notation when it comes to free commutative monoids. In case of the two approaches that we are going to present, we will stick to the notation that is in each case more practical and supports easiest understanding: the first approach will be presented additively, the second one multiplicatively.

We will certainly not use extra expressions like "dual tomonoids" or similarly; we will simply talk about tomonoids $\langle L, \leq, \odot, 1\rangle$ or $\langle L, \leq, \oplus, 0\rangle$, respectively. Only one aspect needs to taken into account. Negativity will be applied only to multiplicatively written tomonoids. Analogously, we call an additively written tomonoid $\langle L, \leq, \oplus, 0\rangle$ *positive* if 0 is the bottom element.

In what follows, a *subtomonoid* of a tomonoid $\boldsymbol{L}$ is a submonoid of $L$ together with the inherited total order. Moreover, a *homomorphism* between tomonoids is defined as usual. Finally, a homomorphism $\chi$ between tomonoids $\boldsymbol{K}$ and $\boldsymbol{L}$ is called *sup-preserving* if, whenever the supremum of elements $a_\iota$, $\iota \in I \neq \emptyset$, exists in $\boldsymbol{K}$, then $\chi(\bigvee_\iota a_\iota)$ is the supremum of the set $\{\chi(a_\iota) \mid \iota \in I\}$, in $\boldsymbol{L}$.

# 3 The partial-algebra method for the representation of totally ordered monoids

## 3.1 The idea

The typical aim of a representation theorem is to describe the structure of an algebra by means of algebras of a simpler type. In the case of residuated lattices the probably most often considered candidates for the basic constituents are lattice-ordered groups, or $\ell$-groups for short. By means of $\ell$-groups it will most likely never be possible to fully understand the structure of residuated lattices. However, for certain subclasses the idea has turned out to be fruitful, as will be exemplified by the results of the first part of the present chapter.

To understand the key idea of what we have in mind, let us see on the basis of a simple example how $\ell$-groups can be used to represent residuated tomonoids. Let $\star\colon [0,1]^2 \to [0,1], \ \ \langle a,b\rangle \mapsto (a+b-1) \vee 0$ be the Łukasiewicz t-norm and consider the tomonoid $\langle [0,1], \leq, \star, 1\rangle$. This is an MV-algebra, and by Mundici's representation theorem we certainly know how it is related to the totally ordered group of reals. Here, however, we want to explain on the basis of the simple case a procedure that is applicable to a class of residuated structures that is more comprehensible than MV-algebras.

Switching to the additive picture, we are led to the algebra $\langle [0,1], \leq, \oplus, 0\rangle$, where $\oplus$ is the truncated sum, that is,

$$a \oplus b \;=\; (a+b) \wedge 1, \quad a,b \in [0,1]. \tag{1}$$

From $\oplus$, we will now define a partial operation, which we denote by $+$. The partial operation $+$ will, where defined, coincide with the total one $\oplus$; in this sense, $+$ will be a restriction of $\oplus$. Our definition goes as follows:

$$a+b \;=\; \begin{cases} a \oplus b & \text{if } a \text{ is the smallest } x \text{ such that } x \oplus b = a \oplus b \\ & \text{and } b \text{ is the smallest } y \text{ such that } a \oplus y = a \oplus b \\ \text{undefined} & \text{otherwise.} \end{cases} \tag{2}$$

Consider now the partial algebra $\langle [0,1], \leq, +, 0\rangle$. If the usual sum of two reals $a, b \in [0,1]$ is strictly greater than 1, their sum $a+b$, according to (2), is undefined. If, however, the usual sum of $a$ and $b$ is at most 1, then $a+b$ stands for what it commonly denotes: the sum of $a$ and $b$ as reals. In short, the partial operation $+$ is the restriction of the usual addition of reals to those pairs whose sum is in $[0,1]$.

Remarkably, we do not lose information when switching to the partial operation $+$. In fact, we can easily recover the original operation: $a \oplus b$ is the maximal element among all defined sums $a' + b'$ such that $a' \leq a$ and $b' \leq b$.

Our aim is to embed $\langle [0,1], \leq, +, 0\rangle$ into $\langle \mathsf{R}, \leq, +, 0\rangle$. How can we construct the totally ordered group of reals from our partial algebra? We first determine the monoid freely generated by $[0,1]$ subject to the condition $a+b=c$ if this holds in the partial algebra. The result is the monoid $\langle \mathsf{R}^+, +, 0\rangle$, the positive reals endowed with the usual addition. Second, we make $\mathsf{R}^+$ the positive cone of a totally ordered Abelian group. The result is $\langle \mathsf{R}, \leq, +, 0\rangle$, and our embedding is complete.

The question remains how the algebra $\langle [0,1], \leq, \oplus, 0\rangle$ with which we started is represented by $\langle \mathsf{R}, \leq, +, 0\rangle$. The situation is as follows. The base set $[0,1]$ is an *interval* of R, consisting of all positive elements below the positive element 1. Furthermore, the total order is inherited from R. Finally, the monoidal operation is given according to (1).

The aim of this section is to show that we can proceed analogously to this example under rather general assumptions.

## 3.2 D.p.r. tomonoids and R-chains

Let us delimit the class of tomonoids to which our method is at present known to apply. Our main assumption is that the order of the tomonoid is the natural one: we require divisibility.

DEFINITION 3.2.1. *A residuated tomonoid $\langle L, \leq, \oplus, 0\rangle$ is called* divisible *if, for any $a, b \in L$ such that $a \leq b$, there are $c, d \in L$ such that $a \oplus c = d \oplus a = b$.*

Hence the structures that we are going to discuss are divisible, positive, residuated tomonoids; we will abbreviate these three properties with "d.p.r.". Note that d.p.r. tomonoids are in a one-to-one correspondence with divisible, integral residuated chains.

We will denote the residuals corresponding to the monoidal operation $\oplus$ of a d.p.r. tomonoid by $\oslash$ and $\oslash$, respectively, and in accordance with our additive notation we will write them in analogy to differences:

$$a \leq b \oplus c \quad \text{iff} \quad a \oslash b \leq c \quad \text{iff} \quad a \oslash c \leq b.$$

Then a positive, residuated tomonoid $L$ is divisible if, for any $a, b \in L$ such that $a \leq b$,

$$a \oplus (b \oslash a) \;=\; (b \oslash a) \oplus a \;=\; b.$$

With a d.p.r. tomonoid, we associate a partial algebra as follows.

DEFINITION 3.2.2. *Let $\langle L, \leq, \oplus, 0\rangle$ be a d.p.r. tomonoid. For $a, b \in L$, let $a+b = a\oplus b$ if $a = (a \oplus b) \oslash b$ and $b = (a \oplus b) \oslash a$; otherwise, let $a + b$ be undefined. Then we call $\langle L, \leq, +, 0\rangle$ the* partial algebra associated with *$\langle L, \leq, \oplus, 0\rangle$.*

Note that this definition is in accordance with the specification (2) in our informal introduction. Namely, let $a$ and $b$ be elements of a d.p.r. tomonoid. Then $(a \oplus b) \oslash b$ is, by definition, the smallest element $x$ such that $x \oplus b = a \oplus b$. Similarly, $(a \oplus b) \oslash a$ is the smallest element $y$ such that $a \oplus y = a \oplus b$.

From the partial algebra that we have associated with a d.p.r. tomonoid in Definition 3.2.2, we can recover the original structure in the following way.

LEMMA 3.2.3. *Let $\langle L, \leq, \oplus, 0\rangle$ be a d.p.r. tomonoid and $\langle L, \leq, +, 0\rangle$ its associated partial algebra. Then, for any $a, b \in L$,*

$$a \oplus b \;=\; \max\,\{a' + b' \mid a' \leq a \text{ and } b' \leq b \text{ such that } a' + b' \text{ is defined}\}. \qquad (3)$$

*Proof.* Assume that $a' \leq a,\ b' \leq b$, and $a'+b'$ is defined. Then $a'+b' = a'\oplus b' \leq a\oplus b$.

Let now $c = a \oplus b$; $a' = c \oslash b$, and $b' = c \oslash a'$. Then $a' \oplus b' = a' \oplus (c \oslash a') = c$. Moreover, $c \oslash b' = c \oslash (c \oslash a') = c \oslash (c \oslash (c \oslash b)) = c \oslash b = a'$. Hence $a' + b'$ exists and equals $a \oplus b$. $\square$

What kind of partial algebras do we get here? The following definition compiles their properties. As regards the existence of partially defined sums, we follow, whenever reasonable, the usual convention: $a+b=c$ means that $a+b$ exists and equals $c$.

DEFINITION 3.2.4. *An* R-chain *is a structure* $\langle L, \leq, +, 0\rangle$ *such that*

(E1) $\langle L, \leq, 0\rangle$ *is a chain with the bottom element* 0,

*and such that* $+$ *is a partial binary operation fulfilling, for any* $a, b, c \in L$, *the following conditions:*

(E2) $(a+b)+c$ *is defined if and only if* $a+(b+c)$ *is defined, and in this case* $(a+b)+c = a+(b+c)$.

(E3) $a+0 = 0+a = a$.

(E4) *If* $a+c$ *and* $b+c$ *are defined, then* $a \leq b$ *if and only if* $a+c \leq b+c$.

*If* $c+a$ *and* $c+b$ *are defined, then* $a \leq b$ *if and only if* $c+a \leq c+b$.

(E5) *If* $a+b$ *is defined, there are* $x, y \in L$ *such that* $a+b = x+a = b+y$.

(E6) *Let* $a \leq b$. *Then there is a largest element* $\bar{a} \leq a$ *such that* $b = \bar{a}+x$ *for some* $x \in L$.

*Similarly, there is a largest element* $\bar{\bar{a}} \leq a$ *such that* $b = y+\bar{\bar{a}}$ *for some* $y \in L$.

(E7) *If* $a \leq c \leq a+b$, *there is an* $x \in L$ *such that* $c = a+x$.

*Similarly, if* $a \leq c \leq b+a$, *there is a* $y \in L$ *such that* $c = y+a$.

The remaining part of this section is devoted to the proof that the partial algebra associated with a d.p.r. tomonoid is in fact an R-chain.

We begin with those properties that are comparably easy to prove.

LEMMA 3.2.5. *Let* $\langle L, \leq, \oplus, 0\rangle$ *be a d.p.r. tomonoid. Then the associated partial algebra* $\langle L, \leq, +, 0\rangle$ *fulfils* (E1), (E3), (E4), (E6), *and* (E7).

*Proof.* (E1) holds $\leq$ coincides by definition with total order of $L$ as a positive tomonoid. For the remaining properties, we show only the first half; the second half will follow in each case dually.

It is easily checked that, for any $a \in L$, $a+0$ and exist and equals $a$. (E3) follows.

Let now $a, b, c \in L$ such that $a+c$ and $b+c$ are defined. If $a \leq b$, then $a+c = a \oplus c \leq b \oplus c = b+c$. Conversely, if $a+c \leq b+c$, then $a = (a \oplus c) \oslash c = (a+c) \oslash c \leq (b+c) \oslash c = (b \oplus c) \oslash c = b$. We have proved (E4).

Next, let $a, b \in L$ such that $a \leq b$. Let $\bar{a} = b \oslash (b \otimes a)$. Then $\bar{a} \leq a$ and $\bar{a}+x = b$, where $x = b \otimes a$. If $a' \leq a$ and $a'+x' = b$, we have $x' = b \otimes a' \geq b \otimes a = x$ and hence $a' = b \oslash x' \leq b \oslash x = \bar{a}$. This shows (E6).

Finally, let $a, b, c \in L$ such that $a \leq c \leq a+b$. We shall show that $a = c \oslash (c \otimes a)$; it will then follow that $a+x = c$, where $x = c \otimes a$. Putting $d = a+b$, we derive from the divisibility of $L$ that $a = d \oslash b = d \oslash (d \otimes a) \geq c \oslash (c \otimes a) \geq (d \oslash (d \otimes c)) \oslash (c \otimes a) = (d \oslash (d \otimes (a \oplus (c \otimes a)))) \oslash (c \otimes a) = (d \oslash ((d \otimes a) \otimes (c \otimes a))) \oslash (c \otimes a) = d \oslash ((c \otimes a) \oplus ((d \otimes a) \otimes (c \otimes a))) = d \oslash (d \otimes a) = a$. (E7) follows. □

We continue with property (E2), the associativity.

LEMMA 3.2.6. *Let* $\langle L, \leq, \oplus, 0\rangle$ *be a d.p.r. tomonoid. Then the associated partial algebra* $\langle L, \leq, +, 0\rangle$ *fulfils* (E2).

*Proof.* Let $a, b, c \in L$ be such that $(a + b) + c$ is defined. Let $e = a + b$ and $d = e + c$. Let $f = d \oslash a$. Then $c = d \oslash e \leq d \oslash a = f \leq d = e + c$, and by (E7), there is a $b' \leq e$ such that $f = b' + c$. Then $b' = f \oslash c = (d \oslash a) \oslash c = (d \oslash c) \oslash a = b$, and we have shown that $f = b + c$.

Next, let $a' = d \oslash f$. Then $d \oslash a' = d \oslash (d \oslash f) = d \oslash (d \oslash (d \oslash a)) = d \oslash a = f$, and it follows $d = a' + f$. Furthermore, $a = (d \oslash c) \oslash b = (d \oslash c) \oslash (f \oslash c) = d \oslash ((f \oslash c) \oplus c) = d \oslash f = a'$. The proof is complete that $(a + b) + c = a + (b + c)$. The other half of (E2) is proved analogously. □

We finally turn to the property (E5). Let us define, for a d.p.r. tomonoid:

$$\begin{aligned} a \preccurlyeq_l b \quad & \text{if there is an } x \in L \text{ such that } b + x \text{ exists and equals } a \\ a \preccurlyeq_r b \quad & \text{if there is an } y \in L \text{ such that } y + b \text{ exists and equals } a, \end{aligned} \tag{4}$$

where the operation $+$ refers to the associated partial algebra. We shall show that $\preccurlyeq_l$ and $\preccurlyeq_r$ are coinciding partial orders; obviously, (E5) will then follow.

LEMMA 3.2.7. *Let* $\langle L, \leq, \oplus, 0\rangle$ *be a d.p.r. tomonoid. Then* $\preccurlyeq_l$ *and* $\preccurlyeq_r$ *are partial orders, both being extended by* $\leq$.

*Proof.* It is clear that $a \preccurlyeq_l b$ or $a \preccurlyeq_r b$ implies $a \leq b$. It further follows that $\preccurlyeq_l$ and $\preccurlyeq_r$ are reflexive and antisymmetric. Finally, the transitivity of $\preccurlyeq_l$ and $\preccurlyeq_r$ follows from (E2). □

LEMMA 3.2.8. *Let* $\langle L, \leq, \oplus, 0\rangle$ *be a d.p.r. tomonoid.*

(i) *Let* $0 < a \leq b \leq c$. *Then* $a \preccurlyeq_l b$ *and* $b \preccurlyeq_l c$ *if and only if* $a \preccurlyeq_l c$. *Similarly,* $a \preccurlyeq_r b$ *and* $b \preccurlyeq_r c$ *if and only if* $a \preccurlyeq_r c$.

(ii) *Let* $a, b \in L$. *Then* $a \preccurlyeq_l b$ *if and only if* $a \preccurlyeq_r b$.

*Proof.* (i) We prove the claim for $\preccurlyeq_r$; the assertion follows for $\preccurlyeq_l$ similarly.

By Lemma 3.2.7, $a \preccurlyeq_r b$ and $b \preccurlyeq_r c$ imply $a \preccurlyeq_r c$.

Let $a \preccurlyeq_r c$. From (E7), we conclude $a \preccurlyeq_r b$. This implies $a = b \oslash (b \oslash a) \geq (c \oslash (c \oslash b)) \oslash (b \oslash a) = c \oslash (c \oslash a) = a$, that is, $(c \oslash (c \oslash b)) \oslash (b \oslash a) = b \oslash (b \oslash a) > 0$. By divisibility, we conclude $c \oslash (c \oslash b) = b$, and it follows $b \preccurlyeq_r c$.

(ii) If $a = 0$ or $a = b$, the claim is clear. Assume $0 < a < b$ and $a \preccurlyeq_r b$.

Let $x$ be such that $b = a + x$. If $x \leq a$, we conclude from $x \leq a < b$ and $x \preccurlyeq_l b$ by part (i) that $a \preccurlyeq_l b$.

Assume $a < x$. From $a < x \leq b$ and $a \preccurlyeq_r b$, we have by part (i) that $x \preccurlyeq_r b$. Let $s$ be such that $b = x + s$. If $s \leq a$, we conclude from $s \preccurlyeq_l b$ and $s \leq a < b$ that $a \preccurlyeq_l b$.

Assume $a < s$. Then $a < s \leq x + s = b = a + x$. Thus there is an $r \leq x$ such that $s = a + r$, and we have $b = x + s = x + a + r$. In particular, $r \preccurlyeq_l b$. If then $r \leq a$, we conclude $a \preccurlyeq_l b$.

Assume $a < r$. Then $a < r \leq x \leq x + a$, and it follows that $r = u + a$ for some $u$. We conclude $b = x + a + r = x + a + u + a$, that is $a \preccurlyeq_l b$.

Analogously, we show that $a \preccurlyeq_l b$ implies $a \preccurlyeq_r b$. □

In what follows, we will denote the coinciding partial orders $\preccurlyeq_l$ and $\preccurlyeq_r$ of the R-chain associated with a d.p.r. tomonoid simply by $\preccurlyeq$.

We have shown:

THEOREM 3.2.9. *Let $\langle L, \leq, \oplus, 0 \rangle$ be a d.p.r. tomonoid. Then the associated algebra $\langle L, \leq, +, 0 \rangle$ is an R-chain.*

### 3.3 Ordinal sum decomposition of R-chains

Having seen that the partial algebras associated with d.p.r. tomonoids are R-chains, we will prove in this section that these partial algebras are ordinal sums of R-chains that are naturally ordered. The latter notion is defined in the expected way.

DEFINITION 3.3.1. *Let $\langle L, \leq, +, 0 \rangle$ be an R-chain. We say that $L$ is* naturally ordered *if, for any $a, b \in L$, $a \leq b$ if and only if there is an $x \in L$ such that $b = a + x$ if and only if there is a $y \in L$ such that $b = y + a$.*

In other words, to be naturally ordered means for R-chains arising from d.p.r. tomonoids that the total order $\leq$ coincides with $\preccurlyeq$.

The notion of an ordinal sum of R-chains is defined as usual.

DEFINITION 3.3.2. *Let $\langle I, \leq \rangle$ be a chain, and for every $i \in I$, let $\langle L_i, \leq, +, 0_i \rangle$ be an R-chain. Put $L = \dot{\bigcup}_{i \in I}(L_i \backslash \{0_i\}) \cup \{0\}$, where $0$ is a new element and $\dot{\cup}$ denotes the disjoint union. For $a, b \in L$, put $a \leq b$ if either $a = 0$, or $a \in L_i$ and $b \in L_j$ such that $i < j$, or $a, b \in L_i$ for some $i$ and $a \leq b$ holds in $L_i$. Similarly, define $a + b$ if either $a = 0$, in which case $a + b = b$, or $b = 0$, in which case $a + b = a$, or $a, b \in L_i$ for some $i$ and $a + b$ is defined in $L_i$, in which case $a + b$ is mapped to the same value as in $L_i$. Then $\langle L, \leq, +, 0 \rangle$ is called the* ordinal sum *of the R-chains $L_i$ w.r.t. $\langle I, \leq \rangle$.*

We easily check:

LEMMA 3.3.3. *An ordinal sum of R-chains is again an R-chain.*

THEOREM 3.3.4. *Let $\langle L, \leq, \oplus, 0 \rangle$ be a d.p.r. tomonoid, and let $\langle L, \leq, +, 0 \rangle$ be the associated R-chain. Then $L$ is the ordinal sum of naturally ordered R-chains.*

*Proof.* By Lemma 3.2.8(i), $L \backslash \{0\} = \dot{\bigcup}_{i \in I} C_i$ for pairwise disjoint convex subsets $C_i$, $i \in I$, of $L$ such that, for $a, b \in L \backslash \{0\}$, the following holds: $a \preccurlyeq b$ if and only if there is an $i \in I$ such that $a, b \in C_i$ and $a \leq b$.

Let $i \in I$. If $a, b \in C_i$ such that $a + b$ exists, we have $a + b \in C_i$ by construction. Consider $C_i \cup \{0\}$ endowed with the restriction of $\leq$ and $+$ to $C_i \cup \{0\}$ as well as with the constant $0$. Then it is easily checked that $\langle C_i \cup \{0\}, \leq, +, 0 \rangle$ fulfils (E1)–(E7). Hence $C_i \cup \{0\}$ is an R-algebra, which by construction is naturally ordered. Furthermore, $L$ is the ordinal sum of the R-algebras $C_i \cup \{0\}$. □

### 3.4 Naturally ordered R-chains

We next turn to the characterisation of naturally ordered R-chains. We will show that any such R-chain can be embedded into the positive cone of totally ordered Abelian group.

LEMMA 3.4.1. *Let $\langle L, \leq, +, 0\rangle$ be a naturally ordered R-chain. For any $a, b, c \in L$, the following holds:*

(i) *If $a + b$ exists, $a_1 \leq a$, and $b_1 \leq b$, then also $a_1 + b_1$ exists.*

(ii) *If $a \leq b + c$, then there are $b_1 \leq b$ and $c_1 \leq c$ such that $a = b_1 + c_1$ exists.*

*Proof.* (i) Let $a + b$ exists, and let $a_1 \leq a$ and $b_1 \leq b$. Then $a_1 \preccurlyeq a$ and $b_1 \preccurlyeq b$, hence there are $x, y$ such that $a = x + a_1$ and $b = b_1 + y$. Thus $a + b = x + a_1 + b_1 + y$, and the claim follows from (E2).

(ii) If $a \leq b$, we put $b_1 = a$ and $c_1 = 0$. If $b < a$, there is by (E7) a $c_1$ such that $a = b + c_1$. By (E4), $c_1 \leq c$; thus we put $b_1 = b$ and we are done. □

We note that the statement of Lemma 3.4.1(ii) is usually refer to as a Riesz decomposition property.

By a scheme of the form (5) in the following lemma to hold, we mean that the sum of any row and any column exists and equals the element to which the respective arrow points to; the order of addition is from left to right or from top to bottom, respectively.

LEMMA 3.4.2. *Let $\langle L, \leq, +, 0\rangle$ be a naturally ordered R-chain and let $a_1, \ldots, a_m$, $b_1, \ldots, b_n \in L$ be such that $a_1 + \cdots + a_m = b_1 + \cdots + b_n$, where $n, m \geq 1$. Then there are $d_{11}, \ldots, d_{mn} \in L$ such that*

$$\begin{array}{ccccc} d_{11} & \ldots & d_{1n} & \rightarrow & a_1 \\ \vdots & & \vdots & & \vdots \\ d_{m1} & \ldots & d_{mn} & \rightarrow & a_m \\ \downarrow & & \downarrow & & \\ b_1 & \ldots & b_n & & \end{array} \tag{5}$$

*and*

$$d_{ik} \wedge d_{jl} = 0 \textit{ for every } 1 \leq i < j \leq m \textit{ and } 1 \leq l < k \leq n. \tag{6}$$

*Proof.* If $m = 1$ or $n = 1$, the assertion is trivial. Let $m = n = 2$; then our assumption is $a_1 + a_2 = b_1 + b_2$. Assume that $a_1 \leq b_1$. Let $d$ be such that $a_1 + d = b_1$; then we put $d_{11} = a_1$, $d_{12} = 0$, $d_{21} = d$, and $d_{22} = b_2$, and we are done. Similarly, we proceed in case $b_1 \leq a_1$.

Assume next that $m \geq 3$ and $n \geq 2$, and that the assertion holds for any pair $m' < m$ and $n' \leq n$. Then, obviously, the assertion holds for the pair $m$ and $n$ as well. □

From a naturally ordered R-chain $\langle L, \leq, +, 0\rangle$ we now construct the free monoid with the elements of $L$ as its generators and with the conditions $a + b = c$, where $a, b, c \in L$ such that $a + b = c$ holds in $L$. We will have to show that the free monoid does not "collapse": the natural embedding of the R-chain is injective.

DEFINITION 3.4.3. *Let $\langle L, \leq, +, 0\rangle$ be a naturally ordered R-chain. We call a finite sequence $\langle a_1, \ldots, a_n\rangle$ of $n \geq 1$ elements of $L$ a* word *of $L$. We denote the set of words of $L$ by $W(L)$, and we define $+: W(L)^2 \to W(L)$ by concatenation.*

*Moreover, let $\sim$ be the smallest equivalence relation on $W(L)$ such that*

$$\langle a_1, \ldots, a_p, a_{p+1}, \ldots, a_n\rangle \sim \langle a_1, \ldots, a_p + a_{p+1}, \ldots, a_n\rangle$$

*holds for any two words in $W(L)$ of the indicated form, where $1 \leq p < n$. We denote the equivalence class of some $\langle a_1, \ldots, a_n\rangle \in W(L)$ by $\langle\!\langle a_1, \ldots, a_n\rangle\!\rangle$ and the set of all equivalence classes by $C(L)$.*

As seen in the next lemma, $C(L)$ is a semigroup under elementwise concatenation, into which $L$, as a semigroup, naturally embeds.

LEMMA 3.4.4. *Let $\langle L, \leq, +, 0\rangle$ be a naturally ordered R-chain.*

(i) *The equivalence relation $\sim$ on $W(L)$ is compatible with $+$. Denoting the induced operation by $+$ again, we have that $\langle C(L), +, \langle\!\langle 0\rangle\!\rangle\rangle$ is a monoid.*

(ii) *Let $a_1, \ldots, a_n, b \in L$, where $n \geq 1$. Then $\langle a_1, \ldots, a_n\rangle \sim \langle b\rangle$ if and only if $a_1 + \ldots + a_n = b$.*

(iii) *Let*

$$\iota\colon L \to C(L), \quad a \mapsto \langle\!\langle a\rangle\!\rangle$$

*be the natural embedding of $L$ into $C(L)$. Then $\iota$ is injective.*

*Furthermore, for $a, b \in L$ the sum $a + b$ is defined and equals $c$ if and only if $\iota(a) + \iota(b) = \iota(c)$.*

*Proof.* (i) Evident.

(ii) For any word $\langle a_1, \ldots, a_n\rangle$ the sum of whose elements exists and equals $b$, the same is true for any word equivalent to $\langle a_1, \ldots, a_n\rangle$.

(iii) The injectivity of $\iota$ follows from part (ii). Let moreover $a, b \in L$. If $a + b = c$, then obviously $\iota(a) + \iota(b) = \iota(c)$. Conversely, $\iota(a) + \iota(b) = \iota(c)$ means $\langle a, b\rangle \sim \langle c\rangle$, that is, $a + b = c$ by part (ii). □

Next, we show that the monoid $\langle C(L), +, \langle\!\langle 0\rangle\!\rangle\rangle$ fulfils the characteristic properties of the positive cone of a partially ordered group. As a preparation, we insert the following generalisation of Lemma 3.4.2.

LEMMA 3.4.5. *Let $\langle L, \leq, +, 0\rangle$ be a naturally ordered R-chain and let $a_1, \ldots, a_m$, $b_1, \ldots, b_n \in L$ such that $\langle a_1, \ldots, a_m\rangle \sim \langle b_1, \ldots, b_n\rangle$, where $n, m \geq 1$. Then there are $d_{11}, \ldots, d_{mn} \in L$ such that* (5) *and* (6) *hold.*

*Proof.* If $m = n$ and $a_1 = b_1, \ldots, a_m = b_m$, the assertion is trivial. Let $a_1, \ldots, b_n$ be arbitrary, and let $d_{11}, \ldots, d_{mn}$ be such that (5) and (6) hold. We shall show how to modify the scheme (5) to preserve both its correctness and the infimum-zero relations (6) when $\langle b_1, \ldots, b_n\rangle$ is replaced (i) by $\langle b_1, \ldots, b_p + b_{p+1}, \ldots, b_n\rangle$ for some $1 \leq p < n$ or (ii) by $\langle b_1, \ldots, b_p^1, b_p^2, \ldots, b_n\rangle$, where $1 \leq p \leq n$ and $b_{p1} + b_{p2} = b_p$.

Ad (i). We replace, for each $i = 1, \ldots, m$, the neighbouring entries $d_{ip}$ and $d_{i,p+1}$ by their sum, which by Lemma 3.4.1(i) exists. Then the sum of the $i$-th row is obviously still $a_i$. To see that the sum of the new column exists and is $b_p + b_{p+1}$, we make repeated use of the fact that two elements one of which is 0 can be interchanged:

$$\begin{aligned} b_p + b_{p+1} &= d_{1p} + \ldots + d_{mp} + d_{1,p+1} + \ldots + d_{m,p+1} \\ &= d_{1p} + d_{1,p+1} + d_{2p} + \ldots + d_{mp} + d_{2,p+1} + \ldots + d_{m,p+1} \\ &= \ldots \\ &= d_{1p} + d_{1,p+1} + d_{2p} + d_{2,p+1} + \ldots + d_{mp} + d_{m,p+1}. \end{aligned}$$

Clearly, the infimum-zero relations still hold.

Ad (ii). We apply Lemma 3.4.2 to the equation $b_p^1 + b_p^2 = d_{1p} + \ldots + d_{mp}$, and replace the column $d_{1p}, \ldots, d_{mp}$ with the new double column. Obviously, in the modified scheme, the rows and columns add up correctly and the required infimum-zero relations hold. □

LEMMA 3.4.6. *Let $\langle L, \leq, +, 0\rangle$ be a naturally ordered R-chain. Then $\langle C(L), +, \langle\langle 0\rangle\rangle\rangle$ is a monoid such that for $\mathfrak{a}, \mathfrak{b}, \mathfrak{c} \in C(L)$:*

(i) *From $\mathfrak{a} + \mathfrak{b} = \langle\langle 0\rangle\rangle$ it follows $\mathfrak{a} = \mathfrak{b} = \langle\langle 0\rangle\rangle$.*

(ii) *From $\mathfrak{a} + \mathfrak{b} = \mathfrak{a} + \mathfrak{c}$ or $\mathfrak{b} + \mathfrak{a} = \mathfrak{c} + \mathfrak{a}$ it follows $\mathfrak{b} = \mathfrak{c}$.*

(iii) *There are $\mathfrak{x}, \mathfrak{y} \in C(L)$ such that $\mathfrak{a} + \mathfrak{b} = \mathfrak{x} + \mathfrak{a} = \mathfrak{b} + \mathfrak{y}$.*

*Proof.* $C(L)$ is a monoid by Lemma 3.4.4(i).

(i) This follows from Lemma 3.4.4(ii).

(ii) We may restrict to the case that $\mathfrak{a} = \langle\langle a\rangle\rangle$ for some $a \in L$. Furthermore, let $\mathfrak{b} = \langle\langle b_1, \ldots, b_m\rangle\rangle$, $\mathfrak{c} = \langle\langle c_1, \ldots, c_n\rangle\rangle$, $m, n \geq 1$, and assume that $\langle a, b_1, \ldots, b_m\rangle \sim \langle a, c_1, \ldots, c_n\rangle$. We will show that $\mathfrak{b} = \mathfrak{c}$; the second part can be proved analogously.

By Lemma 3.4.5, there are elements in $L$ such that

$$\begin{array}{ccccccc} d & d_1 & \ldots & d_n & \rightarrow & a \\ e_1 & e_{11} & \ldots & e_{1n} & \rightarrow & b_1 \\ \vdots & \vdots & & \vdots & & \vdots \\ e_m & e_{m1} & \ldots & e_{mn} & \rightarrow & b_m \\ \downarrow & \downarrow & & \downarrow & & \\ a & c_1 & & c_n, & & \end{array}$$

where any pair of elements one of which is placed further up and further right than the other one, has infimum 0. But the latter condition means $d = a$ and $d_1 = \cdots = d_n = e_1 = \cdots = e_m = 0$. Again using the infimum-zero conditions, we conclude $\langle\langle b_1, \ldots, b_m\rangle\rangle = \langle\langle c_1, \ldots, c_n\rangle\rangle$.

(iii) We only prove the first half of the claim. We may furthermore restrict to the case that $\mathfrak{a} = \langle\langle a \rangle\rangle$ and $\mathfrak{b} = \langle\langle b \rangle\rangle$ for some $a, b \in L$. If $a \leq b$, then $b = x + a$ for some $x \in L$, hence $\langle\langle a, b \rangle\rangle = \langle\langle a, x \rangle\rangle + \langle\langle a \rangle\rangle$. If $a \geq b$, we have $a = b + x = y + b$ for some $x, y \in L$. If then $y \leq x$, we have $x = z + y$ for some $z \in L$ and thus $\langle\langle a, b \rangle\rangle = \langle\langle b, x, b \rangle\rangle = \langle\langle b, z, y, b \rangle\rangle = \langle\langle b, z \rangle\rangle + \langle\langle a \rangle\rangle$. If then $x \leq y$, we have $y = z + x$ for some $z \leq b$ by (E7) and thus $\langle\langle a, b \rangle\rangle = \langle\langle b, x, b \rangle\rangle = \langle\langle b', z, x, b \rangle\rangle = \langle\langle b', y, b \rangle\rangle = \langle\langle b' \rangle\rangle + \langle\langle a \rangle\rangle$, where $b = b' + z$. □

We next establish that the natural order of $C(L)$ is actually a total order.

LEMMA 3.4.7. *Let $\langle L, \leq, +, 0 \rangle$ be a naturally ordered R-chain. Let*

$$\mathfrak{b} \leq \mathfrak{a} \quad \textit{if} \quad \mathfrak{b} + \mathfrak{x} = \mathfrak{a} \textit{ for some } \mathfrak{x} \in C(L)$$

*for $\mathfrak{a}, \mathfrak{b} \in C(L)$. Then $\leq$ is a total order.*

*Proof.* Let $\mathfrak{a} = \langle\langle a_1, \ldots, a_m \rangle\rangle$ and $\mathfrak{b} = \langle\langle b_1, \ldots, b_n \rangle\rangle$ with $m, n \geq 1$. Assume that $a_1 \leq b_1$. Then there is an $x \in L$ such that $b_1 = a_1 + x$, and by Lemma 3.4.6, $\mathfrak{a}$ and $\mathfrak{b}$ are comparable iff so are $\langle\langle a_2, \ldots, a_m \rangle\rangle$ and $\langle\langle x, b_2, \ldots, b_n \rangle\rangle$, where the empty word is identified with $\langle\langle 0 \rangle\rangle$. Similarly, assume $b_1 \leq a_1$. Then there is an $x \in L$ such that $a_1 = b_1 + x$, and $\mathfrak{a}$ and $\mathfrak{b}$ are comparable iff so are $\langle\langle x, a_2, \ldots, a_m \rangle\rangle$ and $\langle\langle b_2, \ldots, b_n \rangle\rangle$. From the fact that $\langle\langle 0 \rangle\rangle$ is comparable with any word, we conclude the assertion by induction. □

We arrive at our main theorem. By an isomorphic embedding of a naturally ordered R-chain $\langle L, \leq, +, 0 \rangle$ into a totally ordered group $\langle G, \leq, +, 0 \rangle$, we mean an injective mapping $\iota \colon L \to G$ such that for $a, b, c \in L$:

$$a \leq b \text{ if and only if } \iota(a) \leq \iota(b)$$
$$a + b \text{ is defined and equals } c \text{ if and only if } \iota(a) + \iota(b) = \iota(c)$$
$$\iota(0) = 0.$$

THEOREM 3.4.8. *Let $\langle L, \leq, +, 0 \rangle$ be a naturally ordered R-chain. Then there exists an isomorphic embedding $\iota$ of the R-chain $\langle L, \leq, +, 0 \rangle$ into a totally ordered group $\langle G, \leq, +, 0 \rangle$. The range of $\iota$ is a convex subset of $G$, whose smallest element is $\langle\langle 0 \rangle\rangle$ and which generates $G$.*

*Proof.* By Lemma 3.4.6 and [19, Chapter II, Theorem 4], there is a totally ordered group $\langle G(L), \leq, +, \langle\langle 0 \rangle\rangle \rangle$ such that $C(L)$ is its positive cone. Moreover, by Lemma 3.4.4(iii), $\iota \colon L \to C(L), \;\; a \mapsto \langle\langle a \rangle\rangle$ is an isomorphic embedding of $L$ into $\langle C(L), \leq, +, \langle\langle 0 \rangle\rangle \rangle$. Extending the range of the mapping $\iota$ to $G(L)$, $\iota$ becomes an isomorphic embedding of $L$ into $\langle G(L), \leq, +, \langle\langle 0 \rangle\rangle \rangle$.

The range of $\iota$ in $G(L)$ is then $\{\langle\langle a \rangle\rangle \mid a \in L\} \subseteq C(L)$. The smallest element of $C(L)$ is $\langle\langle 0 \rangle\rangle$, which consequently is the smallest element of the range of $\iota$. Moreover, let $\mathfrak{g} \in G(L)$ and $a \in L$ such that $\langle\langle 0 \rangle\rangle \leq \mathfrak{g} \leq \langle\langle a \rangle\rangle$. Then $\mathfrak{g} \in C(L)$ and $\mathfrak{g} + \mathfrak{h} = \langle\langle a \rangle\rangle$ for some $\mathfrak{h} \in C(L)$, and it follows that $\mathfrak{g} = \langle\langle b \rangle\rangle$ for some $b \in L$. Thus the range of $\iota$ is a convex subset of $G(L)$, which moreover generates $C(L)$ and consequently $G(L)$. □

### 3.5 The representation of d.p.r. tomonoids

So far, we have associated with a d.p.r. tomonoid an R-chain; we have represented this partial algebra as an ordinal sum of naturally ordered R-chains; and we have shown that each naturally ordered R-chain embeds into a totally ordered Abelian group. Summarising these results, we may now formulate a representation theorem for d.p.r. tomonoids.

For convenience, let us introduce simple notions for our basic constituents.

DEFINITION 3.5.1. *Let $\langle G, \leq, +, 0\rangle$ be a totally ordered group. Then we call the tomonoid $\langle G^+, \leq, +, 0\rangle$ a* group cone.

*Moreover, let $u \in G$ such that $u > 0$. Let $[0, u] = \{g \in G \mid 0 \leq g \leq u\}$, and define*

$$a \oplus b \;=\; (a+b) \wedge u, \quad a, b \in [0, u].$$

*Then we call the tomonoid $\langle [0, u], \leq, \oplus, 0\rangle$ a* group interval.

Note that group cones and group intervals are actually d.p.r. tomonoids.

LEMMA 3.5.2. *Let $\langle L, \leq, \oplus, 0\rangle$ be a d.p.r. tomonoid such that its associated partial algebra is a naturally ordered R-chain. Then $L$ is either a group cone or a group interval.*

*Proof.* Let $\iota \colon L \to G$ the embedding of $L$ into a totally ordered group according to Theorem 3.4.8. We distinguish two cases:

*Case 1.* The addition of $L$ is total. As $G^+$ is generated by $\iota(L)$, it follows $\iota(L) = G^+$, that is, $\langle L, \leq, \oplus, 0\rangle$ is a group cone in this case.

*Case 2.* There are $a, b \in L$ such that $a + b$ is not defined. Let then $u = a \oplus b$; we claim that $u$ is the top element of $L$ and hence $\iota(L) = \{g \in G^+ \mid g \leq \iota(u)\}$.

Assume to the contrary that there is a $v \in L$ such that $u < v$. Let $a' \leq a$ and $b' \leq b$ such that $a' + b' = u$. Then either $a' < a$ or $b' < b$; we assume $b' < b$ and we can proceed similarly in the case $a' < a$. As $L$, as an R-chain, is naturally ordered, there is a $d > 0$ such that $v = u + d$. This means that the sum $a' + b' + d$ is defined; then $b'' = (b' + d) \wedge b > b'$ and $a' \oplus b'' = a \oplus b = u$. However, the sum $a' + b''$ is defined and strictly greater than $a' + b' = a \oplus b = u$. Our claim is proved.

If now $a, b \in L$ such that $a + b$ exists, we have $\iota(a \oplus b) = \iota(a + b) = \iota(a) + \iota(b) \in L$, that is, $\iota(a) + \iota(b) \leq \iota(u)$. If $a + b$ does not exist, $\iota(a) + \iota(b) \notin \iota(L)$, that is $\iota(a) + \iota(b) > \iota(u)$. In this case, $a \oplus b = u$, hence $\iota(a \oplus b) = \iota(u)$. We conclude

$$\iota(a \oplus b) \;=\; (\iota(a) + \iota(b)) \wedge u;$$

hence $L$ is a group interval in this case. □

We next define ordinal sums of d.p.r. tomonoids.

DEFINITION 3.5.3. *Let $\langle I, \leq\rangle$ be a chain, and for each $i \in I$, let $\langle L_i, \leq_i, \oplus_i, 0_i\rangle$ be a d.p.r. tomonoid. Put $L = \dot{\bigcup}_{i \in I}(L_i \backslash \{0_i\}) \cup \{0\}$, where $0$ is a new element and $\dot{\bigcup}$ denotes the disjoint union. For $a, b \in L$, let $a \leq b$ if either $a = 0$, or $a \in L_i$ and $b \in L_j$ such that $i < j$, or $a, b \in L_i$ for some $i$ and $a \leq b$ holds in $L_i$. Moreover, for $a, b \in L$, define $a \oplus 0 = 0 \oplus a = a$; $a \oplus b = a \oplus_i b$ if $a, b \in L_i$ for some $i \in I$; and $a \oplus b = b \oplus a = a$ if $a \in L_i$ and $b \in L_j$ such that $i < j$. Then $\langle L, \leq, \oplus, 0\rangle$ is called the* ordinal sum *of the d.p.r. tomonoids $L_i$ w.r.t. $\langle I, \leq\rangle$.*

We obviously have:

LEMMA 3.5.4. *The ordinal sum of d.p.r. tomonoids is again a d.p.r. tomonoid.*

We can finally state our main result.

THEOREM 3.5.5. *Each d.p.r. tomonoid is the ordinal sum of d.p.r. tomonoids $L_i$, $i \in I$, such that each $L_i$ is either a group cone or a group interval.*

*Proof.* Let $\langle L, \leq, \oplus, 1\rangle$ be a d.p.r. tomonoid. By Theorem 3.2.9, its associated partial algebra $\langle L, \leq, +, 0\rangle$ is an R-chain, and $\oplus$ is determined by $+$ according to (3). Moreover, by Theorem 3.3.4, the R-chain $L$ is the ordinal sum of naturally ordered R-chains $\langle L_i, \leq, +, 0\rangle$, where $i \in I$ and $I$ is a chain.

From (3) we conclude that $L_i$ is closed under $\oplus$. Consequently, $\langle L_i, \leq, \oplus, 0\rangle$ is a tomonoid, in fact a d.p.r. tomonoid. It is furthermore easily seen that $\langle L_i, \leq, +, 0\rangle$ is its associated R-chain. As the latter is naturally ordered, $\langle L_i, \leq, \oplus, 0\rangle$ is by Lemma 3.5.2 a group cone or a group interval. The assertion follows. □

In the dual picture, Theorem 3.5.5 provides a representation of divisible, integral residuated chains, or totally ordered pseudohoops [9]. Adding the assumption that there is a bottom element, we arrive at a representation of totally ordered pseudo-BL algebras. Finally, adding commutativity, we get the well-known representation of totally ordered BL algebras [1]; cf. Chapter V of this Handbook.

## 4 Coextensions of totally ordered monoids

### 4.1 The idea

The second approach that we are going to present in this chapter follows similar aims than the first one; our concern is a better understanding of the structure of residuated chains. However, what now follows could hardly be more different in style from what we have discussed so far.

Our starting point is a simple observation. Recall that the quotients of an integral residuated chain are in a one-to-one correspondence with its filters, and the set of filters is itself totally ordered w.r.t. set-theoretical inclusion. With any integral residuated chain we may hence associate the chain of their quotients. The bottom element of this chain is the trivial algebra, consisting of a single element; and the top element is the algebra under consideration. The intermediate elements may be seen as leading us stepwise from the trivial algebra to more and more fine-grained structures up to the algebra under consideration.

This intuitive picture is certainly easily overridden by the real situation: although the set of quotients cannot be ordered in a completely arbitrary manner, this chain can be very complicated. An example is the Cantor set endowed with its natural order; in such cases we can hardly speak about a stepwise construction process.

Nevertheless it seems to make sense to explore neighbouring elements in the chain, provided that there are any. If two quotients directly follow one another the filter inducing the congruence is Archimedean and accordingly we speak about Archimedean coextensions then. The construction of Archimedean coextensions is again intractable in general, but there is a condition that reduces possibilities drastically, namely, the condition that the congruence classes are order-isomorphic to real intervals.

The fact that we deal with the real line might reveal our original motivation underlying the present study: our ultimate aim has been a classification of left-continuous t-norms. Given the tomonoid based on a t-norm, the detection of filters and their induced quotients may already imply the entanglement of a possibly complicated structure. By this step alone, seemingly exotic cases can often be easily categorised. Moreover, the regular representation of monoids is a convenient geometric tool that accompanies our analysis with a clear intuition.

With regards to the t-norms, our main results implies the following. Consider the tomonoid arising from a left-continuous t-norm and assume further that it possesses an Archimedean filter. Then the t-norm can be described in terms of the quotient induced by this filter and essentially the only information needed is the order type of the congruence classes.

We proceed as follows. The subsequent Section 4.2 introduces the class of totally ordered monoids that we consider this time. The property of divisibility will no longer play a role; but we will deal with the commutative case only and we will assume an order-theoretic completeness condition. In the subsequent Section 4.3, we turn to the chain of quotients of the tomonoids induced by filters.

As a preparation for what follows, and the same time as a visualisation tool, we discuss in Section 4.4 the regular representation of tomonoids, which we call Cayley tomonoids. Section 4.5 contains our main result: a method of constructing from a given tomonoid an Archimedean coextension.

## 4.2 Q.n.c. tomonoids and their quotients

In this second part of the present chapter, totally ordered monoids, or tomonoids for short, serve again as our algebraic framework. This time, however, we will use the multiplicative notation.

Our tomonoids will be assumed to be negative, and we deal with the commutative case only. Moreover, we assume that the tomonoids are almost complete. Here, a poset is called *almost complete* if arbitrary non-empty suprema exist. Finally, we will assume that the multiplication distributes over arbitrary joins. We combine the latter two conditions to one notion named "quantic" because quantales are in fact defined similarly [36].

DEFINITION 4.2.1. *A tomonoid* $\boldsymbol{L} = \langle L, \leq, \odot, 1 \rangle$ *is called* quantic *if* (i) $\boldsymbol{L}$ *is almost complete and* (ii) *for any elements* $a, b_\iota$, $\iota \in I$, *of* $L$ *we have*

$$a \odot \bigvee_\iota b_\iota = \bigvee_\iota (a \odot b_\iota) \quad \text{and} \quad (\bigvee_\iota b_\iota) \odot a = \bigvee_\iota (b_\iota \odot a).$$

The tomonoids that we consider here are quantic, negative, and commutative. We abbreviate these three properties with "q.n.c.". Note that q.n.c. tomonoids are residuated. In fact, q.n.c. tomonoids are in a one-to-one correspondence with almost complete, integral, commutative residuated chains, or almost complete totally ordered basic semihoops.

The motivating examples arise from left-continuous triangular norms.

EXAMPLE 4.2.2. Let [0,1] be the real unit interval endowed with the natural order and let $\star\colon [0,1]^2 \to [0,1]$ be a left-continuous t-norm. Then $\langle [0,1], \leq, \star, 1 \rangle$ is a q.n.c. tomonoid.

We will use in the sequel occasionally the residual of a q.n.c. tomonoid; we denote it by $\to$.

A q.n.c. tomonoid does not necessarily possess a bottom element. If not, we can add an additional element with this role in the usual way.

DEFINITION 4.2.3. *Let $\boldsymbol{L}$ be a q.n.c. tomonoid. Let $\boldsymbol{L}^0 = \boldsymbol{L}$ if $L$ has a bottom element. Otherwise, let $\boldsymbol{L}^0 = \langle L^0, \leq, \odot, 1 \rangle$ arise from $\boldsymbol{L}$ by adding a new element 0; in this case, we extend the total order to $L^0$ such that 0 is the bottom element, and we extend the monoidal operation to $L^0$ such that 0 is absorbing.*

Obviously, for any q.n.c. tomonoid $\boldsymbol{L}$, $\boldsymbol{L}^0$ is again a q.n.c. tomonoid, whose total order is complete and which hence can be seen as a quantale.

In the context of almost complete chains, it makes sense to speak about intervals analogously to the case of reals. An *interval* of a q.n.c. tomonoid $\boldsymbol{L}$ will be a non-empty convex subset of $L$. An interval $J$ of $\boldsymbol{L}$ possesses in $\boldsymbol{L}^0$ an infimum $u$ and a supremum $v$, and we will refer to $J$ by $(u, v)$, $(u, v]$, $[u, v)$, or $[u, v]$, depending on whether or not $u$ and $v$ belong to $J$.

We now turn to quotients of tomonoids. We note that the following definition could be simplified if we included the infimum or supremum to the signature instead of the total order relation.

DEFINITION 4.2.4. *Let $\boldsymbol{L} = \langle L, \leq, \odot, 1 \rangle$ be a q.n.c. tomonoid. An equivalence relation $\sim$ on $L$ is called a* tomonoid congruence *if* (i) *$\sim$ is a congruence of $\boldsymbol{L}$ as a monoid and* (ii) *the $\sim$-classes are convex. We endow then the quotient $[\boldsymbol{L}]_\sim$ with the total order given by*

$$[a]_\sim \leq [b]_\sim \text{ if } a' \leq b' \text{ for some } a' \sim a \text{ and } b' \sim b$$

*for $a, b \in L$, with the induced operation $\odot$, and with the constant $[1]_\sim$. The resulting structure $\langle [\boldsymbol{L}]_\sim, \leq, \odot, [1]_\sim \rangle$ is called a* tomonoid quotient *of $\boldsymbol{L}$.*

Obviously, the congruence classes of a tomonoid quotient are intervals and we have $[a]_\sim < [b]_\sim$ if and only if $a' < b'$ for all $a' \sim a$ and $b' \sim b$.

To describe the totality of quotients of q.n.c. tomonoids is in general difficult. Here, we are interested in only one way of forming quotients: by means of filters.

DEFINITION 4.2.5. *Let $\boldsymbol{L}$ be a q.n.c. tomonoid. Then a* filter *of $\boldsymbol{L}$ is a subtomonoid $\boldsymbol{F} = \langle F, \leq, \odot, 1 \rangle$ of $\boldsymbol{L}$ such that $f \in F$ and $g \geq f$ imply $g \in F$.*

By the *trivial* tomonoid, we mean the one-element tomonoid, consisting of 1 alone. Each q.n.c. tomonoid $\boldsymbol{L}$ possesses the following filters: $\{1\}$, the *trivial* filter, and $\boldsymbol{L}$, the *improper* filter. Thus each non-trivial q.n.c. tomonoid has at least two filters.

As we easily check, a filter of a q.n.c. tomonoid is again a q.n.c. tomonoid. A filter may or may not possess a bottom element; this is actually the reason for which we defined quanticity by requiring an almost complete rather than a complete order.

DEFINITION 4.2.6. *Let $\boldsymbol{F}$ be a filter of a q.n.c. tomonoid $\boldsymbol{L}$. Let $d$ be the infimum of $F$ in $\boldsymbol{L}^0$; then we call $d$ the* boundary *of $\boldsymbol{F}$. If $d$ belongs to $F$, we write $F = d^{\leq}$; if $d$ does not belong to $F$, we write $F = d^{<}$.*

Therefore each filter of a q.n.c. tomonoid $\boldsymbol{L}$ is of the form $d^{<} = (d, 1]$ for some $d \in L^0 \backslash \{1\}$, or $d^{\leq} = [d, 1]$ for some $d \in L$. Each filter $\boldsymbol{F}$ is uniquely determined by its boundary $d$ together with the information whether or not $d$ belongs to $F$. We note that, for some $d \in L$, it is possible that both $d^{<}$ and $d^{\leq}$ are filters.

LEMMA 4.2.7. *Let $\boldsymbol{L}$ be a q.n.c. tomonoid, and let $d \in L$.*

(i) *$d^{\leq}$ is a filter if and only if $d$ is idempotent.*

(ii) *$d^{<}$ is a filter if and only if $d \neq 1$, $d = \bigwedge_{a>d} a$, and $d < a \odot b$ for all $a, b > d$.*

*Proof.* (i) $[d, 1]$ is a filter if and only if $[d, 1]$ is closed under multiplication if and only if $d \odot d = d$, that is, if $d$ is idempotent.

(ii) Let $d^{<}$ be a filter. Then $d < 1$ because each filter contains 1; $d = \inf d^{<} = \inf\,(d, 1] = \bigwedge_{a>d} a$; and $(d, 1]$ is closed under multiplication, that is, $a \odot b > d$ for each $a, b > d$.

Conversely, let $d \neq 1$ such that $d = \bigwedge_{a>d} a$ and $d < a \odot b$ for any $a, b > d$. Then $\{a \in L \mid a > d\}$ is a filter whose infimum is $d$, that is, which equals $d^{<}$. □

In the broader context of residuated lattices, the relevant substructures are convex normal subalgebras; each of the latter induces a quotient and every quotient of a residuated lattice arises in this way [26]. Here we consider a special case of this situation: filters of q.n.c. tomonoids lead to tomonoid quotients. We should, however, be aware of the fact that not all quotients of totally ordered monoids are induced by filters.

DEFINITION 4.2.8. *Let $\boldsymbol{F}$ be a filter of a q.n.c. tomonoid $\boldsymbol{L}$. For $a, b \in L$, let*

$$a \sim_F b \quad \text{if there is an } f \in F \text{ such that } b \odot f \leq a \text{ and } a \odot f \leq b.$$

*Then we call $\sim_F$ the* congruence induced by $\boldsymbol{F}$.

Equivalence relations of this type do not only preserve the tomonoid structure, but also all the three properties that we generally assume here.

LEMMA 4.2.9. *Let $\boldsymbol{L}$ be a q.n.c. tomonoid, and let $\boldsymbol{F}$ be a filter of $\boldsymbol{L}$. Then the congruence induced by $\boldsymbol{F}$ is a tomonoid congruence, and the tomonoid quotient is again quantic, negative, and commutative.*

*Proof.* It is easily checked that $\sim_F$ is compatible with $\odot$ and that the equivalence classes are convex. Clearly, negativity and commutativity are preserved.

Our next aim is to prove that the tomonoid quotient $[\boldsymbol{L}]_{\sim_F}$ is almost complete. For simplification, equivalence classes w.r.t. $\sim_F$ will be denoted by $[\cdot]$. We will prove the following statement, which obviously implies almost completeness:

($\star$) Let $a_\iota \in L$, $\iota \in I$, be such that among $[a_\iota]$, $\iota \in I$, there is no largest element; then

$$\bigvee_\iota [a_\iota] = [\bigvee_\iota a_\iota]. \tag{7}$$

To see ($\star$), let $a_\iota \in L$, $\iota \in I$, and assume that the $[a_\iota]$ do not possess a largest element. Let $a = \bigvee_\iota a_\iota$. Then $[a] \geq [a_\iota]$ for all $\iota$. Moreover, let $b \in L$ be such that $[b] \geq [a_\iota]$ for all $\iota$. Then $b$ is not equivalent to any $a_\iota$, hence $[b] > [a_\iota]$; consequently $b > a_\iota$ for all $\iota$, so that $b \geq a$ and $[b] \geq [a]$. Thus (7) follows.

It remains to show that $\odot$ distributes over suprema in $[\boldsymbol{L}]_{\sim_F}$. Let $b_\iota \in L$, $\iota \in I$, and $a \in L$. Assume first that the elements $[a \odot b_\iota]$, $\iota \in I$, do not possess a maximal element. Then also the $[b_\iota]$ do not possess a maximal element, and ($\star$) implies

$$[a] \odot \bigvee_\iota [b_\iota] = \bigvee_\iota ([a] \odot [b_\iota]). \tag{8}$$

Assume second that the $[a \odot b_\iota]$ possess the maximal element $[a \odot b_\kappa]$, but that the $[b_\iota]$ do not possess a maximal element. Let $\iota \in I$ such that $[b_\iota] > [b_\kappa]$. Then $a \odot b_\iota \sim a \odot b_\kappa$, and we have $a \odot b_\kappa \leq a \odot \inf[b_\iota] = a \odot \bigwedge_{f \in F}(b_\iota \odot f) \leq \bigwedge_{f \in F}(a \odot b_\iota \odot f) = \inf[a \odot b_\iota] = \inf[a \odot b_\kappa] \leq a \odot b_\kappa$. We conclude that $a \odot b_\iota = a \odot b_\kappa$ for any $\iota \in I$ such that $b_\iota > b_\kappa$. Thus $a \odot \bigvee_\iota b_\iota = a \odot b_\kappa$. By ($\star$), $[a] \odot \bigvee_\iota [b_\iota] = [a \odot \bigvee_\iota b_\iota] = [a \odot b_\kappa] = \bigvee_\iota([a] \odot [b_\iota])$, and (8) is proved.

Assume third that the $[b_\iota]$, $\iota \in I$, possess the maximal element $[b_\kappa]$. Then $[a \odot b_\kappa]$ is maximal among the $[a \odot b_\iota]$. Then obviously, (8) holds as well. □

As we will deal in the sequel exclusively with congruences induced by filter, we simplify our notation as follows.

DEFINITION 4.2.10. *Let* $\boldsymbol{L}$ *be a q.n.c. tomonoid, and let* $\boldsymbol{F}$ *be a filter of* $\boldsymbol{L}$*. Let* $\sim_F$ *be the congruence induced by* $\boldsymbol{F}$*. We will refer to the* $\sim_F$*-classes as* $\boldsymbol{F}$-classes *and we denote them by* $[\cdot]_F$*. Similarly, let* $P$ *be the quotient of* $\boldsymbol{L}$ *by* $\sim_F$*. Then we refer to* $P$ *as the* quotient of $\boldsymbol{L}$ by $\boldsymbol{F}$ *and we denote it by* $[\boldsymbol{L}]_F$*.*

*We furthermore call in this case* $\boldsymbol{L}$ *an* coextension of $P$ by $\boldsymbol{F}$*, and we refer to* $\boldsymbol{F}$ *as the* extending *tomonoid.*

### 4.3 The chain of quotients of a q.n.c. tomonoid

Each filter of a q.n.c. tomonoid induces a quotient. Let us now consider the collection of such quotients as a whole. The most basic observation is that, for any two filters, one is included in the other one: the set of all filters is totally ordered by inclusion.

DEFINITION 4.3.1. *Let* $\boldsymbol{L}$ *be a q.n.c. tomonoid. We denote the set of all filters of* $\boldsymbol{L}$ *by* $\mathbb{F}$*, and we endow* $\mathbb{F}$ *with the set-theoretical inclusion* $\subseteq$ *as a total order.*

Again, if we included in our signature the lattice operations instead of the total order, the proof of the next lemma could be kept shorter, as it would follow from the Second Isomorphism Theorem of Universal Algebra [5].

LEMMA 4.3.2. *Let* $\boldsymbol{L}$ *be a q.n.c. tomonoid, and let* $\boldsymbol{F}$ *and* $\boldsymbol{G}$ *be filters of* $\boldsymbol{L}$ *such that* $\boldsymbol{F} \subseteq \boldsymbol{G}$*. Then* $[\boldsymbol{G}]_F$ *is a filter of* $[\boldsymbol{L}]_F$*, and* $[\boldsymbol{L}]_G$ *is isomorphic to the quotient of* $[\boldsymbol{L}]_F$ *by* $[\boldsymbol{G}]_F$*.*

*Proof.* We claim that, for $a, b \in L$, $a \sim_G b$ if and only if $[a]_F \sim_{[G]_F} [b]_F$. Indeed, assume $a \leq b$; then $a \sim_G b$ if and only if there a $g \in G$ such that $b \odot g \leq a$. Since $\boldsymbol{F}$ is a filter contained in $\boldsymbol{G}$, the latter holds if and only if there are a $g \in G$ and an $f \in F$ such that $b \odot g \odot f \leq a$ if and only if $[b \odot g]_F \leq [a]_F$ for some $g \in G$ if and only if $[b]_F \odot [g]_F \leq [a]_F$ for some $[g]_F \in [\boldsymbol{G}]_F$ if and only if $[a]_F \sim_{[G]_F} [b]_F$.

It follows that we can define

$$\varphi\colon [\boldsymbol{L}]_G \to [[\boldsymbol{L}]_F]_{[G]_F}, \quad [a]_G \mapsto [[a]_F]_{[G]_F}$$

and that $\varphi$ is a bijection. Moreover, $\varphi$ preserves $\odot$ and is an order-isomorphism. The lemma follows. □

Lemma 4.3.2 is the basis of our loose statement that a q.n.c. tomonoid is the result of a linear construction process. In general, this process does not proceed in a stepwise fashion. But we can speak about a single step if there is a pair of successive filters; the following definition addresses this case.

For an element $a$ of a tomonoid and $n \geq 1$, we write $a^n$ for $a \odot \ldots \odot a$ ($n$ factors).

DEFINITION 4.3.3. *A q.n.c. tomonoid $\boldsymbol{L}$ is called* Archimedean *if, for each $a, b \in L$ such that $a < b < 1$, we have $b^n \leq a$ for some $n \geq 1$. A coextension of a q.n.c. tomonoid by an Archimedean tomonoid is called* Archimedean.

For two filters $\boldsymbol{F}, \boldsymbol{G} \in \mathbb{F}$, we will write $\boldsymbol{F} \subset\!\cdot\, \boldsymbol{G}$ to express that $\boldsymbol{G}$ is the immediate successor of $\boldsymbol{F}$ in $\mathbb{F}$, that is, $\boldsymbol{G}$ is the next smallest filter to $\boldsymbol{F}$.

THEOREM 4.3.4. *Let $\boldsymbol{L}$ be a q.n.c. tomonoid. Then we have:*

(i) *The largest and smallest elements of $\mathbb{F}$ are $\boldsymbol{L}$ and $\{1\}$, respectively. Moreover, $[\boldsymbol{L}]_L$ is the trivial tomonoid, and $[\boldsymbol{L}]_{\{1\}}$ is isomorphic to $\boldsymbol{L}$.*

(ii) *For each $\boldsymbol{F} \in \mathbb{F} \backslash \{\boldsymbol{L}\}$ such that $\boldsymbol{F}$ is not an immediate predecessor, $\sim_F = \bigcap_{\boldsymbol{G} \supset \boldsymbol{F}} \sim_G$.*

(iii) *For each $\boldsymbol{F} \in \mathbb{F} \backslash \{\{1\}\}$ such that $\boldsymbol{F}$ is not an immediate successor, $\sim_F = \bigcup_{\boldsymbol{G} \subset \boldsymbol{F}} \sim_G$.*

(iv) *For each $\boldsymbol{F}, \boldsymbol{G} \in \mathbb{F}$ such that $\boldsymbol{F} \subset\!\cdot\, \boldsymbol{G}$, $[\boldsymbol{L}]_F$ is an Archimedean coextension of $[\boldsymbol{L}]_G$.*

*Proof.* (i) The largest filter is $\boldsymbol{L}$, and the quotient $[\boldsymbol{L}]_L$ is one-element, that is, trivial. The smallest filter is $\{1\}$, and the quotient $[\boldsymbol{L}]_{\{1\}}$ has singleton classes only, that is, coincides with $\boldsymbol{L}$.

(ii) Let $\boldsymbol{F} \in \mathbb{F}$ such that $\boldsymbol{F}$ is neither $\boldsymbol{L}$ nor the predecessor of another filter. As $\mathbb{F}$ is closed under arbitrary intersections, we then have $F = \bigcap_{\boldsymbol{G} \supset \boldsymbol{F}} G$. Let $a, b \in L$ such that $a \leq b$. We have to show that $a \sim_F b$ if and only if, for each $\boldsymbol{G} \supset \boldsymbol{F}$, $a \sim_G b$. Clearly, $a \sim_F b$ implies $a \sim_G b$ for each $\boldsymbol{G} \supset \boldsymbol{F}$. Conversely, assume $a \sim_G b$ for each $\boldsymbol{G} \supset \boldsymbol{F}$. Then for each $\boldsymbol{G} \supset \boldsymbol{F}$ there is a $g_G \in G$ such that $b \odot g_G \leq a$. It follows $b \odot f \leq a$, where $f = \bigvee_{\boldsymbol{G} \supset \boldsymbol{F}} g_G \in F$, hence $a \sim_F b$.

(iii) Let $\boldsymbol{F} \in \mathbb{F}$ such that $\boldsymbol{F}$ is neither $\{1\}$ nor the successor of another filter. As $\mathbb{F}$ is closed under arbitrary unions, $F = \bigcup_{\boldsymbol{G} \subset \boldsymbol{F}} G$ then. For $a \leq b$, we have $a \sim_F b$ if and only if $b \odot f \leq a$ for some $f \in F$ if and only if $b \odot f \leq a$ for some $f \in G$ such that $\boldsymbol{G} \subset \boldsymbol{F}$ if and only if $a \sim_G b$ for some $\boldsymbol{G} \subset \boldsymbol{F}$.

(iv) Let $\boldsymbol{F}, \boldsymbol{G} \in \mathbb{F}$ such that $\boldsymbol{F} \subset\!\!\cdot\, \boldsymbol{G}$. By Lemma 4.3.2, $[\boldsymbol{L}]_G$ is then isomorphic to the quotient of $[\boldsymbol{L}]_F$ by the filter $[\boldsymbol{G}]_F$. Assume that $[\boldsymbol{G}]_F$ is not Archimedean. Then there is a filter $\boldsymbol{H}$ of $[\boldsymbol{G}]_F$ such that $\{[1]_F\} \subset H \subset [G]_F$. But then $\bigcup H$ is a filter of $\boldsymbol{L}$ such that $F \subset \bigcup H \subset G$, a contradiction. □

## 4.4 The Cayley tomonoid

A monoid can be identified with a monoid under composition of mappings, namely, with the set of mappings acting on the monoid by left (or right) multiplication. This is the regular representation [8], which is due to A. Cayley for the case of groups. If the monoid is commutative, any two of the mappings commute. Moreover, the presence of a compatible total order on the monoid means that the mappings are order-preserving.

Representations of partially ordered monoids by order-preserving mappings have been studied in a more general context under the name $S$-posets [16]. An adaptation of our terminology might be a future issue; for the results presented in this chapter, however, such a step would most likely not improve clarity.

The reason to consider the regular representation of tomonoids is twofold. Most important, it gives us a means to specify tomonoid coextensions in a "modular" way. We will see below that the coextension of a tomonoid splits up into constituents each of which we may specify separately. Second, there is an informal aspect that, in the context of structures that are as theoretical as residuated chains, should not be neglected. The regular representation provides a geometric view on tomonoids that is not to be mixed up with the traditionally used three-dimensional graphs of t-norms. It is rather well in line with the algebraic orientation of this study, visualising quotients in a clear way and, in addition, getting along with two dimensions.

DEFINITION 4.4.1. *Let $\langle R, \leq \rangle$ be a chain, and let $\Phi$ be a set of order-preserving mappings from $R$ to $R$. We denote by $\leq$ the pointwise order on $\Phi$, by $\circ$ the functional composition, and by $\mathrm{id}_R$ the identity mapping on $R$. Assume that* (i) *$\leq$ is a total order on $\Phi$,* (ii) *$\Phi$ is closed under $\circ$, and* (iii) *$id_R \in \Phi$. Then we call $\boldsymbol{\Phi} = \langle \Phi, \leq, \circ, id_R \rangle$ a* composition tomonoid *on $R$.*

In order to characterise the composition tomonoids associated with q.n.c. tomonoids, we introduce the following properties of a composition tomonoid $\boldsymbol{\Phi}$ on a chain $R$:

(C1) $\circ$ is commutative.

(C2) $id_R$ is the top element.

(C3) Every $\lambda \in \Phi$ is sup-preserving.

(C4) Pointwise calculated suprema of non-empty subsets of $\Phi$ exist and are in $\Phi$.

(C5) $R$ has a top element 1, and for each $a \in R$ there is a unique $\lambda \in \Phi$ such that $\lambda(1) = a$.

PROPOSITION 4.4.2. *Let* $\mathbf{\Phi}$ *be a composition tomonoid over a chain* $R$. *Then* $\mathbf{\Phi}$ *is a tomonoid. Furthermore, we have:*

(i) $\mathbf{\Phi}$ *is commutative if and only if* $\mathbf{\Phi}$ *fulfils* (C1).

(ii) $\mathbf{\Phi}$ *is negative if and only if* $\mathbf{\Phi}$ *fulfils* (C2).

(iii) *If* $\mathbf{\Phi}$ *fulfils* (C3) *and* (C4), $\mathbf{\Phi}$ *is quantic.*

*Proof.* The fact that $\mathbf{\Phi}$ is a tomonoid is easily checked, and so are parts (i) and (ii).

Assume (C3) and (C4). Then any non-empty subset of $\Phi$ possesses by (C4) w.r.t. the pointwise order a supremum; that is, $\mathbf{\Phi}$ is almost complete. Furthermore, let $\lambda_\iota, \mu \in \Phi$, $\iota \in I$. Then we have by (C4) for any $r \in R$

$$(\bigvee_\iota \lambda_\iota \circ \mu)(r) = (\bigvee_\iota \lambda_\iota)(\mu(r)) = \bigvee_\iota \lambda_\iota(\mu(r)) = \bigvee_\iota (\lambda_\iota \circ \mu)(r) = (\bigvee_\iota (\lambda_\iota \circ \mu))(r).$$

Moreover, we have by (C3) and (C4) for any $r \in R$

$$(\mu \circ \bigvee_\iota \lambda_\iota)(r) = \mu(\bigvee_\iota \lambda_\iota(r)) = \bigvee_\iota \mu(\lambda_\iota(r)) = \bigvee_\iota (\mu \circ \lambda_\iota)(r) = (\bigvee_\iota (\mu \circ \lambda_\iota))(r).$$

We conclude that $\mathbf{\Phi}$ is quantic. □

By Proposition 4.4.2, each composition tomonoid is a tomonoid. We next recall that, conversely, each tomonoid can be viewed as a composition tomonoid.

PROPOSITION 4.4.3. *Let* $\langle L, \leq, \odot, 1 \rangle$ *be a q.n.c. tomonoid. For each* $a \in L$, *put*

$$\lambda_a \colon L \to L, \quad x \mapsto a \odot x, \tag{9}$$

*and let* $\Lambda = \{\lambda_a \mid a \in L\}$. *Then* $\langle \Lambda, \leq, \circ, \mathrm{id}_L \rangle$ *is a composition tomonoid on* $L$ *fulfilling* (C1)–(C5). *Moreover,*

$$\pi \colon L \to \Lambda, \quad a \mapsto \lambda_a \tag{10}$$

*is an isomorphism of the tomonoids* $\langle L, \leq, \odot, 1 \rangle$ *and* $\langle \Lambda, \leq, \circ, \mathrm{id}_L \rangle$.

DEFINITION 4.4.4. *Let* $\mathbf{L} = \langle L, \leq, \odot, 1 \rangle$ *be a tomonoid. For each* $a \in L$, *the mapping* $\lambda_a$ *defined by* (9) *is called the* (left) translation *by* $a$. *Furthermore, the composition tomonoid* $\langle \Lambda, \leq, \circ, \mathrm{id}_L \rangle$ *assigned to* $\mathbf{L}$ *according to Proposition 4.4.3 is called the* Cayley tomonoid *associated with* $\mathbf{L}$.

Let us state what Proposition 4.4.3 means for t-norms: a left-continuous t-norm corresponds to a monoid under composition of pairwise commuting, order-preserving, and left-continuous mappings from $[0, 1]$ to $[0, 1]$ such that for any $a \in [0, 1]$ exactly one of them maps 1 to $a$.

EXAMPLE 4.4.5. The Cayley tomonoids associated with the three standard t-norms are shown in Figure 1. A selection of translations is indicated in a schematic way.

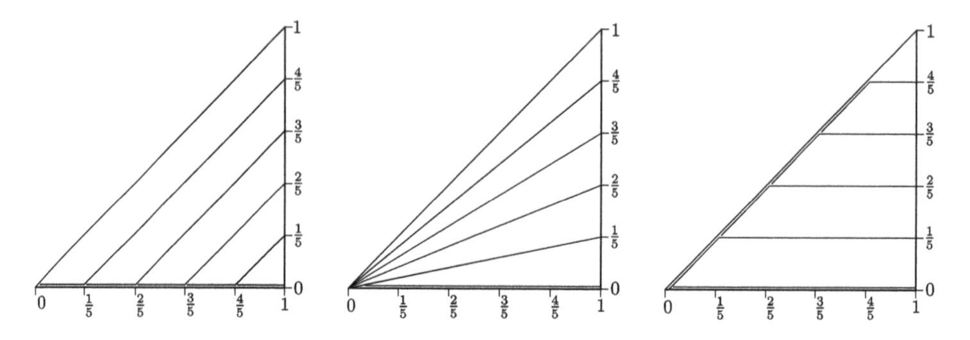


Figure 1. The Cayley tomonoids associated with the tomonoids based on the Łukasiewicz, product, and Gödel t-norm

## Quotients and Cayley tomonoids

Our next aim is to see how quotients of a q.n.c. tomonoid are reflected by its associated Cayley tomonoid.

We will use the following notation and conventions. Let $\boldsymbol{L}$ be a q.n.c. tomonoid and let $\boldsymbol{P}$ be the quotient of $\boldsymbol{L}$ by the filter $\boldsymbol{F}$. Then any $R \in P$ will be considered as a subset of $L$, namely as a class of the congruence on $\boldsymbol{L}$ that yields $\boldsymbol{P}$.

For any $f \in F$, $\lambda_f$ maps $R$ to itself. We write $\lambda_f^R \colon R \to R$ for $\lambda_f$ with its domain and range being restricted to $R$, and we put $\Lambda^R = \{\lambda_f^R \mid f \in F\}$. Note that $\Lambda^F$ is the Cayley tomonoid associated with $\boldsymbol{F}$.

Moreover, let $R \in P$ and $T \in P \backslash \{F\}$, and let $S = R \odot T$. Then for any $t \in T$, $\lambda_t$ maps $R$ to $S$. We write $\lambda_t^{R,S} \colon R \to S$ for $\lambda_t$ with its domain restricted to $R$ and its range restricted to $S$, and we put $\Lambda^{R,S} = \{\lambda_t^{R,S} \mid t \in T\}$.

Finally, we denote a function that maps all values of a set $A$ to the single value $b$ by $c^{A,b}$.

The following lemma describes the sets $\Lambda^R$, where $R$ is an $\boldsymbol{F}$-class; cf. Figure 2.

LEMMA 4.4.6. *Let $\mathbf{L} = \langle L, \leq, \odot, 1 \rangle$ be a q.n.c. tomonoid that possesses the non-trivial filter $\boldsymbol{F}$. Let $\mathbf{P}$ be the quotient of $\mathbf{L}$ induced by $\boldsymbol{F}$.*

(i) *The top element of $\mathbf{P}$ is $F$. Let $u = \inf F \in L^0$; then $u < 1$, and $F$ is one of $(u, 1]$ or $[u, 1]$. Moreover, $\langle \Lambda^F, \leq, \circ, id_F \rangle$ is the Cayley tomonoid of $\boldsymbol{F}$. We have:*

  (a) *Let $f \in F$. If $F = [u, 1]$, $\lambda_f^F(u) = u$; if $F = (u, 1]$, $\bigwedge_{g \in F} \lambda_f^F(g) = u$. Moreover, $\lambda_f^F(1) = f$.*

  (b) *If $F = [u, 1]$, $\Lambda^F$ has the bottom element $c^{F,u}$.*

*Finally,*

$$\pi \colon F \to \Lambda^F, \quad f \mapsto \lambda_f^F$$

*is an isomorphism between $\langle F, \leq, \odot, 1 \rangle$ and $\langle \Lambda^F, \leq, \circ, id_F \rangle$.*

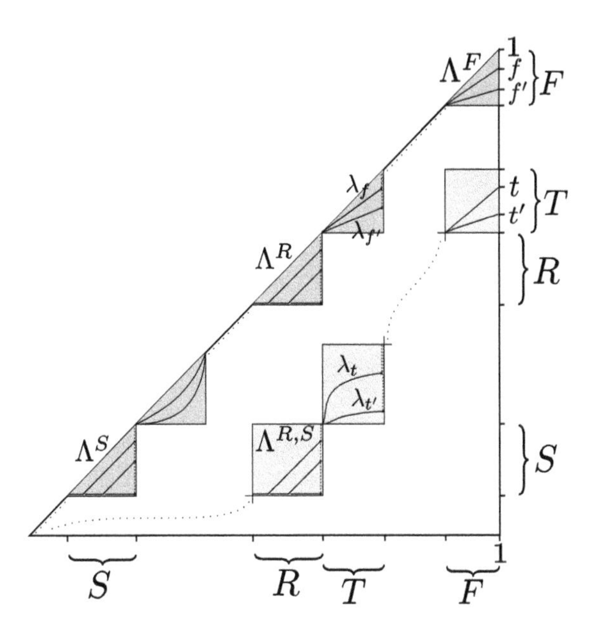


Figure 2. The Cayley tomonoid associated with a q.n.c. tomonoid $\boldsymbol{L}$, which possesses a filter $\boldsymbol{F}$. The translations by the elements $t', t, f', f, 1$ are depicted schematically. It is furthermore indicated how the Cayley tomonoid associated with the quotient of $\boldsymbol{L}$ by $\boldsymbol{F}$ arises. The translation by an element $T$ is shown in light grey, the translation by $F$, which is the identity mapping, is shown in dark grey.

(ii) *Let $R \in P \backslash \{F\}$. Let $u = \inf R$ and $v = \sup R$. If $u = v$, then $R = \{u\}$ and $\lambda_f^R(u) = u$ for any $f \in F$. Assume now $u < v$. Then $R$ is one of $(u, v)$, $[u, v)$, $(u, v]$, or $[u, v]$. Moreover, $\langle \Lambda^R, \leq, \circ, \mathrm{id}_R \rangle$ is a composition tomonoid on $R$ fulfilling* (C1)–(C4) *as well as the following properties:*

(c) *Let $f \in F$. If $u \in R$, $\lambda_f^R(u) = u$; if $u \notin R$, $\bigwedge_{r \in R} \lambda_f^R(r) = u$. Moreover, if $v \notin R$, $\bigvee_{r \in R} \lambda_f^R(r) = \lambda_f(v) = v$.*

(d) *If $R = [u, v]$, then $\Lambda^R$ has the bottom element $c^{R,u}$. If $R = [u, v)$, then $c^{R,u} \notin \Lambda^R$.*

*Finally,*

$$\varrho \colon F \to \Lambda^R, \quad f \mapsto \lambda_f^R \tag{11}$$

*is a surjective sup-preserving homomorphism from $\langle F, \leq, \odot, 1 \rangle$ to $\langle \Lambda^R, \leq, \circ, \mathrm{id}_R \rangle$.*

*Proof.* (i) Here, Proposition 4.4.3 is applied to the q.n.c. tomonoid $\boldsymbol{F}$. (a): Let $f \in F$. If $u \in F$, clearly $\lambda_f^F(u) = u$. If $u \notin F$, we have $u \leq \bigwedge_{g \in F} \lambda_f^F(g) \leq \bigwedge_{g \in F} g = u$, that is, $\bigwedge_{g \in F} \lambda_f^R(g) = u$. Clearly, $\lambda_f(1) = f \odot 1 = f$. (b): If $u \in F$, $\lambda_u^F = c^{F,u}$ is the bottom element of $\Lambda^F$.

(ii) The case that $R$ is a singleton is trivial. Assume $u < v$: The fact that $\Lambda^R$ is a composition tomonoid fulfilling (C1)–(C4) and that $\varrho$, defined by (11), is a sup-preserving and surjective homomorphism follows from Proposition 4.4.3. (c): Let $f \in F$. We see like in the proof of (a) that $\lambda^R_f(u) = u$ if $u \in R$, and $\bigwedge_{r \in R} \lambda^R_f(r) = u$ otherwise. Moreover, if $v \notin R$, then $\lambda_f(v) \notin R$ and consequently $r < \lambda_f(v) \leq v$ for any $r \in R$, that is, $\lambda_f(v) = v$. (d): Assume $u \in R$. If $R$ has a largest element $v$ as well, $v \odot z = u$ for some $z \in F$, and hence $c^{R,u} = \lambda^R_z \in \Lambda^R$. If $R$ does not contain its supremum $v$, then by (c), $\bigvee_{r \in R} \lambda^R_f(r) = v$ for any $f \in F$, and it follows $c^{R,u} \notin \Lambda^R$. □

We next turn to the set $\Lambda^{R,S}$, where $R$ and $S$ are two $\boldsymbol{F}$-classes; cf. again Figure 2. In what follows, we call a pair $A, B$ of elements of the q.n.c. tomonoid $P$ *$\odot$-maximal* if $A = B \to A \odot B$ and $B = A \to A \odot B$.

LEMMA 4.4.7. *Let $\boldsymbol{L} = \langle L, \leq, \odot, 1 \rangle$ be a q.n.c. tomonoid that possesses the non-trivial filter $\boldsymbol{F}$. Let $\boldsymbol{P}$ be the quotient of $\boldsymbol{L}$ induced by $\boldsymbol{F}$. Let $R, T \in P$ such that $T < F$, and let $S = R \odot T$.*

(i) *Let $R, T$ be $\odot$-maximal. Then $S < R$. Let $u = \inf R,\ v = \sup R,\ u' = \inf S$, and $v' = \sup S$. If $u = v$, then $R = \{u\}$, $u' \in S$, and $\lambda^{R,S}_t(u) = u'$ for all $t \in T$. If $u' = v'$, then $S = \{u'\}$ and $\lambda^{R,S}_t = c^{R,u'}$ for all $t \in T$.*

*Assume now $u < v$ and $u' < v'$. If then $u \in R$, we have $u' \in S$. Moreover, $\Lambda^{R,S} = \{\lambda^{R,S}_t \mid t \in T\}$ is a set of mappings from $R$ to $S$ such that:*

(a) *$R$ and $S$ are conditionally complete, and for any $t \in T$, the mapping $\lambda^{R,S}_t$ is sup-preserving.*

(b) *Let $t \in T$. If $u \in R$, $\lambda^{R,S}_t(u) = u'$; if $u \notin R$, $\bigwedge_{r \in R} \lambda^{R,S}_t(r) = u'$.*

(c) *Under the pointwise order, $\Lambda^{R,S}$ is totally ordered.*

(d) *Let $K \subseteq \Lambda^{R,S}$ such that $\bigvee_{\lambda \in K} \lambda(r) \in S$ for all $r \in R$. Then the pointwise calculated supremum of $K$ is in $\Lambda^{R,S}$.*

(e) *If $u' \in S$ and $v \in R$, $\Lambda^{R,S}$ has the bottom element $c^{R,u'}$. If $u' \in S$ and $v \notin R$, then either $\Lambda^{R,S} = \{c^{R,u'}\}$ or $c^{R,u'} \notin \Lambda^{R,S}$. If $v \notin R$ and $v' \in S$, then $u' \in S$ and $\Lambda^{R,S} = \{c^{R,u'}\}$.*

(f) *For any $t \in T$ and $f \in F$, $\lambda^S_f \circ \lambda^{R,S}_t = \lambda^{R,S}_t \circ \lambda^R_f \in \Lambda^{R,S}$.*

*Finally,*

$$\tau\colon T \to \Lambda^{R,S},\quad t \mapsto \lambda^{R,S}_t \tag{12}$$

*is a sup-preserving mapping from $T$ to $\Lambda^{R,S}$ such that, for any $f \in F$ and $t \in T$,*

$$\tau(\lambda^T_f(t)) = \lambda^S_f \circ \tau(t) = \tau(t) \circ \lambda^R_f. \tag{13}$$

(ii) *Let $R, T$ not be $\odot$-maximal. Then $S$ contains a smallest element $u'$, and $\lambda^{R,S}_t = c^{R,u'}$ for all $t \in T$.*

*Proof.* (i) We clearly have $S \leq R$. If $S = R$, the maximal element $Y$ such that $R \odot Y = R \odot T$ would be $F$, in contradiction to the assumptions that $T < F$ and $R, T$ is a $\odot$-maximal pair. Thus $S < R$.

We consider first the case that $R$ is a singleton, that is, $R = \{u\}$. Then $u \odot f = u$ for all $f \in F$. Let $t \in T$; then $\lambda_t^{R,S}(u) \odot f = u \odot t \odot f = u \odot t = \lambda_t^{R,S}(u)$ for any $f \in F$; hence $u' \in S$ and $\lambda_t^{R,S}(u) = u'$.

The case that $S$ is a singleton is trivial. Assume now $u < v$ and $u' < v'$ and let $u \in R$: Then $\lambda_t^{R,S}(u) = u' \in S$ for any $t \in T$. Indeed, we again have $u \odot f = u$ and consequently $\lambda_t^{R,S}(u) \odot f = \lambda_t^{R,S}(u)$ for any $f \in F$.

Claims (a), (c), (d), and the fact that $\tau$, defined by (12), is sup-preserving follow from Proposition 4.4.3. We prove the remaining ones:

(b) Let $t \in T$. If $u \in R$, we have seen above that $\lambda_t^{R,S}(u) = u'$. If $u \notin R$, choose some $\tilde{r} \in R$; then $\bigwedge_{r \in R} \lambda_t^{R,S}(r) = \bigwedge_{f \in F} \lambda_t^{R,S}(\tilde{r} \odot f) = \bigwedge_{f \in F} (\lambda_t^{R,S}(\tilde{r}) \odot f) = \inf S = u'$.

(e) Let $u' \in S$ and $v \in R$. Then, for an arbitrary $\tilde{t} \in T$, $\lambda_{\tilde{t}}(v)$ and $u'$ are both in the congruence class $S$, whose smallest element is $u'$. Thus, for some $f \in F$, we have $\lambda_{\tilde{t}}(v) \odot f = u'$, and consequently $\lambda_t^{R,S} = c^{R,u'}$, where $t = \tilde{t} \odot f \in T$.

Next, let $u' \in S$ and $v \notin R$. For any $t, t' \in T$ such that $t \sim_F t'$, we have $\lambda_t(v) \sim_F \lambda_{t'}(v)$. Consequently, either $\lambda_t(v) \in S$ for all $t \in T$, or $\lambda_t(v) \notin S$ for all $t \in T$. Furthermore, from $v \notin R$ it follows $v \odot f = v$ and thus $\lambda_t(v) \odot f = v \odot t \odot f = v \odot t = \lambda_t(v)$ for all $t \in T$ and $f \in F$. We conclude that, in the former case, $\lambda_t(v) = u'$ for any $t \in T$, that is, $\Lambda^{R,S} = \{c^{R,u'}\}$. In the latter case, $v' \leq \lambda_t(v) = \bigvee_{r \in R} \lambda_t^{R,S}(r) \leq v'$, that is, $\lambda_t(v) = v'$ for all $t \in T$, and $c^{R,u'} \notin \Lambda^{R,S}$.

Finally, let $v \notin R$ and $v' \in S$. Let $t \in T$. Then $\lambda_t(v) = \bigvee_{r \in R} \lambda_t^{R,S}(r) \in S$ and $\lambda_t(v) \odot f = v \odot t \odot f = v \odot t = \lambda_t(v)$ for any $f \in F$; thus $\lambda_t(v) = u' \in S$, that is, $\lambda_t^{R,S} = c^{R,u'}$, and we conclude again $\Lambda^{R,S} = \{c^{R,u'}\}$.

(f) Let $t \in T$, $f \in F$, and $r \in R$. Then we have $(\lambda_f^S \circ \lambda_t^{R,S})(r) = r \odot t \odot f = \lambda_{t \odot f}^{R,S}(r) = \lambda_{f \odot t}^{R,S}(r) = r \odot f \odot t = (\lambda_t^{R,S} \circ \lambda_f^R)(r)$.

Furthermore, $\tau(\lambda_f^T(t))(r) = \lambda_{t \odot f}^{R,S}(r) = r \odot t \odot f$, and also (13) follows.

(ii) Consider first the case that there is an $R' > R$ such that $R' \odot T = S$. Let $r \in R$, $t \in T$, and $r' \in R'$. Then $r < r' \odot f$ for any $f \in F$, and consequently $r \odot t \leq r' \odot f \odot t$ for any $f \in F$. As $r' \odot t \in S$, we conclude that $r \odot t$ is the smallest element of $S$, that is, $\lambda_t^{R,S}(r) = r \odot t = u'$, where $u' = \inf S \in S$.

Similarly, we argue in the case that there is a $T' > T$ such that $R \odot T' = S$. Let $r \in R$, $t \in T$, and $t' \in T'$. Then $t < t' \odot f$ for any $f \in F$, and consequently $r \odot t \leq r \odot t' \odot f$ for any $f \in F$. We conclude again that $u' = \inf S \in S$ and $\lambda_t^{R,S}(r) = r \odot t = u'$. □

Again, let $\boldsymbol{L}$ be a q.n.c. tomonoid, $\boldsymbol{F}$ a filter of $\boldsymbol{L}$, and $\boldsymbol{P}$ the quotient of $\boldsymbol{L}$ by $\boldsymbol{F}$. From an intuitive point of view, we may say with reference to Figure 2 that the Cayley tomonoid associated with $\boldsymbol{L}$ is composed from triangular and rectangular sections, one for each $R \in P$ and for each pair of elements $R, S \in P$, respectively. Lemma 4.4.6(i) deals with top congruence class, the filter $F$, whose associated Cayley tomonoid $\Lambda^F$ is located in the uppermost triangle. Lemma 4.4.6(ii) describes the set $\Lambda^R$ for some $R \in P \backslash \{F\}$, located in one of the remaining triangles. Finally, let $S = R \odot T < R$, where $R, S, T \in P$. Then Lemma 4.4.7 deals with $\Lambda^{R,S}$, located in the rectangular section associated with $R$ and $S$. If $R, T$ is not $\odot$-maximal, $\Lambda^{R,S}$ is trivial by part (ii).

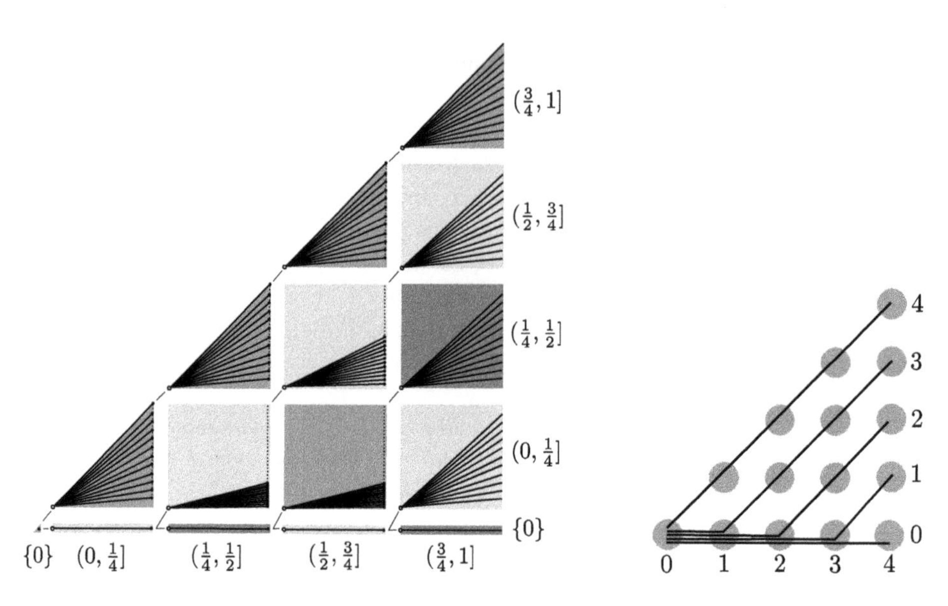


Figure 3. Left: The tomonoid $\langle [0,1], \leq, \star_{\rm H}, 1 \rangle$ (to increase clarity, we have separated the congruence classes by margins). Right: The five-element quotient $\boldsymbol{L_5}$ by the filter $(\frac{3}{4}, 1]$.

We will provide in the sequel some examples of q.n.c. tomonoids based on left-continuous t-norms. Definitions of t-norms are often involved; to keep them as short as possible, we will in general not provide full specifications, but assume commutativity to be used to cover all cases.

EXAMPLE 4.4.8. Let us consider the following t-norm:

$$a \star_{\rm H} b = \begin{cases} 4ab - 3a - 3b + 3 & \text{if } a, b > \frac{3}{4} \\ 4ab - 3a - 2b + 2 & \text{if } \frac{1}{2} < a \leq \frac{3}{4} \text{ and } b > \frac{3}{4} \\ 4ab - 3a - b + 1 & \text{if } \frac{1}{4} < a \leq \frac{1}{2} \text{ and } b > \frac{3}{4} \\ 4ab - 3a & \text{if } a \leq \frac{1}{4} \text{ and } b > \frac{3}{4} \\ 2ab - a - b + \frac{3}{4} & \text{if } \frac{1}{2} < a, b \leq \frac{3}{4} \\ ab - \frac{1}{2}a - \frac{1}{4}b + \frac{1}{8} & \text{if } \frac{1}{4} < a \leq \frac{1}{2} \text{ and } \frac{1}{2} < b \leq \frac{3}{4} \\ 0 & \text{if } a \leq \frac{1}{4} \text{ and } \frac{1}{2} < b \leq \frac{3}{4} \text{ or } a, b \leq \frac{1}{2}. \end{cases} \tag{14}$$

$\star_{\rm H}$ is a modification of a t-norm defined by Hájek in [22]. The tomonoid $\langle [0,1], \leq, \star_{\rm H}, 1 \rangle$ possesses the filter $F = (\frac{3}{4}, 1]$ and the $\boldsymbol{F}$-classes are $\{0\}$, $(0, \frac{1}{4}]$, $(\frac{1}{4}, \frac{1}{2}]$, $(\frac{1}{2}, \frac{3}{4}]$, and $(\frac{3}{4}, 1]$. The quotient by $\boldsymbol{F}$ is isomorphic to $\boldsymbol{L_5}$, the five-element Łukasiewicz chain. An illustration, showing the sets $\Lambda^R$ and $\Lambda^{R,S}$ for all the congruence classes $R$ and $S$, can be found in Figure 3.

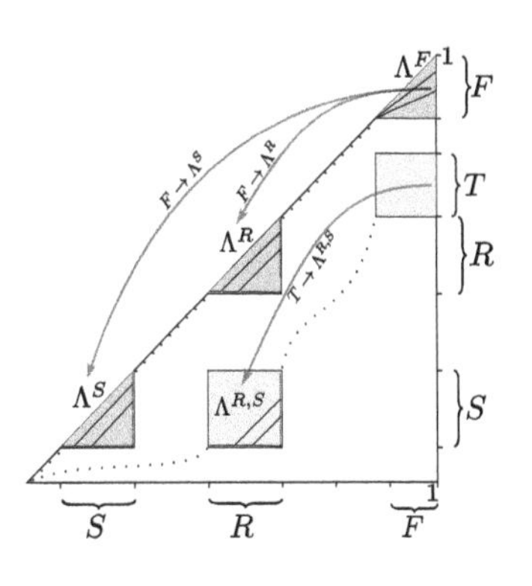


- The Cayley tomonoid $\Lambda^F$ of the extending filter $\boldsymbol{F}$: Proposition 4.5.3.
- For each $R \in P$ such that $R < F$,
    - The composition tomonoid $\Lambda^R$: Theorem 4.5.6
    - The homomorphism $F \to \Lambda^R,\ f \mapsto \lambda^R_f$: Proposition 4.5.7.
- For each pair $R, S \in P$ such that $S = R \odot T$ for some $T \in P \backslash \{F\}$,
    - The set of mappings $\Lambda^{R,S}$: Proposition 4.5.8
    - The mapping $T \to \Lambda^{R,S},\ t \mapsto \lambda^{R,S}_t$: Proposition 4.5.9.

Figure 4. The way we specify a real Archimedean coextension of a q.n.c. tomonoid.

### 4.5 Real Archimedean coextensions

As we have seen in Subsection 4.3, we may associate with a q.n.c. tomonoid the chain of quotients induced by its filters. We shall now have a closer look at the case of two successive elements of this chain. By part (iv) of Theorem 4.3.4, the corresponding coextension is in this case Archimedean.

In what follows, a *real interval* is meant to be a one-element set or one of $(a, b)$, $(a, b]$, $[a, b)$, or $[a, b]$ for $a, b \in \mathrm{R}$ such that $a < b$.

DEFINITION 4.5.1. *Let $\boldsymbol{P}$ be the quotient of the q.n.c. tomonoid $\boldsymbol{L}$ by an Archimedean filter such that each congruence class is order-isomorphic to a real interval. Then we call $\boldsymbol{P}$ a* real Archimedean quotient *of $\boldsymbol{L}$, and we call $\boldsymbol{L}$ a* real Archimedean coextension *of $\boldsymbol{P}$.*

Given a q.n.c. tomonoid $\boldsymbol{P}$, our aim is to describe the real Archimedean coextensions $\boldsymbol{L}$ of $\boldsymbol{P}$. To this end, we will specify, following the lines of Lemmas 4.4.6 and 4.4.7, sectionwise the Cayley tomonoid $\Lambda$ of $\boldsymbol{L}$. That is, for each pair $R$ and $S$ of congruence classes, we specify the translations restricted in domain to $R$ and in range to $S$. Figure 4 may serve as a guide through this section, indicating which section of the Cayley tomonoid is described in which proposition or theorem.

We will use a few auxiliary notions. A *non-minimal* element of a chain $A$ is any $a \in A$ such that $a$ is not the smallest element of $A$. Furthermore, let $\chi$ be an order-preserving mapping from $A$ to another chain $B$. Then we call the set $\{x \in A \mid \chi(x)$ is non-minimal in $B\}$ the *support* of $\chi$. Obviously, the support of $\chi$ is the whole set $A$ if $B$ does not possess a smallest element; and the support of $\chi$ is empty if and only if $B$ possesses a smallest element $u$ and $\chi = c^{A,u}$.

A composition tomonoid $\Phi$ on a chain $R$ will be called *c-isomorphic* to another composition tomonoid $\Psi$ on a chain $S$ if there is an order isomorphism $\iota \colon R \to S$ such that $\Psi = \{\iota \circ \lambda \circ \iota^{-1} \mid \lambda \in \Phi\}$. Note that c-isomorphic composition tomonoids are also isomorphic (as tomonoids); the converse, however, does not in general hold.

We will first be concerned with the sets $\Lambda^R$, where $R$ is an element of the quotient of the tomonoid $\boldsymbol{L}$ that we are going to construct, identifiable with an element of the given tomonoid $\boldsymbol{P}$.

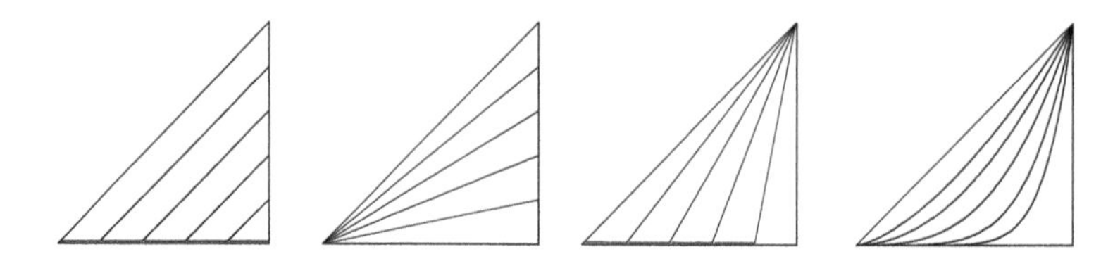

Figure 5. The standard composition tomonoids

DEFINITION 4.5.2.

(i) *Let* $\Phi$ *consist of the functions* $\lambda_t\colon [0,1] \to [0,1],\quad x \mapsto (x+t-1) \vee 0$ *for each* $t \in [0,1]$. *Then* $\langle \Phi, \leq, \circ, id_{[0,1]} \rangle$ *is called the* Łukasiewicz composition tomonoid.

(ii) *Let* $\Phi$ *consist of the functions* $\lambda_t\colon (0,1] \to (0,1],\quad x \mapsto t \cdot x$ *for each* $t \in (0,1]$. *Then* $\langle \Phi, \leq, \circ, id_{(0,1]} \rangle$ *is called the* product composition tomonoid.

(iii) *Let* $\Phi$ *consist of the functions* $\lambda_t\colon [0,1) \to [0,1),\quad x \mapsto \frac{(x+t-1)\vee 0}{t}$ *for each* $t \in (0,1]$. *Then* $\langle \Phi, \leq, \circ, id_{[0,1)} \rangle$ *is called the* reversed product composition tomonoid.

(iv) *Let* $\Phi$ *consist of the functions* $\lambda_t\colon (0,1) \to (0,1),\quad x \mapsto x^{\frac{1}{t}}$ *for each* $t \in (0,1]$. *Then* $\langle \Phi, \leq, \circ, id_{(0,1)} \rangle$ *is called the* power composition tomonoid.

*A composition tomonoid on a chain* $R$ *that is c-isomorphic to one of these four will be called a* standard composition tomonoid.

The four standard composition tomonoids are schematically shown in Figure 5. Note that the key property in which they differ is their base set: the real unit interval with, without the left, right margin.

We start by specifying the uppermost composition tomonoid.

PROPOSITION 4.5.3. *Let* $\boldsymbol{P}$ *be a real Archimedean quotient of the q.n.c. tomonoid* $\boldsymbol{L}$ *by the filter* $\boldsymbol{F}$. *Then* $\langle \Lambda^F, \leq, \circ, id_F \rangle$ *is c-isomorphic to the Łukasiewicz composition tomonoid or to the product composition tomonoid.*

*Proof.* By assumption, $F$ is order-isomorphic to a real interval, which is right-closed and either left-closed or left-open.

Assume first that $F$ possesses a smallest element. Then $\langle F, \leq, \odot, 1 \rangle$ is a q.n.c. tomonoid such that $F$ is order-isomorphic to the real unit interval; that is, $\boldsymbol{F}$ is isomorphic to a tomonoid based on a t-norm. Let $\odot$ be this t-norm. By assumption, $\boldsymbol{F}$ is Archimedean. By [29, Proposition 2.16], $\odot$ is continuous, and due to the Archimedean property, $\odot$ is in fact isomorphic to the Łukasiewicz t-norm. Consequently, $\Lambda^F$ is c-isomorphic to the Łukasiewicz composition tomonoid.

Assume second that $F$ does not have a smallest element. Then $\langle F^0, \leq, \odot, 1 \rangle$ is a q.n.c. tomonoid such that $F^0$ is order-isomorphic to the real unit interval; that is, $\boldsymbol{F}^0$ is again isomorphic to a tomonoid based on a t-norm. Let $\odot$ be this t-norm. Then $(0,1]$ together with the restriction of $\odot$ to $(0,1]$ is an Archimedean tomonoid. Thus we conclude as before from [29, Proposition 2.16] that $\odot$ is continuous, but this time isomorphic to the product t-norm. Consequently, $\Lambda^F$ is in this case c-isomorphic to the product composition tomonoid. □

Our next aim is to characterise the composition tomonoids associated with some of the remaining congruence classes. Several preparations are needed.

For chains $A$ and $B$ that are order-isomorphic to real intervals, *continuity* of a mapping from $A$ to $B$ will be understood in the obvious way.

LEMMA 4.5.4. *Let $\boldsymbol{P}$ be a real Archimedean quotient of the q.n.c. tomonoid $\mathbf{L}$. Let $R \in P$, and assume that $R$ is not a singleton. Then $\langle \Lambda^R, \leq, \circ, \mathrm{id}_R \rangle$ is a composition tomonoid on $R$ fulfilling* (C1)–(C4). *Moreover, the following holds:*

(C6) *Any $\lambda \in \Lambda^R$ is continuous.*

(C7) *For any $\lambda \in \Lambda^R \backslash \{id_R\}$ and any non-minimal element $r$ of $R$, $\lambda(r) < r$.*

*Proof.* As $\boldsymbol{P}$ is assumed to be a real Archimedean quotient, $R$ is order-isomorphic to a real interval. Furthermore, by Proposition 4.5.3, the extending filter $\boldsymbol{F}$ is isomorphic to $\langle [0,1], \star, \leq, 1 \rangle$, where $\star$ is the Łukasiewicz t-norm, or to $\langle (0,1], \cdot, \leq, 1 \rangle$, where $\cdot$ is the product t-norm.

By Lemma 4.4.6, $\Lambda^R$ fulfills (C1)–(C4). Thus we only have to prove (C6) and (C7). We will first show (C7) as well as a strengthened form of (C7) and then (C6).

(C7): Let $f \in F \backslash \{1\}$ and let $r \in R$. Assume that $\lambda_f^R(r) = r \odot f = r$. Then $r \odot f^n = r$ for any $n \geq 1$, and since $\boldsymbol{F}$ is Archimedean, it follows that $r \odot g = r$ for all $g \in F$; thus $r$ is the smallest element of the congruence class $R$. We conclude that if $r$ is not the smallest element of $R$, then $\lambda_f^R(r) < r$.

We next prove:

($\star$) For any $\lambda \in \Lambda^R \backslash \{id_R\}$ and any $r \in R$ that is neither the smallest nor the largest element of $R$, $\bigwedge_{x>r} \lambda(x) < r$.

Let $f \in F \backslash \{1\}$ and let $r \in R$ be neither the smallest nor the largest element of $R$. Let then $g \in F$ be such that $f \leq g^2 < g < 1$. Assume that $\lambda_g^R(x) = x \odot g > r$ for all $x \in R$ such that $x > r$; then $x \odot g^n > r$ for any $n \geq 1$, and since $\boldsymbol{F}$ is Archimedean, it further follows $x \odot h > r$ for all $h \in F$, in contradiction to the fact that $x$ and $r$ are in the same congruence class $R$. Hence there is an $x \in R$ such that $x > r$ and $\lambda_g^R(x) = x \odot g \leq r$. As $r$ is non-minimal and $\lambda_g$ is not the identity, we conclude by (C7) that $\lambda_f^R(x) = x \odot f \leq x \odot g \odot g \leq r \odot g < r$. The proof of ($\star$) is complete.

(C6): Let $f \in F$ and assume that $\lambda_f^R$ is discontinuous at $r \in R$. Note that then $f < 1$ and $r$ is neither the smallest nor the largest element of $R$. Let $p = \lambda_f^R(r)$ and $q = \bigwedge_{x>r} \lambda_f^R(x)$; then $p < q < r$ by ($\star$). By (C4) and (C7), we may choose a $\lambda \in \Lambda^R$ such that $p < \lambda(q) < q$ and $q < \lambda(r) < r$. By ($\star$), there is an $x > r$ such that $\lambda(x) \leq r$. Then $\lambda_f^R(\lambda(x)) \leq \lambda_f^R(r) = p$ and $\lambda(\lambda_f^R(x)) \geq \lambda(q) > p$, a contradiction. □

The proof of the following technical lemma proceeds according to [32].

LEMMA 4.5.5. *For each $k \in \mathrm{N}$, let $g_k$ be an order-automorphism of the open real unit interval $(0,1)$. Assume that, for each $k$,* (i) *$g_k(x) < x$ for $x \in (0,1)$;* (ii) *$g_{k+1}^2 = g_k$; and* (iii) *the functions $g_k$ converge uniformly to $\mathrm{id}_{(0,1)}$. Then there is an order-automorphism $\varphi$ of $(0,1)$ such that $g_k(\varphi(x)) = \varphi(x^{2^{(\frac{1}{2})^k}})$ for each $k \in \mathrm{N}$ and $x \in (0,1)$.*

*Proof.* We have to determine an order-automorphism $\varphi\colon (0,1) \to (0,1)$ such that $(\varphi^{-1} \circ g_k \circ \varphi)(x) = x^{2^{(\frac{1}{2})^k}}$ for each $x \in (0,1)$ and $k \geq 0$. We shall reformulate this problem twice. First, let $y = \ln x \in (-\infty, 0)$ and $\psi\colon (-\infty,0) \to (0,1),\ \ y \mapsto \varphi(e^y)$. Then $\varphi(x) = \psi(y)$, and we have to determine an order isomorphism $\psi\colon (-\infty,0) \to (0,1)$ such that $(\psi^{-1} \circ g_k \circ \psi)(y) = 2^{(\frac{1}{2})^k} y$ for each $y \in (-\infty,0)$ and $k \geq 0$. We next set $z = \ln(-y) \in \mathsf{R}$ and $\chi\colon \mathsf{R} \to (0,1),\ \ z \mapsto \psi(-e^z)$. Then $\psi(y) = \chi(z)$, and our problem is finally to find an order antiisomorphism $\chi\colon \mathsf{R} \to (0,1)$ such that

$$(\chi^{-1} \circ g_k \circ \chi)(z) = z + \tfrac{\ln 2}{2^k}, \quad z \in \mathsf{R},\ k \geq 0. \tag{15}$$

We set $\chi(0) = \frac{1}{2}$, and for any $k \geq 0$ and $n \in \mathsf{Z}$, let $\chi(\frac{n}{2^k} \ln 2) = g_k^n(\frac{1}{2})$. We readily check that this defines $\chi$ unambiguously on the set $R = \{\frac{n}{2^k} \ln 2 \mid k \geq 0,\ n \in \mathsf{Z}\}$. For $z \in R$, (15) is then fulfilled. Moreover, because $g_k(x) < x$ for each $k$ and $x \in (0,1)$, (15) implies that $\chi$ is strictly decreasing.

We next show that $\chi$ can be continuously extended to the whole real line. Let $r \in \mathsf{R}$, and let $(n_k)_k$ be the sequence of natural numbers such that $r \in [\frac{n_k}{2^k} \ln 2, \frac{n_k+1}{2^k} \ln 2)$ for every $k$. We have to prove that the length of the interval $[\chi(\frac{n_k+1}{2^k} \ln 2), \chi(\frac{n_k}{2^k} \ln 2)]$ converges to 0 for $k \to \infty$. But this is the case because $\chi(\frac{n_k+1}{2^k} \ln 2) = g_k(\chi(\frac{n_k}{2^k} \ln 2))$ and $(g_k)_k$ converges uniformly to the identity. Note that the function $\chi\colon \mathsf{R} \to [0,1]$ is decreasing and fulfils (15) because $\chi$ and the $g_k$ are continuous.

It remains to show is that $\chi$ is surjective. Recall that $g_0(x) < x$ for all $x \in (0,1)$, let $u = \bigwedge_n g_0^n(\frac{1}{2})$, and assume $u > 0$. Then, by the continuity of $g_0$, we have $g_0(u) = u$, a contradiction; so $u = 0$. Similarly, we conclude $\bigvee_n g_0^{-n}(\frac{1}{2}) = 1$. So the image of $\chi$ covers the whole interval $(0,1)$. □

New we can characterise the composition tomonoids $\Lambda^R$ for any congruence class $R$.

THEOREM 4.5.6. *Let $\boldsymbol{P}$ be a real Archimedean quotient of the q.n.c. tomonoid $\boldsymbol{L}$. Let $R \in P$, and assume that $R$ is not a singleton. Then $\langle \Lambda^R, \leq, \circ, id_R \rangle$ is a standard composition tomonoid.*

*In fact, if then $R$ has a smallest and a largest element, $\Lambda^R$ is c-isomorphic to the Łukasiewicz composition tomonoid. If $R$ has a largest but no smallest element, $\Lambda^R$ is c-isomorphic to the product composition tomonoid. If $R$ has a smallest but no largest element, $\Lambda^R$ is c-isomorphic to the reversed product composition tomonoid. If $R$ has no smallest and no largest element, $\Lambda^R$ is c-isomorphic to the power composition tomonoid.*

*Proof.* We can assume that $R$ is a real interval with the boundaries 0 and 1.

By Lemma 4.5.4, $\langle \Lambda^R, \leq, \circ, id_R \rangle$ is a composition tomonoid fulfilling (C1)–(C4) and (C6)–(C7). Before proving the four assertions of the second paragraph of the theorem, let us establish the following auxiliary facts (a)–(e).

(a) By (C6), each $\lambda \in \Lambda^R$ is continuous and, by Lemma 4.4.6(ii)(c), if $0 \notin R$ the right limit of $\lambda$ at 0 is 0, and if $1 \notin R$ the left limit of $\lambda$ at 1 is 1. Moreover, $id_R$ is the uniform limit of (any increasing sequence in) $\Lambda^R \backslash \{id_R\}$. In fact, by (C4), $id_R$ is the pointwise supremum of these mappings, and if 0 or 1 are not in $R$, the continuous extension of any $\lambda \in \Lambda^R$ maps 0 to 0 and 1 to 1; thus the claim follows from the compactness of $[0,1]$.

(b) (C7) and (a) imply that each $\lambda \in \Lambda^R$, restricted to its support, is strictly increasing.

(c) If $\lambda(r) = \lambda'(r) > 0$ for some $\lambda, \lambda' \in \Lambda^R$ and $r \in R$, then $\lambda = \lambda'$. Indeed, if then $0 < \lambda(s) < \lambda'(s)$ for some $s \in R$, the pair $\kappa \circ \lambda'$ and $\lambda$ is not comparable for a sufficiently large $\kappa \in \Lambda^R \backslash \{id_R\}$. Thus $\lambda$ and $\lambda'$ coincide on the meet of their supports, and by continuity and monotonicity, we conclude $\lambda = \lambda'$.

(d) Infima of subsets $\Lambda^R$ that possess some lower bound exist and are calculated pointwise. Indeed, let $\lambda_\iota \in \Lambda^R$, $\iota \in I$, be lower bounded. Let $r \in R$ be such that $s = \bigwedge_\iota \lambda_\iota(r) > 0$. Then, for any $\varepsilon > 0$, there is a $\kappa \in \Lambda^R$ such that $s - \varepsilon < \kappa(a) < s$; as $\kappa$ is continuous, there is a $\iota$ such that $s - \varepsilon < (\kappa \circ \lambda_\iota)(r) < s$. We conclude that the supremum of the lower bounds of $\lambda_\iota$, $\iota \in I$, is their pointwise infimum.

(e) For any $r \in R$, $\{\lambda(r) \mid \lambda \in \Lambda^R\} = \{a \in R \mid a \leq r\}$. Indeed, this set is closed under suprema by (C4) and under infima by (d). Moreover, we conclude from (a) and (c) that the set is dense. In view of (c), we see in particular that $1 \in R$ implies that $\Lambda^R$ fulfils (C5).

We now continue the proof of the theorem. Depending on whether or not $R$ possesses a smallest or largest element, we have that $R$ equals $[0,1]$, $(0,1]$, $[0,1)$, or $(0,1)$.

Assume first that $R = [0,1]$. By (e), $\Lambda^R$ fulfils (C1)–(C5) and is thus the Cayley tomonoid associated with a q.n.c. tomonoid. As in the proof of Proposition 4.5.3, we conclude that $\Lambda^R$ is c-isomorphic to the Łukasiewicz composition tomonoid.

Assume second that $R = (0,1]$. We proceed as in the previous case to see that $\Lambda^R$ is c-isomorphic to the product composition tomonoid.

Assume third that $R = [0,1)$. For $\lambda \in \Lambda^R$, let $z_\lambda = \max\{r \in R \mid \lambda(r) = 0\}$; note that this definition is possible because $\lambda(0) = 0$ and $\lim_{r \nearrow 1} \lambda(r) = 1$.

Then, for each $z \in R$, there is exactly one $\lambda \in \Lambda^R$ such that $z = z_\lambda$. Indeed, by (C7) and (a), $\lambda \neq \lambda'$ implies $z_\lambda \neq z_{\lambda'}$. Furthermore, $\Lambda^R$ is closed under suprema and infima, hence $\{z_\lambda \mid \lambda \in \Lambda^R\}$ is a closed subset of $R$, which by (a) is dense and does not possess a largest element and thus equals $R$.

Moreover, for each $\lambda \in \Lambda^R$, we have that $\lambda(r) = 0$ for $r \in [0, z_\lambda]$, strictly increasing on $[z, 1)$, and $\lim_{r \nearrow 1} \lambda(1) = 1$. Thus we can define

$$\tilde{\lambda} \colon (0,1] \to (0,1], \quad x \mapsto \begin{cases} 1 - \lambda^{-1}(1-x) & \text{if } x < 1 \\ 1 - z_\lambda & \text{if } x = 1. \end{cases}$$

It is somewhat tedious, but not difficult to check that $\tilde{\Lambda}^R = \{\tilde{\lambda} \mid \lambda \in \Lambda^R\}$ is a composition tomonoid fulfilling (C1)–(C4). By (f), $\tilde{\Lambda}^R$ fulfils also (C5). We conclude as in the previous case that $\tilde{\Lambda}^R$ is c-isomorphic to the product composition tomonoid. Hence $\Lambda^R$ itself is c-isomorphic to the reversed product composition tomonoid.

Finally, assume that $R = (0,1)$. Then $\Lambda^R$ consists of order-automorphisms of $(0,1)$. Let $\lambda \in \Lambda^R$ and put $\kappa = \bigwedge\{\mu \in \Lambda^R \mid \mu^2 \geq \lambda\}$. We claim that then $\kappa^2 = \lambda$. Indeed, by (d), $\kappa(r) = \bigwedge\{\mu(r) \mid \mu^2 \geq \lambda\}$ for any $r \in R$. By continuity, we calculate $\kappa^2(r) = \bigwedge\{\mu(\mu'(r)) \mid \mu^2, \mu'^2 \geq \lambda\} = \bigwedge\{\mu^2(r) \mid \mu^2 \geq \lambda\} \geq \lambda(r)$. If this inequality was strict, there would be a $\nu \in \Lambda^R \backslash \{id_R\}$ such that $\nu^2(\kappa^2(r)) > \lambda(r)$; but then $(\kappa \circ \nu)^2(r) > \lambda(r)$ although $\kappa \circ \nu < \kappa$, a contradiction. Thus $\kappa^2(r) = \lambda(r)$, that is, $\kappa^2 = \lambda$.

Let now $\lambda_0 \in \Lambda^R \backslash \{id_R\}$, and let $\lambda_k$ be the unique mapping such that $\lambda_k^{2^k} = \lambda_0$.

Then $\lambda_0 < \lambda_1 < \ldots < id$. Moreover, $(\lambda_k)_k$ converges uniformly to *id*. Indeed, let $\lambda = \bigvee_k \lambda_k$; then $\lambda_0 \leq \lambda^k$ for every $k$ and it follows $\lambda = id$. Thus $(\lambda_k)_k$ converges to *id* pointwise and consequently uniformly.

By Lemma 4.5.5, $\Lambda^R$ is c-isomorphic to a composition tomonoid $\tilde{\Lambda}^R$ containing the mappings $\tilde{\lambda}_k \colon (0,1) \to (0,1), \quad r \mapsto r^{2^{(\frac{1}{2})^k}}$ for each $k \in \mathsf{N}$. It follows that the functions $r \mapsto r^q$, where $q = 2^{\frac{m}{2^n}}$ for $m, n \in \mathsf{N}$, are dense in $\tilde{\Lambda}^R$. We conclude that $\tilde{\Lambda}^R$ consist of the mappings $x \mapsto x^q$, where $q \in \{s \in \mathsf{R} \mid s \geq 1\}$; that is, $\tilde{\Lambda}^R$ is the power composition tomonoid. □

Theorem 4.5.6 describes each composition tomonoid $\Lambda^R$ separately. Each element of $\Lambda^R$ is the restriction of a translation $\lambda_f$, where $f$ is an element of the extending filter $\boldsymbol{F}$, to $R$. It remains to determine which mapping in $\Lambda^R$ belongs to which element of $F$. According to our next proposition, the homomorphism $\varrho \colon F \to \Lambda^R, \quad f \mapsto \lambda_f^R$ is already uniquely determined by one non-trivial assignment.

PROPOSITION 4.5.7. *Let $\boldsymbol{\Phi}$ be a standard composition tomonoid on a chain $R$; let $\langle F, \leq, \odot, 1\rangle$ be either the product or the Łukasiewicz tomonoid; let $\tilde{f} \in F \backslash \{1\}$ be non-minimal, and let $\tilde{\lambda} \in \Phi \backslash \{id_R\}$ have a non-empty support. Then there is at most one surjective sup-preserving homomorphism $\varrho \colon F \to \Phi$ such that $\varrho(\tilde{f}) = \tilde{\lambda}$.*

*Proof.* Let $n \geq 1$. As $\boldsymbol{F}$ is the product or the Łukasiewicz tomonoid, and $\tilde{f}$ is a non-minimal element of it, there is a unique $f_n \in F$ such that $f_n^n = \tilde{f}$. Similarly, $\boldsymbol{\Phi}$ is a standard composition tomonoid, and $\tilde{\lambda}$ is a non-minimal element of it; it is readily checked that in each of the four possible cases there is a unique $\lambda_n \in \Phi$ such that $\lambda_n^n = \tilde{\lambda}$.

It follows that any homomorphism mapping $\tilde{f}$ to $\tilde{\lambda}$ must map $f_n$ to $\lambda_n$. As $\varrho$ is supposed to be a sup-preserving homomorphism, the claim follows. □

We now turn to the sets $\Lambda^{R,S}$. We will see that $\Lambda^R$ and $\Lambda^S$ largely determine which mappings can be contained in $\Lambda^{R,S}$. Figure 6 gives an impression of the situation.

PROPOSITION 4.5.8. *Let $\boldsymbol{\Phi}$ be a standard composition tomonoid on the chain $R$ and let $\boldsymbol{\Psi}$ be a standard composition tomonoid on the chain $S$. Furthermore, let $\langle F, \leq, \odot, 1\rangle$ be either the product or the Łukasiewicz tomonoid, and assume that there are surjective homomorphisms $F \to \Phi, \quad f \mapsto \Phi_f$ and $F \to \Psi, \quad f \mapsto \psi_f$. Let*

$$\Xi = \{\xi \colon R \to S \mid \textit{for all } f \in F,\ \xi \circ \varphi_f = \psi_f \circ \xi\}. \tag{16}$$

*Moreover, let $\Xi'$ be a set of mappings from $R$ to $S$ such that* (1) *for any $\xi \in \Xi'$ and $f \in F$, $\psi_f \circ \xi$ and $\xi \circ \varphi_f$ coincide and are in $\Xi'$,* (2) *if the pointwise calculated supremum of a subset of $\Xi'$ exists, it is in $\Xi'$, and* (3) *if $S$ has a smallest element $u'$, $c^{R,u'} \in \Xi'$. Then either $\Xi' = \Xi$ or there is a $\zeta \in \Xi$ such that $\Xi' = \{\xi \in \Xi \mid \xi \leq \zeta\}$.*

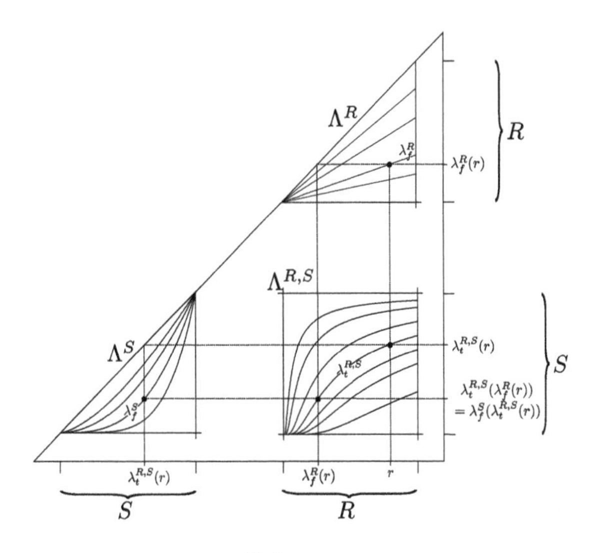


Figure 6. Each mapping contained in $\Lambda^{R,S}$ is uniquely determined by its value at a single point. The figure shows how the value of $\lambda_t^{R,S}$ at $r$ determines its value at $\lambda_f^S(r)$.

*Proof.* If $u' \notin S$ and $\Xi$ is empty, or $u' \in S$ and $\Xi$ contains only $c^{R,u'}$, the assertion is trivial. Let us assume that neither of these possibilities apply. We first show two auxiliary lemmas about $\Xi$.

(a) For any $r \in R$ and any non-minimal $s \in S$, there is at most one $\xi \in \Xi$ such that $\xi(r) = s$.

Indeed, let $\xi, \upsilon \in \Xi$ be such that $\xi(r) = \upsilon(r) = s$. For any $r' < r$, there is an $f \in F$ such that $\varphi_f(r) = r'$, because $\varphi$ is a standard composition tomonoid. Thus $\xi(r') = \xi(\varphi_f(r)) = \psi_f(\xi(r)) = \psi_f(\upsilon(r)) = \upsilon(\varphi_f(r)) = \upsilon(r')$. For any $r' > r$, there an $f \in F$ such that $\varphi_f(r') = r$, thus $\psi_f(\xi(r')) = \xi(\varphi_f(r')) = \xi(r) = \upsilon(r) = \upsilon(\varphi_f(r')) = \psi_f(\upsilon(r'))$, and since $\Psi$ is a standard composition tomonoid and $s$ is non-minimal, it follows $\xi(r') = \upsilon(r')$ again. We conclude $\xi = \upsilon$ and (a) is proved.

(b) Let $\xi, \upsilon \in \Xi$ have a non-empty support. Then there is an $f \in F$ such that either $\xi = \upsilon \circ \varphi_f$ or $\upsilon = \xi \circ \varphi_f$. In particular, $\xi$ and $\upsilon$ are comparable.

Indeed, let $r \in R$ be in the support of both $\xi$ and $\upsilon$. Assume $\xi(r) \leq \upsilon(r)$, and let $f \in F$ be such that $(\upsilon \circ \varphi_f)(r) = \psi_f(\upsilon(r)) = \xi(r)$. Note that $\upsilon \circ \varphi_f \in \Xi$. Then it follows by (a) that $\xi = \upsilon \circ \varphi_f$. Similarly, $\upsilon(r) \leq \xi(r)$ implies that there is an $f \in F$ such that $\upsilon = \xi \circ \varphi_f$. The proof of (b) is complete.

Let $\Xi'$ be a set of functions from $R$ to $S$ such that properties (1)–(3) hold. By (1), $\Xi' \subseteq \Xi$. Assume that $\xi \in \Xi'$, $\upsilon \in \Xi$, and $\upsilon \leq \xi$; we claim that then $\upsilon \in \Xi'$. Indeed, either $S$ has the smallest element $u'$ and $\upsilon = c^{R,u'}$; thus $\upsilon \in \Xi'$ by (3). Or $\upsilon$ has a non-empty support; by (b), then $\upsilon = \xi \circ \varphi_f$ for some $f \in F$; thus $\upsilon \in \Xi'$ by (1).

Assume now that $\Xi'$ is a proper subset of $\Xi$. Because $\Xi$ is totally ordered, any element of $\Xi \backslash \Xi'$ is an upper bound of $\Xi'$; hence the pointwise supremum $\zeta$ of $\Xi'$ exists and is, by (2), in $\Xi'$. Hence also $\zeta \in \Xi$, and we conclude $\Xi' = \{\xi \in \Xi \mid \xi \leq \zeta\}$. □

Again, Proposition 4.5.8 describes the sets $\Lambda^{R,S}$ separately and it remains to determine which mapping in $\Lambda^{R,S}$ belongs to which translation. Similarly as in case of Proposition 4.5.7, the mapping $\tau \colon T \to \Lambda^{R,S}, \; t \mapsto \lambda_t^{R,S}$ is uniquely determined by a single assignment.

PROPOSITION 4.5.9. *Let $R$, $\boldsymbol{\Phi}$, $S$, $\boldsymbol{\Psi}$, $\boldsymbol{F}$, and the mappings $f \mapsto \varphi_f$ and $f \mapsto \psi_f$ be as in Proposition 4.5.8, and let $\Xi$ be defined by* (16). *Let $\boldsymbol{X}$ be a further standard composition tomonoid on the chain $T$, and let $F \to X, \; f \mapsto \chi_f$ be a surjective sup-preserving homomorphism. Let $\tilde{t} \in T$ be non-minimal, and let $\tilde{\xi} \in \Xi$ have a non-empty support. Then there is at most one mapping $\tau \colon T \to \Xi$ such that $\tau(\chi_f(t)) = \psi_f \circ \tau(t)$ for any $t \in T$ and $\tau(\tilde{t}) = \tilde{\xi}$.*

*Proof.* Assume that the mappings $\tau_1, \tau_2 \colon T \to \Xi$ are as indicated.

Let $t > \tilde{t}$ and put $\xi_1 = \tau_1(t), \;\; \xi_2 = \tau_2(t)$. As $\boldsymbol{X}$ is a standard composition tomonoid, there is an $f \in F$ such that $\chi_f(t) = \tilde{t}$. We have $\psi_f \circ \xi_1 = \psi_f \circ \tau_1(t) = \tau_1(\chi_f(t)) = \tau_1(\tilde{t}) = \tilde{\xi}$ and similarly $\psi_f \circ \xi_2 = \tilde{\xi}$. Let $r$ be in the support of $\tilde{\xi}$; then $\psi_f(\xi_1(r)) = \psi_f(\xi_2(r))$ is non-minimal, and we conclude $\xi_1(r) = \xi_2(r)$. We proceed as in the proof of Proposition 4.5.8 to conclude that $\xi_1 = \xi_2$, that is, $\tau_1(t) = \tau_2(t)$.

Let now $t < \tilde{t}$. Then there is an $f \in F$ such that $\chi_f(\tilde{t}) = t$ and we have $\tau_1(t) = \tau_1(\chi_f(\tilde{t})) = \psi_f \circ \tau_1(\tilde{t}) = \psi_f \circ \tau_2(\tilde{t}) = \tau_2(\chi_f(\tilde{t})) = \tau_2(t)$ also in this case.

Hence $\tau_1 = \tau_2$ and the assertion follows. □

This concludes our specification of real Archimedean coextensions. On the basis of the following examples we now demonstrate how Proposition 4.5.3, Theorem 4.5.6 and Propositions 4.5.7, 4.5.8, 4.5.9 can be used to determine the real Archimedean coextensions of a given tomonoid.

EXAMPLE 4.5.10. As a first example of how our theory works, let us get back to Example 4.4.8. Again, let $\boldsymbol{L}_5$ be the five-element Łukasiewicz chain. We are going to determine the real Archimedean coextensions of $\boldsymbol{L}_5$ such that the bottom element is left unaltered and the remaining four elements are expanded to left-open right-closed real intervals.

By Proposition 4.5.3, $\Lambda^F$, where $\boldsymbol{F}$ is the extending tomonoid, is the product or Łukasiewicz composition tomonoid. As $F$ does not possess a smallest element, the former possibility applies.

Furthermore, by Theorem 4.5.6, $\Lambda^{(0,\frac{1}{4}]}, \Lambda^{(\frac{1}{4},\frac{1}{2}]}, \Lambda^{(\frac{1}{2},\frac{3}{4}]}$ are c-isomorphic to the product composition tomonoid as well.

To determine the translations $\lambda_t$, $\frac{3}{4} < t < 1$, it is by Proposition 4.5.7 sufficient to specify one of them. To this end, we choose one element distinct from the identity from each composition tomonoid $\Lambda^{(0,\frac{1}{4}]}, \Lambda^{(\frac{1}{4},\frac{1}{2}]}, \Lambda^{(\frac{1}{2},\frac{3}{4}]}$, and $\Lambda^{(\frac{3}{4},1]}$, and we require that these mappings arise from the same translation.

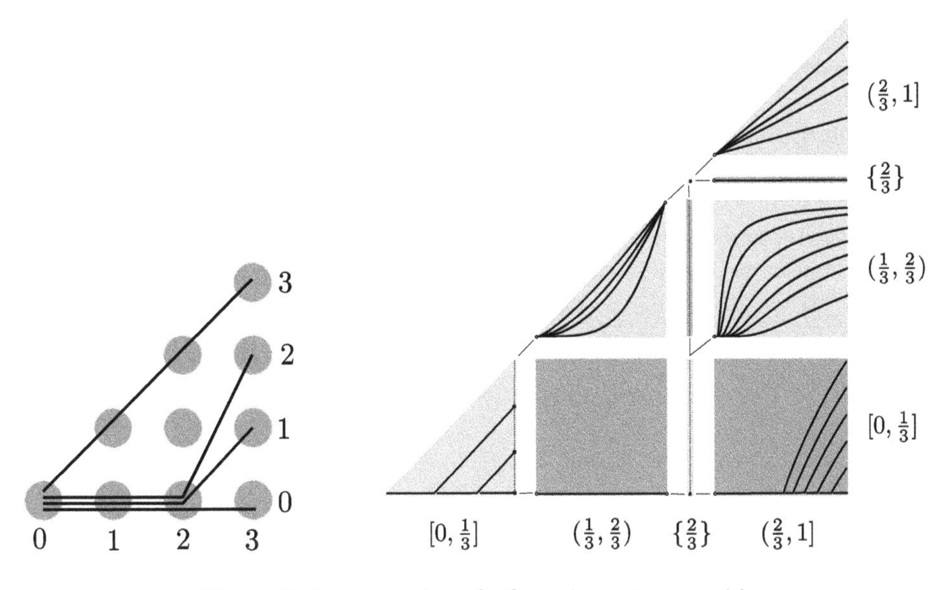


Figure 7. A coextension of a four-element tomonoid

It remains to determine the sets $\Lambda^{R,S}$, where $R$ and $S$ are among $\{0\}$, $(0, \frac{1}{4}]$, $(\frac{1}{4}, \frac{1}{2}]$, $(\frac{1}{2}, \frac{3}{4}]$, $(\frac{3}{4}, 1]$. The case that the singleton $\{0\}$ is involved is trivial and covered by Lemma 4.4.7(ii). Let both $R$ and $S$ be distinct from $\{0\}$; then Proposition 4.5.8 applies. It is straightforward to calculate $\Xi$ according to (16) from $\Lambda^R$ and $\Lambda^S$, which are both product composition tomonoids. The actual set $\Lambda^{R,S}$ results from $\Xi$ by determining a largest element $\zeta$. Note that only in this respect, our construction allows an essential choice.

Still given $R$ and $S$, it remains to determine the mapping $T \to \Lambda^{R,S}, \;\; t \mapsto \lambda_t^{R,S}$, where $T = R \to S$. By Proposition 4.5.9 it is sufficient to make a single assignment. But one assignment is already clear, namely, $\lambda_t^{R,S} = \zeta$, where $t$ is the maximal element of $T$.

The t-norm $\star_{\mathrm{H}}$ is a possible result of this construction; cf. Figure 3 and (14).

EXAMPLE 4.5.11. Next, we construct the real Archimedean coextensions of the four-element drastic tomonoid, that is, the tomonoid specified in Figure 7 (left). We assign to the four elements the real intervals $[0, \frac{1}{3}]$, $(\frac{1}{3}, \frac{2}{3})$, $\{\frac{2}{3}\}$, and $(\frac{2}{3}, 1]$, respectively.

The universe of the extending tomonoid is the left-open interval $(\frac{2}{3}, 1]$ and $\Lambda^{(\frac{2}{3},1]}$ is consequently again the product composition tomonoid.

Moreover, the composition tomonoids $\Lambda^{[0,\frac{1}{3}]}$ and $\Lambda^{(\frac{1}{3},\frac{2}{3})}$ are, according to Theorem 4.5.6, c-isomorphic to the Łukasiewicz and the power composition tomonoid, respectively. By Lemma 4.4.7(ii), $\Lambda^{\{\frac{2}{3}\}}$ consists of the mapping assigning $\frac{2}{3}$ to itself.

We next define an arbitrary translation $\lambda_t$, where $\frac{2}{3} < t < 1$, by selecting from each of the three non-trivial composition tomonoids one mapping different from the identity. Then the translations $\lambda_t$ are uniquely determined also for all remaining $t \in (\frac{2}{3}, 1]$.

We proceed by constructing the set $\Lambda^{(\frac{2}{3},1],(\frac{1}{3},\frac{2}{3})}$ on the basis of (16). The whole set (16) is needed in this case because, by condition (C5), for each $t \in (\frac{1}{3}, \frac{2}{3})$ there must be a translation mapping 1 to $t$. The situation is similar in the case of $\Lambda^{(\frac{2}{3},1],[0,\frac{1}{3}]}$.

Finally, Lemma 4.4.7(i)(e) implies that $\Lambda^{(\frac{1}{3},\frac{2}{3}),[0,\frac{1}{3}]}$ contains the constant 0 mapping only. Again by Lemma 4.4.7(i), $\Lambda^{\{\frac{2}{3}\},[0,\frac{1}{3})}$ consists of the single mapping assigning $\frac{2}{3}$ to 0. Also $\Lambda^{(\frac{2}{3},1],\{\frac{2}{3}\}}$ is trivial, consisting of the constant $\frac{2}{3}$ mapping.

The Cayley tomonoid is thus completely determined. The result is a left-continuous t-norm, for instance the following one:

$$a \star b \;=\; \begin{cases} 3ab - 2a - 2b + 2 & \text{if } a, b > \frac{2}{3} \\ \frac{1}{3}((3a-1)^{\frac{1}{3b-2}} + 1) & \text{if } \frac{1}{3} < a \leq \frac{2}{3} \text{ and } b > \frac{2}{3} \\ (a + \frac{1}{3}\log_2(3b-2)) \vee 0 & \text{if } a \leq \frac{1}{3} \text{ and } b > \frac{2}{3} \\ 0 & \text{if } a, b \leq \frac{2}{3}. \end{cases}$$

# 5 Historical remarks and further reading

## 5.1 Partial algebras

The idea of using partial algebras for the representation of residuated lattices originates from a research field that is led by quite different concerns from those of many-valued logics. Quantum structures were originally meant to be partially ordered algebras occurring in the context of quantum mechanics; an example is the orthomodular lattice of closed subspaces of a complex Hilbert space [33]. Orthomodular lattices have later turned out to be an interesting research object in their own right, and the same applies for their generalisations [10]. Indeed, closed subspaces can be viewed as models of "sharp" measurements, like the position of some object within a certain interval. In contrast, the Hilbert space effects (the positive operators below the identity) correspond to "unsharp" measurements, like the position of some object within a fuzzy set over the reals [7]. The set of effects possesses the internal structure of an effect algebra and the latter structure has turned out to be a rewarding research object [18].

In particular, K. Ravindran studied the relationship between effect algebras and partially ordered groups. He has shown that an effect algebra fulfilling a certain Riesz decomposition property is representable as an interval of a partially ordered group [34]. We note that the underlying technique was the first time employed by R. Baer in a more general context [2]. Among the effect algebras to which the technique is applicable, we do not find the standard effect algebra known from physics. But we do find here the so-called MV-effect algebras, a class of partial algebras that are in a one-to-one correspondence with MV-algebras [17].

We have explored this connection between algebras originating from fuzzy logic and techniques developed for quantum structures in a series of papers [11–14, 37, 38, 40, 41]. The structures under consideration were chosen more and more general. For instance, commutativity and boundedness turned out to be dispensable conditions.

The present exposition treats the most general case with which we have coped so far. However, we have considered the case of a total order only; this restriction has made a remarkable optimisation possible. For a more general framework, we refer to [41].

To summarise, the literature on effect algebras and their various generalisations is rich. Residuated structures viewed as partial algebras, however, have most likely been mainly examined in the afore-mentioned papers. The interested reader can find a more detailed and a more general account than the one given here in particular in [41]. We note that the latter paper is the only one in which the notation of residuated structures is adapted to what is common in logics; a partial multiplication instead of a partial addition is used.

## 5.2 Coextensions of tomonoids

The second part of this chapter addresses various topics and we are not able to provide a comprehensive overview over the related activities. We restrict to several short remarks.

Monoids endowed with a compatible partial order or, more generally, semigroups with a compatible preorder, have been considered in a number of different contexts. As an example, we may mention the paper [3] on compatible preorders and associated decompositions of semigroups. Several particular topics have furthermore been studied; see, e.g., [21], [27], or [28].

The more special totally ordered semigroups appear in the literature less frequently. An early survey is [20]; a comprehensive paper from more recent times is [15]. A property of tomonoids that has recently drawn attention is formal integrality; see, e.g. [24].

A central part of our discussion was devoted to representations of tomonoids by monoids of mappings under composition. We have mostly considered the simplest such representation, the regular representation of tomonoids. Such representations of partially ordered semigroups have been considered in a general context in a series of papers; they are known under the name $S$-posets. The initial paper was [16]; among the newer contributions, we may mention, e.g. [4]. Another viewpoint on the same topic can be found, e.g. in [35].

Quotients of partially ordered monoids in general have apparently not really been considered as a rewarding topic; they are indeed not easy to characterise. In contrast, congruences of residuated tomonoids that preserve the residuals as well are well understood, being identifiable with normal convex subalgebras. See [26] for the general case and [30] also for more special cases like MTL-algebras.

The last topic to be mentioned is a field to which we have not only given a good amount of space in this chapter but which is also very present in the literature: triangular norms. Many results on t-norms are compiled in [29], which is endowed also with a comprehensive bibliography. As a general way of classifying t-norms does not exist, several methods of their construction have been compiled over the years, not all of which are in line with the algebraic structure as exposed in this chapter. For a review of construction methods from a more algebraic perspective, see, e.g. [31]. For an account of MTL-algebras, we refer to [30].

Finally, how tomonoids can be analysed by means of their quotients has been described in our papers [39] and [42]; the former adopts a more geometrical, the latter a more algebraic point of view. In these papers a considerable amount of information can be found in addition to what we have presented here.

**Acknowledgements**

The author acknowledges the support by the Austrian Science Fund (FWF): project I 1923-N25 (New perspectives on residuated posets). Moreover he wishes to thank Gejza Jenca for valuable comments on a draft version of this chapter.

# BIBLIOGRAPHY

[1] Paolo Aglianò and Franco Montagna. Varieties of BL-algebras I: General properties. *Journal of Pure and Applied Algebra*, 181(2–3):105–129, 2003.

[2] Reinhold Baer. Free sums of groups and their generalizations. an analysis of the associative law. *American Journal of Mathematics*, 71:706–742, 1949.

[3] Stojan Bogdanovic and Miroslav Ciric. Quasi-orders and semilattice decompositions of semigroups (a survey). In *Proceedings of Internationals Conference in Semigroups and its Related Topics*, pages 27–57, 1998.

[4] Sydney Bulman-Fleming and Mojgan Mahmoudi. The category of $S$-posets. *Semigroup Forum*, 71(3): 443–461, 2005.

[5] Stanley Burris and H.P. Sankappanavar. *A Course in Universal Algebra*, volume 78 of *Graduate Texts in Mathematics*. Springer, 1981.

[6] Manuela Busaniche and Franco Montagna. Hájek's logic BL and BL-algebras. In Petr Cintula, Petr Hájek, and Carles Noguera, editors, *Handbook of Mathematical Fuzzy Logic - Volume 1*, volume 37 of *Studies in Logic, Mathematical Logic and Foundations*, pages 355–447. College Publications, London, 2011.

[7] Paul Busch, Marian Grabowski, and Pekka Johannes Lahti. *Operational Quantum Physics*, volume 31. Springer, 1995.

[8] Alfred Hoblitzelle Clifford and Gordon Bambord Preston. *The Algebraic Theory of Semigroups*, volume 7. American Mathematical Society, Providence, RI, 1967.

[9] Anatolij Dvurečenskij. Aglianò-Montagna type decomposition of linear pseudo hoops and its applications. *Journal of Pure and Applied Algebra*, 211(3):851–861, 2007.

[10] Anatolij Dvurečenskij and Sylvia Pulmannová. *New Trends in Quantum Structures*, volume 516 of *Mathematics and Its Applications*. Kluwer and Ister Science Ltd., Dordrecht, Bratislava, 2000.

[11] Anatolij Dvurečenskij and Thomas Vetterlein. Generalized pseudo-effect algebras. In *Lectures on Soft Computing and Fuzzy Logic*, pages 89–111. Springer, 2001.

[12] Anatolij Dvurečenskij and Thomas Vetterlein. Pseudoeffect algebras. I: Basic properties. *International Journal of Theoretical Physics*, 40(3):685–701, 2001.

[13] Anatolij Dvurečenskij and Thomas Vetterlein. Pseudoeffect algebras. II: Group representations. *International Journal of Theoretical Physics*, 40(3):703–726, 2001.

[14] Anatolij Dvurečenskij and Thomas Vetterlein. Algebras in the positive cone of po-groups. *Order*, 19(2): 127–146, 2002.

[15] Katie Evans, Michael Konikoff, James J. Madden, Rebecca Mathis, and Gretchen Whipple. Totally ordered commutative monoids. *Semigroup Forum*, 62(2):249–278, 2001.

[16] Syed M. Fakhruddin. On the category of $S$-posets. *Acta Scientiarum Mathematicarum*, 52(1–2):85–92, 1988.

[17] David J. Foulis. MV and Heyting effect algebras. *Foundations of Physics*, 30(10):1687–1706, 2000.

[18] David J. Foulis and Mary K. Bennett. Effect algebras and unsharp quantum logics. *Foundations of Physics*, 24(10):1331–1352, 1994.

[19] László Fuchs. *Partially Ordered Algebraic Systems*. Pergamon Press, Oxford, 1963.

[20] E. Ya. Gabovich. Fully ordered semigroups and their applications. *Russian Mathematical Surveys*, 31(1):147–216, 1976.

[21] Victoria Gould and Lubna Shaheen. Perfection for pomonoids. *Semigroup Forum*, 81(1):102–127, 2010.

[22] Petr Hájek. Observations on the monoidal t-norm logic. *Fuzzy Sets and Systems*, 132(1):107–112, 2002.

[23] Rostislav Horčík. *Algebraic Properties of Fuzzy Logics*. PhD thesis, Czech Technical University, Faculty of Electrical Egineering, Prague, 2005.

[24] Rostislav Horčík. Solutions to some open problems on totally ordered monoids. *Journal of Logic and Computation*, 20(4):977–983, 2010.

[25] Rostislav Horčík. Algebraic semantics: Semilinear FL-algebras. In Petr Cintula, Petr Hájek, and Carles Noguera, editors, *Handbook of Mathematical Fuzzy Logic - Volume 1*, volume 37 of *Studies in Logic, Mathematical Logic and Foundations*, pages 283–353. College Publications, London, 2011.

[26] Peter Jipsen and Constantine Tsinakis. A survey of residuated lattices. In Jorge Martínez, editor, *Ordered Algebraic Structures*, pages 19–56. Kluwer, Dordrecht, 2002.
[27] Zsofia Juhasz and Alexei S. Vernitskiĭ. Filters in (quasiordered) semigroups and lattices of filters. *Communications in Algebra*, 39(11):4319–4335, 2011.
[28] Lila Kari and Gabriel Thierrin. Languages and compatible relations on monoids. In Gheorghe Păun, editor, *Mathematical Linguistics and Related Topics*, pages 212–220. Editura Academiei Române, Bucharest, 1995.
[29] Erich Peter Klement, Radko Mesiar, and Endre Pap. *Triangular Norms*, volume 8 of *Trends in Logic*. Kluwer, Dordrecht, 2000.
[30] Carles Noguera, Francesc Esteva, and Joan Gispert. On some varieties of MTL-algebras. *Logic Journal of the Interest Group of Pure and Applied Logic*, 13(4):443–466, 2005.
[31] Carles Noguera, Francesc Esteva, and Lluís Godo. Generalized continuous and left-continuous t-norms arising from algebraic semantics for fuzzy logics. *Information Sciences*, 180(8):1354–1372, 2010.
[32] Andrei D. Polyanin and Alexander V. Manzhirov. *Handbook of Integral Equations*. Chapman and Hall/CRC, Boca Raton, FL, second edition, 2008.
[33] Pavel Pták and Sylvia Pulmannová. *Orthomodular Structures as Quantum Logics: Intrinsic Properties, State Space and Probabilistic Topics*, volume 44. Springer, 1991.
[34] Kuppusamy Ravindran. *On a Structure Theory of Effect Algebras*. PhD thesis, Kansas State University, 1996.
[35] Vladimir B. Repnitskiĭ and Alexei S. Vernitskiĭ. Semigroups of order-preserving mappings. *Communications in Algebra*, 28(8):3635–3641, 2000.
[36] Kimmo I. Rosenthal. *Quantales and their Applications*, volume 234. Longman Scientific & Technical Harlow, 1990.
[37] Thomas Vetterlein. BL-algebras and effect algebras. *Soft Computing*, 9(8):557–564, 2005.
[38] Thomas Vetterlein. Partial algebras for Łukasiewicz Logic and its extensions. *Archive for Mathematical Logic*, 44(7):913–933, 2005.
[39] Thomas Vetterlein. Regular left-continuous t-norms. *Semigroup Forum*, 77(3):339–379, 2008.
[40] Thomas Vetterlein. Weak effect algebras. *Algebra Universalis*, 58(2):129–143, 2008.
[41] Thomas Vetterlein. Pseudo-BCK algebras as partial algebras. *Information Sciences*, 180(24):5101–5114, 2010.
[42] Thomas Vetterlein. Totally ordered monoids based on triangular norms. *Communications in Algebra*, 43:1–37, 2015.

THOMAS VETTERLEIN
Department of Knowledge-Based Mathematical Systems
Johannes Kepler University Linz
Altenberger Straße 69
4040 Linz, Austria
Email: Thomas.Vetterlein@jku.at

# Chapter XIII: Semantic Games for Fuzzy Logics

CHRISTIAN G. FERMÜLLER

## 1 Introduction

Deductive fuzzy logics nowadays come in many variants and types. Clearly there is no single logical system that is adequate for all applications and contexts. This fact imparts significance to the problem of *justifying particular logics* with respect to *specific principles of reasoning*. In other words, one is challenged to motivate the choice of a concrete logic with respect to basic models of reasoning that go beyond the mere presentation of some set of truth functions or of some proof system. Among the various models of fuzzy logics that have been proposed in this vein are Lawry's voting semantics [37], Paris's acceptability semantics [49], re-randomising semantics [34], and approximation semantics [6, 50]. This chapter addresses the challenge by presenting *semantic games* for some important fuzzy logics, in particular for Łukasiewicz logic, but also, e.g., for Product and for Gödel logic. Semantic games, sometimes also called evaluation games, characterize the evaluation of a given formula with respect to a given assignment of truth values to atomic formulas by a game between two players, that assume the roles of the proponent (or defender) and the opponent (or attacker) of the formula, respectively. This alternative to Tarski-style semantics was introduced by Jaako Hintikka in the late 1960s for classical logic [31]. Independently of Hintikka, Robin Giles suggested in the 1970s a game based interpretation of Łukasiewicz logic [25]. As will get clear in this chapter, both Hintikka's and Giles's games are starting points for a whole range of different game based characterizations of various important fuzzy logics.

A related enterprise, namely the connection between Ulam–Rényi games and t-norm based fuzzy logics, is presented in Chapter XIV of this volume of this Handbook. The reader should also be aware of the fact that there are many other kinds of logical games, that have at least partly been extended to many-valued logics as well: for example Lorenzen-style dialogue games, model comparison games (in particular Ehrenfreucht–Fraïssé games), and various forms of model construction games; however, here, we strictly focus on semantic games.

The chapter is structured as follows: We start by revisiting Hintikka's classical semantic game in Section 2 and observe that by simply placing this basic game into a many-valued setting one obtains a characterization of the so-called weak fragment of Łukasiewicz logic, also known as Kleene–Zadeh logic KZ. We also show that straightforward variations of Hintikka's game rules do not lead beyond connectives that are already definable in KZ. Section 3 presents a generalization of Hintikka's game to full Łukasiewicz logic, that requires the explicit reference to a truth value at any state of the game. In Section 4 we review Giles's game for Łukasiewicz logic, which is based on a more general concept of game states, but does not explicitly involve truth values during evaluation. Section 5 looks at Giles's game from a more abstract point of view and formulates dialogue as well payoff principles that should be maintained in generalizations and variants of that game if one aims at extracting truth functions from optimal strategies. We also show how games for other logics arise in this manner. Section 6 takes up a topic that has already been discussed by George Metcalfe in Chapter III (on proof theory) of this Handbook: the connection between Giles's game and logical rules for a particular type of hypersequent system. We present this concept from a somewhat different point of view here, and work out some details not for the original case of Łukasiewicz logic, but rather for Abelian logic instead. Section 7 describes another type of generalization of Hintikka's game that employs a stack of game states in addition to the current formula and role distribution. It is shown how Łukasiewicz, Gödel, as well as Product logic can be characterized in this manner. Section 8 presents yet another variant of semantic games based on introducing random choices, in addition to the choices made by the two players of a Hintikka-style game. In Section 9 the idea of considering random choices is lifted from the propositional level to rules for certain types of semi-fuzzy quantifiers in the context of a Giles-style game. Section 10 consists in a very concise synopsis of the various semantic games discussed in the previous sections. We close with historical remarks and hints on the sources in the final Section 11.

## 2 Hintikka's game — from classical to many-valued logics

### 2.1 Hintikka's classical semantic game

We will call Hintikka's classic game for characterizing truth in a given model the $\mathcal{H}$-game. Like in all semantic games that we will consider in this chapter, there are two players, called *myself* (or simply I) and *you*, here, who can both act either in the role of the *proponent* **P** or of the *opponent* **O**[1] of a given classical first-order formula $\varphi$. Throughout this chapter, instead of referring explicitly to variable assignments, we will assume that there is a constant in the language for each domain element of the interpretation with respect to which the formula is to be evaluated. For simplicity, we will identify the constant with the corresponding domain element. Initially I act as **P** and you act as **O**. My aim is to show that the initial formula is true in a given interpretation $\mathcal{M}$. More generally, in any state of the game, it is **P**'s aim to show that the formula in focus at the given state, called *current formula*, is true in $\mathcal{M}$. The game proceeds according to the following rules. Note that these rules only refer to the players' roles (*role distribution*) and to the outermost connective of the current formula.

$(R^{\mathcal{H}}_{\wedge})$ If the current formula is $\varphi \wedge \psi$ then **O** chooses whether the game continues with $\varphi$ or with $\psi$.

[1] See Section 11 for a remark on alternative names for roles and players.

($R^{\mathcal{H}}_{\vee}$) If the current formula is $\varphi \vee \psi$ then **P** chooses whether the game continues with $\varphi$ or with $\psi$.

($R^{\mathcal{H}}_{\neg}$) If the current formula is $\neg\varphi$, the game continues with $\varphi$, except that the roles of the players are switched: the player who is currently acting as **P**, acts as **O** at the the next state, and vice versa for the current **O**.

($R^{\mathcal{H}}_{\forall}$) If the current formula is $\forall x\varphi(x)$ then **O** chooses an element $c$ of the domain of $\mathcal{M}$ and the game continues with $\varphi(c)$.

($R^{\mathcal{H}}_{\exists}$) If the current formula is $\exists x\varphi(x)$ then **P** chooses an element $c$ of the domain of $\mathcal{M}$ and the game continues with $\varphi(c)$.

Except for rule $R^{\mathcal{H}}_{\neg}$, the players' roles remain unchanged. The game ends when an atomic formula $p$ is hit. The player who is currently acting as **P** *wins* and the other player, acting as **O**, *loses* if $p$ is true in the given interpretation $\mathcal{M}$. We associate payoff 1 with winning and payoff 0 with losing. We also include the truth constants $\top$ and $\bot$, with their usual interpretation, among the atomic formulas. The game starting with formula $\varphi$ is called the *$\mathcal{H}$-game for $\varphi$ under $\mathcal{M}$.*

THEOREM 2.1.1 (Hintikka[2]). *I have a winning strategy in the $\mathcal{H}$-game for $\varphi$ under the (classical) interpretation $\mathcal{M}$ iff $\varphi$ is true in $\mathcal{M}$ (in symbols: $v_{\mathcal{M}}(\varphi) = 1$).*

## 2.2 Hintikka's game in a fuzzy logic setting

Recall that in game theory one is usually not just talking about winning or losing a game, but rather about the players' strategies for maximizing their payoffs. Since we have identified winning or losing the $\mathcal{H}$-game with receiving the payoff 0 or 1, respectively, this perspective also covers the $\mathcal{H}$-game: instead of talking about a winning strategy, we may refer to a strategy that guarantees payoff 1. With the exception of the explicit evaluation game in Section 3 we will employ this more general perspective here and identify truth values with payoff values that lie between 0 and 1. In this manner the $\mathcal{H}$-game can be straightforwardly generalized to a many-valued setting that is sometimes simply referred to as '(the) fuzzy logic', e.g. in the well known textbook [47]. This logic is occasionally also called the 'weak fragment of Łukasiewicz logic' or just 'weak Łukasiewicz logic' (see, e.g., [22]). Following [1], we prefer to call this logic *Kleene–Zadeh logic*, or KZ for short, here.

At the propositional level the semantics of KZ is specified by extending a given interpretation $\mathcal{M}$, i.e., an assignment $v_{\mathcal{M}}(\cdot)$ of propositional variables to truth values in $[0, 1]$, to arbitrary formulas as follows:

$$
\begin{aligned}
v_{\mathcal{M}}(\varphi \wedge \psi) &= \min\{v_{\mathcal{M}}(\varphi), v_{\mathcal{M}}(\psi)\}\\
v_{\mathcal{M}}(\varphi \vee \psi) &= \max\{v_{\mathcal{M}}(\varphi), v_{\mathcal{M}}(\psi)\}\\
v_{\mathcal{M}}(\neg\varphi) &= 1 - v_{\mathcal{M}}(\varphi)\\
v_{\mathcal{M}}(\bot) &= 0\\
v_{\mathcal{M}}(\top) &= 1.
\end{aligned}
$$

[2] A proof for Theorem 2.1.1 can easily be extracted from the more general case of Theorem 2.2.2, below.

Interestingly, neither the rules nor the notion of a state in an $\mathcal{H}$-game have to be changed in order to characterize the logic KZ. We only have to generalize the possible payoff values for the $\mathcal{H}$-game from $\{0,1\}$ to the unit interval $[0,1]$, as already indicated above. More precisely, the payoff for the player who is in the role of **P** when a game under $\mathcal{M}$ ends with the atomic formula $p$ is $v_{\mathcal{M}}(p)$, while the value for **O** is $1 - v_{\mathcal{M}}(p)$. If the payoffs are modified in this manner we will speak of an $\mathcal{H}$-mv-game *under* $\mathcal{M}$.

KZ can be straightforwardly extended to predicate logic. At the first-order level an interpretation $\mathcal{M}$ includes a non-empty set $D$ as domain. With respect to our convention identifying domain elements with constants, the semantics of the universal and the existential quantifier is given by

$$v_{\mathcal{M}}(\forall x \varphi(x)) = \inf\{v_{\mathcal{M}}(\varphi(c)) \mid c \in D\}$$
$$v_{\mathcal{M}}(\exists x \varphi(x)) = \sup\{v_{\mathcal{M}}(\varphi(c)) \mid c \in D\}.$$

A slight complication arises for quantified formulas in $\mathcal{H}$-mv-games: there might be no element $c$ in the domain of $\mathcal{M}$ such that $v_{\mathcal{M}}(\varphi(c)) = \inf\{v_{\mathcal{M}}(\varphi(c)) \mid c \in D\}$ or no domain element $d$ such that $v_{\mathcal{M}}(\varphi(d)) = \sup\{v_{\mathcal{M}}(\varphi(d)) \mid d \in D\}$. A simple way to deal with this fact is to restrict attention to so-called witnessed models [29], where constants that witness all arising infima and suprema are assumed to exist. In other words: infima are minima and suprema are maxima in witnessed models. For KZ and for Łukasiewicz logic validity is not affected by restricting to witnessed models. However, for other logics like Gödel logic the set of formulas that always evaluate to 1 increases if only witnessed models are considered. In any case, we are not so much interested in validity here, but rather in concrete valuations. Therefore we adopt a more general solution that refers to optimal payoffs up to some $\epsilon$.

DEFINITION 2.2.1. *Suppose that, for every $\epsilon > 0$, player **X** has a strategy that guarantees her a payoff of at least $w - \epsilon$, while her opponent has a strategy that ensures that **X**'s payoff is at most $w + \epsilon$, then $w$ is called the* value for **X** of the game.

This notion, which corresponds to that of an $\epsilon$-equilibrium as known from game theory, allows us to state the following generalization of Theorem 2.1.1.

THEOREM 2.2.2. *The value for myself of the $\mathcal{H}$-mv-game for $\varphi$ under the interpretation $\mathcal{M}$ is $w$ iff $v_{\mathcal{M}}(\varphi) = w$.*

*Proof.* We argue by induction on the complexity of $\varphi$. The induction hypothesis generalizes the statement in the theorem by referring to (the player in the role of) **P**, instead of just to myself, and by including that the value for **O** is $1 - v_{\mathcal{M}}(\varphi)$.

If $\varphi$ is an atomic formula then the claim follows directly from Definition 2.2.1.

If $\varphi = \varphi_1 \vee \varphi_2$, then by rule $R^{\mathcal{H}}_{\vee}$ **P** can choose whether to continue the game with the current formula $\varphi_1$ or $\varphi_2$. By the induction hypothesis the value of the $\mathcal{H}$-mv-game for $\varphi_i$ for the player in role **P** is $v_{\mathcal{M}}(\varphi_i)$ for $i = \{1,2\}$. To maximize her payoff **P** will therefore choose $\varphi_i$ for $i \in \{1,2\}$ such that $v_{\mathcal{M}}(\varphi_i) = \max\{v_{\mathcal{M}}(\varphi_1), v_{\mathcal{M}}(\varphi_2)\}$. This guarantees that the value for the game for $\varphi$ is $\max\{v_{\mathcal{M}}(\varphi_1), v_{\mathcal{M}}(\varphi_2)\} = v_{\mathcal{M}}(\varphi)$ as required. Moreover, it follows from the induction hypothesis that the value for **O**, after **P**'s choice of $\varphi_i$, is $\min\{1 - v_{\mathcal{M}}(\varphi_1), 1 - v_{\mathcal{M}}(\varphi_2)\} = 1 - \max\{v_{\mathcal{M}}(\varphi_1), v_{\mathcal{M}}(\varphi_2)\} = 1 - v_{\mathcal{M}}(\varphi)$, as required.

The case for $\varphi = \varphi_1 \wedge \varphi_2$ is like the one for $\varphi = \varphi_1 \vee \varphi_2$ with **P** and **O** switched and the case for negation follows directly from the induction hypothesis.

If $\varphi = \exists x \varphi'(x)$, then by the induction hypothesis and by Definition 2.2.1, we obtain that for every $\epsilon > 0$ player **P** has a strategy that guarantees her a payoff of at least $v_{\mathcal{M}}(\varphi'(c)) - \epsilon$ in the game for $\varphi'(c)$, for some domain element $c$. Therefore, for every $\delta > 0$ player **P** can pick a $d \in D$ such that her payoff in the game for $\varphi'(d)$ is not less than $\sup\{v_{\mathcal{M}}(\varphi'(d)) \mid d \in D\} - \delta = v_{\mathcal{M}}(\exists x \varphi'(x)) - \delta$. Analogously, we conclude that for each domain element $d$ and for every $\delta > 0$ player **O** has a strategy to ensure that **P**'s payoff is at most $v_{\mathcal{M}}(\exists x \varphi'(x)) + \delta$ in a game for $\varphi'(d)$. Therefore the value for **P** of the game for $\varphi = \exists x \varphi'(x)$ is $\sup\{v_{\mathcal{M}}(\varphi'(d)) \mid d \in D\}$, as required. Taking into account that the value for **O** remains inverse to the one for **P**, we conclude that the induction hypothesis also holds for $\varphi$. □

## 2.3 Limits of Hintikka-style games

Note that there is no rule for implication in the $\mathcal{H}$-game or the $\mathcal{H}$-mv-game. Of course, we could simply define $\varphi \to \psi$ as $\neg\varphi \vee \psi$, like in classical logic. However this does not work for the (standard) implication of full Łukasiewicz logic Ł nor for other t-norm based fuzzy logics. As a consequence, at least at a first glimpse, the possibilities for extending $\mathcal{H}$-mv-games to logics more expressive than KZ look very limited if we insist on *Hintikka's principle* that a state of the game is fully determined by a formula and a distribution of the two roles (**P** and **O**) to the two players. There are only three elementary building blocks on which rules can be based: choices by **P**, choices by **O**, and role switch. By combining these building blocks one is led to a more general concept of propositional game rules, related to those described in [17] for connectives defined by arbitrary finite deterministic and non-deterministic matrices. In order to facilitate a concise specification of all rules of that type, we introduce the following technical notion.

DEFINITION 2.3.1. *An* $n$-selection *is a non-empty subset* $S$ *of* $\{1, \ldots, n\}$*, where each element of* $S$ *may additionally be marked by a* switch sign.

A game rule for an $n$-ary connective $\diamond$ in a *generalized* $\mathcal{H}$*-mv-game* is specified by a non-empty set $\{S_1, \ldots, S_m\}$ of $n$-selections. According to this concept, a round in a generalized $\mathcal{H}$-mv-game consists of two phases. The scheme for the corresponding game rule specified by $\{S_1, \ldots, S_m\}$ is as follows:

**(Phase 1):** If the current formula is $\diamond(\varphi_1, \ldots, \varphi_n)$ then **O** chooses an $n$-selection $S_i$ from $\{S_1, \ldots, S_m\}$.

**(Phase 2):** **P** chooses an element $j \in S_i$. The game continues with formula $\varphi_j$, where the roles of the players are switched if $j$ is marked by a switch sign.

REMARK 2.3.2. *A variant of this scheme arises by letting* **P** *choose the* $n$*-selection* $S_i$ *in phase 1 and* **O** *choose* $j \in S_i$ *in phase 2. But note that playing the game for* $\diamond(\varphi_1, \ldots, \varphi_n)$ *according to that role inverted scheme is equivalent to playing the game for* $\neg \diamond (\neg\varphi_1, \ldots, \neg\varphi_n)$ *using the exhibited scheme.*

REMARK 2.3.3. *The rules $R^{\mathcal{H}}_{\wedge}$, $R^{\mathcal{H}}_{\vee}$, and $R^{\mathcal{H}}_{\neg}$ can be understood as instances of the above scheme:*
*- $R^{\mathcal{H}}_{\wedge}$ is specified by $\{\{1\},\{2\}\}$,*
*- $R^{\mathcal{H}}_{\vee}$ is specified by $\{\{1,2\}\}$, and*
*- $R^{\mathcal{H}}_{\neg}$ is specified by $\{\{1^*\}\}$, where the asterisk is used as switch mark.*

THEOREM 2.3.4. *In a generalized $\mathcal{H}$-mv-game, each rule of the type described above corresponds to a connective that is definable in the logic* KZ.

*Proof.* The argument for the adequateness of all semantic games considered in this paper proceeds by backward induction on the game tree.

For (generalized) $\mathcal{H}$-mv-games the base case is trivial: by definition **P** receives payoff $v_{\mathcal{M}}(p)$ and **O** receives payoff $1 - v_{\mathcal{M}}(p)$ if the game ends with the atomic formula $p$.

For the inductive case assume that the current formula is $\diamond(\varphi_1, \ldots, \varphi_n)$ and that the rule for $\diamond$ is specified by the set $\{S_1, \ldots, S_m\}$ of $n$-selections, where $S_i = \{j(i,1), \ldots, j(i,k(i))\}$ for $1 \leq i \leq m$ and $1 \leq k(i) \leq n$. Remember that the elements of $S_i$ are numbers $\in \{1, \ldots, n\}$, possibly marked by a switch sign. For sake of clarity let us first assume that there are no switch signs, i.e., no role switches occur. Let us say that a player **X** *can force payoff* $w$ if **X** has a strategy that guarantees her a payoff $\geq w$ at the end of the game. By the induction hypothesis, **P** can force payoff $v_{\mathcal{M}}(B)$ for herself and **O** can force payoff payoff $1 - v_{\mathcal{M}}(B)$ for himself if $B$ is among $\{\varphi_1, \ldots, \varphi_n\}$ and does indeed occur at a successor state to the current one; in other words, if $B = \varphi_{j(i,\ell)}$ for some $i \in \{1, \ldots, m\}$ and $\ell \in \{1, \ldots, k(i)\}$. Since **O** chooses the $n$-selection $S_i$, while **P** chooses an index number in $S_i$, **P** can force payoff

$$\min_{1 \leq i \leq m} \max_{1 \leq \ell \leq k(i)} v_{\mathcal{M}}(\varphi_{j(i,\ell)})$$

at the current state, while **O** can force payoff

$$\max_{1 \leq i \leq m} \min_{1 \leq \ell \leq k(i)} (1 - v_{\mathcal{M}}(\varphi_{j(i,\ell)})) = 1 - \min_{1 \leq i \leq m} \max_{1 \leq \ell \leq k(i)} v_{\mathcal{M}}(\varphi_{j(i,\ell)}).$$

If both players play optimally these payoff values are actually achieved. Therefore the upper expression corresponds to the truth function for $\diamond$. Both expressions have to be modified by uniformly substituting $1 - v_{\mathcal{M}}(\varphi_{j(i,\ell)})$ for $v_{\mathcal{M}}(\varphi_{j(i,\ell)})$ whenever $j(i,\ell)$ is marked by a switch sign in $S_1$ for $1 \leq i \leq m$ and $1 \leq k(i) \leq n$.

To infer that the connective $\diamond$ is definable in logic KZ it suffices to observe that its truth function, described above, can be composed from the functions $\lambda x(1-x)$, $\lambda x, y \min\{x,y\}$, and $\lambda x, y \max\{x,y\}$. But these functions are the truth functions for $\neg$, $\wedge$, and $\vee$, respectively, in KZ. □

Theorem 2.3.4 confirms our initial observation that there are hardly any options for generalizing game semantics to more expressive fuzzy logics, if we insist on Hintikka's principle that a game state should be fully determined by a formula and one of the two possible role distributions. We thus have to look for non-trivial augmentations of the $\mathcal{H}$-mv-game. A number of quite different such extensions will be considered in the succeeding sections.

# 3 An explicit evaluation game for Łukasiewicz logic

From the point of view of continuous t-norm based fuzzy logics, as popularized by Petr Hájek [28] and amply documented in this handbook, Kleene–Zadeh logic KZ is unsatisfying: while min is a t-norm, its residuum, which corresponds to implication in Gödel logic, is not expressible in KZ. If we define implication by $\varphi \to \psi =_{def} \neg\varphi \vee \psi$, in analogy to classical logic, then $\varphi \to \varphi$ is not valid, i.e., $v_{\mathcal{M}}(\varphi \to \varphi)$ is not true in all interpretations. In fact, formulas that do not contain truth constants are never valid in KZ.

The most important t-norm based fuzzy logic extending KZ is (full) Łukasiewicz logic Ł. The language of Ł extends that of KZ by implication $\to$, strong conjunction &, and strong disjunction $\oplus$. The semantics of these connectives is given by

$$v_{\mathcal{M}}(\varphi \to \psi) = \min\{1, 1 - v_{\mathcal{M}}(\varphi) + v_{\mathcal{M}}(\psi)\}$$
$$v_{\mathcal{M}}(\varphi \,\&\, \psi) = \max\{0, v_{\mathcal{M}}(\varphi) + v_{\mathcal{M}}(\psi) - \})$$
$$v_{\mathcal{M}}(\varphi \oplus \psi) = \min\{1, v_{\mathcal{M}}(\varphi) + v_{\mathcal{M}}(\psi)\}.$$

All other propositional connectives could be defined in Ł, e.g., from $\to$ and $\bot$, or from & and $\neg$, alone. However, neither $\to$ nor & nor $\oplus$ can be defined in KZ. For this reason KZ is sometimes called the weak fragment of Ł, as mentioned in Section 2.

The increased expressiveness of Ł over KZ is particularly prominent at the first-order level: while in KZ there are only trivially valid formulas, which involve the truth constants in an essential manner, the set of valid first-order formulas in Ł is not even recursively enumerable, due to a classic result of Scarpellini [51].

As pointed out at the end of last section, it seems to be impossible to characterize full Łukasiewicz logic Ł by a trivial extension of the $\mathcal{H}$-game, comparable to the smooth shift from $\mathcal{H}$-games to $\mathcal{H}$-mv-games. However by introducing an explicit reference to a value $\in [0,1]$ at every state of the game we may define an *explicit evaluation game* or, shortly, $\mathcal{E}$-game for Ł.

Like above, we call the players *myself* (*I*) and *you*, and the roles **P** and **O**. In addition to the role distribution and the current formula, also a *current value* $\in [0,1]$ is included in the specification of a game state. We will thus denote $\mathcal{E}$-game states as pairs $\langle \varphi, r \rangle$, where $\varphi$ is the current formula and $r$ the current value. If $\langle \varphi, r \rangle$ is the initial state we speak of the $\mathcal{E}$-game *for* $\langle \varphi, r \rangle$.

Atomic formulas correspond to tests, like in the classical $\mathcal{H}$-game. If the current state is $\langle \alpha, r \rangle$, where $\alpha$ is an atomic formula, then the game ends and (the current) **P** wins if $v_{\mathcal{M}}(\alpha) \geq r$, otherwise **O** wins.

The rules for weak conjunction and disjunction remain essentially as in the $\mathcal{H}$-game, except for the additional reference to a value $r \in [0,1]$. This value however does not change in these moves.

($R^{\mathcal{E}}_{\wedge}$) If the current state $\langle \varphi \wedge \psi, r \rangle$ then **O** chooses whether the game continues with $\langle \varphi, r \rangle$ or with $\langle \psi, r \rangle$.

($R^{\mathcal{E}}_{\vee}$) If the current state is $\langle \varphi \vee \psi, r \rangle$ then **P** chooses whether the game continues with $\langle \varphi, r \rangle$ or with $\langle \psi, r \rangle$.

The rule for strong disjunction consists of two actions. First **P** divides the value of the current formula between the disjuncts; then **O** chooses one of the disjuncts (with the corresponding value) for the next state of the game.

($R^{\mathcal{E}}_{\oplus}$) If the current state is $\langle \varphi \oplus \psi, r\rangle$, then **P** chooses $r_\varphi, r_\psi \geq 0$ such that $r_\varphi + r_\psi = r$ and **O** chooses whether the game continues with $\langle \varphi, r_\varphi\rangle$ or with $\langle \psi, r_\psi\rangle$.

Note that the rule $R^{\mathcal{E}}_{\vee}$ for weak disjunction can be seen as a restricted case of the rule $R^{\mathcal{E}}_{\oplus}$ for strong disjunction, where either $r_\varphi = r$ and $r_\psi = 0$ or, conversely, $r_\psi = r$ and $r_\varphi = 0$.

Negation involves the switch of roles, as in the $\mathcal{H}$-game. However it is not sufficient to simply switch the roles **P** and **O** and to continue the game with the inverse value for the unnegated formula. The reason for this that we want to consistently interpret the statement to be defended by the player in role **P** at a state $\langle \psi, r\rangle$ as the claim that $v_{\mathcal{M}}(\psi) \geq r$. For $\psi = \neg\varphi$ this implies that the player has to defend $v_{\mathcal{M}}(\varphi) \leq 1 - r$ at the next state. After role switch the player who previously denied this claim acts in role **P** and therefore has to defend $v_{\mathcal{M}}(\varphi) > 1 - r$. To avoid the use of $>$ (or of $\leq$) instead of $\geq$, we reformulate the relevant condition as follows. If **O** denies **P**'s claim that $v_{\mathcal{M}}(\neg\varphi) \geq r$ then she asserts that $v_{\mathcal{M}}(\neg\varphi) < r$. This is equivalent to the claim that $v_{\mathcal{M}}(\neg\varphi) \leq r'$ for some $r'$ strictly smaller than $r$, which in turn amounts to claiming $v_{\mathcal{M}}(\varphi) \geq 1 - r'$.

($R^{\mathcal{E}}_{\neg}$) If the current state is $\langle \neg\varphi, r\rangle$ then **O** chooses an $r'$, where $0 \leq r' < r$, and the game continues with $\langle \varphi, 1 - r'\rangle$ with the roles of players switched.

The rule for the strong conjunction is dual to the one of strong disjunction. It again refers to two actions: modification of the value by **P** and a choice by **O**.

($R^{\mathcal{E}}_{\&}$) If the current state is $\langle \varphi \,\&\, \psi, r\rangle$ then **P** chooses a value $\bar{r}$, where $0 \leq \bar{r} \leq 1 - r$; then **O** chooses whether to continue the game with $\langle \varphi, r + \bar{r}\rangle$ or with $\langle \psi, 1 - \bar{r}\rangle$.

The universal quantifier rule is analogous to the one for the $\mathcal{H}$-game. The state $\langle \forall x \varphi(x), r\rangle$ corresponds to **P**'s claim that $\inf\{v_{\mathcal{M}}(\varphi(c)) \mid c \in D\} \geq r$. **O** has to provide a counterexample, i.e., to find a $d$ such that $v_{\mathcal{M}}(\varphi(d)) < r$. Clearly the choice of a counterexample is independent of the (non)existence of an witnessing element for the infimum.

($R^{\mathcal{E}}_{\forall}$) If the current state is $\langle \forall x \varphi(x), r\rangle$ then **O** chooses some $c$ in the domain $D$ of the interpretation $\mathcal{M}$ and the game continues with $\langle \varphi(c), r\rangle$.

The situation is different for the existential quantifier. Now **P** has to provide a witness for the existential claim, i.e., for $\sup\{v_{\mathcal{M}}(\varphi(c)) \mid c \in D\} \geq r$. But as mentioned in Section 2, if the supremum is not a maximum, this poses a problem. It can happen that **P**'s claim is true, but that nevertheless there does not exist a witnessing element that would directly show this. The solution for the case of such non-witnessed models is similar to the one from Section 2. We relax the winning condition and allow the player who is currently in the role **P** to select a witness for which the value of the formula may not be equal to $r$, but is arbitrarily close. To this aim we let **O** decrease the value of the formula (where, of course, it is in **O**'s interest to decrease it as little as possible) and only *then* require **P** to find a witness (for the decreased value). Note that this does not affect **O**'s winning condition. If in the state $\langle \exists x\varphi(x), r\rangle$ the value $r$ is strictly greater than $\sup\{v_{\mathcal{M}}(\varphi(c)) \mid c \in D\}$ then **O** can always win by choosing an $\epsilon$ that lies between the supremum and $r$. The just discussed rule can actually be stated formally without explicit reference to $\epsilon$ as follows.

($R^{\mathcal{E}}_{\exists}$) If the current state is $\langle \exists x\varphi(x), r\rangle$ then **O** chooses $r' < r$ and **P** chooses $c \in D$; the game continues with $\langle \varphi(c), \max\{0, r'\}\rangle$.

Compared to Theorems 2.1.1 and 2.2.2, the adequateness theorem for the $\mathcal{E}$-game reveals a somewhat less direct correspondence to the standard semantics of Ł.

THEOREM 3.0.1. *I, initially acting as* **P**, *have a winning strategy in the* $\mathcal{E}$*-game for* $\langle \varphi, r\rangle$ *under an* Ł*-interpretation* $\mathcal{M}$ *iff* $v_{\mathcal{M}}(\varphi) \geq r$.

*Proof.* We prove the claim by induction on complexity of $\varphi$. In the induction hypothesis we actually do not care whether I or you are initially in the role of **P**.

The base case, where $\chi$ is an atomic formula, is obvious.

If $\varphi = \varphi_1 \vee \varphi_2$ we argue similarly to the corresponding case of the proof of Theorem 2.2.2. By the induction hypothesis **P** has a winning strategy for $\langle \varphi_i, r_i\rangle$ iff $v_{\mathcal{M}}(\varphi_i) \geq r_i$ for $i \in \{1, 2\}$. **P**'s winning strategy for $\langle \varphi_1 \vee \varphi_2, r\rangle$ is obtained by the choice of $\varphi_i$ such that $v_{\mathcal{M}}(\varphi_i) = \max\{v_{\mathcal{M}}(\varphi_1), v_{\mathcal{M}}(\varphi_2)\}$. Conversely a winning strategy for **P** for $\langle \varphi_1 \vee \varphi_2, r\rangle$ contains either one for $\langle \varphi_1, r\rangle$ or one for $\langle \varphi_2, r\rangle$. Therefore we obtain $\max\{v_{\mathcal{M}}(\varphi_1), v_{\mathcal{M}}(\varphi_2)\} \geq r$ by the induction hypothesis.

The cases for $\varphi = \varphi_1 \wedge \varphi_2$ is analogous.

Consider $\varphi = \neg\varphi_1$: we have $v_{\mathcal{M}}(\varphi) \geq r$ iff $v_{\mathcal{M}}(\varphi_i) \leq 1 - r$ iff $v_{\mathcal{M}}(\varphi_i) < 1 - r'$ for every $r' < r$. Now note that, since we are dealing with a finite game of perfect information, **O** has a winning strategy iff **P** does not have a winning strategy. Therefore the induction hypothesis implies that for every $r' < r$ **O** has a winning strategy in the game for $\langle \varphi_1, 1 - r'\rangle$ iff $v_{\mathcal{M}}(\neg\varphi_1) \geq r$. Since the rule $R^{\mathcal{E}}_{\neg}$ entails a role switch we obtain that **P** has a winning strategy for $\langle \varphi, r\rangle$ iff $v_{\mathcal{M}}(\varphi) \geq r$.

For $\varphi = \varphi_1 \oplus \varphi_2$ suppose that **P** has a winning strategy for $\langle \varphi, r\rangle$. By the rule $R^{\mathcal{E}}_{\oplus}$ this means that **P** has a winning strategy for $\langle \varphi_1, r_1\rangle$ as well as for $\langle \varphi_1, r_2\rangle$ for some $r_1$ and $r_2$ satisfying $r_1 + r_2 = r$. By the induction hypothesis we obtain that $v_{\mathcal{M}}(\varphi_1) \geq r_1$ and $v_{\mathcal{M}}(\varphi_2) \geq r_2$. But this implies that $v_{\mathcal{M}}(\varphi) = v_{\mathcal{M}}(\varphi_1 \oplus \varphi_2) = \min\{1, v_{\mathcal{M}}(\varphi_1) + v_{\mathcal{M}}(\varphi_2)\} \geq r_1 + r_2 = r$, as required. Conversely, suppose that $v_{\mathcal{M}}(\varphi) = v_{\mathcal{M}}(\varphi_1 \oplus \varphi_2) \geq r$. By the induction hypothesis **P** has a winning strategy for $(\varphi_i, r_i)$ whenever $v_{\mathcal{M}}(\varphi_i) \geq r_i$, for $i \in \{1, 2\}$. Since $v_{\mathcal{M}}(\varphi_1 \oplus \varphi_2) = \min\{1, v_{\mathcal{M}}(\varphi_1) + v_{\mathcal{M}}(\varphi_2)\}$, **P** can choose $r_1 = v_{\mathcal{M}}(\varphi_1)$ and $r_2 = v_{\mathcal{M}}(\varphi_2)$ and combine the winning strategies for $\langle \varphi_1, r_1\rangle$ and $\langle \varphi_2, r_2\rangle$ into one for $\langle \varphi_1 \oplus \varphi_2, r_1 + r_2\rangle = \langle \varphi, r\rangle$.

For $\varphi = \varphi_1 \,\&\, \varphi_2$ suppose that **P** has a winning strategy for $\langle \varphi, r\rangle$. By the rule $R^{\mathcal{E}}_{\&}$ this means that for some non-negative $\bar{r} \leq 1 - r$ **P** has a winning strategy for $\langle \varphi_1, r + \bar{r}\rangle$ as well as for $\langle \varphi_2, 1 - \bar{r}\rangle$. By the induction hypothesis we obtain that $v_{\mathcal{M}}(\varphi_1) \geq r + \bar{r}$ and $v_{\mathcal{M}}(\varphi_2) \geq 1 - \bar{r}$. Joining these facts we obtain that $v_{\mathcal{M}}(\varphi) = v_{\mathcal{M}}(\varphi_1 \,\&\, \varphi_2) =$

$\max\{0, v_{\mathcal{M}}(\varphi_1){+}v_{\mathcal{M}}(\varphi_2){-}1\} \geq \max\{0, r{+}\bar{r}{+}1{-}\bar{r}{-}1\} = r$ as required. Conversely, suppose that $v_{\mathcal{M}}(\varphi) = v_{\mathcal{M}}(\varphi_1 \,\&\, \varphi_2) \geq r$. Let **P** choose $\bar{r} = 1 - v_{\mathcal{M}}(\varphi_2)$. Then we have $v_{\mathcal{M}}(\varphi_2) = 1 - \bar{r}$. Because of $v_{\mathcal{M}}(\varphi_1 \,\&\, \varphi_2) = \max\{0, v_{\mathcal{M}}(\varphi_1) + v_{\mathcal{M}}(\varphi_2) - 1\}$, we moreover have $v_{\mathcal{M}}(\varphi_1) = v_{\mathcal{M}}(\varphi_1 \,\&\, \varphi_2) + 1 - v_{\mathcal{M}}(\varphi_2) = v_{\mathcal{M}}(\varphi_1 \,\&\, \varphi_2) + \bar{r}$. Therefore, by the induction hypothesis, there exist winning strategies for **P** for the game $\langle \varphi_2, 1 - \bar{r} \rangle$ and for the game $\langle \varphi_1, r + \bar{r} \rangle$. By combining these strategies we obtain **P**'s winning strategy for $\langle \varphi, r \rangle$.

For $\varphi = \forall x \varphi_1(x)$ the rule $R^{\mathcal{E}}_{\forall}$ entails that **P** has a winning strategy for $\langle \varphi, r \rangle$ iff she has a winning strategy for $\langle \varphi_1(c), r \rangle$ for all $c \in D$. By the induction hypothesis the latter is equivalent to $v_{\mathcal{M}}(\varphi_1(c)) \geq r$ for all $c \in D$. But this in turn is equivalent to $v_{\mathcal{M}}(\forall x \varphi_1(x)) \geq r$.

For $\varphi = \exists x \varphi_1(x)$ let us once more check the two directions of the equivalence separately. First suppose that **P** has a winning strategy for $\langle \varphi, r \rangle$. By the rule $R^{\mathcal{E}}_{\exists}$ this means that for all $r' < r$ **P** can find a $c \in D$ such that she has a winning strategy for $\langle \varphi_1(c), \max\{0, r'\} \rangle$. By the induction hypothesis this implies $v_{\mathcal{M}}(\varphi_1(c)) \geq \max\{0, r'\}$ for all $r' < r$, and therefore $v_{\mathcal{M}}(\varphi) = v_{\mathcal{M}}(\exists x \varphi_1(x)) = \sup\{v_{\mathcal{M}}(\varphi_1(c)) \mid c \in D\} \geq \sup\{r' \mid r' < r\} = r$. Conversely, suppose $v_{\mathcal{M}}(\varphi) = v_{\mathcal{M}}(\exists x \varphi_1(x)) \geq r$. This implies that for every $r' < r$ there is $c \in D$ such that $v_{\mathcal{M}}(\varphi_1(c)) \geq r'$. By the induction hypothesis this implies that for every $r' < r$ there is a $c \in D$ such that **P** has a winning strategy for $\langle \varphi_1(c), r' \rangle$. According to rule $R^{\mathcal{E}}_{\exists}$ these winning strategies can be combined into one for $\langle \exists x \varphi_1(x), r \rangle = \langle \varphi, r \rangle$. □

## 4 Giles's game for Łukasiewicz logic

Already in the 1970s Robin Giles [25] introduced a game that was intended to provide 'tangible meaning' to reasoning about statements with dispersive semantic tests as they appear in physics. For the logical rules of his game Giles referred not to Hintikka or Henkin, but rather to the dialogue game based semantics for intuitionistic logic by Lorenzen [39, 40]. In particular, Giles proposed the following rule for implication:

($R^{\mathcal{G}}_{\to}$) He who asserts $\varphi \to \psi$ agrees to assert $\psi$ if his opponent will assert $\varphi$.

Like we did in Sections 2 and 3, above, Giles refers to the players as I and you, respectively. In contrast to $\mathcal{H}$-games, the rule $R^{\mathcal{G}}_{\to}$ introduces game states, where more than one formula may be currently asserted by each player. Since, in general, it matters whether we assert the same statement just once or more often, game states are now denoted as pairs of multisets of formulas. Following Giles, we call these multisets *my tenet* and *your tenet*, respectively. Formally we denote a state as

$$[\varphi_1, \ldots, \varphi_n \mid \psi_1, \ldots, \psi_m],$$

where $[\varphi_1, \ldots, \varphi_n]$ is your tenet and $[\psi_1, \ldots, \psi_m]$ is my tenet.

The payoff at a final game state, where all currently asserted formulas are atomic, is defined in terms of expected risks of payments to be made to the opposing player whenever an atomic assertion made by a player turns out to be false. More precisely, a *binary experiment* $\mathsf{E}_p$ is associated with each atomic formula $p$. 'Binary' here means that $\mathsf{E}_p$ either *fails* or *succeeds*. The special experiment $\mathsf{E}_\bot$ always fails. We stipulate that I have to pay 1€ to you for each of my assertions of $\mathsf{E}_p$, where a corresponding trial of $\mathsf{E}_p$ fails. Likewise, you have to pay 1€ to me for each of your assertions that does not pass the associated test. The central feature of Giles's payoff scheme is that that each experiment $\mathsf{E}_p$ may be *dispersive*, meaning that $\mathsf{E}_p$ may yield different results when repeated. But a fixed *failure probability* $\pi(\mathsf{E}_p)$ is known to the players for each $p$; we call this probability the risk value $\langle p \rangle$ of $p$. Remember that it matters whether we assert the same proposition just once or more often. For a final game state

$$[p_1, \ldots, p_n \mid q_1, \ldots, q_m] .$$

we can therefore specify the expected total amount of money (in €) that I have to pay to you at the exhibited state by

$$\langle p_1, \ldots, p_n \mid q_1, \ldots, q_m \rangle = \sum_{1 \leq i \leq m} \langle q_i \rangle \; - \; \sum_{1 \leq j \leq n} \langle p_j \rangle .$$

We call this number briefly my *risk* associated with that state. Note that the risk can be negative, i.e., the risk values of the relevant propositions may be such that I expect an (average) net payment by you to myself.

As an example consider the state $[p, p \mid q]$, where you have asserted $p$ twice and I have asserted $q$ once. Three trials of experiments are involved in the corresponding evaluation: two trials of $\mathsf{E}_p$, one for each of your assertions, and one trial of $\mathsf{E}_q$ to test my assertion. If $\langle p \rangle = 0.2$, i.e., if the probability that the experiment $\mathsf{E}_p$ fails is 0.2 and $\langle q \rangle = 0.5$ then $\langle p, p \mid q \rangle = 0.1$. This means that my expected loss of money according to the outlined betting scheme is 0.1€. On the other hand, if $\langle p \rangle = \langle q \rangle = 0.5$, then $\langle p, p \mid q \rangle = -0.5$, which means that I expect an (average) gain of 0.5€.

In the context of fuzzy logic, one may interpret this setup as a model of reasoning under vagueness. As linguists and philosophers of language have repeatedly pointed out, competent language users, in concrete dialogues, either (momentarily and provisionally) accept or don't accept utterances upon receiving them. No 'degrees of truth' enter the picture at this level; vagueness rather consists in a certain brittleness or dispersiveness of such highly context dependent decisions (see, e.g, [53]). One imagines that the dialogue partners repeatedly solicit answers to the question "Do you accept $p$?" from competent speakers who are familiar with the given context of assertion, but who may have different standards of acceptance of $p$, reflecting its vagueness. With respect to the terminology introduced above, the experiment $\mathsf{E}_p$ consists in asking this question; $\mathsf{E}_p$ fails if the answer is negative. To arrive at a 'degree of truth' for $p$ we assume that the players have a particular expectation for $\mathsf{E}_p$ to fail or to succeed, that only depends on $p$. We may thus arrive at a many-valued interpretation by stipulating that $v_{\mathcal{M}}(p) = 1 - \langle p \rangle$.

Like in the $\mathcal{H}$-game and the $\mathcal{E}$-game, we distinguish the roles of a proponent **P** and of an opponent **O** for any occurrence of a complex formula. I act as **P** and you as **O** for any formula in my tenet, while you act as **P** and I act as **O** for any formula in your tenet. Following the terminology of Lorenzen for logical dialogue games, one also refers to a (semi-)move by a player in role **O** as an *attack* and calls the corresponding reaction of the other player (in role **P**) a *defense* of the attacked formula occurrence.

The rules $R^{\mathcal{H}}_{\wedge}$, $R^{\mathcal{H}}_{\vee}$, $R^{\mathcal{H}}_{\forall}$, and $R^{\mathcal{H}}_{\exists}$ defined in Section 2 basically remain unchanged for $\mathcal{G}$-games. However, we reformulate these rules as well as Giles's original implication rule, stated above, to better reflect the context in which these rules apply in a $\mathcal{G}$-game.

($R^{\mathcal{G}}_{\wedge}$) If the current formula is $\varphi \wedge \psi$ then the game continues in a state where the indicated occurrence of $\varphi \wedge \psi$ in **P**'s tenet is replaced by either $\varphi$ or by $\psi$, according to **O**'s choice.

($R^{\mathcal{G}}_{\vee}$) If the current formula is $\varphi \vee \psi$ then the game continues in a state where the indicated occurrence of $\varphi \vee \psi$ in **P**'s tenet is replaced by either $\varphi$ or by $\psi$, according to **P**'s choice.

($R^{\mathcal{G}}_{\rightarrow}$) If the current formula is $\varphi \rightarrow \psi$ then the indicated occurrence of $\varphi \rightarrow \psi$ is removed from **P**'s tenet and **O** chooses whether to continue the game at the resulting state or whether to add $\varphi$ to **O**'s tenet and $\psi$ to **P**'s tenet before continuing the game.

($R^{\mathcal{G}}_{\forall}$) If the current formula is $\forall x \varphi(x)$ then **O** chooses an element $c$ of the domain of $\mathcal{M}$ and the game continues in a state where the indicated occurrence of $\forall x \varphi(x)$ in **P**'s tenet is replaced by $\varphi(c)$.

($R^{\mathcal{G}}_{\exists}$) If the current formula is $\exists x \varphi(x)$ then **P** chooses an element $c$ of the domain of $\mathcal{M}$ and the game continues in a state where the indicated occurrence of $\exists x \varphi(x)$ in **P**'s tenet is replaced by $\varphi(c)$.

For later reference, we point out that $R^{\mathcal{G}}_{\rightarrow}$ contains a hidden *principle of limited liability*: referring to an occurrence of $\varphi \rightarrow \psi$, the player in role **O** may, instead of asserting $\varphi$ in order to elicit **P**'s assertion of $\psi$, explicitly choose not to attack $\varphi \rightarrow \psi$ at all. This option results in a branching of the game tree. The state $[\Gamma \mid \Delta, \varphi \rightarrow \psi]$, where $\Gamma$ and $\Delta$ are multisets of sentences asserted by you and me, respectively, and where the exhibited occurrence indicates that you currently refer to my assertion of $\varphi \rightarrow \psi$, has the two possible successor states $[\varphi, \Gamma \mid \Delta, \psi]$ and $[\Gamma \mid \Delta]$. In the latter state you have chosen to limit your liability in the following sense. Attacking an assertion by the other player should never incur an expected (positive) loss, which were the case if the risk associated with asserting $\varphi$ is higher than that for asserting $\psi$. In such cases a rational player in role **O** will explicitly renounce an attack on $\varphi \rightarrow \psi$. For all other logical connectives the principle is ensured by the fact that the rules of the $\mathcal{G}$-game ensure that each occurrence of a formula can be attacked at most once: the attacked occurrence is removed from the state in the transition to a corresponding successor state.

Another form of the principle of limited liability arises for defense moves. In defending any sentence $\varphi$, **P** has to be able to hedge her possible loss associated with the assertions made in defense of $\varphi$ to at most 1€. This is already the case for all logical rules considered so far. However, as shown in [14, 19], by making this principle explicit we arrive at a rule for strong conjunction, that is missing in Giles [25, 26]:

($R^{\mathcal{G}}_{\&}$) If the current formula is $\varphi \,\&\, \psi$ then **P** chooses whether to continue the game at a state where the indicated occurrence of $\varphi \,\&\, \psi$ is replaced by $\varphi$ as well as $\psi$ in **P**'s tenet, or by a single occurrence of $\bot$, instead.

The above description might yet be too informal to see in which sense every $\mathcal{G}$-game, just like an $\mathcal{H}$-game, constitutes an ordinary two-person zero-sum extensive game of finite depth with perfect information. For this purpose one should be a bit more precise than Giles and specify for each non-final state which player is to move next and which of the formulas in **P**'s tenet is the "current formula". For this purpose we introduce the notion of a *regulation*, which is a function $\rho$ that maps every non-final game state $[\Gamma \mid \Delta]$ into an occurrence of some non-atomic formula in either your tenet $\Gamma$ or in my tenet $\Delta$; $\rho([\Gamma \mid \Delta])$ is called the *current formula.* When the current formula in a state has to be made explicit, we will underline it. If the initial state of a game is $[\mid \varphi]$ we speak of a *$\mathcal{G}$-game for $\varphi$.*

Given any Ł-interpretation $\mathcal{M}$ we define a corresponding risk value assignment by $\langle p \rangle_{\mathcal{M}} = 1 - v_{\mathcal{M}}(p)$ for every propositional variable $p$. Rather than to refer to the minimal upper bound of risks associated with the final states that I can enforce if we both play rationally, we want to talk about the *value* of a game, as defined in Section 2. To be able to apply Definition 2.2.1, we therefore stipulate that my risk at a final state (as defined above) is your payoff, while my payoff is the inverse of this risk, entailing that the game is zero-sum. These conventions allow us to formulate the characterization of Łukasiewicz logic Ł by $\mathcal{G}$-games as follows.

THEOREM 4.0.1. *The value for myself of a $\mathcal{G}$-game for $\varphi$ under the risk value assignment $\langle \cdot \rangle_{\mathcal{M}}$ and an arbitrary regulation $\rho$ is $v_{\mathcal{M}}(\varphi)$.*

*Proof.* Note that every run of a $\mathcal{G}$-game is finite; therefore we can once more 'solve' the game by backward induction. Recall that for every final state $[p_1, \ldots, p_m \mid q_1, \ldots, q_n]$, we have defined my associated risk as

$$\langle p_1, \ldots, p_n \mid q_1, \ldots, q_m \rangle = \sum_{1 \leq i \leq m} \langle q_i \rangle \; - \sum_{1 \leq j \leq n} \langle p_j \rangle .$$

To calculate the value of the game, i.e., the minimal upper bound of the final risk that I can enforce at a given non-final state $S$ we have have to take into account two rationality principles, that arise since the $\mathcal{G}$-game is zero-sum:

1. If you can choose the successor state to $S$ then my final risk at $S$ is the maximum over all risks associated with successor states to $S$.

2. If, on the other hand, I can choose the successor state to $S$ then my final risk at $S$ is the minimum over all risks associated with the successor states.

Correspondingly, if the current formula is an implication, then the rule $R^{\mathcal{G}}_{\to}$ requires us to show that the notion of my risk $\langle \cdot \mid \cdot \rangle$ can be extended from final states to arbitrary states in a manner that guarantees that the following conditions are satisfied:

$$\langle \Gamma \mid \underline{\varphi \to \psi}, \Delta \rangle \;=\; \max\{\langle \Gamma \mid \Delta \rangle, \langle \Gamma, \varphi \mid \psi, \Delta \rangle\} \tag{1}$$

$$\langle \Gamma, \underline{\varphi \to \psi} \mid \Delta \rangle \;=\; \min\{\langle \Gamma \mid \Delta \rangle, \langle \Gamma, \psi \mid \varphi, \Delta \rangle\}. \tag{2}$$

To connect risk for arbitrary states with truth value assignments (Ł-interpretations) we extend the semantics of Ł from formulas to multisets $\Gamma$ of formulas as follows:

$$v_{\mathcal{M}}(\Gamma) = \sum_{\varphi \in \Gamma} v_{\mathcal{M}}(\varphi).$$

Risk value assignments are in one to one correspondence with truth value assignments via $\langle p \rangle = \langle p \rangle_{\mathcal{M}} = 1 - v_{\mathcal{M}}(p)$ for all propositional variables $p$, which extends to

$$\langle p_1, \ldots, p_m \mid q_1, \ldots, q_n \rangle_{\mathcal{M}} = n - m + v_{\mathcal{M}}([p_1, \ldots, p_m]) - v_{\mathcal{M}}([q_1, \ldots, q_n]).$$

Correspondingly, we define the following function for arbitrary states:

$$\langle \Gamma \mid \Delta \rangle_{\mathcal{M}} = |\Delta| - |\Gamma| + v_{\mathcal{M}}(\Gamma) - v_{\mathcal{M}}(\Delta).$$

Note that this in particular entails

$$\langle \mid \varphi \rangle_{\mathcal{M}} = 1 - v_{\mathcal{M}}([\varphi]) = 1 - v_{\mathcal{M}}(\varphi). \tag{3}$$

It remains to show that $\langle \cdot \mid \cdot \rangle_{\mathcal{M}}$ indeed specifies my final risk at any state if I play rationally. For final states this is immediate. If the current formula selected by the regulation $\rho$ is an implication in my tenet then we have to check that $\langle \cdot \mid \cdot \rangle_{\mathcal{M}}$ satisfies condition (1):

$$\begin{aligned}
\langle \Gamma \mid \underline{\varphi \to \psi}, \Delta \rangle_{\mathcal{M}} &= |\Delta| + 1 - |\Gamma| + v_{\mathcal{M}}(\Gamma) - v_{\mathcal{M}}(\Delta) - v_{\mathcal{M}}(\varphi \to \psi) \\
&= \langle \Gamma \mid \Delta \rangle_{\mathcal{M}} + 1 - v_{\mathcal{M}}(\varphi \to \psi) \\
&= \langle \Gamma \mid \Delta \rangle_{\mathcal{M}} + 1 - \min\{1, 1 - v_{\mathcal{M}}(\varphi) + v_{\mathcal{M}}(\psi)\} \\
&= \langle \Gamma \mid \Delta \rangle_{\mathcal{M}} - \min\{0, v_{\mathcal{M}}(\psi) - v_{\mathcal{M}}(\varphi)\} \\
&= \langle \Gamma \mid \Delta \rangle_{\mathcal{M}} + \max\{0, v_{\mathcal{M}}(\varphi) - v_{\mathcal{M}}(\psi)\} \\
&= \langle \Gamma \mid \Delta \rangle_{\mathcal{M}} + \max\{0, \langle \varphi \mid \psi \rangle_{\mathcal{M}}\} \\
&= \max\{\langle \Gamma \mid \Delta \rangle_{\mathcal{M}}, \langle \Gamma, \varphi \mid \psi, \Delta \rangle_{\mathcal{M}}\}.
\end{aligned}$$

For states where the current formula is an implication in your tenet condition (2) can be checked as follows:

$$\begin{aligned}
\langle \Gamma, \underline{\varphi \to \psi} \mid \Delta \rangle_{\mathcal{M}} &= |\Delta| - |\Gamma| - 1 + v_{\mathcal{M}}(\Gamma) + v_{\mathcal{M}}(\varphi \to \psi) - v_{\mathcal{M}}(\Delta) \\
&= \langle \Gamma \mid \Delta \rangle_{\mathcal{M}} - 1 + v_{\mathcal{M}}(\varphi \to \psi) \\
&= \langle \Gamma \mid \Delta \rangle_{\mathcal{M}} - 1 + \min\{1, 1 - v_{\mathcal{M}}(\varphi) + v_{\mathcal{M}}(\psi)\} \\
&= \langle \Gamma \mid \Delta \rangle_{\mathcal{M}} - 1 + \min\{1, 1 + \langle \psi \mid \varphi \rangle_{\mathcal{M}}\} \\
&= \langle \Gamma \mid \Delta \rangle_{\mathcal{M}} + \min\{0, \langle \psi \mid \varphi \rangle_{\mathcal{M}}\} \\
&= \min\{\langle \Gamma \mid \Delta \rangle_{\mathcal{M}}, \langle \Gamma, \psi \mid \varphi, \Delta \rangle_{\mathcal{M}}\}.
\end{aligned}$$

If the current formula is a strong conjunction, the following conditions arise from the rationality principles and the rule $R^{\mathcal{G}}_{\&}$:

$$\langle \Gamma \mid \underline{\varphi \,\&\, \psi}, \Delta \rangle = \min\{\langle \Gamma \mid \Delta, \bot \rangle, \langle \Gamma \mid \Delta, \varphi, \psi \rangle\} \tag{4}$$

$$\langle \Gamma, \underline{\varphi \,\&\, \psi} \mid \Delta \rangle = \max\{\langle \Gamma, \bot \mid \Delta \rangle, \langle \Gamma, \varphi, \psi \mid \Delta \rangle\}. \tag{5}$$

The corresponding arguments are as follows:

$$\begin{array}{rcl}
\langle\Gamma \mid \underline{\varphi \,\&\, \psi}, \Delta\rangle_{\mathcal{M}} & = & |\Delta| + 1 - |\Gamma| + v_{\mathcal{M}}(\Gamma) - v_{\mathcal{M}}(\Delta) - v_{\mathcal{M}}(\varphi \,\&\, \psi) \\
& = & \langle\Gamma \mid \Delta\rangle_{\mathcal{M}} + 1 - v_{\mathcal{M}}(\varphi \,\&\, \psi) \\
& = & \langle\Gamma \mid \Delta\rangle_{\mathcal{M}} + 1 - \max\{0, v_{\mathcal{M}}(\varphi) + v_{\mathcal{M}}(\psi) - 1\} \\
& = & \langle\Gamma \mid \Delta\rangle_{\mathcal{M}} + \min\{1, (1 - v_{\mathcal{M}}(\varphi)) + (1 - v_{\mathcal{M}}(\psi))\} \\
& = & \langle\Gamma \mid \Delta\rangle_{\mathcal{M}} + \min\{1, \langle \mid \varphi, \psi\rangle_{\mathcal{M}}\} \\
& = & \min\{\langle\Gamma \mid \Delta, \bot\rangle_{\mathcal{M}}, \langle\Gamma \mid \Delta, \varphi, \psi\rangle_{\mathcal{M}}\}. \\
\langle\Gamma, \underline{\varphi \,\&\, \psi} \mid \Delta\rangle_{\mathcal{M}} & = & |\Delta| - |\Gamma| - 1 + v_{\mathcal{M}}(\Gamma) + v_{\mathcal{M}}(\varphi \,\&\, \psi) - v_{\mathcal{M}}(\Delta) \\
& = & \langle\Gamma \mid \Delta\rangle_{\mathcal{M}} - 1 + v_{\mathcal{M}}(\varphi \,\&\, \psi) \\
& = & \langle\Gamma \mid \Delta\rangle_{\mathcal{M}} - 1 + \max\{0, v_{\mathcal{M}}(\varphi) + v_{\mathcal{M}}(\psi) - 1\} \\
& = & \max\{\langle\Gamma, \bot \mid \Delta\rangle_{\mathcal{M}}, \langle\Gamma, \varphi, \psi \mid \Delta\rangle_{\mathcal{M}}\}.
\end{array}$$

Analogous conditions corresponding to the rules $R^{\mathcal{G}}_{\vee}$, $R^{\mathcal{G}}_{\wedge}$, $R^{\mathcal{G}}_{\forall}$, and $R^{\mathcal{G}}_{\exists}$, respectively, can be checked straightforwardly. But note that, just like in the corresponding cases of the $\mathcal{H}$-mv-game, the quantifier rules entail a reference to some 'margin of error' $\epsilon$, as indicated in Definition 2.2.1 of the value of a game. □

REMARK 4.0.2. *Note that the above proof of Theorem 4.0.1 can be read as justification of Łukasiewicz logic with respect to Giles's game based model of approximate reasoning. Rather than imposing the truth functions for the various connectives in the first place, they are derived from the rules and the payoff scheme of the game in conjunction with the general concept of rationality that underlies game theory.*

At a first glimpse, Giles's game looks very different from the $\mathcal{H}$-mv-game. However, in a sense, it may actually be viewed as closer in spirit to the $\mathcal{H}$-mv-game (and therefore Hintikka's classic $\mathcal{H}$-game) than the $\mathcal{E}$-game described in Section 3. The main point here is that, in contrast to the $\mathcal{E}$-game, no explicit reference to truth values is made in the $\mathcal{G}$-game. Instead, like in the $\mathcal{H}$-mv-game, there is a direct match between (optimal) payoff for myself and truth values. Giles's motivation of payoff values in terms of bets on expected results of dispersive binary experiments seems to put his version of game semantics apart from Hintikka's. However note that the offered interpretation of truth values as inverted risk values is in fact completely independent from the semantic game itself. Therefore we may choose to ignore that part of Giles's semantic altogether and simply speak of assignments of values $\in [0, 1]$ to atomic formulas, just like for the $\mathcal{H}$-mv-game. Conversely, we may add Giles's betting scenario to the $\mathcal{H}$-mv-game and interpret the value assigned to an atomic formula as the inverted risk of having to pay 1€ when the claim that a particular dispersive experiment, characterized by a given failure probability, succeeds turns out to false. In other words, the only remaining essential difference between the $\mathcal{G}$-game and the $\mathcal{H}$-mv-game is that more than one formula occurrence has to be taken into account in general in the former case. In Sections 7 and 8 we will investigate variants of semantic games that address this issue. But before doing so, we will investigate (in Section 5) a generalization of the $\mathcal{G}$-game that leaves its notion of a game state unchanged and discuss (in Section 6) the relation of the $\mathcal{G}$-game and some of its variants to hypersequent based proof systems.

# 5 Generalizing Giles's game

In this section we review a general framework for Giles-style games at the propositional level. In contrast to the previous sections, we will not talk about specific rules for particular logical connectives, but rather specify a general rule format appropriate for this type of game. Moreover we will look at the evaluation of final (atomic) game states from a wider perspective that is neither dependent on particular motivations regarding reasoning in physics (as in Giles) nor on the presence of vagueness (as indicated in the Section 4). The main result recorded in this section is that truth functions over the reals—and in this sense: fuzzy logics—can be recovered for any concrete instance of this general game based framework.

We stick with the notion, introduced in Section 4, of a game state as consisting of two multisets of formulas: my tenet and your tenet. We will denote atomic tenets by $\gamma$ or $\delta$, possibly primed, and arbitrary tenets by upper Greek letters $\Gamma, \Delta, \ldots$. Moreover, we write $[\Gamma, \Delta]$ to denote the union of the multisets $\Gamma$ and $\Delta$ as well as $[\Gamma, \phi]$ instead of $[\Gamma, [\phi]]$, etc.

## 5.1 General payoff principles

Giles's story about risking money to be paid when losing bets on dispersive experiments might be intriguing from a philosophical point of view, however, mathematically, it boils down to the definition of a particular ordinary payoff function in the usual game theoretic sense, i.e., an assignment of real numbers to all final states of the game. This observation motivates the formulation of general principles for assigning payoff values to atomic states. As in previous sections, we will only be interested in payoff for myself and thus simply speak of 'the payoff' associated with an atomic state. (More precisely, we can think of your payoff for the same state as directly inverse to mine. In other words, the game is zero-sum. This is codified in the *Payoff Principle* 2, below.

DEFINITION 5.1.1 (Payoff). *A* payoff function *assigns a real number to every atomic game state. The payoff of the atomic game state $[\gamma \mid \delta]$ is denoted as $\langle \gamma \mid \delta \rangle$.*

**Payoff Principle 1** (Context independence)**.** *A payoff function $\langle \cdot \mid \cdot \rangle$ is* context independent *if for all atomic tenets $\gamma, \delta, \gamma', \delta', \gamma''$, and $\delta''$ the following holds: If $\langle \gamma' \mid \delta' \rangle = \langle \gamma'' \mid \delta'' \rangle$ then $\langle \gamma, \gamma' \mid \delta', \delta \rangle = \langle \gamma, \gamma'' \mid \delta'', \delta \rangle$.*

Context independence entails that the payoff for a state $[\gamma, \gamma' \mid \delta, \delta']$ is solely determined by the payoffs of its sub-states $[\gamma \mid \delta]$ and $[\gamma' \mid \delta']$. This property is crucial for recovering a truth functional (compositional) semantics for all our games.

PROPOSITION 5.1.2. *Let $\langle \cdot \mid \cdot \rangle$ be a context independent payoff function and let $G = [\gamma, \gamma' \mid \delta, \delta']$ be an atomic game state. Then there exists an associative and commutative binary operation $\oplus$ on $\mathbb{R}$ such that $\langle G \rangle = \langle \gamma \mid \delta \rangle \oplus \langle \gamma' \mid \delta' \rangle$.*

*Proof.* Assume that $\langle \gamma \mid \delta \rangle = \langle \gamma'' \mid \delta'' \rangle = x$ and $\langle \gamma' \mid \delta' \rangle = \langle \gamma''' \mid \delta''' \rangle = y$. Then $\langle \gamma'', \gamma''' \mid \delta'', \delta''' \rangle = \langle \gamma, \gamma''' \mid \delta, \delta''' \rangle = \langle \gamma, \gamma' \mid \delta, \delta' \rangle$ by applying context independence twice. Thus we may write $\langle \gamma, \gamma' \mid \delta, \delta' \rangle = x \oplus y$. Associativity and commutativity of $\oplus$ directly follow from the fact that tenets are multisets. □

REMARK 5.1.3. *We will call $\oplus$ as specified in Proposition 5.1.2 the* aggregation function corresponding to $\langle\cdot \mid \cdot\rangle$. *In Giles's original game the function $\oplus$ is ordinary addition, which motivates our notation.*

**Payoff Principle 2** (Symmetry). *A payoff function $\langle\cdot \mid \cdot\rangle$ is* symmetric *if $\langle\gamma \mid \delta\rangle = -\langle\delta \mid \gamma\rangle$ for all atomic tenets $\gamma$ and $\delta$.*

If $\langle\cdot \mid \cdot\rangle$ is context independent and symmetric then the payoff of an arbitrary atomic game state can be decomposed as follows:

$$\begin{aligned}\langle p_1, \ldots, p_n \mid q_1, \ldots, q_m\rangle &= \langle p_1 \mid \rangle \oplus \cdots \oplus \langle p_n \mid \rangle \oplus \langle \mid q_1\rangle \oplus \cdots \oplus \langle \mid q_m\rangle \\ &= -\langle \mid p_1\rangle \oplus \cdots \oplus -\langle \mid p_n\rangle \oplus \langle \mid q_1\rangle \oplus \cdots \oplus \langle \mid q_m\rangle .\end{aligned}$$

Note that symmetry implies that $\langle\gamma \mid \gamma\rangle = 0$. In other words, the payoff is 0 in any atomic state where your tenet is identical to mine. Moreover, this shows that one could focus on single tenets instead of two-sided states.

PROPOSITION 5.1.4. *Let $\langle\cdot \mid \cdot\rangle$ be a context independent and symmetric payoff function. Then*

*(i) $-$ distributes over the corresponding aggregation function $\oplus$, i.e., for all payoff values $x$ and $y$, $-(x \oplus y) = -x \oplus -y$.*

*(ii) $-$ is inverse to $\oplus$, i.e., $x \oplus -x = 0$ holds for all values $x$.*

*Proof. (i)* Let $[\gamma_1 \mid \delta_1]$ and $[\gamma_2 \mid \delta_2]$ be two atomic states where $\langle\gamma_1 \mid \delta_1\rangle = x$ and $\langle\gamma_2 \mid \delta_2\rangle = y$. Then

$$\begin{aligned}-(x \oplus y) &= -(\langle\gamma_1 \mid \delta_1\rangle \oplus \langle\gamma_1 \mid \delta_2\rangle) && \text{by definition of } x, y \\ &= -\langle\gamma_1, \gamma_2 \mid \delta_1, \delta_2\rangle && \text{by Proposition 5.1.2} \\ &= \langle\delta_1, \delta_2 \mid \gamma_1, \gamma_2\rangle && \text{by Payoff Principle 1 (symmetry)} \\ &= \langle\delta_1 \mid \gamma_1\rangle \oplus \langle\delta_2 \mid \gamma_2\rangle && \text{by Proposition 5.1.2} \\ &= -\langle\gamma_1 \mid \delta_1\rangle \oplus -\langle\gamma_2 \mid \delta_2\rangle && \text{by Payoff Principle 1 (symmetry)} \\ &= -x \oplus -y && \text{by definition of } x, y.\end{aligned}$$

*(ii)* Let $[\gamma \mid \delta]$ be an atomic game state such that $\langle\gamma \mid \delta\rangle = x$. Then

$$\begin{aligned}x \oplus -x &= \langle\gamma \mid \delta\rangle \oplus -\langle\gamma \mid \delta\rangle && \text{by definition of } x \\ &= \langle\gamma \mid \delta\rangle \oplus \langle\delta \mid \gamma\rangle && \text{by Payoff Principle 1 (symmetry)} \\ &= \langle\gamma, \delta \mid \gamma, \delta\rangle && \text{by Proposition 5.1.2} \\ &= \langle\gamma \mid \gamma\rangle \oplus \langle\delta \mid \delta\rangle && \text{by Proposition 5.1.2} \\ &= 0 \oplus 0 && \text{by Payoff Principle 1 (symmetry)} \\ &= 0 && \text{by Proposition 5.1.2.}\end{aligned}$$

□

Note that every context independent and symmetric payoff function induces via its aggregation function a totally ordered Abelian group with (some subset of) the reals R as base set and with 0 as neutral element.

Given Proposition 5.1.4 we can rewrite the decomposition of the payoff for an atomic state $[p_1, \ldots, p_n \mid q_1, \ldots, q_m]$ as

$$\langle p_1, \ldots, p_n \mid q_1, \ldots, q_m\rangle = \bigoplus_{1 \le i \le m} \langle \mid q_i\rangle \oplus -\bigoplus_{1 \le j \le n} \langle \mid p_i\rangle .$$

**Payoff Principle 3** (Monotonicity)**.** *A payoff function* $\langle \cdot \mid \cdot \rangle$ *is* monotone *if for all tenets* $\gamma$,$\delta$,$\gamma'$, $\delta'$, $\gamma''$, *and* $\delta''$ *the following holds: if* $\langle \gamma' \mid \delta' \rangle \leq \langle \gamma'' \mid \delta'' \rangle$ *then* $\langle \gamma, \gamma' \mid \delta', \delta \rangle \leq \langle \gamma, \gamma'' \mid \delta'', \delta \rangle$.

PROPOSITION 5.1.5. *Let* $\langle \cdot \mid \cdot \rangle$ *be a monotone and context independent payoff function and* $\oplus$ *the corresponding aggregation function. Then for all payoff values* $x$, $y$, $z$*:*

*(i) If* $y \leq z$, *then* $x \oplus y \leq x \oplus z$.

*(ii)* min *and* max *distribute over* $\oplus$, *i.e.,* $\min\{x \oplus y, x \oplus z\} = x \oplus \min\{y, z\}$ *and* $\max\{x \oplus y, x \oplus z\} = x \oplus \max\{y, z\}$.

*Proof. (i)* Let $G = [\gamma \mid \delta]$, $\psi' = [\gamma' \mid \delta']$, and $\psi'' = [\gamma'' \mid \delta'']$ be three atomic states such that $\langle \psi \rangle = x$, $\langle \psi' \rangle = y$, and $\langle \psi'' \rangle = z$. Then the premise $y \leq z$ amounts to $\langle \gamma' \mid \delta' \rangle \leq \langle \gamma'' \mid \delta'' \rangle$ and $x \oplus y \leq x \oplus z$ to $\langle \gamma \mid \delta \rangle \oplus \langle \gamma' \mid \delta' \rangle \leq \langle \gamma \mid \delta \rangle \oplus \langle \gamma'' \mid \delta'' \rangle$ or, equivalently, to $\langle \gamma, \gamma' \mid \delta, \delta' \rangle \leq \langle \gamma, \gamma'' \mid \delta, \delta'' \rangle$, which is just an instance of Payoff Principle 3.

*(ii)* We only consider the equation for min; the argument for max is analogous. Assume that $y \leq z$ holds. Then, by *(i)*, $x \oplus y \leq x \oplus z$ holds for all $x$ and thus also $\min\{x \oplus y, x \oplus z\} = x \oplus y = x \oplus \min\{y, z\}$. On the other hand, if $z \leq y$ then $x \oplus z \leq x \oplus y$ and thus also $\min\{x \oplus y, x \oplus z\} = x \oplus z = x \oplus \min\{y, z\}$. □

We combine the three payoff principles discussed in this section in the following notion, that will be central for Theorem 5.3.1 and Corollary 5.3.3 of Section 5.3.

DEFINITION 5.1.6 (Discriminating Payoff Function). *We call a payoff function* $\langle \cdot \mid \cdot \rangle$ discriminating *if it is context independent, symmetric, and monotone.*

## 5.2 Dialogue principles for logical connectives

We now turn our attention to logical connectives and look for dialogue rules that regulate the stepwise reduction of states with logically complex assertions to final atomic states. We assume perfect information, which in particular implies that the two players have common knowledge of the payoff values. Since we strive for full generality, we will not consider conjunction, disjunction, implication, etc., separately, but rather specify a generic format of dialogue rules for arbitrary $n$-ary connectives ($n \geq 1$). It turns out that two simple and general *dialogue principles*, in combination with discriminating payoff functions, suffice to guarantee that a truth functional semantics can be extracted from the corresponding game.

**Dialogue Principle 1** (Decomposition)**.** *A (dialogue) rule for an* $n$*-ary connective* $\diamond$ *is* decomposing *whenever in any corresponding round of the game exactly one occurrence of a compound formula* $\diamond(\varphi_1, \ldots, \varphi_n)$ *is removed from the current state and (possibly zero) occurrences of* $\varphi_1, \ldots, \varphi_n$ *and of truth constants are added to obtain the successor state. (Below we will give a step-by-step description of what is meant by 'round' here.)*

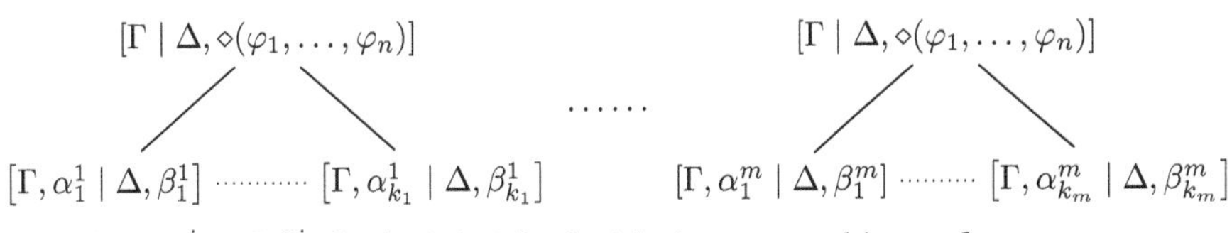

where $\alpha^i_j$ and $\beta^i_j$, for $1 \leq j \leq k_i$, $1 \leq i \leq m$, are multisets of zero or more occurrences of the formulas $\varphi_1, \ldots, \varphi_n$ and of truth constants.

Figure 1. Generic dialogue rule for your attack of my assertion of $\diamond(\varphi_1, \ldots, \varphi_n)$

Note that the decomposition principle entails that each occurrence of a formula can be attacked at most once: it is simply removed from the state in the corresponding round of the game. Moreover, an attack (i.e., a move by the player who is currently in the role of the opponent **O**) may or may not involve sub-formulas of the attacked formula occurrence (and/or truth constants) to be asserted by the attacking player. For example, remember that in Giles's $\mathcal{G}$-game, according to rule $R^{\mathcal{G}}_{\to}$, attacking $\varphi \to \psi$ requires the attacker to assert $\varphi$ (see Section 4). We require the reply to any attack to follow at once. In our example of an attack to $\varphi \to \psi$ in the $\mathcal{G}$-game this means that an assertion of $\psi$ will be added to the tenet of the attacked player. In general, the attacking player may choose between one of several available forms of attacking a particular formula, as witnessed by the rule for (weak) conjunction in the original game. Likewise, as exemplified in Giles's rule $R^{\mathcal{G}}_{\vee}$ for disjunction, a rule may also involve a choice on the side of the defending player. Consequently, every round of the game may be thought of as consisting of a sequence of three consecutive moves. (We only consider the case where you attack one of the formulas asserted by myself, the other case is dual.)

1. You pick an occurrence of a compound formula $\diamond(\psi_1, \ldots, \psi_n)$ from my current tenet for attack (or possibly for dismissal, see below).
2. You choose the form of attack (if there is more than one form available).
3. I choose the way in which I want to defend, i.e., to reply to the given attack on the indicated occurrence of $\diamond(\psi_1, \ldots \psi_n)$ (if such a choice is possible).

The corresponding rule may be depicted as shown in Figure 1. That there is a forest rather than a single tree rooted in $[\Gamma \mid \Delta, \diamond(\psi_1, \ldots, \psi_n)]$ reflects the fact that you may choose between different forms of attack for formulas of the form $\diamond(\psi_1, \ldots, \psi_n)$. In contrast, the branching in the trees corresponds to *my* possible choices in defending against your particular attack.

Recall from Section 4 that we have appealed to two forms of the *principle of limited liability* to explain the form of rules $R^{\mathcal{G}}_{\to}$ and $R^{\mathcal{G}}_{\&}$, respectively. In our current context it is appropriate to formulate this two-fold principle in the following more abstract manner:

**Limited liability for defense (LLD):** A player can always choose to assert $\bot$ in reply to an attack by her opponent.

**Limited liability for attack (LLA):** A player can always declare not to attack a particular occurrence of a formula that has been asserted by her opponent.

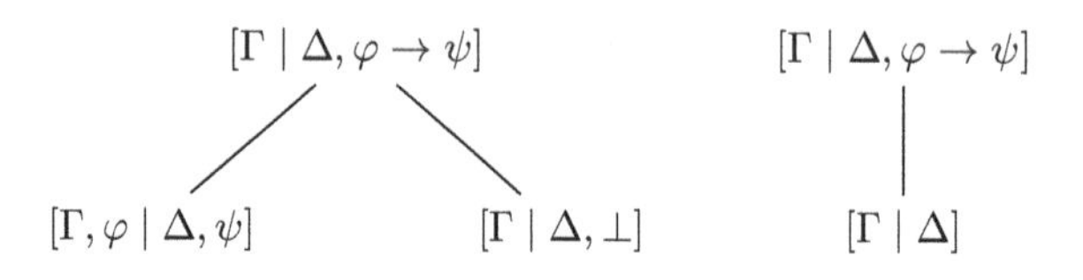

Figure 2. Implication rule (your attack) with two-fold principle of limited liability

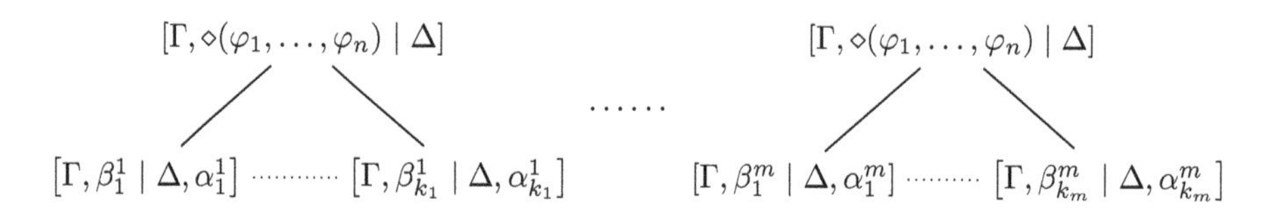

Figure 3. Generic dialogue rule dual to that in Figure 1; i.e., for my attack on your assertion of $\diamond(\varphi_1, \ldots, \varphi_n)$ ($m$, $k_i$, $\alpha^i_j$, and $\beta^i_j$ are like in Figure 1).

To illustrate the above dialogue rule format by a concrete example, consider the case of your attack on my assertion of $\varphi \rightarrow \psi$ in a variant of Giles's game where both forms of the principle of limited liability, LLD and LLA, are imposed. The resulting version of the implication rule is depicted in Figure 2.
The right (degenerate) tree in Figure 2 corresponds to your declaration not to attack the exhibited occurrence of $\varphi \rightarrow \psi$ at all. We treat this case as a special form of attack, where the 'attacked' formula occurrence (current formula) is simply removed to obtain the successor state. The first tree indicates a choice by me (i.e., the defending player): I may either according to LLD assert $\bot$ in reply to your attack or else assert $\psi$ in exchange for your assertion of $\varphi$.

The second principle that we want to maintain in generalizing Giles's game is player neutrality, i.e., role duality: you and me have the very same obligations and rights in attacking or defending a particular type of formula.

**Dialogue Principle 2** (Duality). *Dialogue rule $\delta_\diamond$ for my (your) assertion of a formula of the form $\diamond(\varphi_1, \ldots, \varphi_n)$ is called* dual *to the rule $\delta'_\diamond$ for your (my) assertion of $\diamond(\varphi_1, \ldots, \varphi_n)$ if $\delta_\diamond$ is obtained from $\delta'_\diamond$ by just switching the roles of the players.*

*We will say that a dialogue game* has dual rules *whenever for every dialogue rule of the game there is a dual rule.*

Figure 3 depicts the generic dialogue that is dual to that in Figure 1. Note that now *I* am the one who, in attacking your assertion of $\diamond(\varphi_1, \ldots, \varphi_n)$, is free to pick a tree of the forest, whereas the branching in the tree now refers to *your* choices when defending against my attack.

Note that since the format of decomposing rules allows for a choice between different types of attacks as well as corresponding replies, we may speak without loss of generality of *the* dialogue rule for a connective $\diamond$ if the game has dual rules.

### 5.3 Extracting truth functions

Following the well known game theoretic principle of backward induction, that we have already seen at play in previous sections, the maximal payoff value that I can enforce at a game state $S$—for short: my *enforceable payoff* at $S$—amounts to the minimum of enforceable payoffs at the successor states of $S$ if it is *your* turn to move at $S$ as well as to the maximum of enforceable payoffs at the successor states if it is *my* turn to move at $S$. Correspondingly, the function $\langle \cdot \mid \cdot \rangle$ that denotes my enforceable payoff at an arbitrary state in our dialogue games (where a round involves a move by both of us in turn) is induced by the corresponding payoff function for atomic game states and by the following *min-max conditions* for non-atomic game states:

$$\langle \Gamma \mid \diamond(\varphi_1, \dots \varphi_n), \Delta \rangle = \min_{1 \leq i \leq m} \max_{1 \leq j \leq k_i} \langle \Gamma, \alpha^i_j \mid \Delta, \beta^i_j \rangle \quad (6)$$

$$\langle \diamond(\varphi_1, \dots \varphi_n), \Gamma \mid \Delta \rangle = \max_{1 \leq i \leq m} \min_{1 \leq j \leq k_i} \langle \Gamma, \beta^i_j \mid \Delta, \alpha^i_j \rangle, \quad (7)$$

where $m, k_i, \alpha^i_j$, and $\beta^i_j$ are defined as in Figure 1. We call this function the *extended payoff* function.[3]

Above, we have defined context independence, symmetry, and monotonicity for payoff functions which, by definition, refer only to atomic game states. However, by inspecting Definitions 1, 2, and 3 it is obvious that neither these properties, nor those expressed in Propositions 5.1.2, 5.1.4, and 5.1.5 depend on the atomicity of the formulas in a corresponding tenet. Therefore we can speak without ambiguity of context independence, symmetry, and monotonicity for arbitrary functions from general states to real numbers, not just for proper payoff functions.

THEOREM 5.3.1. *Let $\eth$ be a dialogue game with a discriminating payoff function and decomposing dual rules. Then the extended payoff function denoting my enforceable payoff is context independent, symmetric, and monotone.*

*Proof.* Given a discriminating payoff function $\langle \cdot \mid \cdot \rangle$ with corresponding aggregation function $\oplus$, we define a function $v$ from (arbitrary) game states to the real numbers inductively as follows:

$$(a) \quad v([\,\mid p]) = \langle\, \mid p \rangle$$

$$(b) \quad v([\,\mid \Delta]) = \bigoplus_{\psi \in \Delta} v([\,\mid \psi]))$$

$$(c) \quad v([\Gamma \mid \Delta]) = v([\,\mid \Delta]) \oplus -v([\,\mid \Gamma])$$

$$(d) \quad v([\,\mid \diamond(\varphi_1, \dots \varphi_n)]) = \min_{1 \leq i \leq m} \max_{1 \leq j \leq k_i} v(\left[\alpha^i_j \mid \beta^i_j\right]),$$

where $m$, $k_i$, $\alpha^i_j$, and $\beta^i_j$ are defined as in Figure 1.

[3] It can easily be checked that the above min-max conditions define a unique extension of any discriminating payoff function to arbitrary game states if the dialogue rules are dual and discriminating. As pointed out in [19] (for Giles's game) this fact implies that the order of rule applications is irrelevant: we arrive at the same enforceable payoff, independently of the specific formula occurrence that is picked by you or myself for attack at any given state.

We prove that $v$ indeed calculates my enforceable payoff, i.e., it coincides with $\langle \cdot \mid \cdot \rangle$ on atomic states and fulfills the min-max conditions. Moreover we show that it is context independent, symmetric, and monotone.

It is straightforward to check that $v([\gamma \mid \delta])$ indeed coincides with $\langle \gamma \mid \delta \rangle$ for all atomic states $[\gamma \mid \delta]$. Following this observation we will from now on usually write $\langle \Gamma \mid \Delta \rangle$ instead of $v([\Gamma \mid \Delta])$, even if the tenets $\Gamma$ and $\Delta$ are not atomic.

The symmetry of $v([\cdot \mid \cdot])$ immediately follows from its definition, where (here as well as further on) we freely exploit the commutativity and associativity of $\oplus$.

$$\begin{aligned} -v([\Gamma \mid \Delta]) = - \langle \Gamma \mid \Delta \rangle &= -(\langle \mid \Delta \rangle \oplus - \langle \mid \Gamma \rangle) && \text{by definition of } v \text{ (c)} \\ &= - \langle \mid \Delta \rangle \oplus \langle \mid \Gamma \rangle && \text{by Proposition 5.1.4(i)} \\ &= \langle \Delta \mid \Gamma \rangle && \text{by definition of } v \text{ (c).} \end{aligned}$$

Note that the definition of $v$ directly entails that, just like the payoff at atomic states, the enforceable payoff at arbitrary states can also be obtained from the enforceable payoffs for sub-states by applying $\oplus$: we will refer to *merging* of and *partitioning*, respectively. More precisely:

$$\begin{aligned} \langle \Gamma, \Gamma' \mid \Delta', \Delta \rangle &= \langle \mid \Delta', \Delta \rangle \oplus - \langle \mid \Gamma, \Gamma' \rangle && \text{by definition of } v \text{ (c)} \\ &= (\langle \mid \Delta' \rangle \oplus \langle \mid \Delta \rangle) \oplus -(\langle \mid \Gamma' \rangle \oplus \langle \mid \Gamma \rangle) && \text{by definition of } v \text{ (b)} \\ &= \langle \mid \Delta' \rangle \oplus \langle \mid \Delta \rangle \oplus - \langle \mid \Gamma' \rangle \oplus - \langle \mid \Gamma \rangle && \text{by Proposition 5.1.4} \\ &= \langle \Gamma' \mid \Delta' \rangle \oplus \langle \Gamma \mid \Delta \rangle && \text{by definition of } v \text{ (c).} \end{aligned}$$

Given this fact, it is easy to see that $\langle \cdot \mid \cdot \rangle$ is context independent. Let $[\Gamma' \mid \Delta']$, $[\Gamma'' \mid \Delta'']$ be two game states such that $\langle \Gamma' \mid \Delta' \rangle = \langle \Gamma'' \mid \Delta'' \rangle$. Then for arbitrary tenets $\Gamma$ and $\Delta$:

$$\begin{aligned} \langle \Gamma, \Gamma' \mid \Delta', \Delta \rangle &= \langle \Gamma' \mid \Delta' \rangle \oplus \langle \Gamma \mid \Delta \rangle && \text{by partitioning} \\ &= \langle \Gamma'' \mid \Delta'' \rangle \oplus \langle \Gamma \mid \Delta \rangle && \text{by assumption} \\ &= \langle \Gamma, \Gamma'' \mid \Delta'', \Delta \rangle && \text{by merging.} \end{aligned}$$

Monotonicity also straightforwardly carries over from atomic to arbitrary game states. Let $[\Gamma' \mid \Delta']$, $[\Gamma'' \mid \Delta'']$ be two game states such that $\langle \Gamma' \mid \Delta' \rangle \leq \langle \Gamma'' \mid \Delta'' \rangle$. Then for arbitrary tenets $\Gamma$ and $\Delta$:

$$\begin{aligned} \langle \Gamma, \Gamma' \mid \Delta', \Delta \rangle &= \langle \Gamma' \mid \Delta' \rangle \oplus \langle \Gamma \mid \Delta \rangle && \text{by partitioning} \\ &\leq \langle \Gamma'' \mid \Delta'' \rangle \oplus \langle \Gamma \mid \Delta \rangle && \text{by assumption and Proposition 5.1.5(i)} \\ &= \langle \Gamma, \Gamma'' \mid \Delta, \Delta'' \rangle && \text{by merging.} \end{aligned}$$

It remains to check that the min-max conditions are satisfied. For states of the form $[\Gamma \mid \Delta, \diamond(\varphi_1, \ldots \varphi_n)]$ we obtain min-max condition (6) as follows:

$$\begin{aligned} &\langle \Gamma \mid \Delta, \diamond(\varphi_1, \ldots \varphi_n) \rangle \\ &\quad = \langle \Gamma \mid \Delta \rangle \oplus \langle \mid \diamond(\varphi_1, \ldots \varphi_n) \rangle && \text{by partitioning} \\ &\quad = \langle \Gamma \mid \Delta \rangle \oplus \min_{1 \leq i \leq m} \max_{1 \leq j \leq k_i} \left( \langle \alpha^i_j \mid \beta^i_j \rangle \right) && \text{by definition of } v \text{ (d)} \\ &\quad = \min_{1 \leq i \leq m} \max_{1 \leq j \leq k_i} \left( \langle \Gamma \mid \Delta \rangle \oplus \langle \alpha^i_j \mid \beta^i_j \rangle \right) && \text{by Proposition 5.1.5(ii)} \\ &\quad = \min_{1 \leq i \leq m} \max_{1 \leq j \leq k_i} \left( \langle \Gamma, \alpha^i_j \mid \beta^i_j, \Delta \rangle \right) && \text{by merging.} \end{aligned}$$

The dual min-max condition (7) exploits the symmetry of $\langle \cdot \mid \cdot \rangle$:

$$\begin{aligned}
&\langle \Gamma, \diamond(\varphi_1, \dots \varphi_n) \mid \Delta \rangle \\
&\quad = - \langle \Delta \mid \Gamma, \diamond(\varphi_1, \dots \varphi_n) \rangle && \text{by symmetry} \\
&\quad = - \min_{1 \le i \le m} \max_{1 \le j \le k_i} (\langle \Delta, \alpha_j^i \mid \beta_j^i, \Gamma \rangle) && \text{by min-max condition (6)} \\
&\quad = \max_{1 \le i \le m} \min_{1 \le j \le k_i} (- \langle \Delta, \alpha_j^i \mid \beta_j^i, \Gamma \rangle) && \text{by Proposition 5.1.5(ii)} \\
&\quad = \max_{1 \le i \le} \min_{1 \le j \le k_i} (\langle \beta_j^i, \Gamma \mid \Delta, \alpha_j^i \rangle) && \text{by symmetry,}
\end{aligned}$$

where $m$, $k_i$, $\alpha_j^i$, and $\beta_j^i$ are defined as in Figure 1. □

REMARK 5.3.2. *The duality of dialogue rules is used only indirectly in the above proof: it is reflected in the corresponding duality of the two min-max conditions and in the symmetry of the extended payoff function.*

COROLLARY 5.3.3. *Let $\mathfrak{D}$ be a game with discriminating payoff function and decomposing dual rules. Then for each connective $\diamond$ there is a function $f_\diamond$ such that $\langle \mid \diamond(\varphi_1, \dots \varphi_n) \rangle = f_\diamond(\langle \mid \varphi_1 \rangle, \dots, \langle \mid \varphi_n \rangle)$ for all formulas $\varphi_1, \dots, \varphi_n$, where $\langle \cdot \mid \cdot \rangle$ denotes the extended payoff function of Theorem 5.3.1.*

*Proof.* Applying min-max condition (6) as well as context independence and symmetry, we obtain

$$\begin{aligned}
\langle \mid \diamond(\varphi_1, \dots, \varphi_n) \rangle &= \min_{1 \le i \le m} \max_{1 \le j \le k_i} \langle \alpha_j^i \mid \beta_j^i \rangle \\
&= \min_{1 \le i \le m} \max_{1 \le j \le k_i} (\langle \mid \beta_j^i \rangle \oplus \langle \alpha_j^i \mid \rangle) \\
&= \min_{1 \le i \le m} \max_{1 \le j \le k_i} (\langle \mid \beta_j^i \rangle \oplus - \langle \mid \alpha_j^i \rangle) \\
&= \min_{1 \le i \le m} \max_{1 \le j \le k_i} \left( \bigoplus_{\beta \in \beta_j^i} \langle \mid \beta \rangle \oplus - \bigoplus_{\alpha \in \alpha_j^i} \langle \mid \alpha \rangle \right),
\end{aligned}$$

where $\oplus$ is the aggregation function corresponding to $\langle \cdot \mid \cdot \rangle$; $m$, $k_i$, $\beta_j^i$, and $\alpha_j^i$ obviously again refer to the dialogue rule for $\diamond(\varphi_1, \dots \varphi_n)$ as exhibited in Figure 1. Note that the $\alpha_j^i$s and $\beta_j^i$s are multisets containing only the formulas $\varphi_1, \dots, \varphi_n$ and truth constants, which of course are evaluated to constant real numbers. Therefore that last expression defines the required function $f_\diamond$. □

REMARK 5.3.4. *To emphasize that $f_\diamond$ is of type $\mathsf{R}^n \to \mathsf{R}$ it can be rewritten as*

$$f_\diamond(x_1, \dots, x_m) = \min_{1 \le i \le n} \max_{1 \le j \le k_i} \left( \bigoplus_{y \in \overline{\beta_j^i}} y \oplus - \bigoplus_{x \in \overline{\alpha_j^i}} x \right),$$

*where $\overline{\beta_j^i}$ is a multiset of real numbers defined with respect to the multiset of formulas $\beta_j^i$ as follows: $\overline{\beta_j^i} = \{\overline{\varphi} \mid \varphi \in \beta_j^i\}$, where $\overline{\varphi} = x_i$ when $\varphi = \varphi_i$ for $1 \le i \le n$ and $\overline{\varphi} = \langle \mid \varphi \rangle$ whenever $\varphi$ is a truth constant.*

Note that the duality of the rules entails

$$\langle \diamond(\varphi_1, \dots \varphi_n) \mid \rangle = - \langle \mid \diamond(\varphi_1, \dots \varphi_n) \rangle = -f_\diamond(\langle \mid \varphi_1 \rangle, \dots, \langle \mid \varphi_n \rangle).$$

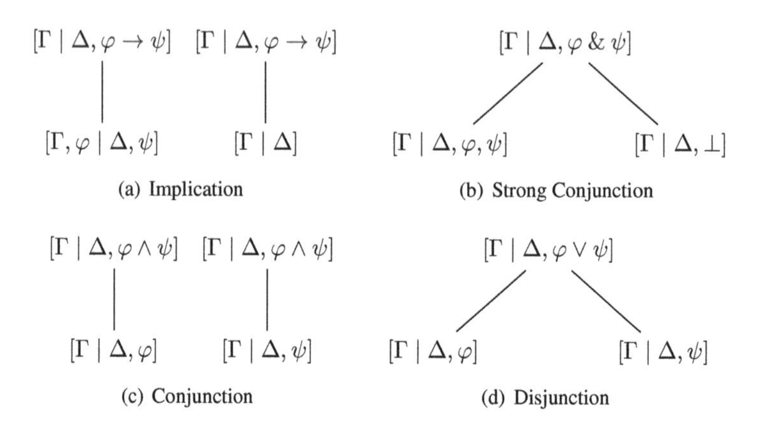


Figure 4. Rules of the $\mathcal{G}$-game for myself as **P** and you as **O**

By identifying payoff values with truth values we may thus claim to have extracted a unique truth function for $\diamond$ from a given payoff function and any decomposing dialogue rule for $\diamond$. However, as we will see in the next section, standard truth functions for many affected logics usually are based on different sets of truth values. To obtain those truth functions from an appropriate game we have to use certain bijections between payoff values and truth values, as we will explain below.

### 5.4 Revisiting the game for Łukasiewicz logic

To illustrate the emergence of concrete logics as instances of the general framework for games presented in Sections 5.1, 5.2, and 5.3, we should first check whether Giles's original $\mathcal{G}$-game for Łukasiewicz logic Ł is indeed covered. While the assignment of risk $\langle\cdot \mid \cdot\rangle$ to atomic states, as defined for the $\mathcal{G}$-game in Section 4, amounts to a discriminating payoff function (according to Definition 5.1.6), the connection to the standard truth functional semantics for Ł becomes clearer when we convert risk, that is to be minimized, to payoff, that is to be maximized, and set

$$\begin{aligned}
\langle p_1, \ldots, p_n \mid q_1, \ldots, q_m \rangle &= -\langle p_1, \ldots, p_n \mid q_1, \ldots, q_m \rangle \\
&= -\sum_{1 \le i \le m} \langle q_i \rangle + \sum_{1 \le j \le n} \langle p_j \rangle \\
&= -\sum_{1 \le i \le m} -\langle \mid q_i \rangle + \sum_{1 \le j \le n} -\langle \mid p_j \rangle \\
&= \sum_{1 \le i \le m} \langle \mid q_i \rangle - \sum_{1 \le j \le n} \langle \mid p_j \rangle .
\end{aligned}$$

Clearly, the aggregation function corresponding to $\langle\cdot \mid \cdot\rangle$ is ordinary addition. Figure 4 presents the dialogue rules in the format defined in Figures 1 and 3. Because of duality—which is obvious from Giles's generic presentation of the rules—we only have to consider your attacks on my assertions explicitly.

Note that discriminating payoff functions have 0 as neutral element. If we want to match the functions $f_{\rightarrow}$, $f_{\&}$, $f_{\wedge}$, and $f_{\vee}$ extracted from these dialogue rules according to Corollary 5.3.3 with standard truth functions over $[0, 1]$ we still have to add 1 to the payoff. It is straightforward to check that, modulo that transformation, the functions extracted from the rules in Figure 4 indeed coincide with the standard truth functions for Ł, reviewed in Section 4. We only illustrate the case for implication. From the rule for my assertion of $\varphi \rightarrow \psi$, which gives you a choice between asserting $\varphi$ to force me to assert $\varphi$ or else to declare that you will not attack this assertion at all, we obtain the following instance of min-max condition (1):

$$\langle \mid \varphi \rightarrow \psi \rangle = \min\{\langle \varphi \mid \psi \rangle, \langle \mid \rangle\} = \min\{0, \langle \mid \psi \rangle - \langle \mid \varphi \rangle\}.$$

Adding 1 yields the truth function $v(\varphi \rightarrow \psi) = 1 + \langle \mid \varphi \rightarrow \psi \rangle = \min\{1, 1 + \langle \mid \psi \rangle + 1 - (\langle \mid \varphi \rangle + 1)\} = \min\{1, 1 - v(\varphi) + v(\psi)\}$. The truth function for the other connectives are obtained in the same manner.

## 5.5 Finitely-valued Łukasiewicz logics

Instead of considering arbitrary risk (and therefore also arbitrary truth values) from $[0, 1]$, one may restrict the set of permissible risk values (equivalently: truth values) to $V_n = \{\frac{i}{n-1} \mid 1 \leq i < n\}$, for some $n \geq 2$. Since $V_n$ is closed with respect to addition, subtraction, as well as min and max, truth functions for all *finitely-valued* Łukasiewicz logics $Ł_n$ are obtained just like those for Ł.

Note that by this observation we have also covered classical logic, which coincides with $Ł_2$. This means that classical logic can be modeled by a version of Giles's game where the experiments that determine the payoffs are not dispersive: every atomic proposition $p$ is simply true or false, entailing a determinate payment of 1€ for every assertion of $p$ in case it is false. For every assignment of risk values 0 or 1 to atomic formulas I have a strategy for avoiding (net) payment in a game starting with my assertion of a formula $\varphi$, if $\varphi$ is true under that assignment; on the other hand, if $\varphi$ is false, my best strategy limits my payment to you to 1€.

## 5.6 Cancellative hoop logic

A more interesting case is cancellative hoop logic CHL [10]. The truth value set of CHL is $(0, 1]$; correspondingly the truth constant $\bot$, along with negation ($\neg$) is removed from the language. The truth functions for implication and strong conjunction are given as

$$\begin{aligned} v(\varphi \,\&\, \psi) &= v(\varphi) \cdot v(\psi) \\ v(\varphi \rightarrow \psi) &= \begin{cases} \frac{v(\psi)}{v(\varphi)} & \text{if } v(\varphi) > v(\psi) \\ 1 & \text{else.} \end{cases} \end{aligned}$$

At first sight it is unclear how to obtain these truth functions from dialogue rules in our framework. However remember that in the game for Łukasiewicz logics—assuming that Giles's "risk values" have already been translated into payoff values by multiplying with $-1$—we still had to shift payoff values by 1 to obtain the standard truth function $\tilde{\diamond}$ from the function $\varphi_{\diamond}$ that can be extracted from the dialogue rule for the connective $\diamond$. It will be helpful to visualize the general form of this relation, as follows:

$$\begin{array}{ccc} \mathcal{V}_{\text{payoff}} & \xrightarrow{f_\diamond} & \mathcal{V}_{\text{payoff}} \\ \mu\uparrow & & \downarrow\sigma \\ \mathcal{V}_{\text{truth}} & \xrightarrow{\tilde{\diamond}} & \mathcal{V}_{\text{truth}} \end{array}$$

In the case of Ł we have $\mathcal{V}_{\text{truth}} = [0,1]$, $\mathcal{V}_{\text{payoff}} = [-1,0]$, $\mu(x) = x-1$, and $\sigma(x) = x+1$. In CHL we have $\mathcal{V}_{\text{truth}} = (0,1]$. If we set $\mu(x) = \log(x)$ and accordingly $\mathcal{V}_{\text{payoff}} = (-\infty, 0]$ and $\rho(x) = \exp(x)$, then the implication rule of Giles's game (see Figure 4) yields the truth function for implication in CHL. In the same manner addition $(+)$ over $(-\infty, 0]$ maps into multiplication $(\cdot)$ over $(0,1]$. However, the function $f_\&$ extracted from the dialogue rule for & of Giles's game (with risk inverted into payoff) is $\&(x,y) = \max\{-1, x-1+y-1\}$ rather than the required $+$. (Note that the Łukasiewicz t-norm that models & in the standard semantics for Ł is obtained by adding $+1$, i.e., by applying $\sigma$, as explained above.) To obtain a dialogue rule for & such that $f_\& = +$, we have to drop the option to reply to an attack on $\varphi\&\psi$ by asserting $\bot$, instead of asserting $\varphi$ and $\psi$. In other words we simply drop the principle of limited liability LLD from the original rule for strong conjunction.

## 5.7 Abelian logic

So far we have only considered logics where the set of truth values is a proper subset of R and where we had to explicitly transform payoff values into truth values and vice versa. But there is an interesting and well studied logic, namely Slaney and Meyer's Abelian logic A [23, 46] which coincides with one of Casari's logics for modeling comparative reasoning in natural language [7], where arbitrary real-valued payoffs in a Giles-style game can be directly interpreted as truth values. The truth value set of A indeed is R. The truth functions for implication $(\rightarrow)$ is subtraction and the truth function for strong conjunction (&) is addition over R. In addition, max and min serve as truth functions for disjunction $(\vee)$ and weak conjunction $(\wedge)$, respectively.

The game based characterization of A is particularly simple: just drop both forms of the principle of limited liability, LLA and LLD, from Giles's game. In other words: every assertion made by the opposing player, including those of the form $\varphi \rightarrow \psi$, has to be attacked, moreover the only permissible reply to attack an $\varphi\&\psi$ is to assert both $\varphi$ and $\psi$. (The latter rule has already been used for CHL, above.) The functions that can be extracted from the resulting dialogue rules according to Corollary 5.3.3 are precisely those mentioned above: $f_\rightarrow = -$, $f_\& = +$, $f_\wedge = \min$, and $f_\vee = \max$.

We will revisit Abelian logic and present the corresponding game in greater detail in Section 6.

## 5.8 Alternative aggregation functions

In all the above examples, the aggregation function $\oplus$ corresponding to the respective payoff function has been addition $(+)$. This raises the question, whether in fact $\oplus$ always has to be $+$. This question is of some interest, since every truth function that can be directly extracted from a Giles-style game is built up from $\oplus$, $-$, min, max, and constant real numbers corresponding to truth constants. (By 'directly extracted' we mean:

disregarding further transformations—like $+1$ for Ł, and exp for CHL—that we may want to apply to map payoffs into standard truth values for particular logics.)

To settle this question in the negative it suffices to check that for any assignment $v$ of reals to atomic propositions

$$\langle \gamma \mid \delta \rangle = \sqrt[3]{\sum_{q\in\delta} v(q)^3} - \sqrt[3]{\sum_{p\in\gamma} v(p)^3}$$

is a discriminating payoff function with $\oplus(x,y) = \sqrt[3]{x^3+y^3}$ as corresponding aggregation function. However, we do not know of any many-valued logic in the literature where definitions of truth functions involve this or other possible aggregation functions different from $+$.

The above observations trigger the question whether for any aggregation function the ordered group $G = \langle \mathsf{R}; \leq, \oplus, 0, - \rangle$ is isomorphic to $\langle \mathsf{R}; \leq, +, 0, - \rangle$. A partly positive answer is provided by noting that $\psi$ is Archimedean. This is essentially due to monotonicity (Payoff Principle 3) and the standard order $\leq$ on the base set R. Therefore Hölder's Theorem [36] entails that $G$ is isomorphic to a *sub*group of $\langle \mathsf{R}; \leq, +, 0, - \rangle$.

# 6 Giles's game and hypersequents — the case of Abelian logic

Hypersequent systems are an important generalization of Gentzen's well known sequent framework for classical and intuitionistic logic. Roughly speaking, a hypersequent is a finite collection of sequents, viewed disjunctively. Chapter III of this Handbook not only demonstrates that hypersequents are a versatile tool for defining analytic proof systems for a wide variety of fuzzy logics, but also contains a section on Giles's game in this context. In particular, it is explained there in which sense the rules of the $\mathcal{G}$-game for Łukasiewicz logic Ł correspond to the logical rules of the hypersequent calculus GŁ. Obviously, this topic is also of central importance to the current handbook chapter. The relation between the $\mathcal{G}$-game and system GŁ is certainly the most important example of a direct connection between semantic games and cut-free inference systems for fuzzy logics; in fact, it is the only case of this kind that has been worked out in some detail in the literature so far. However, rather than repeating the material presented in Section 5.2 of [43], we want to address the issue from a slightly different angle and will work out the closely related case of Abelian logic A in some detail here.

We start with the Giles-style game for A that was briefly indicated in Section 5.7 and show that this semantic game (i.e., this game for *checking graded truth* in a given model) can be converted into a game for *checking validity* for formulas of A. The crucial step is to abstract away from concrete evaluations and to record the possible choices of the proponent **P** of the game in so-called *disjunctive states*. Different choices by the opponent **O** still correspond to different branches of the game tree, like in all games considered in the chapter. The only difference is that this game now consists in a tree of disjunctive states. At the leave nodes, where all formulas are atomic, we do not any longer just calculate my value or risk with respect to a particular interpretation, but rather have to check whether for *every interpretation* at least one of the disjunctive components of the final disjunctive state is a winning state in the sense of the original semantic game.

The just sketched disjunctive state scenario may be interpreted in two different manners: either as a new game in its own right, where the players have to move in ignorance of concrete risk value assignments (or corresponding interpretations), or, alternatively, simply as a device for *uniformly* computing winning strategies for **P** in the original semantic game. Here "uniformly" refers to the fact that we take into account all possible interpretations at every possible state of the game at once, rather than proceeding in a cases by case manner. (These two interpretations are explained in more detail at the end of Subsection 6.3.) Whatever interpretation one prefers, it turns out that at the level of strategies for disjunctive states (disjunctive strategies) the rules of the game for logic A directly correspond to the logical rules of a hypersequent system GA that is sound and cut-free complete for logic A.

### 6.1 Abelian logic revisited

The formulas of Abelian logic A are built up from propositional variables and the truth constant $t$ using the connectives $\neg$, $\vee$, $\wedge$, $\rightarrow$, and &.[4] The semantics of A is often specified with respect to arbitrary lattice ordered Abelian groups (Abelian $\ell$-groups). However, in our context it is more appropriate to make use of the fact that it suffices to consider the particular $\ell$-group $\langle \mathsf{R}, +, \max, -, 0 \rangle$. More precisely, the set of real numbers $\mathsf{R}$ is taken as set of truth values and any corresponding interpretation $\mathcal{M}$ extends an assignment $v_{\mathcal{M}}$ of reals to propositional variables to arbitrary formulas as follows.

$$\begin{aligned} v_{\mathcal{M}}(t) &= 0 \\ v_{\mathcal{M}}(\varphi \vee \psi) &= \max\{v_{\mathcal{M}}(\varphi), v_{\mathcal{M}}(\psi)\} \\ v_{\mathcal{M}}(\varphi \wedge \psi) &= \min\{v_{\mathcal{M}}(\varphi), v_{\mathcal{M}}(\psi)\} \\ v_{\mathcal{M}}(\neg\varphi) &= -v_{\mathcal{M}}(\varphi) \\ v_{\mathcal{M}}(\varphi \rightarrow \psi) &= v_{\mathcal{M}}(\psi) - v_{\mathcal{M}}(\varphi) \\ v_{\mathcal{M}}(\varphi \,\&\, \psi) &= v_{\mathcal{M}}(\varphi) + v_{\mathcal{M}}(\psi). \end{aligned}$$

A formula $\varphi$ is called *valid* in A iff $v_{\mathcal{M}}(\varphi) \geq 0$ for every interpretation $\mathcal{M}$.

### 6.2 A Giles-style game for Abelian logic

We have already indicated in Section 5.7 how a semantic game can be obtained for logic A that is similar to Giles's $\mathcal{G}$-game for Ł. In particular, just like in the $\mathcal{G}$-game, each state in the corresponding $\mathcal{G}$A-game consists of a pair of multisets (tenets) of formulas denoted as

$$[\varphi_1, \ldots, \varphi_n \mid \psi_1, \ldots, \psi_m],$$

where $[\varphi_1, \ldots, \varphi_n]$ is your tenet and $[\psi_1, \ldots, \psi_m]$ is my tenet. Also recall that in every concrete instance of the game some *regulation* picks an occurrence of a non-atomic formula at any given state; with respect to this *current formula*, I act as proponent **P** and you act as opponent **O** if the occurrence is in my tenet. The roles are switched if the current formula is in your tenet. We state the rules of the $\mathcal{G}$A-game explicitly; in the notation introduced in Section 5, we can depict these rules as in Figure 5:

[4] Strong conjunction (&) for Abelian logic is often denoted by $+$ and negation $\neg$ as $-$. Moreover 0 and $e$ are alternative signs for the truth constant denoted by $t$ here.

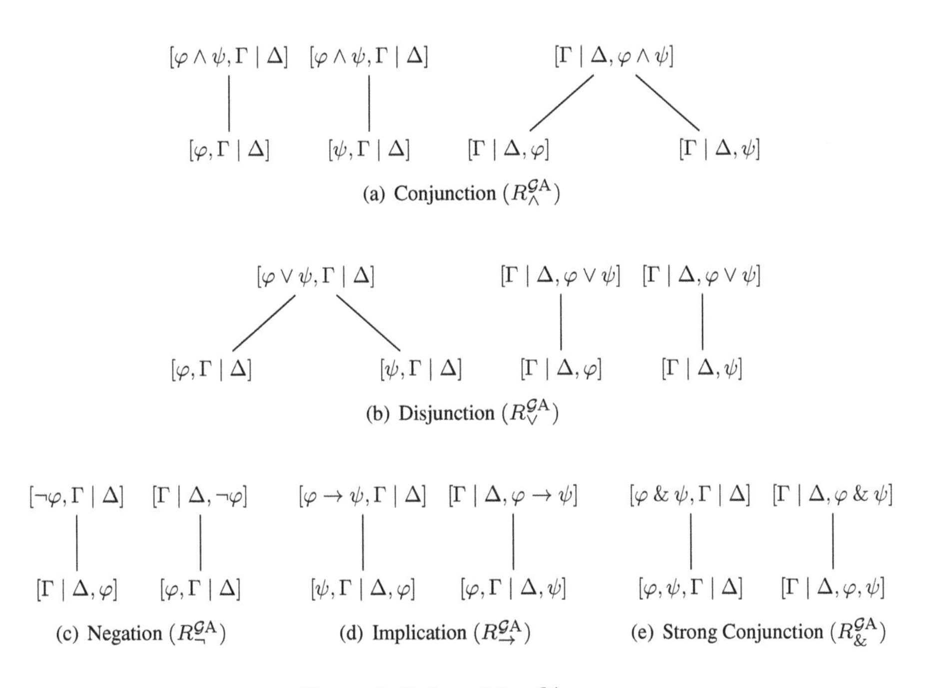


Figure 5. Rules of the $\mathcal{G}$A-game

($R^{\mathcal{G}\mathrm{A}}_{\wedge}$) If the current formula is $\varphi \wedge \psi$ then the game continues in a state where the indicated occurrence of $\varphi \wedge \psi$ in **P**'s tenet is replaced by either $\varphi$ or by $\psi$, according to **O**'s choice.

($R^{\mathcal{G}\mathrm{A}}_{\vee}$) If the current formula is $\varphi \vee \psi$ then the game continues in a state where the indicated occurrence of $\varphi \vee \psi$ in **P**'s tenet is replaced by either $\varphi$ or by $\psi$, according to **P**'s choice.

($R^{\mathcal{G}\mathrm{A}}_{\neg}$) If the current formula is $\neg\varphi$ then the game continues in a state where the indicated occurrence of $\neg\varphi$ is removed from **P**'s tenet and an occurrence of $\varphi$ is added to **O**'s tenet.

($R^{\mathcal{G}\mathrm{A}}_{\to}$) If the current formula is $\varphi \to \psi$ then the game continues in a state where the indicated occurrence of $\varphi \to \psi$ is replaced by $\psi$ in **P**'s tenet and an occurrence of $\varphi$ is added to **O**'s tenet.

($R^{\mathcal{G}\mathrm{A}}_{\&}$) If the current formula is $\varphi \,\&\, \psi$ then the game continues in a state where the indicated occurrence of $\varphi \,\&\, \psi$ in **P**'s tenet is removed and an occurrence of $\varphi$ as well as an occurrence of $\psi$ are added to **P**'s tenet.

REMARK 6.2.1. *It is instructive to compare the rules of the $\mathcal{G}$A-game with those of the $\mathcal{G}$-game. The $\mathcal{G}$A-game-rules for $\wedge$ and $\vee$ are exactly as in the $\mathcal{G}$-game, and therefore basically like already in Hintikka's $\mathcal{H}$-game. Rule $R_{\to}^{\mathcal{G}\mathrm{A}}$ differs from the implication rule $R_{\to}^{\mathcal{G}}$ of the $\mathcal{G}$-game, since the latter rule gives **O** the option not to attack **P**'s assertion of the current formula and consequently to have it removed without replacement from **P**'s tenet. This option, which is an instance of the principle of limited liability for attack (LLA), referred to in Sections 4 and 5, is missing in $R_{\to}^{\mathcal{G}\mathrm{A}}$: every occurrence of an implication has to be attacked. Similarly, rule $R_{\&}^{\mathcal{G}\mathrm{A}}$ differs from rule $R_{\&}^{\mathcal{G}}$, since the latter rule gives **P** the option to assert $\bot$ instead of the two conjuncts. In other words in the $\mathcal{G}$-game **P** can invoke the principle of limited liability for defense (LLD), whereas this option is missing in the $\mathcal{G}$A-game. We did not formulate a negation rule for the $\mathcal{G}$-game, but rather pointed out that negation for Łukasiewic logic is defined by $\neg\varphi = \varphi \to \bot$. Actually, negation for Abelian logic can be defined analogously by $\neg\varphi = \varphi \to t$ and therefore we could also have omitted rule $R_{\neg}^{\mathcal{G}\mathrm{A}}$, in principle.*

At a final game state where $[p_1, \ldots, p_n]$ is your tenet and $[q_1, \ldots, q_m]$ is my tenet, the payoff for myself in a $\mathcal{G}$A-game with respect to an interpretation $\mathcal{M}$ is specified as

$$\langle p_1, \ldots, p_n \mid q_1, \ldots, q_m \rangle = \sum_{1 \le i \le m} v_{\mathcal{M}}(q_i) - \sum_{1 \le j \le n} v_{\mathcal{M}}(p_j).$$

The following theorem follows from the general results of Section 5. It can also be shown directly in analogy to the proof of Theorem 4.0.1.

THEOREM 6.2.2. *In any $\mathcal{G}$A-game for $\varphi$ the value for myself under a given* A-*interpretation $\mathcal{M}$ and an arbitrary regulation $\rho$ is $v_{\mathcal{M}}(\varphi)$.*

### 6.3 Disjunctive states

Note that Theorem 6.2.2 does not refer to validity in logic A, but rather to 'graded truth' in a given interpretation, like all other semantic games for fuzzy logics. However, given the definition of validity in A in Section 6.1 above, we obtain the following.

COROLLARY 6.3.1. *A formula $\varphi$ is valid in* A *iff for every* A-*interpretation $\mathcal{M}$ and every regulation $\rho$ the value for myself of the corresponding $\mathcal{G}$A-game for $\varphi$ is $\geq 0$.*

My optimal strategies for a $\mathcal{G}$A-game starting in state $[\mid \varphi]$, or in fact any non-final state for that matter, will of course depend on the given interpretation $\mathcal{M}$, in general. An inspection of the rules in Figure 5 reveals that there are only two cases where I have to make a choice: (1) if the current formula is an occurrence of $\varphi \vee \psi$ on my tenet then I, acting as **P**, have to decide whether to replace it by $\varphi$ or by $\psi$; similarly, (2) if the current formula is an occurrence of $\varphi \wedge \psi$ on your tenet then I, acting as **O**, get to decide whether it should be replaced it by $\varphi$ or by $\psi$. Recall that choices to be made by you amount to branching in any tree that represents a strategy for myself. To be able to keep track also of my own options we introduce the notion of a *disjunctive state*.[5] By this we just mean a finite multiset of ordinary states, written as

[5] Disjunctive states are also referred to as 'state disjunctions' (see [19, 43]). This notational ambiguity may be understood to reflect the two different interpretations of the corresponding rule system indicated at the end of this subsection.

$$[\varphi_1^1, \ldots, \varphi_{n_1}^1 \mid \psi_1^1, \ldots, \psi_{m_1}^1] \bigvee \cdots \bigvee [\varphi_1^k, \ldots, \varphi_{n_k}^k \mid \psi_1^k, \ldots, \psi_{m_k}^k].$$

We now replace in part (a) of Figure 5

$$\begin{array}{cc} [\varphi \wedge \psi, \Gamma \mid \Delta] & [\varphi \wedge \psi, \Gamma \mid \Delta] \\ | & | \\ [\varphi, \Gamma \mid \Delta] & [\psi, \Gamma \mid \Delta] \end{array} \quad \text{by} \quad \begin{array}{c} [\varphi \wedge \psi, \Gamma \mid \Delta] \\ | \\ [\varphi, \Gamma \mid \Delta] \bigvee [\psi, \Gamma \mid \Delta] \end{array}$$

and in part (b) of Figure 5

$$\begin{array}{cc} [\Gamma \mid \Delta, \varphi \vee \psi] & [\Gamma \mid \Delta, \varphi \vee \psi] \\ | & | \\ [\Gamma \mid \Delta, \varphi] & [\Gamma \mid \Delta, \psi] \end{array} \quad \text{by} \quad \begin{array}{c} [\Gamma \mid \Delta, \varphi \vee \psi] \\ | \\ [\Gamma \mid \Delta, \varphi] \bigvee [\Gamma \mid \Delta, \psi]. \end{array}$$

We finally add an initial '$\mathcal{D} \bigvee$' to each state exhibited in Figure 5 and above, where $\mathcal{D}$ is a meta-variable for an arbitrary (possibly empty) disjunctive state. The resulting rule system can be interpreted in two different ways:

1. As a new game, where the states are now disjunctive states. For this new game we are not interested in payoff values at the final (disjunctive) states. Rather we declare that such a final disjunctive state, where all occurrences of formulas are atomic, is a *winning* state for myself iff for every interpretation $\mathcal{M}$ there is at least one disjunct $[p_1, \ldots, p_n \mid q_1, \ldots, q_m]$, such that $\langle p_1, \ldots, p_n \mid q_1, \ldots, q_m \rangle = \sum_{1 \leq i \leq m} v_{\mathcal{M}}(q_i) - \sum_{1 \leq j \leq n} v_{\mathcal{M}}(p_j) \geq 0$. Clearly, Corollary 6.3.1 implies that I have a winning strategy in this new game starting with the (single disjunct) state $[\mid \varphi]$ iff $\varphi$ is valid in Abelian logic.

2. We may alternatively view the new rule system as a calculus for the systematic and uniform construction of optimal strategies for myself in the original $\mathcal{G}A$-game. To obtain such an optimal strategy for a given interpretation $\mathcal{M}$ one selects in each final disjunctive state a disjunct ($\mathcal{G}A$-game state) $[p_1, \ldots, p_n \mid q_1, \ldots, q_m]$, where $\langle p_1, \ldots, p_n \mid q_1, \ldots, q_m \rangle$ is maximal and removes all other disjuncts. Every remaining (non-disjunctive) final state can be traced to a unique parent $\mathcal{G}A$-game state (disjunct) in the parent node (disjunctive state) in the strategy tree. All disjuncts except this $\mathcal{G}A$-game state are removed from the parent state. This procedure is iterated until we arrive at an ordinary $\mathcal{G}A$-game strategy.

### 6.4 A hypersequent system for Abelian logic

As already mentioned at the beginning of this section, a hypersequent is a multiset of sequents, interpreted as disjunctions of sequents, and written as

$$\Gamma_1 \vdash \Delta_1 \mid \ldots \mid \Gamma_n \vdash \Delta_n.$$

Sequents are understood here as pairs of multisets of formulas, just like states in Giles-style games. Consequently disjunctive states correspond to hypersequents.

Like an ordinary sequent calculus, hypersequent systems consist of axioms (initial hypersequents), structural rules, and logical rules. The following variant of a hypersequent system for Abelian logic can be found in [44].

*Axioms:*

$$(Ax) \quad \mathcal{G} \mid \Gamma, t^n \vdash \Gamma, t^m \quad \text{where } t^k \text{ denotes } k \geq 0 \text{ occurrences of } t$$

*Structural rules:*

$$\frac{\mathcal{G} \mid \Gamma \vdash \Delta \mid \Gamma \vdash \Delta}{\mathcal{G} \mid \Gamma \vdash \Delta}\ (EC) \qquad \frac{\mathcal{G} \mid \Gamma_1, \Gamma_2 \vdash \Delta_1, \Delta_2}{\mathcal{G} \mid \Gamma_1 \vdash \Delta_1 \mid \Gamma_2 \vdash \Delta_2}\ (Split)$$

*Logical rules:*

$$\frac{\mathcal{G} \mid \varphi, \Gamma \vdash \Delta \mid \psi, \Gamma \vdash \Delta}{\mathcal{G} \mid \varphi \wedge \psi, \Gamma \vdash \Delta}\ (\wedge, l) \qquad \frac{\mathcal{G} \mid \Gamma \vdash \Delta, \varphi \qquad \mathcal{G} \mid \Gamma \vdash \Delta, \psi}{\mathcal{G} \mid \Gamma \vdash \Delta, \varphi \wedge \psi}\ (\wedge, r)$$

$$\frac{\mathcal{G} \mid \varphi, \Gamma \vdash \Delta \qquad \mathcal{G} \mid \psi, \Gamma \vdash \Delta}{\mathcal{G} \mid \varphi \vee \psi, \Gamma \vdash \Delta}\ (\vee, l) \qquad \frac{\mathcal{G} \mid \Gamma \vdash \Delta, \varphi \mid \Gamma \vdash \Delta, \psi}{\mathcal{G} \mid \Gamma \vdash \Delta, \varphi \vee \psi}\ (\wedge, r)$$

$$\frac{\mathcal{G} \mid \Gamma \vdash \Delta, \varphi}{\mathcal{G} \mid \neg\varphi, \Gamma \vdash \Delta}\ (\neg, l) \qquad \frac{\mathcal{G} \mid \varphi, \Gamma \vdash \Delta}{\mathcal{G} \mid \Gamma \vdash \Delta, \neg\varphi}\ (\neg, r)$$

$$\frac{\mathcal{G} \mid \psi, \Gamma \vdash \Delta, \varphi}{\mathcal{G} \mid \varphi \to \psi, \Gamma \vdash \Delta}\ (\to, l) \qquad \frac{\mathcal{G} \mid \varphi, \Gamma \vdash \Delta, \psi}{\mathcal{G} \mid \Gamma \vdash \Delta, \varphi \to \psi}\ (\to, r)$$

$$\frac{\mathcal{G} \mid \varphi, \psi, \Gamma \vdash \Delta}{\mathcal{G} \mid \varphi \,\&\, \psi, \Gamma \vdash \Delta}\ (\&, l) \qquad \frac{\mathcal{G} \mid \Gamma \vdash \Delta, \varphi, \psi}{\mathcal{G} \mid \Gamma \vdash \Delta, \varphi \,\&\, \psi}\ (\&, r)$$

Note that the rule system for disjunctive states described in Section 6.3 can be obtained directly for the above logical rules by reading them bottom-to-top and replacing '$\mathcal{G}$' by '$\mathcal{D}$', '$|$' by '$\bigvee$', and '$\cdot \vdash \cdot$' by '$[\cdot \mid \cdot]$'. Conversely, we may view the logical hypersequent rules as directly derived from the rules for disjunctive game states.

Regarding the axioms and structural rules, it is important to realize that the hypersequent calculus remains complete for A if *EC* (external contraction) and *Split* are only applied to atomic hypersequents and if all instances of axioms are atomic. In fact this observation makes clear that the rule *EC* is redundant altogether. Since the truth constant $t$ is interpreted by 0, every atomic instance of an axiom $(Ax)$ satisfies the winning condition for the corresponding to final disjunctive game states. The rule *Split* has no direct correspondence in the rule system for disjunctive states. It may be seen as a device that allows one to reduce checking whether a given atomic hypersequent corresponds to a final disjunctive state satisfying the winning condition to the simpler case of $Ax$, where only the form of a single disjunct (sequent) is relevant.

Finally note that the particular order of applications of logical rules in a hypersequent derivation corresponds to a particular regulation in the game. Since every regulation leads to the same result (see Theorem 6.2.2 and Corollary 6.3.1), we conclude that systematic proof search in the hypersequent system is possible without backtracking at the level of logical rules.

# 7 Backtracking games

We have pointed out in Sections 3 and 4 that the $\mathcal{E}$-game as well as the $\mathcal{G}$-game deviate from what we called Hintikka's Principle, according to which a game state is fully determined by a single formula and a role distribution, i.e., the information whether myself or you are currently 'defending' the formula as the proponent **P**, while the other player is in the role of the opponent **O**. In this chapter we consider a variant of semantic games, where, unlike in the $\mathcal{E}$-game, no explicit reference to truth values is needed, and where, in contrast to the $\mathcal{G}$-game, the focus at each state is on a single formula. This is achieved by allowing the players to 'backtrack' to previous states of the game. Strictly speaking, the resulting *backtracking games* also do not satisfy Hintikka's Principle, since a stack of game states that are available for backtracking is now needed to fully describe a given state of a game. However the introduction of a stack for backtracking allows one to characterize not just Łukasiewicz logic Ł, but also Gödel logic G, and Product logic Π. In all previous sections it was always clear from the context to which logic we refer and thus we did not have to overload the notation for a corresponding evaluation function with an explicit reference to the logic in question. But in this section it is more appropriate to use $v^{Ł}_{\mathcal{M}}$, $v^{G}_{\mathcal{M}}$, and $v^{\Pi}_{\mathcal{M}}$ to refer the three corresponding valuation functions that specify the semantics of Ł, G, and Π, respectively. ($v^{Ł}_{\mathcal{M}}$ has been defined in Section 3. For the other two logics we will recall the relevant clauses for extending truth value assignments to compound formulas in the corresponding Subsections 7.2 and 7.4, respectively.)

REMARK 7.0.1. *In the following we will focus on the propositional level. However, we emphasize that adding Hintikka's original quantifier rules $R^{\mathcal{H}}_{\forall}$ and $R^{\mathcal{H}}_{\exists}$ results in characterizations of the corresponding first order logics, analogously to the case of the $\mathcal{H}$-mv-game for* KZ *and Giles's game for* Ł.

## 7.1 A backtracking game for Łukasiewicz logic

The backtracking game for Łukasiewicz logic may be viewed as a 'sequentialized' version of the $\mathcal{G}$-game. As announced, we introduce a *stack* on which information about an alternative state is stored (in a last-in first-out manner) when making particular moves. Initially the stack is empty. Upon reaching an atomic formula the game only ends if the stack is empty. Otherwise, the game *backtracks* to the state (formula and role distribution) indicated by the top element of the stack. That stack element is thereby removed from the stack.

In addition to the stack, we need to keep track of the *preliminary payoff* $\sigma_{\mathbf{P}}$ for the player that is currently acting as **P**. The preliminary payoff $\sigma_{\mathbf{O}}$ for **O** is $-\sigma_{\mathbf{P}}$ throughout the game. When the game ends, the preliminary payoff becomes final. Initially, $\sigma_{\mathbf{P}} = 1$. We call the resulting game variant the *backtracking game for* Ł or *$\mathcal{B}$Ł-game* for short.

The rules $R^{\mathcal{H}}_{\wedge}$, $R^{\mathcal{H}}_{\vee}$, and $R^{\mathcal{H}}_{\neg}$ of the $\mathcal{H}$-game of Section 2 are directly taken over into the $\mathcal{B}$Ł-game; no reference to the game stack or to $\sigma_{\mathbf{P}}$ and $\sigma_{\mathbf{O}}$ is needed. This implies that the $\mathcal{B}$Ł-game (for Ł) actually is an extension of the $\mathcal{H}$-mv-game for Kleene–Zadeh logic KZ (see Section 2). The rules for strong conjunction and implication are as follows ($\neg\varphi$ is treated as $\varphi \to \bot$):

($R^{\mathcal{B}Ł}_{\&}$) If the current formula is $\varphi \& \psi$ then **P** can choose either (1) to continue the game with $\varphi$ and to put $\psi$ together with the current role distribution on the stack, or (2) to continue the game with $\bot$.

($R^{\mathcal{B}Ł}_{\to}$) If the current formula is $\varphi \to \psi$ then **O** can choose either (1) to put $\psi$ on the stack with the current role distribution and to continue the game with $\varphi$ and inverted roles, or (2) to continue the game with the top element of the stack. If the stack is empty, the game ends.

($R^{\mathcal{B}Ł}_{\text{at}}$) If the current formula is an atom $p$ then $v_{\mathcal{M}}(p) - 1$ is added to $\sigma_{\mathbf{P}}$ and the same value is subtracted from $\sigma_{\mathbf{O}}$. The game ends if the stack is empty and is continued with the top element of the stack otherwise.

Again, we speak of the *$\mathcal{B}Ł$-game for $\varphi$ under $\mathcal{M}$* if the game starts with the current formula $\varphi$ where initially I am **P** and you are **O**.

THEOREM 7.1.1. *The value for myself of the $\mathcal{B}Ł$-game for $\varphi$ under the Ł-interpretation $\mathcal{M}$ is $w$ iff $v^{Ł}_{\mathcal{M}}(\varphi) = w$.*

*Proof.* We generalize to $\mathcal{B}Ł$-games that may start with any formula, role distribution, preliminary payoffs $\sigma_{\mathbf{P}} = -\sigma_{\mathbf{O}}$ and any stack content. We use $\mathcal{S}^I$ to denote the multiset of $|\mathcal{S}^I|$ formulas on the stack where I am assigned the role of **P**, and $\mathcal{S}^Y$ to denote the multiset of $|\mathcal{S}^Y|$ formulas on the stack where you are assigned the role of **P**. (Note that we ignore the order of stack elements, but not the number of occurrences of the same formula on the stack.) We define $s(\varphi) = 1$ if $\varphi$ is atomic, $s(\neg\varphi) = s(\varphi) + 1$, and $s(\varphi \circ \varphi') = s(\varphi) + s(\varphi') + 1$ for $\circ \in \{\vee, \wedge, \&, \to\}$. We prove the following by induction on $n = s(\varphi) + \sum_{\psi \in \mathcal{S}^I \cup \mathcal{S}^Y} s(\psi)$: $u$ is the value for myself of the $\mathcal{B}Ł$-game under interpretation $\mathcal{M}$ that starts with formula $\varphi$ and with myself as **P** iff

$$u = \sigma_{\mathbf{P}} + v^{Ł}_{\mathcal{M}}(\varphi) - 1 + \sum_{\psi \in \mathcal{S}^I} (v^{Ł}_{\mathcal{M}}(\psi) - 1) - \sum_{\psi \in \mathcal{S}^Y} (v^{Ł}_{\mathcal{M}}(\psi) - 1).$$

The theorem follows for $\sigma_{\mathbf{P}} = 1$ and $|\mathcal{S}^I| = |\mathcal{S}^Y| = 0$. For the case where I am initially in the role of **O** we have

$$u = \sigma_{\mathbf{O}} - v^{Ł}_{\mathcal{M}}(\varphi) + 1 + \sum_{\psi \in \mathcal{S}^I} (v^{Ł}_{\mathcal{M}}(\psi) - 1) - \sum_{\psi \in \mathcal{S}^Y} (v^{Ł}_{\mathcal{M}}(\psi) - 1).$$

At the base case, $n = 1$, the stack is empty and $\varphi$ is atomic. Therefore $v^{Ł}_{\mathcal{M}}(\varphi) - 1$ is added to $\sigma_{\mathbf{P}}$. The game ends at that state and $\sigma_{\mathbf{P}} + v^{Ł}_{\mathcal{M}}(\varphi) - 1$ is the payoff for myself as well as the value of the game, as required.

For the induction step we distinguish the following cases:

*$\varphi$ is atomic, but $n > 1$:* $v^{Ł}_{\mathcal{M}}(\varphi) - 1$ is added to $\sigma_{\mathbf{P}}$ or is subtracted from $\sigma_{\mathbf{O}}$, depending on whether I am acting as **P** or as **O**. The game continues with the formula and role distribution at the top of the stack. Clearly, the induction hypothesis is preserved.

*$\varphi = \varphi' \wedge \varphi''$:* We continue either with the game where $\varphi'$ or with the game where $\varphi''$ is the initial formula, according to **O**'s choice. Therefore we have to replace $v^{Ł}_{\mathcal{M}}(\varphi)$ by $\min\{v^{Ł}_{\mathcal{M}}(\varphi'), v^{Ł}_{\mathcal{M}}(\varphi'')\}$ to obtain the value for **P** of the original game from the values for **P** of the two possible succeeding games. This clearly matches the truth function for $\wedge$.

$\varphi = \varphi' \vee \varphi''$: like the case for $\varphi = \varphi' \wedge \varphi''$, except, since now **P** herself can choose the successor game, for replacing $v^{Ł}_{\mathcal{M}}(\varphi)$ by $\max\{v^{Ł}_{\mathcal{M}}(\varphi'), v^{Ł}_{\mathcal{M}}(\varphi'')\}$ in the value for **P**.

$\varphi = \varphi' \,\&\, \varphi''$: By the induction hypothesis, if $v^{Ł}_{\mathcal{M}}(\varphi') + v^{Ł}_{\mathcal{M}}(\varphi'') - 1$ is $\geq 0$ then the value is maximized for **P** by choosing option (1): continue with $\varphi'$, while putting $\varphi''$ on the stack. However, if $v^{Ł}_{\mathcal{M}}(\varphi') + v^{Ł}_{\mathcal{M}}(\varphi'') - 1$ is below 0 then **P** is better off by continuing the game with $\varphi$ as new initial formula, i.e., choosing option (2) of rule $R^{\mathcal{B}Ł}_{\&}$. Therefore, putting the two options together, the value for **P** of the original game results from the values of the two possible succeeding games when we replace by $v^{Ł}_{\mathcal{M}}(\varphi)$ by $\max\{0, v^{Ł}_{\mathcal{M}}(\varphi') + v^{Ł}_{\mathcal{M}}(\varphi'') - 1\}$. This matches the truth function for &.

$\varphi = \varphi' \to \varphi''$: If $v^{Ł}_{\mathcal{M}}(\varphi') > v^{Ł}_{\mathcal{M}}(\varphi'')$ then the (negative) contribution of $\varphi'$ to the value of the game for **O** is higher than the (positive) contribution of $\varphi''$ for **O** and therefore **O** will choose option (1) of rule $R^{\mathcal{B}Ł}_{\to}$ and let the game continue with $\varphi'$ and inverted roles, while $\varphi''$ is put on the stack. If, on the other hand, $v^{Ł}_{\mathcal{M}}(\varphi') \leq v^{Ł}_{\mathcal{M}}(\varphi'')$ then **O** will choose option (2) and discard $\varphi$ altogether. In the latter case the game continues with the next formula/role distribution pair on the stack, unless the stack is empty and the game ends. Combining the two options we obtain the value for **P** from her values of the possible succeeding games as given by the induction hypothesis: we replace by $v^{Ł}_{\mathcal{M}}(\varphi)$ by $\min\{1, 1 - v^{Ł}_{\mathcal{M}}(\varphi'') + v^{Ł}_{\mathcal{M}}(\varphi')\}$. This matches the truth function for $\to$. □

REMARK 7.1.2. *An alternative way of proving Theorem 7.1.1 consists of transforming Giles's game into a $\mathcal{B}Ł$-game and vice versa.*

## 7.2 A backtracking game for Gödel logic

Like in KZ, but unlike in Ł, we only have to consider min and max as truth functions for conjunction and disjunction, respectively, for Gödel logic G. Recall that the semantics of implication in G is specified by $v^{\mathrm{G}}_{\mathcal{M}}(\varphi \to \psi) = v^{\mathrm{G}}_{\mathcal{M}}(\psi)$ if $v^{\mathrm{G}}_{\mathcal{M}}(\varphi) > v^{\mathrm{G}}_{\mathcal{M}}(\psi)$ and $v^{\mathrm{G}}_{\mathcal{M}}(\varphi \to \psi) = 1$ otherwise. Negation is again defined by $\neg\varphi =_{def} \varphi \to \bot$ and therefore does not need separate consideration. To obtain a backtracking game for G, called $\mathcal{B}$G-game, we define the following rule:

($R^{\mathcal{B}\mathrm{G}}_{\to}$) If the current formula is $\varphi \to \psi$ then the game is continued with $\psi$ in the current role distribution and $\varphi$ is put on the stack together with the inverse role distribution.

Note that no choice of the players is involved in this rule. Below, we will present an alternative implication rule with choice. Here, however, choices remain restricted to conjunctions and disjunctions, for which the rules $R^{\mathcal{H}}_{\wedge}$ and $R^{\mathcal{H}}_{\vee}$ of the $\mathcal{H}$-game remain in place.

($R^{\mathcal{B}\mathrm{G}}_{\mathrm{at}}$) If the current formula is atomic then the game ends if the stack is empty and is continued with the top element of the stack otherwise.

Keeping track of payoff values is more involved in the $\mathcal{B}$G-game than in the $\mathcal{B}Ł$-game. An (ordered) tree $\tau$ of all formula occurrences visited during the game is built up for that purpose. At a state where the current formula $\varphi$ is a conjunction or a disjunction the subformula of $\varphi$ chosen by **O** or **P**, respectively, is attached to $\tau$ as successor node to $\varphi$. If the current formula $\varphi$ is an implication $\varphi' \to \varphi''$ then $\varphi'$ and $\varphi''$ are attached to $\tau$ as the right and left successor node to $\varphi$, respectively. When an atomic formula $p$ is reached then the corresponding leaf node $p$ is labeled by $v^{\mathrm{G}}_{\mathcal{M}}(p)$. To compute the payoff at the end of a game, the values (labels) at the leaf nodes are finally propagated upwards in $\tau$ as follows. Let $\varphi$ be the non-atomic formula at an internal node of $\tau$, where each successor node has already been labeled by a value:

If $\varphi = \varphi' \to \varphi''$, then $\varphi$ is labeled by 1 if $f' \leq f''$ and by $f''$ if $f' > f''$, where $f'$ and $f''$ are the values that label $\varphi'$ and $\varphi''$, respectively.

If $\varphi = \varphi' \vee \varphi''$ or $\varphi = \varphi' \wedge \varphi''$, then the same value that labels the successor node of $\varphi$ also labels $\varphi$ itself.

The $\mathcal{B}$G-*game for* $\varphi$ *under* $\mathcal{M}$ starts with an empty stack, the current formula $\varphi$ (that is also the initial tree $\tau$) and the role distribution where I am **P** and you are **O**. The payoff for myself in that game is given by the label $f$ of $\varphi$ in $\tau$ (computed as explained above, once the game has ended). The payoff for you is $-f$. (In other words: the $\mathcal{B}$G-game is a zero sum game.)

THEOREM 7.2.1. *The value for myself of the* $\mathcal{B}$G-*game for* $\varphi$ *under the* G-*interpretation* $\mathcal{M}$ *is* $w$ *iff* $v^{\mathrm{G}}_{\mathcal{M}}(\varphi) = w$.

*Proof.* If $\varphi$ does not contain $\wedge$ or $\vee$ then the tree $\tau$ of the game is just the tree of *all* occurrences $\underline{\psi}$ of subformulas $\psi$ of $\varphi$, where $\underline{\psi}$ is labeled by $v^{\mathrm{G}}_{\mathcal{M}}(\psi)$. In particular, the payoff for myself, and therefore the value of the game for $\varphi$ coincides with $v^{\mathrm{G}}_{\mathcal{M}}(\varphi)$.

It remains to check that the values labeling formulas of the form $\varphi' \wedge \varphi''$ and $\varphi' \vee \varphi''$ correspond to $\min\{v^{\mathrm{G}}_{\mathcal{M}}(\varphi'), v^{\mathrm{G}}_{\mathcal{M}}(\varphi'')\}$ and $\max\{v^{\mathrm{G}}_{\mathcal{M}}(\varphi'), v^{\mathrm{G}}_{\mathcal{M}}(\varphi'')\}$, respectively. To this aim, we refer to the *polarity* $\pi_\varphi(\underline{\psi}) \in \{+, -\}$ of an occurrence $\underline{\psi}$ of a subformula in $\varphi$, defined in a top down manner, as follows:

- $\pi_\varphi(\underline{\varphi}) = +$,
- $\pi_\varphi(\underline{\underline{\psi} \circ \underline{\psi'}}) = \pi_\varphi(\underline{\psi}) = \pi_\varphi(\underline{\psi'})$ for $\circ \in \{\wedge, \vee\}$,
- $\pi_\varphi(\underline{\underline{\psi} \to \underline{\psi'}}) = \pi_\varphi(\underline{\psi'})$; $\pi_\varphi(\underline{\psi}) = -$ if $\pi_\varphi(\underline{\underline{\psi} \to \underline{\psi'}}) = +$ and $\pi_\varphi(\underline{\psi}) = +$ if $\pi_\varphi(\underline{\underline{\psi} \to \underline{\psi'}}) = -$.

It is straightforwardly checked by induction that I am **P** and you are **O** in a state with current formula $\psi$ iff $\pi_\varphi(\underline{\psi}) = +$. For $\psi = \varphi' \vee \varphi''$ this implies that I (as **P**) will choose a subformula labeled by the maximal (truth) value. On the other hand, for $\psi = \varphi' \wedge \varphi''$ you (as **O**) will choose a subformula labeled by the minimal value. The case where $\pi_\varphi(\underline{\psi}) = -$ is dual: I (as **O**) will choose the subformula of $\psi = \varphi' \wedge \varphi''$ that minimizes your (i.e., **P**'s) payoff and therefore maximizes my (**O**'s) own payoff. Likewise, for $\psi = \varphi' \vee \varphi''$ you (as **P**) will choose a subformula with the maximal value. □

REMARK 7.2.2. *If we retain the rule* $R^{\mathcal{H}}_{\neg}$ *for negation in addition to the (here different) negation defined by* $\neg\varphi =_{def} \varphi \to \bot$ *we obtain a game for the logic* $\mathrm{G}_{\sim}$*, which is Gödel logic augmented by involutive negation.*

### 7.3 An implicit backtracking game for Gödel logic

The $\mathcal{B}$G-game presented above is unsatisfying in a few aspects. As we have already mentioned, no choice of either player is involved in the rule $R_{\to}^{\mathcal{B}\mathrm{G}}$. In fact, if we focus on formulas where implication is the only binary connective, the $\mathcal{B}$G-game can be viewed as just a particular implementation of the evaluation algorithm for G-formulas. Thus a lot of the appeal of game semantics is lost. Another drawback is the comparatively complex way of computing the payoff. In this section we seek to address these worries by defining an alternative semantic game for G where backtracking and thus the use of a stack is left *implicit* in the very same way as a stack for backtracking is implicit in recursive programs: the stack only gets explicit when the recursion is unraveled.

We use $\mathcal{I}\mathrm{G}(\varphi,\omega)$ to denote the *implicit backtracking game* for the logic G ($\mathcal{I}$G-game) for the formula $\varphi$, starting with role distribution $\omega$ and use $\langle\mathcal{I}\mathrm{G}(\varphi,\omega)\rangle_{\mathbf{P}}$ to denote the value for **P** of that game. (Depending on $\omega$, **P** is either myself or you.) Of course, $\mathcal{I}\mathrm{G}(\varphi,\omega)$ also refers to a given interpretation $\mathcal{M}$. However we prefer to keep that reference implicit in order to simplify notation. Like all other games described in this paper, the $\mathcal{I}$G-game is zero-sum. Modulo this clarification, it is sufficient to mention only the payoff for **P** in the following: the payoff for **O** is always inverse to that for **P**.

The rule for implication in the $\mathcal{I}$G-game is as follows.

($R_{\to}^{\mathcal{I}\mathrm{G}}$) In $\mathcal{I}\mathrm{G}(\varphi\to\psi,\omega)$ **P** chooses whether (1) to continue the game as $\mathcal{I}\mathrm{G}(\varphi,\omega)$ or (2) to play, in addition to $\mathcal{I}\mathrm{G}(\psi,\omega)$, also $\mathcal{I}\mathrm{G}(\varphi,\widehat{\omega})$, where $\widehat{\omega}$ denotes the role distribution that is inverse to $\omega$. In the latter case the payoff for **P** is $1$ if $\langle\mathcal{I}\mathrm{G}(\varphi,\omega)\rangle_{\mathbf{P}}\geq\langle\mathcal{I}\mathrm{G}(\psi,\widehat{\omega})\rangle_{\mathbf{P}}$ and $-1$ otherwise.

REMARK 7.3.1. *While the formulation of $R_{\to}^{\mathcal{I}\mathrm{G}}$ looks quite different from that of the rules for the $\mathcal{B}$Ł- or the $\mathcal{B}$G-game, the difference lies only in the fact that in $R_{\to}^{\mathcal{I}\mathrm{G}}$ we hide details of implementation. If in choice (2) we insist in playing $\mathcal{I}\mathrm{G}(\varphi,\widehat{\omega})$ first and consequently in putting G with $\widehat{\omega}$ on a stack, we obtain a version of the rule that is analogous to those of the earlier games.*

($R_{\mathrm{at}}^{\mathcal{I}\mathrm{G}}$) The payoff for **P** at $\mathcal{I}\mathrm{G}(\varphi,\omega)$ is $v_{\mathcal{M}}^{\mathrm{G}}(\varphi)$.

Note that we do not insist that the game ends upon reaching an atomic formula. Indeed, the payoff may be preliminary since it may only refer to a sub-game of the overall game, as indicated in rule $R_{\to}^{\mathcal{I}\mathrm{G}}$.

The rules for conjunction and disjunction in the $\mathcal{I}$G-game are virtually identical to $R_{\wedge}^{\mathcal{H}}$ and $R_{\vee}^{\mathcal{H}}$ and can be formulated as follows:

($R_{\wedge}^{\mathcal{I}\mathrm{G}}$) In $\mathcal{I}\mathrm{G}(\varphi\wedge\psi,\omega)$ **O** chooses whether to continue the game as $\mathcal{I}\mathrm{G}(\varphi,\omega)$ or as $\mathcal{I}\mathrm{G}(\psi,\omega)$.

($R_{\vee}^{\mathcal{I}\mathrm{G}}$) In $\mathcal{I}\mathrm{G}(\varphi\vee\psi,\omega)$ **P** chooses whether to continue the game as $\mathcal{I}\mathrm{G}(\varphi,\omega)$ or as $\mathcal{I}\mathrm{G}(\varphi,\omega)$.

Remember that no rule for negation is needed because we have $\neg\varphi =_{def} \varphi\to\bot$.

THEOREM 7.3.2. *The value for myself of the $\mathcal{I}$G-game for $\varphi$ under the G-interpretation $\mathcal{M}$ is $w$ iff $v_{\mathcal{M}}^{\mathrm{G}}(\varphi)=w$.*

*Proof.* We show by induction on the complexity of $\varphi$ that the value $\langle \mathcal{I}\mathrm{G}(\varphi,\omega)\rangle_{\mathbf{P}}$ for **P** of $\mathcal{I}\mathrm{G}(\varphi,\omega)$ is $v^{\mathrm{G}}_{\mathcal{M}}(\varphi)$ for every role distribution $\omega$. (The theorem clearly follows for the role distribution $\omega$ where I am **P** and you are **O**.)

According to the rule $R^{\mathcal{I}\mathrm{G}}_{\mathrm{at}}$ the payoff for **P** is $v^{\mathrm{G}}_{\mathcal{M}}(\varphi)$ if $\varphi$ is atomic. Therefore we have $\langle \mathcal{I}\mathrm{G}(\varphi,\omega)\rangle_{\mathbf{P}} = v^{\mathrm{G}}_{\mathcal{M}}(\varphi)$ in this case.

For the induction step we distinguish the following cases:

$\varphi = \varphi' \wedge \varphi''$: Since **O** can choose whether to continue the game as $\mathcal{I}\mathrm{G}(\varphi',\omega)$ or as $\mathcal{I}\mathrm{G}(\varphi'',\omega)$ and since the payoff for **O** is inverse to that of **P** we obtain $\langle \mathcal{I}\mathrm{G}(\varphi,\omega)\rangle_{\mathbf{P}} = \min\{\langle \mathcal{I}\mathrm{G}(\varphi',\omega)\rangle_{\mathbf{P}}, \langle \mathcal{I}\mathrm{G}(\varphi'',\omega)\rangle_{\mathbf{P}}\}$ and therefore, by the induction hypothesis, $\mathcal{I}\mathrm{G}(\varphi,\omega) = \min\{v^{\mathrm{G}}_{\mathcal{M}}(\varphi'), v^{\mathrm{G}}_{\mathcal{M}}(\varphi'')\}$, as required.

$\varphi = \varphi' \vee \varphi''$: This case is analogous to that for conjunction, except that now the player currently in role **P** can choose how to continue the game. Consequently we obtain $\langle \mathcal{I}\mathrm{G}(\varphi,\omega)\rangle_{\mathbf{P}} = \max\{\langle \mathcal{I}\mathrm{G}(\varphi',\omega)\rangle_{\mathbf{P}}, \langle \mathcal{I}\mathrm{G}(\varphi'',\omega)\rangle_{\mathbf{P}}\}$ and thus $\mathcal{I}\mathrm{G}(\varphi,\omega) = \max\{v^{\mathrm{G}}_{\mathcal{M}}(\varphi'), v^{\mathrm{G}}_{\mathcal{M}}(\varphi'')\}$, as required.

$\varphi = \varphi' \rightarrow \varphi''$: If $\langle \mathcal{I}\mathrm{G}(\varphi'',\omega)\rangle_{\mathbf{P}} \geq \langle \mathcal{I}\mathrm{G}(\varphi',\widehat{\omega})\rangle_{\mathbf{P}}$ then by rule $R^{\mathcal{I}\mathrm{G}}_{\rightarrow}$ the payoff for **P** and therefore also $\langle \mathcal{I}\mathrm{G}(F,\omega)\rangle_{\mathbf{P}}$ is 1, i.e., optimal for **P**. Consequently **P** will choose to continue the game with the two sub-games $\mathcal{I}\mathrm{G}(\varphi'',\omega)$ and $\mathcal{I}\mathrm{G}(\varphi',\widehat{\omega})$. By the induction hypothesis we have $v^{\mathrm{G}}_{\mathcal{M}}(\varphi') \leq v^{\mathrm{G}}_{\mathcal{M}}(\varphi'')$ in this case, implying $v^{\mathrm{G}}_{\mathcal{M}}(\varphi) = 1$, as required. If, on the other hand, $\langle \mathcal{I}\mathrm{G}(\varphi'',\omega)\rangle_{\mathbf{P}} < \langle \mathcal{I}\mathrm{G}(\varphi',\widehat{\omega})\rangle_{\mathbf{P}}$ then **P** will maximize her payoff by continuing the game as $\mathcal{I}\mathrm{G}(\varphi'',\omega)$. In this case the induction hypothesis implies that $v^{\mathrm{G}}_{\mathcal{M}}(\varphi') > v^{\mathrm{G}}_{\mathcal{M}}(\varphi'')$ and therefore $\langle \mathcal{I}\mathrm{G}(\varphi,\omega)\rangle_{\mathbf{P}} = \langle \mathcal{I}\mathrm{G}(\varphi'',\omega)\rangle_{\mathbf{P}} = v^{\mathrm{G}}_{\mathcal{M}}(\varphi'') = v^{\mathrm{G}}_{\mathcal{M}}(\varphi)$, again as required. □

### 7.4 An implicit backtracking game for Product logic

Recall that the semantics of Product logic $\Pi$ is specified by extending a given assignment $v_{\mathcal{M}}$ of values in $[0,1]$ to atomic formulas as follows:

$$v^{\Pi}_{\mathcal{M}}(\varphi \,\&\, \psi) = v^{\Pi}_{\mathcal{M}}(\varphi) \cdot v^{\Pi}_{\mathcal{M}}(\psi)$$

$$v^{\Pi}_{\mathcal{M}}(\varphi \rightarrow \psi) = \begin{cases} 1 & \text{if } v^{\Pi}_{\mathcal{M}}(\varphi) \leq v^{\Pi}_{\mathcal{M}}(\psi) \\ v^{\Pi}_{\mathcal{M}}(\psi)/v^{\Pi}_{\mathcal{M}}(\varphi) & \text{otherwise.} \end{cases}$$

Negation is treated as a defined connective via $\neg\varphi =_{def} \varphi \rightarrow \bot$, where $v^{\Pi}_{\mathcal{M}}(\bot) = 0$.

One could define a semantic game with explicit backtracking for $\Pi$ that is very similar to the $\mathcal{B}$Ł-game defined at the beginning of this Section. Roughly speaking one only needs to change the propagation of preliminary payoffs when reaching atomic formulas: instead of addition and subtraction we have to use multiplication and division, respectively. However, as for Gödel logic G above, we prefer to present such a game at a more abstract and compact level that leaves the reference to a game stack and to preliminary payoffs implicit.

The implicit backtracking game for $\Pi$ ($\mathcal{I}\Pi$-game) for a formula $\varphi$ starting with role distribution $\omega$ is denoted by $\mathcal{I}\Pi(\varphi,\omega)$. By $\langle \mathcal{I}\Pi(\varphi,\omega)\rangle_{\mathbf{P}}$ we denote the value for **P** of that game. Again, we suppress the reference to the underlying interpretation $\mathcal{M}$. Once more, we describe a zero-sum game and thus it is sufficient to specify only the payoff for **P** explicitly.

The implication rule of the $\mathcal{I}\Pi$-game is as follows.

($R^{\mathcal{I}\Pi}_{\to}$) In $\mathcal{I}\Pi(\varphi \to \psi, \omega)$ **O** chooses whether (1) to end the game immediately and accept payoff 1 for **P** and $-1$ for herself or (2) to continue by playing $\mathcal{I}\Pi(\psi, \omega)$ as well as $\mathcal{I}\Pi(F, \widehat{\omega})$, where $\widehat{\omega}$ denotes the role distribution that inverts $\omega$. In this case we have the payoff $\langle \mathcal{I}\Pi(\varphi \to \psi, \omega)\rangle_{\mathbf{P}} = \langle \mathcal{I}\Pi(\psi, \omega)\rangle_{\mathbf{P}} \,/\, \langle \mathcal{I}\Pi(\varphi, \widehat{\omega})\rangle_{\mathbf{P}}$.

For strong conjunction &, product is used in logic $\Pi$and therefore the following rule will come as no surprise:

($R^{\mathcal{I}\Pi}_{\&}$) In $\mathcal{I}\Pi(\varphi \,\&\, \psi, \omega)$ the game splits into the sub-games $\mathcal{I}\Pi(\varphi, \omega)$ and $\mathcal{I}\Pi(\psi, \omega)$, with total payoff $\langle \mathcal{I}\Pi(\varphi \,\&\, \psi, \omega)\rangle_{\mathbf{P}} = \langle \mathcal{I}\Pi(\varphi, \omega)\rangle_{\mathbf{P}} \cdot \langle \mathcal{I}\Pi(\psi, \omega)\rangle_{\mathbf{P}}$.

Negation is left implicit by $\neg\varphi =_{def} \varphi \to \bot$.

THEOREM 7.4.1. *The value for myself of the $\mathcal{I}\Pi$-game for $\varphi$ under the $\Pi$-interpretation $\mathcal{M}$ is $w$ iff $v^{\Pi}_{\mathcal{M}}(\varphi) = w$.*

*Proof.* The proof is very similar to that of Theorem 7.3.2; we show by induction that the value $\langle \mathcal{I}\Pi(F, \omega)\rangle_{\mathbf{P}}$ for **P** of the game $\mathcal{I}\Pi(F, \omega)$ is $v^{\Pi}_{\mathcal{M}}(\varphi)$ for every role distribution $\omega$.

If $\varphi$ is atomic then the payoff for **P** is $v^{\Pi}_{\mathcal{M}}(\varphi)$ and therefore $\langle \mathcal{I}\Pi(\varphi, \omega)\rangle_{\mathbf{P}} = v^{\Pi}_{\mathcal{M}}(\varphi)$. The induction step for implication and strong conjunction is as follows (the cases for $\varphi = \varphi' \wedge \varphi''$ and for $\varphi = \varphi' \vee \varphi''$ are exactly as in Theorem 7.3.2):

$\varphi = \varphi' \to \varphi''$: If $\langle \mathcal{I}\Pi(\varphi'', \omega)\rangle_{\mathbf{P}} > \langle \mathcal{I}\Pi(\varphi', \widehat{\omega})\rangle_{\mathbf{P}}$, then $\langle \mathcal{I}\Pi(\varphi'', \omega)\rangle_{\mathbf{P}} \,/\, \langle \mathcal{I}\Pi(\varphi', \widehat{\omega})\rangle_{\mathbf{P}}$ is greater than 1. This implies that in this case **O** achieves a higher payoff by choosing option (1) in rule $R^{\mathcal{I}\Pi}_{\to}$ and the game ends with the payoff 1 for **P** and thus $-1$ for **O** herself. On the other hand, if $\langle \mathcal{I}\Pi(\varphi'', \omega)\rangle_{\mathbf{P}} > \langle \mathcal{I}\Pi(\varphi', \widehat{\omega})\rangle_{\mathbf{P}}$, then **O** will choose option (2) and we obtain $\langle \mathcal{I}\Pi(\varphi, \omega)\rangle_{\mathbf{P}} = \langle \mathcal{I}\Pi(\varphi'', \omega)\rangle_{\mathbf{P}} \,/\, \langle \mathcal{I}\Pi(\varphi', \widehat{\omega})\rangle_{\mathbf{P}}$. Finally, if $\langle \mathcal{I}\Pi(\varphi'', \omega)\rangle_{\mathbf{P}} = \langle \mathcal{I}\Pi(\varphi', \widehat{\omega})\rangle_{\mathbf{P}}$ then the choice of **O** is immaterial since the payoff for **P** will always be 1. Clearly, the induction hypothesis yields $\langle \mathcal{I}\Pi(\varphi, \omega)\rangle_{\mathbf{P}} = v^{\Pi}_{\mathcal{M}}(\varphi)$ in all three cases.

$\varphi = \varphi' \,\&\, \varphi''$: By rule $R^{\mathcal{I}\Pi}_{\&}$ we obtain $\langle \mathcal{I}\Pi(\varphi, \omega)\rangle_{\mathbf{P}} = \langle \mathcal{I}\Pi(\varphi', \omega)\rangle_{\mathbf{P}} \cdot \langle \mathcal{I}\Pi(\varphi'', \omega)\rangle_{\mathbf{P}}$ and therefore $\langle \mathcal{I}\Pi(\varphi, \omega)\rangle_{\mathbf{P}} = v^{\Pi}_{\mathcal{M}}(\varphi') \cdot v^{\Pi}_{\mathcal{M}}(\varphi'') = v^{\Pi}_{\mathcal{M}}(\varphi)$ by the induction hypothesis. □

REMARK 7.4.2. *We have only treated propositional logics in this section, but we want to emphasize that all backtracking games presented here can straightforwardly be generalized to the first-order level by adding the rules $R^{\mathcal{H}}_{\forall}$ and $R^{\mathcal{H}}_{\exists}$ defined for the $\mathcal{H}$-game and the $\mathcal{H}$-mv-game in Section 2. As discussed there, this entails making use of the general definition of the value of a game, which refers to optimal payoffs only up to some $\epsilon$.*

## 8 Propositional random choice games

Following Giles, we have introduced the idea of expected payoffs in a randomized setting in Section 4. However, Giles applied this idea only to the interpretation of *atomic* formulas. For the interpretation of logical connectives and quantifiers in any of the semantic games mentioned so far it does not matter whether the players seek to maximize expected or a certain payoff or, equivalently, try to minimize either expected or certain payments to the opposing player. In [21, 22] it has been shown that considering random choices of witnessing constants in quantifier rules for *Giles-style* games, allows one to model certain (semi-)fuzzy quantifiers that properly extend first-order Łukasiewicz logic. We will take up this idea in Section 9. However in this section we want to explore the consequences of introducing random choices in rules for propositional connectives in the context of *Hintikka-style* games, i.e., games that respect Hintikka's principle, as explained in Section 2.

The results of Section 2 show that, in order to go beyond logic KZ with Hintikka-style games, a new variant of rules has to be introduced. As already indicated, a particularly simple type of new rule, that does not entail any change in the structure of game states, arises from randomization. So far we have only considered rules where either **P** or **O** chooses the sub-formula of the current formula to continue the game with. In game theory one often introduces *Nature* as a special kind of additional player, who does not care what the next state looks like, when it is her time to move and therefore is modeled by a uniformly random choice between all moves available to *Nature* at that state. As we will see below, introducing *Nature* leads to increased expressive power of semantic games. In fact, to keep the presentation of the games simple, we prefer to leave the role of *Nature* only implicit and just speak of random choices, without attributing them officially to a third player. The most basic rule of the indicated type refers to a new propositional connective $\pi$ and can be formulated as follows.[6]

($R_\pi^{\mathcal{R}}$) If the current formula is $\varphi\pi\psi$ then a uniformly random choice determines whether the game continues with $\varphi$ or with $\psi$.

REMARK 8.0.1. *Note that no role switch is involved in the above rule: the player acting as **P** remains in this role at the succeeding state; likewise for **O**.*

We call the $\mathcal{H}$-mv-game augmented by rule $R_\pi^{\mathcal{R}}$ the *(basic) $\mathcal{R}$-game*. We claim that the new rule gives raise to the following truth function, to be added to the semantics of logic KZ:

$$v_{\mathcal{M}}(\varphi\pi\psi) = (v_{\mathcal{M}}(\varphi) + v_{\mathcal{M}}(\psi))/2.$$

KZ($\pi$) denotes the logic arising from KZ by adding $\pi$. To assist a concise formulation of the adequateness claim for the $\mathcal{R}$-game we have to adapt Definition 2.2.1 by replacing 'payoff' with 'expected payoff'. In fact, since we restrict attention to the propositional level here, we can use the following simpler definition.

DEFINITION 8.0.2. *If player **X** has a strategy that leads to an expected payoff for her of at least $w$, while her opponent has a strategy that ensures that **X**'s expected payoff is at most $w$, then $w$ is called the* expected value for **X** *of the game.*

THEOREM 8.0.3. *A propositional formula $F$ evaluates to $v_{\mathcal{M}}(\varphi) = w$ in a KZ($\pi$)-interpretation $\mathcal{M}$ iff the basic $\mathcal{R}$-game for $F$ with payoffs matching $\mathcal{M}$ has expected value $w$ for myself.*

[6] A similar rule is considered in [54] in the context of partial logic.

*Proof.* Taking into account that $v_{\mathcal{M}}(\varphi)$ coincides with the value of the $\mathcal{H}$-mv-game matching $\mathcal{M}$ if $\varphi$ does not contain the new connective $\pi$, we only have to add the case for a current formula of the form $\varphi\pi\psi$ to the usual backward induction argument. However, because of the random choice involved in rule $R_{\pi}^{\mathcal{R}}$, it is now her *expected* payoff that **P** seeks to maximize and **O** seeks to minimize.

Suppose the current formula is $\varphi\pi\psi$. By the induction hypothesis, at the successor state $\sigma_{\varphi}$ with current formula $\varphi$ (the player who is currently) **P** can force[7] an expected payoff $v_{\mathcal{M}}(\varphi)$ for herself, while **O** can force an expected payoff $1 - v_{\mathcal{M}}(\varphi)$ for himself. Therefore the expected value for **P** for the game starting in $\sigma_{\varphi}$ is $v_{\mathcal{M}}(\varphi)$ for **P**. The same holds for $\psi$ instead of $\varphi$. Since the choice between the two successor states $\sigma_{\varphi}$ and $\sigma_{\psi}$ is uniformly random, we conclude that the expected value for **P** for the game starting with $G\pi H$ is the average of $v_{\mathcal{M}}(\varphi)$ and $v_{\mathcal{M}}(\psi)$, i.e., $(v_{\mathcal{M}}(\varphi) + v_{\mathcal{M}}(\psi))/2$. The theorem thus follows from the fact that I am the initial **P** in the relevant $\mathcal{R}$-game. □

Note that the function $(x + y)/2$ cannot be composed solely from the functions $1 - x$, $\min\{x, y\}$, and $\max\{x, y\}$ and the values 0 and 1. Therefore we can make the following observation.

PROPOSITION 8.0.4. *The connective $\pi$ is not definable in logic* KZ.

But also the following stronger fact holds.

PROPOSITION 8.0.5. *The connective $\pi$ is not definable in Łukasiewicz logic* Ł.

*Proof.* By McNaughton's Theorem [42] a function $f\colon [0,1]^n \to [0,1]$ corresponds to a formula of propositional Łukasiewicz logic iff $f$ is piecewise linear, where every linear piece has integer coefficients. But clearly the coefficient of $(x + y)/2$ is not an integer. □

REMARK 8.0.6. *We may also observe that, in contrast to Ł, not only $\overline{0.5} =_{def} \bot\pi\top$, but in fact every rational number in $[0,1]$ with a finite (terminating) expansion in the binary number system is definable as a truth constant in logic* $\mathrm{KZ}(\pi)$.

Conversely to Proposition 8.0.5 we also have the following.

PROPOSITION 8.0.7. *None of the connectives $\&$, $\oplus$, $\to$ of Ł can be defined in* $\mathrm{KZ}(\pi)$.

*Proof.* Let $\Psi$ denote the set of all interpretations $\mathcal{M}$, where $0 < v_{\mathcal{M}}(p) < 1$ for all propositional variables $p$. The following claim can be straightforwardly checked by induction.

*Claim:* For every formula $\varphi$ of $\mathrm{KZ}(\pi)$ one of the following holds:

(1) $0 < v_{\mathcal{M}}(\varphi) < 1$ for all $\mathcal{M} \in \Psi$,

(2) $v_{\mathcal{M}}(\varphi) = 1$ for all $\mathcal{M} \in \Psi$, or

(3) $v_{\mathcal{M}}(\varphi) = 0$ for all $\mathcal{M} \in \Psi$.

Clearly this claim does not hold for Ł-formulas of the form $\varphi \,\&\, \psi$, $\varphi \oplus \psi$, and $\varphi \to \psi$. Therefore the connectives $\&$, $\oplus$, $\to$ cannot be defined in $\mathrm{KZ}(\pi)$. □

[7] We re-use the terminology introduced in the proof of Theorem 2.3.4, but applied to *expected* payoffs here.

In light of the above propositions, the question arises whether one can come up with further game rules, that, like $R^{\mathcal{R}}_{\pi}$, do not sacrifice what we above called *Hintikka's principle*, i.e., the principle that game state is determined solely by a formula and a role distribution. An obvious way to generalize rule $R^{\mathcal{R}}_{\pi}$ is to allow for a (potentially) biased random choice:

($R^{\mathcal{R}}_{\pi^p}$) If the current formula is $\varphi\pi^p\psi$ then the game continues with $\varphi$ with probability $p$, but continues with $\psi$ with probability $1-p$.

Clearly, $\pi$ coincides with $\pi^{0.5}$. But for other values of $p$ we obtain a new connective. It is straightforward to check that Proposition 8.0.7 also holds if we replace $\pi$ by $\pi^p$ for any $p \in [0,1]$.

Interestingly, there is a fairly simple game based way to obtain a logic that properly extends Łukasiewicz logic by introducing a unary connective D that signals that the payoff values for **P** is to be doubled (capped to 1, as usual) at the end of the game.

($R^{\mathcal{R}}_{\mathrm{D}}$) If the current formula is $\mathrm{D}\varphi$ then the game continues with $\varphi$, but with the following changes at the final state. The payoff, say $x$, for **P** is changed to $\min\{1,2x\}$, while the the payoff $1-x$ for **O** is changed to $1-\min\{1,2x\}$.

REMARK 8.0.8. *Instead of explicitly capping the modified payoff for* **P** *to* 1 *one may equivalently give* ***O*** *the opportunity to either continue that game with doubled payoff for* **P** *(and inverse payoff for* ***O*** *herself) or to simply end the game at that point with payoff* 1 *for* **P** *and payoff* 0 *for* ***O*** *herself.*

Let us use KZ(D) for the logic obtained from KZ by adding the connective D with the following truth function to KZ:

$$v_{\mathcal{M}}(\mathrm{D}\varphi) = \min\{1, 2\cdot v_{\mathcal{M}}(\varphi)\}.$$

Moreover, we use KZ($\pi$, D) to denote the extension of KZ with both $\pi$ and D and call the $\mathcal{R}$-game augmented by rule $R^{\mathcal{R}}_{\mathrm{D}}$ the D-*extended* $\mathcal{R}$*-game.*

THEOREM 8.0.9. *A propositional formula $\varphi$ evaluates to $v_{\mathcal{M}}(\varphi) = w$ in a* KZ($\pi$, D)*-interpretation $\mathcal{M}$ iff the* D*-extended $\mathcal{R}$-game for $\varphi$ with payoffs matching $\mathcal{M}$ has expected value $w$ for myself.*

*Proof.* The proof of Theorem 8.0.3 is readily extended to the present one by considering the additional inductive case of $\mathrm{D}\psi$ as current formula. By the induction hypothesis, the expected value for **P** of the game for $G$ (under the same interpretation $\mathcal{M}$) is $v_{\mathcal{M}}(\psi)$. Therefore rule $R^{\mathcal{R}}_{\mathrm{D}}$ entails that the expected value for **P** of the game for $\mathrm{D}\psi$ is $\min\{1, 2\cdot v_{\mathcal{M}}(\psi)\}$. □

Given Proposition 8.0.7 and Theorem 8.0.9 the following simple observation is of some significance.

PROPOSITION 8.0.10. *The connectives &, $\oplus$ and $\rightarrow$ of Ł are definable in* KZ($\pi$, $D$).

*Proof.* It is straightforward to check that the following definitions of $\oplus$, &, and $\rightarrow$ as derived connectives in KZ($\pi$, $D$) match the corresponding truth functions for logic Ł: $\varphi \oplus \psi =_{def} \mathrm{D}(\varphi\pi\psi)$, $\varphi \,\&\, \psi =_{def} \neg\mathrm{D}(\neg\varphi\pi\neg\psi)$, and $\varphi \rightarrow \psi =_{def} \mathrm{D}(\neg\varphi\pi\psi)$. □

REMARK 8.0.11. *Note that Proposition 8.0.10 jointly with Theorem 8.0.9 entails that one can provide game semantics for (an extension of) Łukasiewicz logic Ł without dropping Hintikka's principle as in $\mathcal{E}$-games and in $\mathcal{G}$-games.*

REMARK 8.0.12. *The definitions mentioned in the proof of Proposition 8.0.10 give rise to corresponding additional rules for the* D*-extended $\mathcal{R}$-game. In particular, for strong disjunction we obtain:*

($R^{\mathcal{R}}_{\oplus}$) *If the current formula is $\varphi \oplus \psi$ then a random choice determines whether to continue the game with $\varphi$ or with $\psi$. But in any case the payoff for* **P** *is doubled (capped to* 1*), while the payoff for* **O** *remains inverse to that for* **P**.

*By further involving role switches similar rules for strong conjunction and for implications are readily obtained. It remains to be seen whether these rules can assist in arguing for the plausibility of the corresponding connective in intended application scenarios. But in any case, it is clear that, compared to the sole specification of truth functions, the game interpretation provides an additional handle for assessing the adequateness of the Łukasiewicz connectives for formalizing reasoning with graded notions and vague propositions.*

Like $R^{\mathcal{R}}_{\pi}$, also the rule $R^{\mathcal{R}}_{\mathrm{D}}$ can be generalized in an obvious manner:

($R^{\mathcal{R}}_{\mathrm{M}_c}$) If the current formula is $\mathrm{M}_c\varphi$ then the game continues with $\varphi$, but with the following changes at the final state. The payoff, say $x$, for **P** is changed to $\min\{1, c \cdot x\}$, while the the payoff $1 - x$ for **O** is changed to $1 - \min\{1, c \cdot x\}$.

Adding further instances of $\pi^p$ and $\mathrm{M}_c$ to $\mathrm{KZ}(\pi, \mathrm{D})$ leads to more expressive logics, related to Rational Łukasiewicz Logic and to divisible MV-algebras [24].[8]

REMARK 8.0.13. *Like in Section 7, we have restricted our attention to propositional logics in this section. However, once more, we straightforwardly obtain corresponding first order logics by extending $\mathcal{R}$-game s with Hintikka's original rules $R^{\mathcal{H}}_{\forall}$ and $R^{\mathcal{H}}_{\exists}$ for universal and existential quantification.*

## 9 Random choice rules for semi-fuzzy quantifiers

As we have seen in the last section, randomization provides a powerful tool for characterizing extensions of logic KZ, i.e., of the 'weak' fragment of Łukasiewicz logic at the propositional level. In this section we will show that allowing for random choices is also useful for characterizing certain types of generalized quantifiers. The simplest quantifier rule of the indicated type can be formulated in direct analogy to the two quantifier rules of the $\mathcal{H}$-game as follows.

($R^{\mathcal{H}}_{\Pi}$) If the current formula is $\Pi x \varphi(x)$ an element $c$ of the domain of $\mathcal{M}$ is chosen randomly and the game continues with $\varphi(c)$.

[8] The following observation (by Petr Hájek) is relevant here: If one adds the truth constant $0.5$ to Ł then *all* rational numbers are expressible. Therefore $\mathrm{KZ}(\pi, \mathrm{D})$ extends not only Ł, but also Rational Pavelka Logic, where all rationals truth constants are added to Ł (see [28]). On the other hand, neither, e.g., $\pi^{1/3}$ or $\mathrm{M}_3$ seem to be expressible in $\mathrm{KZ}(\pi, \mathrm{D})$.

Of course, it is not yet clear what exactly we mean by a random choice of a domain element, in particular if the domain is infinite. Fortunately, the intended application of modeling natural language semantics justifies the focus on finite domains: throughout the whole section we will therefore assume that the domain is finite. Moreover we will assume that 'random' means 'uniformly random'. Note that the latter assumption is imperative if we insist that $\Pi$ is a *logical* quantifier: the meaning of a logical quantifier should be independent of any particular order of domain elements.

While we could add rule $(R^{\mathcal{H}}_{\Pi})$ to the $\mathcal{H}$-game or to the $\mathcal{H}$-mv-game introduced in Section 2, we prefer to switch right away to the more general setting of the $\mathcal{G}$-game (Giles's game) for Łukasiewicz logic Ł as presented in Section 4. In that context rule $R^{\mathcal{H}}_{\Pi}$ has to be reformulated as follows.

$(R^{\mathcal{G}}_{\Pi})$ If the current formula is $\Pi x \varphi(x)$ then the game continues in a state where the indicated occurrence of $\Pi x \varphi(x)$ is replaced by $\varphi(c)$ for some randomly chosen domain element $c$.

REMARK 9.0.1. *Note that rule $R^{\mathcal{G}}_{\Pi}$ does not refer to the two players (myself and you) at all. Like for the case of the random choice connective $\pi$ of Section 8, we may think of a third player* Nature *as responsible for the choice. In any case, the rule applies independently of whether $\Pi x \varphi(x)$ is in my or in your tenet.*

It turns out that $\Pi$ corresponds to a fuzzy quantifier or, more precisely, proportionality quantifier as introduced by Zadeh [55] to model natural language expressions like *few*, *most*, *about a half*, *about ten*, etc. In Zadeh's approach any function from $[0, 1]$ into $[0, 1]$ may serve as a truth function for a (monadic) fuzzy quantifier that may be applied to an arbitrary formula of some underlying fuzzy logic, in principle. However, as has been repeatedly pointed out in the literature (see, e.g., [22] or the monograph [27]) quite fundamental problems of interpretation arise for quantifiers where not only the quantified formula itself may take an intermediary truth value, but where also the scope of the quantifier occurrence may be fuzzy. Such quantifiers are called type IV quantifiers by Liu and Kerre [38]. To get an idea of the problems alluded to above, consider a formalization of the statement *About half of the visitors are tall* in two different scenarios: in the first scenario half of the visitors are clearly tall and half of them are clearly short, whereas in the second scenario all visitors are of the same height, which is borderline between tall and not tall. Intuitively, we want to accept the statement—i.e, judge it as clearly true—in the first case, but not in the second. However, we cannot distinguish properly between the two indicated scenarios if all that is to be taken into account for computing the truth value of the quantified statement is the 'average' truth value of the instances of the scope formula. For this reason we avoid type IV quantifiers and focus on so-called semi-fuzzy quantifiers (type III quantifiers in the classification of [38]), where the scope is always a crisp, i.e., classical, formula and not just any Ł-formula. Formally, we specify the language for an extension Ł$(Qs)$, of first order Łukasiewicz logic Ł, where $Qs$ is a list of (unary) quantifier symbols other than $\forall$ or $\exists$, as follows:

$$\gamma ::= \bot \mid \hat{P}(\vec{t}) \mid \neg\gamma \mid (\gamma \vee \gamma) \mid (\gamma \wedge \gamma) \mid \forall v \gamma \mid \exists v \gamma$$

$$\varphi ::= \gamma \mid \tilde{P}(\vec{t}) \mid \neg\varphi \mid (\varphi \vee \varphi) \mid (\varphi \wedge \varphi) \mid (\varphi \to \varphi) \mid (\varphi \,\&\, \varphi) \mid \forall v \varphi \mid \exists v \varphi \mid \mathrm{Q} v \gamma,$$

where $\hat{P}$ and $\tilde{P}$ are meta-variables for classical and for general (i.e., possibly fuzzy) predicate symbols, respectively, $\mathrm{Q} \in Qs$; $v$ is our meta-variable for object variables; $\vec{t}$ denotes a sequence of terms, i.e., either object variable or constant symbol, matching the arity of the preceding predicate symbol. Note the scope of the additional quantifiers from $Qs$ is always a classical formula. Otherwise the syntax is as for Ł itself.

The following notion supports a crisp specification of truth functions for semi-fuzzy proportionality quantifiers over finite interpretations.

DEFINITION 9.0.2. *Let $\psi(x)$ be a classical formula and $v_{\mathcal{M}}(\cdot)$ a corresponding evaluation function over the finite domain $D$. Then*

$$\mathrm{Prop}_x\, \psi(x) = \frac{\sum_{c \in D} v_{\mathcal{M}}(\psi(c))}{|D|}.$$

$\mathrm{Prop}_x\, \psi(x)$ thus denotes the proportion of all elements in $D$ satisfying the classical formula $\psi$. Remember that we stipulated above that all random choices are made with respect to a uniform probability distribution over $D$. Therefore $\mathrm{Prop}_x\, \psi(x)$ denotes the probability that a randomly chosen element satisfies $\psi$.

The following theorem generalizes Theorem 4.0.1 and states that rule $R_{\Pi}^{\mathcal{G}}$ matches the extension of the valuation function for Ł to Ł$(\Pi)$ by

$$v_{\mathcal{M}}(\Pi x \psi(x)) = \mathrm{Prop}_x\, \psi(x).$$

THEOREM 9.0.3. *A Ł$(\Pi)$-sentence $\varphi$ is evaluated to $v_{\mathcal{M}}(\varphi) = w$ in an interpretation $\mathcal{M}$ iff the $\mathcal{G}$-game for $\varphi$ augmented by rule $R_{\Pi}^{\mathcal{G}}$ has value $1 - w$ for myself under risk value assignment $\langle\cdot\rangle_{\mathcal{M}}$.*

Theorem 9.0.3 is an instance of Theorem 9.2.2, proved below.

## 9.1 A note on binary quantifiers

Natural language quantifiers are usually binary, as in *About half of the students are present*, rather than unary as in *About half* [*of the elements in the domain of discourse*] *are present*. However, binary quantifiers like *about half*, *many*, *at least a third*, etc, are extensional. This means that, like in the above example, the first argument of the binary quantifier—its range—is only used to restrict the universe of discourse. More formally, let $\hat{\psi}$ denote the set of domain elements that satisfy the (crisp) predicate expressed by the classical formula $\psi(x)$. If Q is a unary quantifier, then ${}^{\psi}\mathrm{Q}x\varphi(x)$ is a quantified statement defined by $v_{\mathcal{M}}({}^{\psi}\mathrm{Q}x\varphi(x)) = v_{\mathcal{M}'}(\mathrm{Q}x\varphi(x))$, where $\mathcal{M}'$ denotes the interpretation that results from $\mathcal{M}$ by restricting the domain of $\mathcal{M}$ to $\hat{\psi}$. This reduces extensional binary quantification to unary quantification, here illustrated for $\Pi$ as follows:

($R_{\Pi^2}^{\mathcal{G}}$) Asserting ${}^{\psi}\Pi x\varphi(x)$ reduces to asserting $\varphi(c)$ where $c$ is a (uniformly) randomly chosen element of $\hat{\psi}$: $\varphi(c)$ replaces ${}^{\psi}\Pi x\varphi(x)$ in the corresponding tenet.

If the classical formula $\psi(x)$ is atomic then it is clear what it means to randomly choose an element of $\hat{\psi}$ (unless $\hat{\psi}$ is empty). However, if $\psi(x)$ is of arbitrary logical complexity, then we may remain within our game semantic framework by employing the $\mathcal{H}$-game of Section 2 to find an appropriate random witness element as follows:

1. Choose a random domain element $c$.
2. Initiate an $\mathcal{H}$-game where a Proponent **P** defends $\psi(c)$ against an Opponent **O**.
3. If **P** wins the $\mathcal{H}$-game, then the main $\mathcal{G}$-game is continued with the constant $c$, i.e., with $\varphi(c)$ replacing ${}^{\psi}\Pi x\varphi(x)$. Otherwise, return to 1.

Note that it is important to keep the objectives of the players **P** and **O** in the $\mathcal{H}$-game independent from the objectives of the players in the $\mathcal{G}$-game. By Theorem 2.1.1 **P** wins the $\mathcal{H}$-game against the rational Opponent **O** if and only if the classical formula $\psi(c)$ is true, i.e., if $c \in \psi$. Note that the indicated procedure and therefore the main $\mathcal{G}$-game will fail to terminate if the range $\psi$ is empty. This is in accordance with the above definition that leaves $v_{\mathcal{M}}({}^{\psi}\mathrm{Q}x\varphi(x))$ undefined if the range is empty. According to the classic linguistic paper [5] by Barwise and Cooper this matches intuitions about natural language quantifiers applied to an empty range.

### 9.2 Blind choice quantifiers

Remember that in the context of the $\mathcal{G}$-game we have considered three types of challenges to the defender **X** of a quantified sentence $\mathrm{Q}x\psi(x)$. In each case **X** has to assert $\psi(c)$, but the constant (domain element) is either

(A) chosen by the attacker (i.e., by the current opponent **O**),

(D) chosen by the defender (i.e., by the current proponent **P**), or

(R) chosen randomly (i.e., by *Nature*).

We will speak of a challenge of type A, D, or R, respectively. With respect to a formula $\mathrm{Q}x\psi(x)$ we will say that the two players (myself and you) either bet *for* or *against* $\psi(c)$. Betting for $\psi(c)$ simply means to assert $\psi(c)$, betting against $\psi(c)$ is equivalent to betting for $\neg\psi(c)$ and thus amounts to an assertion of $\psi$ in exchange for an assertion of $\psi(c)$ by the opposing player. We interpret the latter bet as follows: **X** pays 1€ for a ticket that entitles her to receive whatever payment by her opponent **Y** is due for **Y**'s assertion of $\psi(c)$ according to the results of associated dispersive experiments made at the end of the game.

Like in Section 5 for proportional rules in Giles's game, we will speak of a *round* of a game as consisting of a player's *attack* of an assertion made by the other player, followed by a *defense* of that latter player. Moreover, we recall from Section 5 that by the principle of limited liability for defense (LLD) asserting $\perp$ is always a valid defense move. Moreover, by the other form of the principle of limited liability (LLA), the opponent, instead of attacking an assertion in some specific way, may grant the assertion which will consequently be deleted from the current state of the game. In general, when an assertion of $\mathrm{Q}x\psi(x)$ is attacked, the round results in a state where both players are placing certain numbers of bets for or against various instances of $\psi(x)$, where the challenge determining the constants replacing $x$ can be of type A, D, or R. In this manner we arrive at a rich set of possible quantifier rules. However, here we are only interested in type R challenges. We will call $\psi(c)$ a random instance of $\psi(x)$ if $c$ has been chosen randomly.

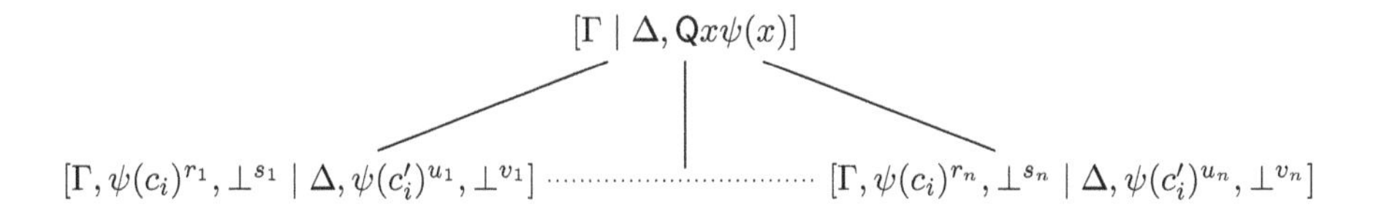

Figure 6. Schematic blind choice quantifier rule — my possible defenses to a particular attack by you.

We first investigate the family of *blind choice quantifiers*, defined as follows.

DEFINITION 9.2.1. Q *is a* (semi-fuzzy) blind choice quantifier *if it can be specified by a game rule satisfying the following two conditions:*

*(i) Only challenges of type R are allowed: an attack on* $Qx\psi(x)$ *followed by a defense move results in a state where both players have placed a certain number (possibly zero) of bets* for *and* against *random instances of* $\psi(x)$.

*(ii) The identity of the random constants is revealed to the players only at the end of the round, i.e., after an attack has been chosen by the one player and a corresponding defense move has been chosen by the other player.*

*Like in all other game rules the occurrence of the attacked formula is removed from the corresponding tenet.*

Figure 6 depicts possible state transitions involved in the application of a blind choice quantifier rule, where I am the proponent (defender) of the formula $Qx\psi(x)$. $\Gamma$ and $\Delta$ denote arbitrary multisets of formulas; the classical formula $\psi(x)$ forms the scope of $Qx\psi(x)$ asserted by myself and attacked by you; $\bot^k$ denotes $k$ occurrences of $\bot$; and $\psi(c_i)^k$ is used as an abbreviation for the $k$ assertions of random instances $\psi(c_1), \ldots, \psi(c_k)$. Note that in general there is more than one way in which you may attack my assertion of $Qx\psi(x)$. Figure 6 only shows the scheme for one particular attack. A presentation of a full rule consists of a finite number of instances of this scheme. The root of all these trees is labeled by $[\Gamma \mid \Delta, Qx\psi(x)]$, which means that the effect of the different attacks is shown only at the end of a full round, i.e., only after also a corresponding defense has been chosen. The principle of limited liability LLA implies that you, the attacker, may choose to simply remove the exhibited occurrence of $Qx\psi(x)$ from the state. In other words, every rule includes an instance of Figure 6 that consists of only one branch ($n = 1$), where $r_1 = s_1 = u_1 = v_1 = 0$. For myself as defender, the principle of limited liability LLD implies that in any other instance of the schematic tree there is a branch $i$ with $r_i = s_i = u_i = 0$ and $v_i = 1$, i.e., where I reply to your attack by asserting $\bot$.

We assume that for every rule for my assertion of a formula, there is a corresponding rule for your assertion of the same formula, that arises by switching our roles. It therefore suffices to investigate explicitly only those rules, where I am in role **P**, i.e., for occurrences of quantified formulas in my tenet.[9]

[9] Formulating a rule only for myself in the role of **P** has the advantage that we do not have to consider two cases when we specify the game states that result from applying the rule.

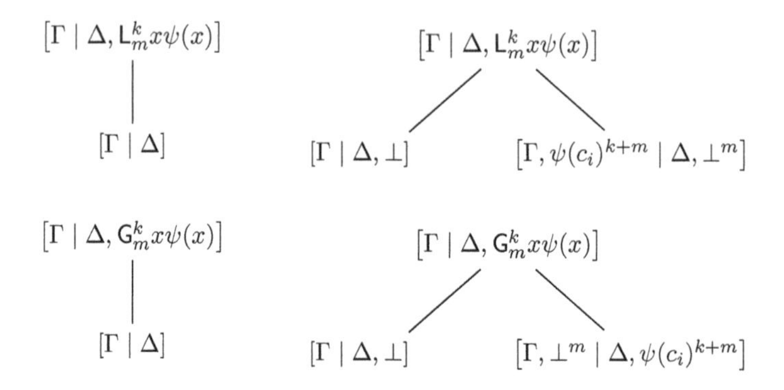

Figure 7. Games rules $R_{\mathsf{L}^k_m}$ and $R_{\mathsf{G}^k_m}$

We specify two concrete families of blind choice quantifiers, $\mathsf{L}^k_m$ and $\mathsf{G}^k_m$, for every $k, m \geq 1$, by the following rules:

($R_{\mathsf{L}^k_m}$) If I assert $\mathsf{L}^k_m x\psi(x)$ then you may attack by betting for $k$ random instances of $\psi(x)$, while I bet against $m$ random instances of $\psi(x)$.

($R_{\mathsf{G}^k_m}$) If I assert $\mathsf{G}^k_m x\psi(x)$ then you may attack by betting against $m$ random instances of $\psi(x)$, while I bet for $k$ random instances of $\psi(x)$.

We insist on condition *(ii)* of Definition 9.2.1: the random constants used to obtain the mentioned instances of $\psi(x)$ are only revealed to the players after they have placed their bets. Moreover, although not explicitly mentioned, the principle of limited liability remains in force in both forms (LLA and LLD). Therefore, by LLA, I as the defender (i.e., in role **P**) may also respond to your attack by asserting $\bot$. On the other hand, by LLD, you may grant the formula occurrence in question without attacking it. It is then simply removed from my tenet. However, if none of the players invokes the principle of limited liability the following successor game states are reached:

$$\text{for } \mathsf{L}^k_m x\psi(x): \quad [\Gamma, \psi(c_i)^{k+m} \mid \Delta, \bot^m]$$

$$\text{for } \mathsf{G}^k_m x\psi(x): \quad [\Gamma, \bot^m \mid \Delta, \psi(c_i)^{k+m}].$$

Consequently, the rules for my assertion of $\mathsf{L}^k_m x\psi(x)$ or of $\mathsf{G}^k_m x\psi(x)$ can be depicted as shown in Figure 7. The rules for your assertion of $\mathsf{L}^k_m x\psi(x)$ or of $\mathsf{G}^k_m x\psi(x)$ are analogous.

We claim that these rules match the extension of Ł to Ł($\mathsf{L}^k_m$, $\mathsf{G}^k_m$) by

$$v_{\mathcal{M}}(\mathsf{L}^k_m x\psi(x)) = \min\{1, \max\{0, 1+k-(m+k)\,\mathrm{Prop}_x\,\psi(x)\}\} \tag{8}$$

$$v_{\mathcal{M}}(\mathsf{G}^k_m x\psi(x)) = \min\{1, \max\{0, 1-k+(m+k)\,\mathrm{Prop}_x\,\psi(x)\}\}. \tag{9}$$

THEOREM 9.2.2. *A* Ł($\mathsf{L}^k_m$, $\mathsf{G}^k_m$)*-sentence $\varphi$ is evaluated to $v_{\mathcal{M}}(\varphi) = x$ in an interpretation $\mathcal{M}$ iff every $\mathcal{G}$-game for $\varphi$ augmented by the rules $R_{\mathsf{L}^k_m}$ and $R_{\mathsf{G}^k_m}$ is has value $1-x$ for me under risk value assignment $\langle\cdot\rangle_{\mathcal{M}}$.*

*Proof.* Relative to the proof of Theorem 4.0.1 (see [19, 25, 26]) we only have to consider states of the form $\left[\Gamma \mid \Delta, \mathsf{L}^k_m x\psi(x)\right]$ and $\left[\Gamma \mid \Delta, \mathsf{G}^k_m x\psi(x)\right]$. (I.e., we only consider situations where my assertion of an $\mathsf{L}^k_m$- or $\mathsf{G}^k_m$-quantified sentences is considered for attack by you. The cases for your assertions of $\mathsf{L}^k_m x\psi(x)$ or $\mathsf{G}^k_m x\psi(x)$ are dual.) In fact, since $\mathsf{G}^k_m$ is treated analogously to $\mathsf{L}^k_m$, we may focus on states of the form $\left[\Gamma \mid \Delta, \mathsf{L}^k_m x\psi(x)\right]$ without loss of generality. Like for the other connectives, we obtain the total risk at such a state as the sum of the risk for the exhibited assertion and of the risk for the rest of the state:

$$\left\langle \Gamma \mid \Delta, \mathsf{L}^k_m x\psi(x)\right\rangle = \langle \Gamma \mid \Delta\rangle + \left\langle \mid \mathsf{L}^k_m x\psi(x)\right\rangle.$$

It remains to show that the reduction of the exhibited quantified formula to instances according to rule $(R_{\mathsf{L}^k_m})$ results in a risk that corresponds to the specified truth function if we play rationally. According to Figure 7 the three possible successor states are $\left[\psi(c_i)^{k+m} \mid \bot^m\right]$, $[\mid]$, and $[\mid \bot]$, respectively. In the first case, revealing the constants to the players also reveals the amount of money I have to pay, since only classical formulas are involved: I have to pay $m$€ to you for my $m$ assertions of $\bot$, while for each of your $k+m$ assertions you have to pay me either 0€ or 1€. In total I have to pay to you between $-k$€ and $m$€, depending on the random constants $c_i$. The risk value of the game state *before* the identities of the constants are revealed to the players is therefore calculated as the *expected* value for this amount. It is binomially distributed and readily computed as

$$m - \sum_{i=0}^{k+m} i \cdot (\mathrm{Prop}_x\, \psi(x))^{k+m-i}(1 - \mathrm{Prop}_x\, \psi(x))^i \binom{k+m}{i} =$$

$$m - (k+m)(1 - \mathrm{Prop}_x\, \psi(x)) = -k + (k+m)\,\mathrm{Prop}_x\, \psi(x)).$$

The second case (state $[\mid]$, carrying risk 0) arises if you choose to grant my assertion of the formula, which for you is the rational choice if the above expression is below 0. The third case (state $[\mid \bot]$, carrying risk 1) arises if I invoke the principle of limited liability LLD to hedge my expected loss. Thus we obtain

$$\left\langle \mid \mathsf{L}^k_m x\psi(x)\right\rangle = \min\{1, \max\{0, -k + (k+m)\,\mathrm{Prop}_x\, \psi(x)\}\} = 1 - v_{\mathcal{M}}(\mathsf{L}^k_m x\psi(x)),$$

which means that the claimed correspondence between the truth function and the risk resulting from playing rationally holds. □

**At least about a third.** For illustration, let us take a closer look at quantifiers of the form $\mathsf{G}^s_{2s}$. We argue that these quantifiers can be used to model the natural language expression *at least about a third.* Note that the attacker of $\mathsf{G}^s_{2s} x\psi(x)$ is supposed to believe that $\psi(x)$ holds for clearly less than a third of all domain elements (otherwise she would grant the assertion). Consequently she is willing to place $2s$ bets *against* random instances of $\psi(x)$ if the defender places $s$ bets *for* such random instances. Figure 8 shows the resulting truth functions for sample sizes $(2s+s)$ $3, 6$, and $9$, where the horizontal axis corresponds to $\mathrm{Prop}_x\, \psi(x)$ and the vertical axis to $v_{\mathcal{M}}(\mathsf{G}^s_{2s} x\psi(x))$. Functions like these are routinely suggested to represent natural language quantifiers like *at least about*

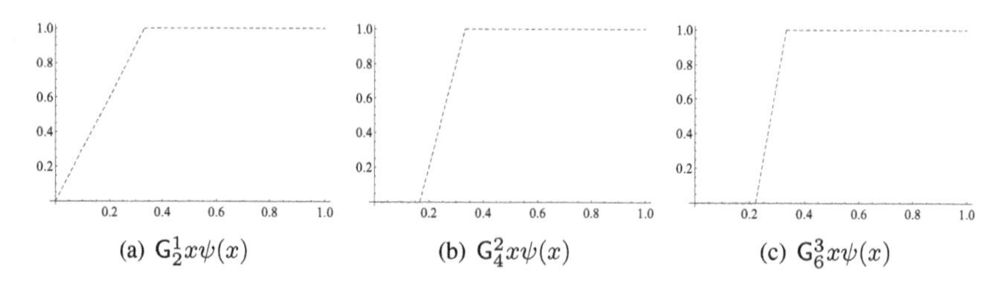

(a) $\mathsf{G}_2^1 x\psi(x)$ (b) $\mathsf{G}_4^2 x\psi(x)$ (c) $\mathsf{G}_6^3 x\psi(x)$

Figure 8. Truth functions for $\mathsf{G}_{2s}^s x\psi(x)$

*a third* in the fuzzy logic literature.[10] However no justification beyond intuitive plausibility is usually given. In contrast, our model allows one to extract such truth functions from an underlying semantic principle: namely the willingness to bet that sufficiently many randomly chosen witnesses support or put into doubt the relevant statement.

As noted above, the quantifiers $\mathsf{L}_m^k$ and $\mathsf{G}_m^k$ are only (very restricted) examples of blind choice quantifiers. Nevertheless, they turn out to be expressive enough to define *all* blind choice quantifiers in the context of Kleene–Zadeh logic KZ (weak Łukasiewicz logic):

THEOREM 9.2.3. *All blind choice quantifiers can be expressed using quantifiers of the form* $\mathsf{L}_m^k$ *and* $\mathsf{G}_m^k$, *conjunction* $\wedge$, *disjunction* $\vee$, *and* $\bot$.

*Proof.* As illustrated in Figure 6 above, the game state resulting from an attack and a corresponding defense of my assertion of a blind choice quantifiers is always of the form $[\Gamma, \psi(c_i)^r, \bot^s \mid \Delta, \psi(c_i')^u, \bot^v]$. Analogously to the proof of Theorem 9.2.2, the associated risk before the identities of the constants are revealed is computed as

$$\langle \Gamma \mid \Delta \rangle + v - s + (u - r)(1 - \mathrm{Prop}_x\, \psi(x)).$$

Remember that $\psi(c_i)^k$ is short hand notation for $k$ (in general) different random instances of $\psi(x)$. As a first step towards a simplified uniform presentation of arbitrary blind choice quantifiers, note the following. Instead of picking $u + r$ random constants we can rather investigate the game state $[\Gamma, \psi(c)^r, \bot^s \mid \Delta, \psi(c)^u, \bot^v]$ where only one random constant $c$ is picked, since this modification does not change the *expected* risk. As a further step, note that game states where assertions of $\psi(c)$ are made by *both* players show redundancies in the sense that there are equivalent game states where $\psi(c)$ occurs only in one of the two multisets of assertions that represent a state. Likewise for game states with assertions of $\bot$ made by both players. Depending on $v, s, u$, and $r$, an equivalent game state is given by:

(1) $\left[\Gamma, \psi(c)^{r-u} \mid \Delta, \bot^{v-s}\right]$ if $v > s$ and $r > u$

(2) $\left[\Gamma, \psi(c)^{r-u}, \bot^{s-v} \mid \Delta\right]$ if $v \leq s$ and $r > u$

(3) $\left[\Gamma \mid \Delta, \psi(c)^{u-r}, \bot^{v-s}\right]$ if $v > s$ and $r \leq u$

(4) $\left[\Gamma, \bot^{s-v} \mid \Delta, \psi(c)^{u-r}\right]$ if $v \leq s$ and $r \leq u$.

[10] For example in [27] trapezoidal functions like the ones in Figure 8 are explicitly suggested for natural language quantifiers of this kind.

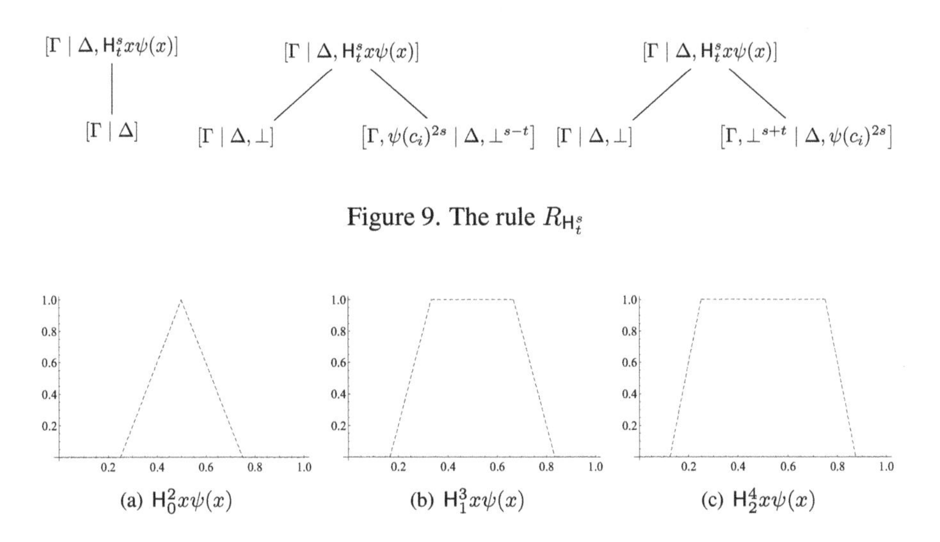


Figure 9. The rule $R_{\mathsf{H}^s_t}$

(a) $\mathsf{H}^2_0 x\psi(x)$ (b) $\mathsf{H}^3_1 x\psi(x)$ (c) $\mathsf{H}^4_2 x\psi(x)$

Figure 10. Truth functions for $\mathsf{H}^s_t x\psi(x)$

Note that states of type (2) are redundant: Playing rationally, you will invoke the principle of limited liability LLA and grant the quantified formula, rather than make an assertion without being compensated by any assertions made by myself. The resulting state is $[\Gamma \mid \Delta]$ in this case. On the other hand, states of type (3) reduce to state $[\Gamma \mid \Delta, \bot]$, since I may invoke the principle of liability LLD. For states of type (1) I will invoke principle LLD if $v - s > r - u$. Similarly, you will invoke principle LLA to ensure that only those states of type (4) have to be considered where $s - v \leq u - r$. For appropriate choices of $k$ and $m$, this leaves us with states that result from the rules for either $\mathsf{L}^k_m x\psi(x)$ or for $\mathsf{G}^k_m x\psi(x)$.

Finally observe that all of my defenses to your attack on $\mathsf{Q}x\psi(x)$ lead to successor states which are reached also by suitable instances of $\mathsf{G}^k_m x\psi(x)$, of $\mathsf{L}^k_m x\psi(x)$ or of $\bot$. Hence my risk for that attack amounts to the minimum of the risk values for these successor states, which in turn equals the risk value for asserting the disjunction of these instances. Similarly, since you can choose between several attacks on $\mathsf{Q}x\psi(x)$ in the first place, my risk for $\mathsf{Q}x\psi(x)$ amounts to the maximum of the risks for these attacks. Hence it is equal to the risk of the conjunction of these disjunctions. □

**About half.** As an example consider the family of quantifiers $\mathsf{H}^s_t$, defined by the game rule depicted in Figure 9.

We suggest that $\mathsf{H}^s_t$ induces plausible fuzzy models for the natural language quantifier *about half*. Figure 10 shows the truth functions for three different quantifiers of this family, where the horizontal axis corresponds to $\mathrm{Prop}_x\, \psi(x)$ and the vertical axis to $v_{\mathcal{M}}(\mathsf{H}^s_t x\psi(x))$.

The two parameters of $\mathsf{H}^s_t$ can be interpreted as follows: $s$ determines the sample size (i.e., the number of random instances involved in reducing the quantified formula), while $t$ may be called the *tolerance*, since the smaller $t$ gets, the closer $\mathrm{Prop}_x\, \psi(x)$ has to be to $1/2$ if $\mathsf{H}^s_t x\psi(x)$ is to be evaluated as perfectly true. If $t = 0$ (zero tolerance) then $v_{\mathcal{M}}(\mathsf{H}^s_0 x\psi(x)) = 1$ if only if $\mathrm{Prop}_x\, \psi(x) = 1/2$ in $M$. By increasing $t$ (while maintaining the same sample size $s$) the range of values for $\mathrm{Prop}_x\, \psi(x)$ that guarantee $v_{\mathcal{M}}(\mathsf{H}^s_0 x\psi(x)) = 1$ grows symmetrically around $1/2$.

As an instance of Theorem 9.2.3 we obtain that $\mathsf{H}^s_t x\psi(x)$ is equivalent to the formula $\mathsf{G}^{s-t}_{s+t} x\psi(x) \wedge \mathsf{L}^{s+t}_{s-t} x\psi(x)$. The tree at the center of Figure 9 corresponds to the rule for $\mathsf{G}^{s-t}_{s+t}$ and the one at the right hand side corresponds to the rule for $\mathsf{L}^{s+t}_{s-t}$. The tree at the left hand side corresponds to the fact that the attacker may choose to grant the formula.

Next we show how arbitrary blind choice quantifiers can be reduced to the quantifier $\Pi$ introduced at the beginning of this section if implication, negation, and strong disjunction, but also truth constants that evaluate to certain rational numbers are available in the language. (This actually corrects an error in [22]).

THEOREM 9.2.4. *The blind choice quantifiers $\mathsf{L}^k_m$ and $\mathsf{G}^k_m$ for all $m, k \geq 1$ can be expressed in Ł($\Pi$) enriched by certain truth constants via the following reductions:*

$$v_{\mathcal{M}}(\mathsf{L}^k_m x\psi(x)) = v_{\mathcal{M}}([\neg(\overline{(1+k)/(m+k)} \rightarrow \Pi x\psi(x))]^{m+k}_{\oplus})$$
$$v_{\mathcal{M}}(\mathsf{G}^k_m x\psi(x)) = v_{\mathcal{M}}([\neg(\Pi x\psi(x) \rightarrow \overline{(k-1)/(m+k)})]^{m+k}_{\oplus})$$

*where $[\phi]^n_{\oplus}$ denotes $\phi \oplus \ldots \oplus \phi$, $n$ times, and $\overline{a}$ denotes the truth constant for $a \in [0,1]$.*

*Proof.* Note that the truth functions of $\mathsf{G}^k_m x\psi(x)$ and $\mathsf{L}^k_m x\psi(x)$ depend only on the value of $\mathrm{Prop}_x\, \psi(x)$, while the random choice quantifier $\Pi$ is directly represented by the truth function $\mathrm{Prop}_x\, \psi(x)$. Hence the equivalences can easily be checked by computing the truth value of the respective right hand side formula and comparing it to the truth function for the corresponding quantifier. □

COROLLARY 9.2.5. *All blind choice quantifiers can be expressed in* Ł($\Pi$) *enriched by rational truth constants.*

The corollary follows directly from Theorems 9.2.3 and 9.2.4.

### 9.3 Deliberate choice quantifiers

In the previous section we surveyed the family of blind choice quantifiers and concluded that these quantifiers all amount to piecewise linear truth functions. A much more general class of quantifiers arises by dropping condition *(ii)* of Definition 9.2.1. As an example of this class we investigate the family of so-called *deliberate choice quantifiers*, specified by the following schematic game rule, where $\psi$ is a classical formula:

($R_{\Pi^k_m}$) If I assert $\Pi^k_m x\psi(x)$ and you decide to attack, then $k+m$ (not necessarily different) constants are chosen randomly and I have to pick $k$ of those constants, say $c_1, \ldots, c_k$, and bet for $\psi(c_1), \ldots, \psi(c_k)$, while simultaneously betting against $\psi(c'_1), \ldots, \psi(c'_m)$, where $c'_1, \ldots, c'_m$ are the remaining $m$ random constants.

(Recall that the scheme for your assertions of $\Pi^k_m x\psi(x)$ arises from switching our roles.) Although not mentioned explicitly, we emphasize that principle of limited liability remains in place: after the constants are chosen, by LLD, I may assert $\perp$ (i.e., agree to pay 1€) instead of betting as indicated above. Therefore I have $1 + \binom{k+m}{k}$ possible defenses to your attack on my assertion of $\Pi^k_m x\psi(x)$: either I choose to hedge my loss by asserting $\bot$ or I pick $k$ out of the $k + m$ random constants to proceed as indicated.

We claim that this rule matches the extension of Ł to Ł$(\Pi^k_m)$ by

$$v_{\mathcal{M}}(\Pi^k_m \psi(x)) = \binom{k+m}{k} (\mathrm{Prop}_x\, \psi(x))^k (1 - \mathrm{Prop}_x\, \psi(x))^m.$$

THEOREM 9.3.1. *A Ł$(\Pi^k_m)$-sentence $\varphi$ is evaluated to $v_{\mathcal{M}}(\varphi) = x$ in interpretation $\mathcal{M}$ iff every $\mathcal{G}$-game for $\varphi$ augmented by rule $R_{\Pi^k_m}$ has value $1 - x$ for myself under risk value assignment $\langle\cdot\rangle_{\mathcal{M}}$.*

*Proof.* Like in the proof of Theorem 9.2.2, we only have to consider states of the form $\left[\Gamma \mid \Delta, \Pi^k_m x\psi(x)\right]$. Again, we can separate the risk for the exhibited assertion from the risk for the remaining assertions:

$$\left\langle \Gamma \mid \Delta, \Pi^k_m x\psi(x)\right\rangle = \langle \Gamma \mid \Delta\rangle + \left\langle \mid \Pi^k_m x\psi(x)\right\rangle.$$

It remains to show that my optimal way to reduce the exhibited quantified formula to instances as required by rule $(R_{\Pi^k_m})$ results in a risk that corresponds to the specified truth function. For the following argument remember that the principle of limited liability is in place. Moreover remember that $\psi(x)$ is classical. This means that I either finally have to pay 1€ for my assertion of $\Pi^k_m x\psi(x)$ or do not have to pay anything at all for it. The latter is only the case if all my bets for $\psi(c_1), \ldots, \psi(c_k)$, as well as all my bets against $\psi(c'_1), \ldots, \psi(c'_m)$, for $c_1, \ldots, c_k$, $c'_1, \ldots, c'_m$ as specified in rule $(R_{\Pi^k_m})$, succeed. Let the random variable $K$ denote the number of chosen elements $c$ on which my bet is successful, i.e., where $\langle\psi(c)\rangle = 0$. Then $K$ is binomially distributed and the probability that this event obtains (the inverse of my associated risk) is readily calculated to be

$$\binom{k+m}{k} \mathrm{Prop}_x\, \psi(x)^k (1 - \mathrm{Prop}_x\, \psi(x))^m. \qquad \square$$

At a first glance, the deliberate choice quantifier $\Pi^k_m$ might seem suitable for modeling the natural language quantifier *about k out of* $m + k$. However, a look at the corresponding graph for $\left\langle\Pi^1_1 x\psi(x)\right\rangle$ (Figure 11) reveals that the risk for asserting $\Pi^1_1 x\psi(x)$ is always larger than 0.5. Therefore the statement is never more than just 'half-true'. This is clearly not in accordance with intuitions about the truth conditions for statements like *About half of the doors are locked* if, say, 49 out of 100 are locked.

An additional mechanism is needed to obtain more appropriate models of natural quantifier expressions like *about half.* While there are many ways to achieve the desired effect, we confine ourselves here to a particularly simple operator that nicely fits our semantic framework, since it arises by simply multiplying involved bets. Given a number $n \geq 2$ and a semi-fuzzy quantifier Q we specify the semi-fuzzy quantifier $\mathsf{W}_n(\mathsf{Q})$ by the following rule.

($\mathsf{W}_n(\mathsf{Q})x\psi(x)$) If I assert $\mathsf{W}_n(\mathsf{Q})x\psi(x)$ then you have to place $n$ bets *against* $\mathsf{Q}x\psi(x)$ while I have to bet *for* $\mathsf{Q}x\psi(x)$ just once. (Analogously for your assertion of $\mathsf{W}_n(\mathsf{Q})x\psi(x)$.)

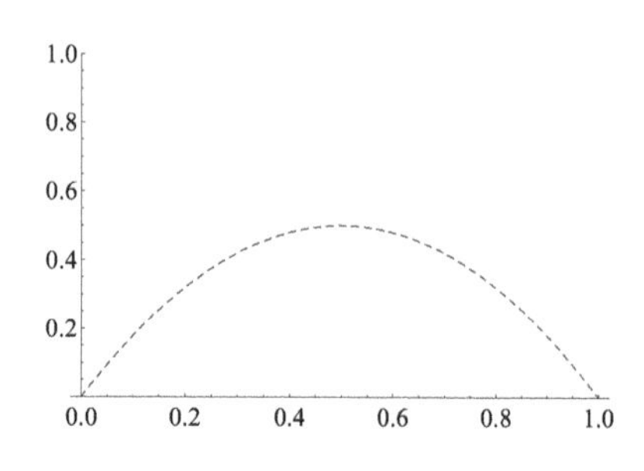


Figure 11. Truth value for $\Pi^1_1 x\psi(x)$ (depending on $p$)

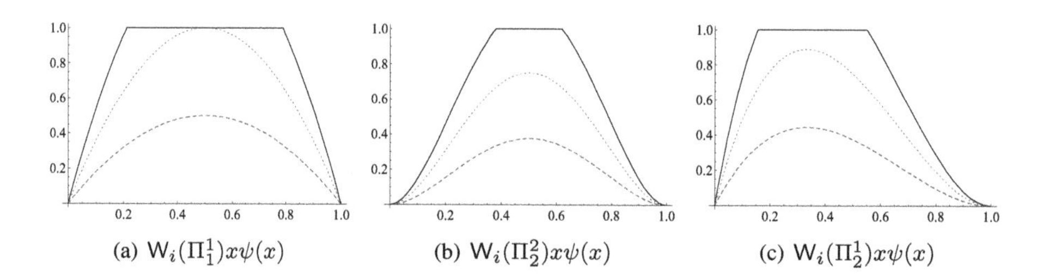


(a) $\mathsf{W}_i(\Pi^1_1)x\psi(x)$ (b) $\mathsf{W}_i(\Pi^2_2)x\psi(x)$ (c) $\mathsf{W}_i(\Pi^1_2)x\psi(x)$

Figure 12. $\mathsf{W}_i$-modified proportionality quantifiers; the graphs correspond to the cases $i = 1$, $i = 2$, and $i = 3$ from bottom to top in each diagram.

Note that $\mathsf{W}_n$ is acting here as a *quantifier modifier*; for any semi-fuzzy quantifier Q, $\mathsf{W}_n(\mathsf{Q})$ still denotes a semi-fuzzy quantifier. Furthermore the principle of limited liability remains in place, hence the game state $\langle \Gamma \mid \Delta, \mathsf{W}_n(\mathsf{Q})x\psi(x)\rangle$ is reduced to $\langle \Gamma, \bot^n \mid \Delta, \mathsf{Q}x\psi(x)^{n+1}\rangle$ or to $\langle \Gamma \mid \Delta\rangle$, depending on whether it is is preferable from the attacker's point of view to attack or to grant the assertion of $\mathsf{Q}x\psi(x)^{n+1}$. (The defender never has to invoke the principal of limited liability in optimal strategies.) Moreover, similarly as in Theorem 9.2.4, $\mathsf{W}_n$ can be expressed using negation and strong conjunction by

$$v_{\mathcal{M}}(\mathsf{W}_n(\mathsf{Q})x\psi(x)) = v_{\mathcal{M}}(\neg(\neg \mathsf{Q}x\psi(x))^{n+1}).$$

For the the truth functions for some of the quantifiers of type $\mathsf{W}_n(\Pi^k_m)$ see Figure 12.

The quantifier $\mathsf{W}_3(\Pi^2_2)$ may be considered as formal fuzzy counterpart of the informal expression *about half*. Likewise, $\mathsf{W}_3(\Pi^1_2)$ may be understood as model of *about a third*. Moreover, $\mathsf{W}_3(\Pi^1_1)$ might serve as a model of *very roughly half*, whereas $\mathsf{W}_2(\Pi^1_1)$ might be appropriate as fuzzy model of the (unhedged) determiner *half*.

In a similar manner, deliberate choice quantifiers can be used to generate plausible candidate models for the proportional reading of *many*. In particular, consider a model where asserting (the formal counterpart of) *Many* [*domain elements*] *are* $\psi$ is expressed by a willingness to place a certain number of bets for random instances of $\psi(x)$. This amounts to considering the family of quantifiers $\Pi^i_0$. The corresponding truth functions are depicted in Figure 13.

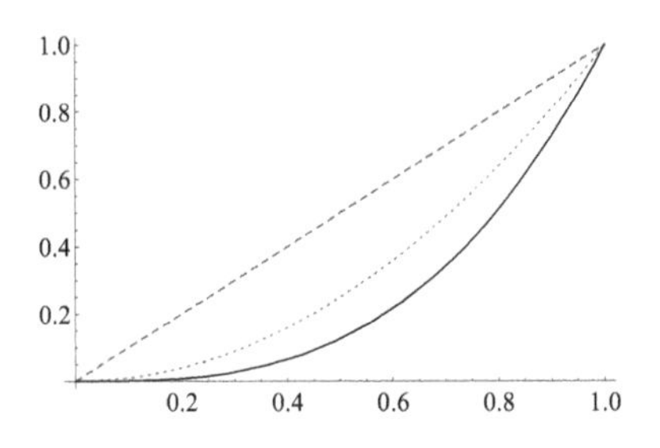


Figure 13. Truth functions for $\Pi_0^i x\psi(x)$ for $i = 1, 2, 3$ from top to bottom

Like for *about half* etc, above, one may want to evaluate *Many* [*domain elements*] *are* $\psi$ as perfectly true (truth value 1) even if $\mathrm{Prop}_x\, \psi(x)$ is somewhat smaller than 1. Again, this can be achieved by employing the $\mathsf{W}_n$-operator, which requires the attacker to place several bets against the contended assertion.

## 10 A brief synopsis of semantic games

To assist a comparison between the various semantic games that we have presented in Sections 2 to 8 we summarize their main characteristics in a table:

| game | state determined by | payoffs | logic(s) | section |
|---|---|---|---|---|
| $\mathcal{H}$-game | single formula + role distribution | bivalent (win/lose) | CL | 2 |
| $\mathcal{H}$-mv-game | single formula + role distribution | many-valued | KZ | 2 |
| $\mathcal{E}$-game | single formula + role distribution + value | bivalent (win/lose) | Ł | 3 |
| $\mathcal{G}$-game(s) | two multisets (tenets) of formulas | many-valued (expected risk) | Ł, $Ł_n$, A CHL, ... | 4, 5 |
| $\mathcal{B}$-games | single formula + role distribution + stack | many-valued | Ł, G, Π | 7 |
| $\mathcal{R}$-game(s) | single formula + role distribution | many-valued (expected value) | $\mathrm{KZ}(\pi)$ $\mathrm{KZ}(\pi, \mathrm{D})$ | 8 |

## 11 Historical remarks and further reading

There are a lot of links between logic and games. We refer to [35] for an overview, including a brief history that goes back to antiquity. For our specific topic—*semantic games*, also known as *evaluation games*—Leon Henkin's [30] is an important precursor. Henkin pointed out that the game based understanding of universal and existential quantifiers may be applied to situations where Tarski's classic definition of truth in model fails. In the late 1960s Jaakko Hintikka [31] started to explore semantic games as an alternative to Tarski-style semantics for classical logic. Important references are [32] and the handbook chapter [33], written jointly with Gabriel Sandu.[11] This approach allowed to consider effects of incomplete information in semantic games under the name of IF (independence friendly) logic [41] (see also the hints on related topics, below). Independently, Rohit Parikh [48] also characterized classical and intuitionistic truth in terms of games. More importantly for our specific context, Robin Giles [25, 26], also already in the 1970s, suggested a game based interpretation of Łukasiewicz logic that is quite different form Hintikka's game, at least at a first glance. Giles seemingly was not aware of Hintikka's (or Parikh's) semantic games, but rather referred to the work of Paul Lorenzen and Kuno Lorenz as an inspiration. Lorenzen had suggested to model constructive validity by logical dialogue games already in the late 1950s [40]. This has then been taken up by Lorenzen's student Lorenz, e.g. in [39]. While Giles's rules for (weak) conjunction, (weak) disjunction, and the standard quantifiers are indeed close to Lorenzen's so-called particle rules for **P**-**O**-dialogues, Giles's game should be classified as a semantic game, since it characterizes (graded) truth with respect to a given interpretation. As explained in Section 4, for Giles an interpretation was specified by an assignment of probabilities to dispersive experiments associated with atomic formulas. However, at least with hindsight, it is clear that Giles's game implicitly refers to standard interpretations (i.e., interpretations over the real unit interval) of Łukasiewicz logic.

Cintula and Majer [9] generalized Hintikka's game to an 'evaluation game' for Łukasiewicz logic. We have called this game explicit evaluation game or $\mathcal{E}$-game, here, in order to distinguish it from other semantic games. In fact the rules presented here in Section 3 slightly deviate from those of [9], but are easily checked to be equivalent.

The 'smooth transition' from Hintikka's classical game to a many-valued setting presented in Section 2.2 is due to [21] and [22]. The limits for Hintikka-style games described in Section 2.3 have originally been presented in [17].

Presentations of Giles's game, that are close to that in Section 4 can be found, e.g., in [13, 22]. The connection between hypersequent systems and Giles-style dialogue games for t-norm based fuzzy logics has first been sketched in [8] and is explained in some more detail in [14]. Giles did not consider strong (t-norm) conjunction. Such a rule has been motivated and defined in [19].

The generalization(s) of Giles's game presented in Section 5 have largely been lifted from the paper [22].

As already pointed out in Section 6, the connection between Giles's game and a hypersequent system for Łukasiewicz logic has already been taken up in Chapter III of this Handbook [43] by George Metcalfe, based on the paper [19]. A related treatment can be found in [15]. The material on Abelian logic presented in Section 6 is new in principle. But, given the closeness of the respective hypersequents for Łukasiewicz logic and Abelian logic (see [44, 45]), it amounts to a straightforward exercise relative to the above mentioned sources. In fact, we have deliberately refrained from a more

[11] Hintikka uses *Myself* and *Nature* as names for the players and *Verifier* and *Falsifier* for the two roles. Sometimes the identity of the players is left implicit in semantic games and the Verifier is called *Eloise* or *∃loise* or simply ∃, while the Falsifier is called *Abelard* or *∀belard* or ∀. However, as already pointed out in Section 2, we refer to the players of all semantic games considered in this chapter as *myself (I)* and *you*, distinguishing between the role of an proponent **P** and an opponent **O**, respectively. This terminology can be traced back to Giles [25] and Lorenzen [40], respectively.

thorough formalization of the corresponding game along the lines of [19], here, in order to keep the presentation focused on essential features that emphasize the close relation to Giles's game for Łukasiewicz logic.

The backtracking games ($\mathcal{B}$-games) of Section 7 have originally been presented in [18]. The logic $G_{\sim}$, briefly mentioned at the end of Subsection 7.2, is considered at various places in the literature (see, e.g., [11]), but has not yet been considered from a game semantic point of view.

Propositional random choice games ($\mathcal{R}$-games), that are the topic of Section 8, are due to [16] (where a synopsis similar to that presented here as Section 10 can be found).

The presentation of random choice rules for semi-fuzzy quantifiers in Section 9 closely follows that of [22]. The idea of using random choices of witness constants for modeling the evaluation of formulas involving semi-fuzzy quantifiers in the context of Łukasiewicz logic has first been presented in [21].

We finally mention some related topics that have not been covered in this chapter:

- As already mentioned above, Lorenzen [40] defined a dialogue game that was intended to characterize intuitionistic validity. In [12] it is shown how parallel dialogue games that model different synchronization mechanisms between underlying Lorenzen-style dialogues result in characterizations of various intermediate logics, among them Gödel logic. These parallel dialogue games are closely related to hypersequent systems and in particular lead to a game based interpretation of Avron's so-called communication rule [2].

- A game for Gödel logic that could be classified as a semantic game is implicitly presented in [20]. It proceeds by reducing claims about the relative order of truth degrees of complex formulas to claims only involving atomic formulas. Evaluating the resulting claims with respect to a given interpretation yields a semantic game. However the intended use of the game in [20] was to check the validity of formulas. For this purpose the game continues after reduction to atomic order claim in such a way that the opponent **O** wins the game if the initial formula is not valid. The game is related to the sequent-of-relations system introduced in [4].

- We have not dealt with finitely-valued logics in a systematic manner here. In fact, it is rather straightforward to describe explicit evaluation games for all logics that are specified by finite truth tables (matrices). As explained in [17], this in turn can be generalized to all logics characterized by so-called Nmatrices (nondeterministic) matrices, introduced in [3].

- A particular interesting topic that has not yet been fully explored is the relation between the game based equilibrium semantics for IF logic (independence friendly logic, see [41, 52]) and many-valued logics. IF logic results from extending Hintikka's (perfect information) game for classical logic to incomplete information, i.e., to situations where at some states not all previous moves are known to both players. The truth functions min, max, and $1 - x$ over $[0, 1]$ for conjunction, disjunction, and negation, respectively, naturally arise in this context if we look for Nash equilibria in such games. As pointed out, e.g., in [16, 54], further propositional connectives, including the operator $\pi$, treated in Section 8, can be modeled by semantic games of incomplete information.

## Acknowledgements

Work on this chapter has been supported Austrian Science Fund (FWF) projects P25417-G15 (LOGFRADIG) and I1897-N25 (MoVaQ).

The author thanks Jesse Alama, George Metcalfe, and Ondrej Majer for valuable comments on a draft version of this chapter. Also the assistance of Matthias Hofer and Christoph Roschger in catching and repairing errors is gratefully acknowledged.

## BIBLIOGRAPHY

[1] Stefano Aguzzoli, Brunella Gerla, and Vicenzo Marra. Algebras of fuzzy sets in logics based on continuous triangular norms. In *Proceedings of the 10th European Conference on Symbolic and Quantitative Approaches to Reasoning with Uncertainty (ECSQARU 2009)*, pages 875–886. Springer, 2009.

[2] Arnon Avron. Hypersequents, logical consequence and intermediate logics for concurrency. *Annals of Mathematics and Artificial Intelligence*, 4(3–4):225–248, 1991.

[3] Arnon Avron and Iddo Lev. Non-deterministic multiple-valued structures. *Journal of Logic and Computation*, 15(3):241–261, 2005.

[4] Matthias Baaz and Christian G. Fermüller. Analytic calculi for projective logics. In Neil V. Murray, editor, *Proceedings of TABLEAUX 1999*, volume 1617 of *Lecture Notes in Computer Science*, pages 36–50, Berlin, 1999. Springer.

[5] Jon Barwise and Robin Cooper. Generalized quantifiers and natural language. *Linguistics and Philosophy*, 4(2):159–219, 1981.

[6] A.D.C. Bennett, Jeff B. Paris, and Alena Vencovska. A new criterion for comparing fuzzy logics for uncertain reasoning. *Journal of Logic, Language and Information*, 9(1):31–63, 2000.

[7] Ettore Casari. Comparative logics and Abelian $\ell$-groups. In *Logic Colloquium '88 (Padova, 1988)*, volume 127 of *Studies in Logic and the Foundations of Mathematics*, pages 161–190. North-Holland, Amsterdam, 1989.

[8] Agata Ciabattoni, Christian G. Fermüller, and George Metcalfe. Uniform rules and dialogue games for fuzzy logics. In *Proceedings of LPAR 2004*, volume 3452 of *Lecture Notes in Artificial Intelligence*, pages 496–510. Springer, 2005.

[9] Petr Cintula and Ondrej Majer. Towards evaluation games for fuzzy logics. In Ondrej Majer, Ahti-Veikko Pietarinen, and Tero Tulenheimo, editors, *Games: Unifying Logic, Language, and Philosophy*, volume 15 of *Logic, Epistemology, and the Unity of Science*, pages 117–138. Springer, 2009.

[10] Francesc Esteva, Lluís Godo, Petr Hájek, and Franco Montagna. Hoops and fuzzy logic. *Journal of Logic and Computation*, 13(4):532–555, 2003.

[11] Francesc Esteva, Lluís Godo, Petr Hájek, and Mirko Navara. Residuated fuzzy logics with an involutive negation. *Archive for Mathematical Logic*, 39(2):103–124, 2000.

[12] Christian G. Fermüller. Parallel dialogue games and hypersequents for intermediate logics. In Marta Cialdea Mayer and Fiora Pirri, editors, *Proceedings of TABLEAUX 2003*, pages 48–64, Rome, 2003.

[13] Christian G. Fermüller. Dialogue games for many-valued logics — an overview. *Studia Logica*, 90(1): 43–68, 2008.

[14] Christian G. Fermüller. Revisiting Giles's game. In Ondrej Majer, Ahti-Veikko Pietarinen, and Tero Tulenheimo, editors, *Games: Unifying Logic, Language, and Philosophy*, volume 15 of *Logic, Epistemology, and the Unity of Science*, pages 209–227. Springer, 2009.

[15] Christian G. Fermüller. On Giles style dialogue games and hypersequent systems. In Hykel Hosni and Franco Montagna, editors, *Probability, Uncertainty and Rationality*, pages 169–195. Springer, Frankfurt, 2010.

[16] Christian G. Fermüller. Hintikka-style semantic games for fuzzy logics. In Christoph Beierle and Carlo Meghini, editors, *Foundations of Information and Knowledge Systems, Proceedings of the 8th International Symposium, FoIKS 2014, Bordeaux, 2014*, volume 8367 of *Lecture Notes in Computer Science*, pages 193–210. Springer, 2014.

[17] Christian G. Fermüller. On matrices, Nmatrices and games. To appear in *Journal of Logic and Computation,* doi: 10.1093/logcom/ext024, 2014.

[18] Christian G Fermüller. Semantic games with backtracking for fuzzy logics. In *44th International Symposium on Multiple-Valued Logic (ISMVL 2014)*, pages 38–43. IEEE, 2014.

[19] Christian G. Fermüller and George Metcalfe. Giles's game and proof theory for Łukasiewicz logic. *Studia Logica*, 92(1):27–61, 2009.

[20] Christian G Fermüller and Norbert Preining. A dialogue game for intuitionistic fuzzy logic based on comparison of degrees of truth. In *Proceedings of InTech'03, International Conference on Intelligent Technologies*, pages 142–151. Institute for Science and Technology Research and Development, Chiang Mai University, 2003.

[21] Christian G. Fermüller and Christoph Roschger. Randomized game semantics for semi-fuzzy quantifiers. In Salvatore Greco, Bernadette Bouchon-Meunier, Giulianella Coletti, Mario Fedrizzi, Benedetto Matarazzo, and Ronald R. Yager, editors, *Advances in Computational Intelligence - 14th International Conference IPMU 2012*, volume 300 of *Communications in Computer and Information Science*, pages 632–641. Springer, 2012.

[22] Christian G. Fermüller and Christoph Roschger. Randomized game semantics for semi-fuzzy quantifiers. *Logic Journal of the Interest Group of Pure and Applied Logic*, 22(3):413–439, 2014.

[23] Adriana Galli, Renato A. Lewin, and Marta Sagastume. The logic of equilibrium and Abelian lattice ordered groups. *Archive for Mathematical Logic*, 43(2):141–158, 2004.

[24] Brunella Gerla. Rational Łukasiewicz's logic and DMV-algebras. *Neural Network World*, 11(6): 579–594, 2001.

[25] Robin Giles. A non-classical logic for physics. *Studia Logica*, 33(4):399–417, 1974.

[26] Robin Giles. A formal system for fuzzy reasoning. *Fuzzy Sets and Systems*, 2:233–257, 1979.

[27] Ingo Glöckner. *Fuzzy Quantifiers: A Computational Theory*, volume 193 of *Studies in Fuzziness and Soft Computing*. Springer, 2006.

[28] Petr Hájek. *Metamathematics of Fuzzy Logic*, volume 4 of *Trends in Logic*. Kluwer, Dordrecht, 1998.

[29] Petr Hájek. On witnessed models in fuzzy logic. *Mathematical Logic Quarterly*, 53(1):66–77, 2007.

[30] Leon Henkin. Some remarks on infinitely long formulas. In *Proceedings of the Symposium on Foundations of Mathematics, Warsaw, 1959*, pages 167–183, 1961.

[31] Jaakko Hintikka. Language-games for quantifiers. In Nicholas Rescher, editor, *Studies in Logical Theory*, pages 46–72. Blackwell, Oxford, 1968. Reprinted in [32] pp. 53–82.

[32] Jaakko Hintikka. *Logic, Language-Games and Information: Kantian Themes in the Philosophy of logic*. Clarendon Press, Oxford, 1973.

[33] Jaakko Hintikka and Gabriel Sandu. Game-theoretical semantics. In Johan van Benthem and Alice ter Meulen, editors, *Handbook of Logic and Language*, pages 361–410. Elsevier and the MIT Press, Oxford/Shannon/Tokio/Cambridge (Massachusetts), 1997.

[34] Ellen Hisdal. Are grades of membership probabilities? *Fuzzy Sets and Systems*, 25(3):325–348, 1988.

[35] Wilfrid Hodges. Logic and games. In Edward N. Zalta, editor, *The Stanford Encyclopedia of Philosophy*. Spring 2013 edition, 2013.

[36] Otto Hölder. Die Axiome der Quantität und die Lehre vom Mass. *Berichte uber die Verhandlungen der Koeniglich Sachsischen Gesellschaft der Wissenschaften zu Leipzig, Mathematisch-Physikaliche Klasse*, 53:1–64, 1901.

[37] Jonathan Lawry. A voting mechanism for fuzzy logic. *International Journal of Approximate Reasoning*, 19(3-4):315–333, 1998.

[38] Yaxin Liu and Etienne E. Kerre. An overview of fuzzy quantifiers (I): Interpretations. *Fuzzy Sets and Systems*, 95(1):1–21, 1998.

[39] Kuno Lorenz. Dialogspiele als semantische Grundlage von Logikkalkülen. *Archiv für mathemathische Logik und Grundlagenforschung*, 11:32–55, 73–100, 1968.

[40] Paul Lorenzen. Logik und Agon. In *Atti Congr. Internaz. di Filosofia (Venezia, 12-18 Settembre 1958), Vol. IV*, pages 187–194. Sansoni, 1960.

[41] Allen L. Mann, Gabriel Sandu, and Merlijn Sevenster. *Independence-Friendly Logic: A Game-Theoretic Approach*. Cambridge University Press, 2011.

[42] Robert McNaughton. A theorem about infinite-valued sentential logic. *Journal of Symbolic Logic*, 16(1):1–13, 1951.

[43] George Metcalfe. Proof theory for mathematical fuzzy logic. In Petr Cintula, Petr Hájek, and Carles Noguera, editors, *Handbook of Mathematical Fuzzy Logic - Volume 1*, volume 37 of *Studies in Logic, Mathematical Logic and Foundations*, pages 209–282. College Publications, London, 2011.

[44] George Metcalfe, Nicola Olivetti, and Dov Gabbay. Sequent and hypersequent calculi for Abelian and Łukasiewicz logics. *ACM Transactions on Computational Logic*, 6(3):578–613, 2005.

[45] George Metcalfe, Nicola Olivetti, and Dov M. Gabbay. *Proof Theory for Fuzzy Logics*, volume 36 of *Applied Logic Series*. Springer, 2008.

[46] Robert K. Meyer and John K. Slaney. Abelian logic from A to Z. In Graham Priest, Richard Routley, and Jean Norman, editors, *Paraconsistent Logic: Essays on the Inconsistent*, pages 245–288. Philosophia Verlag, 1989.
[47] Hung T. Nguyen and Elbert A. Walker. *A First Course in Fuzzy Logic*. Chapman and Hall/CRC, Boca Raton, FL, second edition, 2000.
[48] Rohit Parikh. D-structures and their semantics. *Notices of the American Mathematical Society*, 19:A329, 1972.
[49] Jeff Paris. A semantics for fuzzy logic. *Soft Computing*, 1:143–147, 1997.
[50] Jeff Paris. Semantics for fuzzy logic supporting truth functionality. In Vilém Novák and Irina Perfilieva, editors, *Discovering the World with Fuzzy Logic*, volume 57 of *Studies in Fuzziness and Soft Computing*, pages 82–104. Springer, Heidelberg, 2000.
[51] Bruno Scarpellini. Die Nichtaxiomatisierbarkeit des unendlichwertigen Prädikatenkalküls von Łukasiewicz. *Journal of Symbolic Logic*, 27(2):159–170, 1962.
[52] Merlijn Sevenster and Gabriel Sandu. Equilibrium semantics of languages of imperfect information. *Annals of Pure and Applied Logic*, 161(5):618–631, 2010.
[53] Stewart Shapiro. *Vagueness in Context*. Oxford University Press, Oxford, 2006.
[54] Xuefeng Wen and Shier Ju. Semantic games with chance moves revisited: From IF logic to partial logic. *Synthese*, 190(9):1605–1620, 2013.
[55] Lotfi A. Zadeh. A computational approach to fuzzy quantifiers in natural languages. *Computers and Mathematics*, 9:149–184, 1983.

CHRISTIAN G. FERMÜLLER
Theory and Logic Group 185.2
Vienna University of Technology
Favoritenstraße 9-11
1040 Vienna, Austria
Email: chrisf@logic.at

# Chapter XIV: Ulam–Rényi Game Based Semantics for Fuzzy Logics

FERDINANDO CICALESE AND FRANCO MONTAGNA

## 1 Introduction

Several game semantics for (classical or non-classical) logics are based on the following idea (see for instance [28]): given a logic L, we associate a game $G_\phi$ between two players I and II with each formula $\phi$ of L, where I tries to defend $\phi$ and II tries to attack $\phi$. The game (more precisely, the set of all games of the form $G_\phi$) is appropriate for the logic L whenever I has a winning strategy precisely in those games $G_\phi$, where $\phi$ is a valid formula according to L. In this way, formulas are interpreted as games and proofs are interpreted as winning strategies. This idea has been carried further also for Fuzzy Logic: e.g. in [8], the authors introduce three complete game semantics for Łukasiewicz, Gödel and product logic, based on Giles's game [22]. A cut-free proof system is associated with these games and the proof search turns out to be in Co-NP. Very interestingly, the rules are common to the three logics, only the axioms change.

A different approach is pursued, e.g. in [15], where the authors discuss games for evaluating formulas with respect to given models in Łukasiewicz logic. Such 'semantic games' generalize Hintikka's game semantic for classical logic [27] and are in fact the topic of Chapter XIII of this Handbook. Although this line of research is certainly interesting from a logical point of view, it is probably less interesting from a game-theoretic perspective, since the corresponding games are specifically tailored to characterize the logics in question.

The Ulam–Rényi game, see [39, 42] constitutes a quite different type of game that is of interest also quite independent of any logical concerns. Here, one of the players chooses a number from a discrete finite set and the other player tries to identify the first player's number by asking yes-no questions. In up to a fixed number of answers the first player can lie and only the upper bound on the number of lies allowed is known to the player asking the questions. As will become clear, the process of recording the information in these games obeys many-valued logic. But the game is quite interesting *per se* as it is also strongly related to the theory of error correcting codes (see [4, 13, 29]). Its classical version (without lies) constitutes a complete game semantics for classical logic, while its version with a finite number of lies constitutes a complete semantics for Łukasiewicz logic, see [31] and [32]. In the case of Ulam–Rényi games, not only an interesting logic has an interpretation by means of an interesting game, but the logic itself is also useful to find good strategies to win the game.

The goal of this chapter is twofold: (i) to illustrate the Ulam–Rényi game with lies, its strategies, and its relationships to the handling of reliable transmission of information and (ii) to show that the Ulam–Rényi game and its variants constitute an appropriate game semantics for the four main fuzzy logics, Łukasiewicz, Gödel, product, and BL.

The paper is organized as follows: in Section 2 we introduce all necessary logical and algebraic background. In Section 3 we illustrate the Ulam–Rényi game and prove that it constitutes an appropriate game semantics for Łukasiewicz logic. In Section 4 we discuss several game strategies and their costs. In any strategy, the number of questions cannot be smaller than the theoretical minimum given by the *sphere packing bound*. Remarkably enough, for all sufficiently large search spaces, this theoretical minimum can be attained. Then we introduce the variants of the game with asymmetric or half lies, where Responder can lie if the correct answer to the question is YES. The combinatorics of the Ulam–Rényi game with half-lies is also discussed.

In the last sections we discuss variants of the Ulam–Rényi game which constitute an appropriate game semantics for Gödel, product, and BL logics. We first note that the asymmetric Ulam–Rényi game can be analyzed in terms of a multichannel variant of the classical problem of reliable communication: in this multichannel variant, signals travel through different channels with different levels of reliability and different costs. We present the multichannel Ulam–Rényi game proposed in [13], which constitutes a game semantics for BL. Next we investigate game semantics for Gödel and product logic. Although BL is a common sublogic of Łukasiewicz, Gödel, and product logics, the multichannel semantics proposed in [13] extends the game semantics for Łukasiewicz logic but does not include a semantics for Gödel logic or for product logic as a special case. The basic idea for the game semantics of Gödel logic is that all channels are either always reliable or always unreliable, but the probability for a channel to be reliable depends on the channel (in particular, there is a channel which is reliable with probability 1). For product logic, we use an asymmetric channel in which only positive answers may be distorted with a given probability, depending on the price paid by Questioner. Finally, we present a variant of the Cicalese–Mundici multichannel game for BL, which avoids the restriction of increasingly noisy channels and has the semantics for Gödel, product, and Łukasiewicz logics as special cases.

## 2 Preliminaries

The logics investigated in this chapter, i.e., BL, SBL, Gödel logic G, Łukasiewicz logic Ł, and product logic Π, have been introduced in Chapter I of this Handbook . We now review their algebraic semantics, consisting of the variety of BL-algebras, SBL-algebras, Gödel algebras, Wajsberg algebras, and product algebras, respectively, for details see Chapter V of this Handbook.

DEFINITION 2.0.1. *A* BL -algebra *is an algebra* $\mathbf{A} = \langle A, \cdot, \rightarrow, \vee, \wedge, 0, 1\rangle$ *such that:*

*(1)* $\langle A, \cdot, 1\rangle$ *is a commutative monoid.*

*(2)* $\langle L, \vee, \wedge\rangle$ *is a lattice with bottom* 0 *and top* 1.

*(3)* $\rightarrow$ *is a binary operation satisfying the residuation, divisibility, and prelinearity properties:*

| | |
|---|---|
| *(res)* | $x \cdot y \leq z$ *iff* $x \leq y \to z$ |
| *(div)* | $x \cdot (x \to y) = x \wedge y$ |
| *(prel)* | $(x \to y) \vee (y \to x) = 1.$ |

*A* BL-*algebra is said to be:*

- *An* SBL-algebra *if it satisfies ($\neg x$ is an abbreviation for $x \to 0$):*

  *(sn)* $\neg x \vee \neg\neg x = 1.$

- *A* product algebra *if it satisfies:*

  ($\Pi$) $\neg x \vee ((x \to (x \cdot y)) \to y) = 1.$

- *A* Gödel algebra *if it satisfies:*

  *(contr)* $x \cdot x = x.$

- *A* Wajsberg algebra *if it satisfies:*

  *(dn)* $\neg\neg x = x.$

In a BL-algebra, meet and join are definable in terms of $\cdot$ and $\to$ by $x \wedge y = x \cdot (x \to y)$ and $x \vee y = ((x \to y) \to y) \wedge ((y \to x) \to x)$. Hence, we may refer only to operations $\cdot$ and $\to$ and to the constant 0. Wajsberg algebras are term equivalent to MV-*algebras*, that is, algebras $\langle A, \oplus, \neg, 0\rangle$ where:

(1) $\langle A, \oplus, 0\rangle$ is a commutative monoid and

(2) $\neg\neg x = x$, $x \oplus \neg 0 = \neg 0$ and $y \oplus (\neg(\neg x \oplus y)) = x \oplus (\neg(\neg y \oplus x))$.

modulo the following translations: $\neg x = x \to 0$, $x \oplus y = \neg x \to y$, and $x \cdot y = \neg(\neg x \oplus \neg y)$, $x \to y = \neg x \oplus y$, respectively. Hence, in the following we identify MV-algebras and Wajsberg algebras. Let us also note that in the literature, the monoid operation of an MV-algebra is often denoted by $\odot$ instead of $\cdot$.

The classes of BL-algebras, SBL-algebras, MV-algebras (or Wajsberg algebras), product algebras, and Gödel algebras constitute varieties of algebras, which will be denoted by $\mathbb{BL}$, $\mathbb{SBL}$ $\mathbb{MV}$, $\mathbb{P}$, and $\mathbb{G}$, respectively.

It is well known that if $\boldsymbol{A}$ is a member of any subvariety $\mathbb{V}$ of $\mathbb{BL}$, then $\boldsymbol{A}$ is a subdirect product of totally ordered algebras in $\mathbb{V}$ (also called $\mathbb{V}$-*chains* in the sequel).

Moreover the varieties $\mathbb{MV}$, $\mathbb{P}$, and $\mathbb{G}$ are generated as quasivarieties by the algebra $[0,1]_Ł = \langle [0,1], \cdot_Ł, \to_Ł, \max, \min, 0, 1\rangle$, $[0,1]_\Pi = \langle [0,1], \cdot_\Pi, \to_\Pi, \max, \min, 0, 1\rangle$, and $[0,1]_G = \langle [0,1], \cdot_G, \to_G, \max, \min, 0, 1\rangle$, respectively [23, 25], where:

- $x \cdot_Ł y = \max\{x + y - 1, 0\}$ and $x \to_Ł y = \min\{1 - x + y, 1\}$.
- $\cdot_\Pi$ is the ordinary product on $[0,1]$ and $x \to_\Pi y = 1$ if $x \leq y$ and $x \to_\Pi y = \frac{y}{x}$ otherwise.
- $x \cdot_G y = \min\{x, y\}$ and $x \to_G y = 1$ if $x \leq y$ and $x \to_G y = y$ otherwise.

Hence, every quasiequation which is true in $[0,1]_Ł$ (in $[0,1]_\Pi$ or in $[0,1]_G$ respectively) is true in every Wajsberg algebra (resp., in product algebra or Gödel algebra).

For every natural number $n$, there is a unique (up to isomorphism) Wajsberg chain with $n+1$ elements, namely the subalgebra of $[0,1]_{Ł}$ with domain $\{0, \frac{1}{n}, \ldots, \frac{n-1}{n}, 1\}$. This algebra will be denoted by $\mathbf{Ł}_n$. Note that $\mathbf{Ł}_n$ is a subalgebra of $\mathbf{Ł}_m$ if and only if $n$ divides $m$.

Given a BL-algebra $\boldsymbol{A}$, a *valuation* into $\boldsymbol{A}$ is a homomorphism $v$ from the algebra of formulas of BL into $\boldsymbol{A}$. A formula $\phi$ is said to be *valid* in a class $\mathbb{K}$ of BL-algebras iff for every $\boldsymbol{A} \in \mathbb{K}$ and for every valuation $v$ into $\boldsymbol{A}$, $v(\phi) = 1$. Then we may define BL, SBL, Ł, G, and Π, as the logics whose theorems are precisely the formulas valid in all BL-algebras (resp., SBL-algebras, Wajsberg algebras, Gödel algebras and product algebras). Note that all these logics are algebraizable in the sense of [6]. This allows us to work inside the varieties $\mathbb{BL}$, $\mathbb{SBL}$, $\mathbb{MV}$, $\mathbb{P}$, and $\mathbb{G}$ instead of working in the corresponding logics. For instance, if L is any of BL, SBL, Ł, G, or Π, then L is complete with respect to a class $\mathbb{K}$ of L-algebras iff $\mathbb{K}$ generates the whole variety of L-algebras.

DEFINITION 2.0.2. *A* Wajsberg hoop *is an algebra* $\boldsymbol{H} = \langle H, \cdot, \rightarrow, 1\rangle$ *such that:*

*(1)* $\langle H, \cdot, 1\rangle$ *is a commutative monoid.*

*(2) Letting* $x \leq y$ *iff* $x \rightarrow y = 1$, $x \wedge y = x \cdot (x \rightarrow y)$, *and* $x \vee y = (x \rightarrow y) \rightarrow y$, $(H, \vee, \wedge)$ *is a lattice with top element* 1, *whose associated order is* $\leq$.

*(3) The residuation property* $x \cdot y \leq z$ *iff* $x \leq y \rightarrow z$ *holds.*

*A* cancellative hoop *is a Wajsberg hoop satisfying* $x \rightarrow (x \cdot y) = y$.

In [5] it is shown that Wajsberg hoops are precisely the subreducts of Wajsberg algebras relative to the language $\{\cdot, \rightarrow, 1\}$. Moreover, a Wajsberg hoop is cancellative iff it satisfies the cancellation law: $x \cdot y = x \cdot z$ implies $y = z$. Finally, a finite Wajsberg hoop is the reduct of a finite Wajsberg algebra. The hoop reduct of the $(n+1)$-element Wajsberg chain, $\mathbf{Ł}_n$, will be denoted by $\boldsymbol{W}_n$.

DEFINITION 2.0.3. *Let* $\langle I, \leq\rangle$ *be a finite and totally ordered set with minimum* $i_0$. *For all* $i \in I \backslash \{i_0\}$, *let* $\boldsymbol{A}_i$ *be a totally ordered Wajsberg hoop and let* $\boldsymbol{A}_{i_0}$ *be a totally ordered Wajsberg algebra. Suppose further that for* $i \neq j \in I$, $A_i \cap A_j = \{1\}$ *and let* 0 *be the minimum of* $\boldsymbol{A}_{i_0}$ *(if this condition is not satisfied we replace* $\boldsymbol{A}_i$ *by an isomorphic copy satisfying it). Then* $\bigoplus_{i\in I} \boldsymbol{A}_i$ *(the* ordinal sum of the family $\langle \boldsymbol{A}_i\rangle_{i\in I}$*) is the structure whose base set is* $\bigcup_{i\in I} A_i$, *whose bottom is* 0, *whose top is* 1, *and whose operations are*

$$x \rightarrow y = \begin{cases} x \rightarrow_{\boldsymbol{A}_i} y & \text{if } x, y \in A_i \\ y & \text{if } \exists i > j\ (x \in A_i \text{ and } y \in A_j) \\ 1 & \text{if } \exists i < j\ (x \in A_i \setminus \{1\} \text{ and } y \in A_j) \end{cases}$$

*and*

$$x \cdot y = \begin{cases} x \cdot_{\boldsymbol{A}_i} y & \text{if } x, y \in A_i \\ x & \text{if } \exists i < j\ (x \in A_i \backslash \{1\},\ y \in A_j) \\ y & \text{if } \exists i < j\ (y \in A_i \backslash \{1\},\ x \in A_j). \end{cases}$$

*The algebras* $\boldsymbol{A}_i$ *are called the* Wajsberg components *of* $\bigoplus_{i\in I} \boldsymbol{A}_i$ *and* $\boldsymbol{A}_{i_0}$ *is called* the first component *of* $\bigoplus_{i\in I} \boldsymbol{A}_i$.

THEOREM 2.0.4. *Every* BL*-chain is the ordinal sum of a family* $\langle \boldsymbol{A}_i \rangle_{i \in I}$*, where* $I$ *is a totally ordered set with minimum* $i_0$*, for* $i \in I \setminus \{i_0\}$*,* $\boldsymbol{A}_i$ *is a totally ordered Wajsberg hoop, and* $\boldsymbol{A}_{i_0}$ *is a totally ordered Wajsberg algebra.*

*In particular, (i) any product chain is the ordinal sum of the two-element Wajsberg algebra and of a totally ordered cancellative hoop and (ii) any Gödel chain is the ordinal sum whose first component is (isomorphic to) the two-element Wajsberg algebra* $\mathbf{Ł}_1$ *and whose other components are isomorphic to* $\boldsymbol{W}_1$ *(the Wajsberg hoop with two elements).*

THEOREM 2.0.5.

- BL *is sound and complete with respect to the class* $\mathbb{BL}_{\rm fin}$ *of all finite ordinal sums of the form* $\mathbf{Ł}_{n_0} \oplus \boldsymbol{W}_{n_1} \oplus \ldots \oplus \boldsymbol{W}_{n_k}$ *where* $\mathbf{Ł}_{n_0}$ *is a finite Wajsberg chain and other components are finite totally ordered Wajsberg hoops. Hence, any formula* $\phi$ *is a theorem of* BL *iff for every valuation* $v$ *into any algebra* $\boldsymbol{A} \in \mathbb{BL}_{\rm fin}$ *one has* $v(\phi) = 1$.
- SBL *is complete with respect to the class* $\mathbb{SBL}_{\rm fin}$ *of all finite ordinal sums of the form* $\mathbf{Ł}_1 \oplus \boldsymbol{W}_{n_1} \oplus \ldots \oplus \boldsymbol{W}_{n_k}$*, where* $\mathbf{Ł}_1$ *is the two-element Wajsberg algebra and the other components are totally ordered finite Wajsberg hoops.*
- G *is complete with respect to the class* $\mathbb{G}_{\rm fin}$ *of all finite Gödel chains.*
- Ł *is complete with respect to the class* $\mathbb{W}_{\rm fin}$ *of all finite Wajsberg chains.*
- $\Pi$ *is complete with respect to any class of product algebras containing an infinite product chain.*

REMARK 2.0.6. *If* $\boldsymbol{A}_i$ *and* $\boldsymbol{B}_i$ *are* BL*-algebras* ($i = 0, \ldots, n$)*, then from* $\boldsymbol{A}_i \subseteq \boldsymbol{B}_i$ *we easily get that* $\bigoplus_{i=0}^{n} \boldsymbol{A}_i \subseteq \bigoplus_{i=0}^{n} \boldsymbol{B}_i$*. It follows that any finite* BL*-chain of the form* $\mathbf{Ł}_{n_0} \oplus \boldsymbol{W}_{n_1} \oplus \ldots \oplus \boldsymbol{W}_{n_k}$ *is a subalgebra of* $\mathbf{Ł}_{m_0} \oplus \boldsymbol{W}_{m_1} \oplus \ldots \oplus \boldsymbol{W}_{m_k}$*, where* $m_0 = \max\{n_0, 2\}$ *and for* $0 < i \leq k$*,* $m_i = \prod_{j \leq i} \max\{n_j, 2\}$*. As a consequence,* BL *is complete with respect to the class of finite ordinal sums of the form* $\mathbf{Ł}_{m_0} \oplus \boldsymbol{W}_{m_1} \oplus \ldots \oplus \boldsymbol{W}_{m_k}$ *with* $m_0 < m_1 < \ldots < m_k$.

REMARK 2.0.7. *Actually,* BL *is also complete with respect to the class* $\mathbb{BL}_{\rm fin\Pi}$ *of finite ordinal sums whose components are either finite totally ordered Wajsberg hoops or product chains, thought of as ordinal sums of a two-element Wajsberg hoop and a totally ordered cancellative hoop. Likewise,* SBL *is complete with respect to the class of finite ordinal sums whose first component is the two-element Wajsberg chain and whose other components are either finite and totally ordered Wajsberg hoops or product chains. This semantics includes a complete semantics for* Ł *(when only the first component is present), for* $\Pi$ *(when only a product component is present, thought of as the ordinal sum of* $\mathbf{2}$ *and a totally ordered cancellative hoop), and for* G *(when all components, including the first component, have two elements).*

# 3 Ulam–Rényi game

## 3.1 Ulam–Rényi game without lies

The game is played by two contestants: *Questioner* and *Responder*. At the outset, a natural number $N > 1$ is fixed and known to both players. Then, Responder chooses one of the numbers in the set $S = \{1, 2, \ldots, N\}$. $S$ is referred to as the search or secret space and the number chosen by Responder is referred to as the secret and denoted by $x_{\text{secret}}$. As suggested by the name, $x_{\text{secret}}$ is not revealed to Questioner. In fact, Questioner's role is to identify $x_{\text{secret}}$ by only asking YES–NO questions. The game without lies is known as the *Twenty Question game*. It is familiar to all our readers.

As a warm up, let us analyze some simple strategies for Questioner.

**Strategy 1.** Ask: "Is $x_{\text{secret}} = 1$?". If the answer is YES, the game is finished. If the answer is NO, then ask: "Is $x_{\text{secret}} = 2$?". If the answer is YES, the game is finished. If the answer is NO, then ask: "Is $x_{\text{secret}} = 3$?", etc. In the worst case, when the secret is $N$, exactly $N$ questions are asked. Moreover, the expected value of the number of questions necessary to guess the secret—when each number is equally likely to be the secret—is

$$\frac{1}{N}(1 + 2 + \cdots + N) = \frac{N+1}{2}.$$

A better strategy is easy to discover. Let $\Omega$ denote the set of numbers which are candidates to be the secret because of the answers already received. Before the first question is asked we have $\Omega = S = \{1, \ldots, N\}$.

**Strategy 2.** Questioner partitions the set of candidates into two sets $X_1$ and $X_2$ as evenly as possible, i.e., $||X_1| - |X_2|| \leq 1$ (we can also assume for simplicity that $N$ is a power of 2, hence we can guarantee $|X_1| = |X_2|$ throughout the game) and asks the question "*Is* $x_{\text{secret}} \in X_1$*?*". If the answer is YES, then $\Omega = X_1$, otherwise $\Omega = X_2$. Note that as a result of this question (independently of the answer received) the set of numbers which are now candidates to be the secret has halved. After repeating the above steps $\lceil \log(N) \rceil$ times, there remains at most one candidate left—since $\frac{N}{2^{\lceil \log(N) \rceil}} \leq 1$—and Questioner can guess the secret.

It is easily shown that this strategy is the best one with respect to the maximum[1] number of questions that are asked before the secret is identified. In addition, the same strategy is also the one for which the expected value of the number of questions is minimum when the secret number $x_{\text{secret}}$ is chosen at random with uniform distribution. Indeed, this is the strategy which minimizes the expected value of the cardinality of the updated search space: If $N$ is the cardinality of the search space and Questioner asks "Is $x_{\text{secret}} \in X$?" where $X$ has cardinality $k$, the expected value of the cardinality of the updated search space is $E(k) = k \cdot \frac{k}{N} + (N-k)\frac{N-k}{N} = \frac{k^2+(N-k)^2}{N}$, which is minimal when $k = \frac{N}{2}$.

Let us now consider possible logical interpretation of the Ulam–Rényi game without lies. Each question can be identified with the set of numbers $Q \subseteq S$ for which the answer to the question is YES. Therefore, by a *question* we understand a subset of the secret space. We can then speak of question $Q$ for any $Q \subseteq S$.

If a question $Q$ is asked by Questioner and Responder answers YES (respectively NO) we will say that each $x \in Q$ (respectively $x \notin Q$) satisfies the answer to $Q$ and each $x \notin Q$ (respectively $x \in Q$) falsifies the answer. From this, we also immediately

[1] The maximum is computed over all possible choices of the secret; this analysis is referred to as *worst case analysis*.

get that a number $x$ satisfies (respectively falsifies) the answer YES (respectively NO) to a question $Q$ if and only if $x$ satisfies (respectively falsifies) the answer NO (respectively YES) to the opposite question $S \setminus Q$. This also shows that for each $Q \subseteq S$, questions $Q$ and $S \setminus Q$ are equivalent.

For the sake of a simplified description, in order to define a representation of the information exchanged by the players at any point during the game, we can focus on the answers. Each answer divides the current search space in two parts, the numbers that might be the secret and the numbers which cannot. Let $\Omega = \Omega(\overline{a})$ be the updated search space resulting from a sequence of answers $\overline{a}$. Then the information content of the answers may be expressed as follows: any elements of $\Omega$ can be the secret, any other element cannot.

Suppose, for instance, that $N = 16$ and that $\Omega = \{2, 6, 10, 14\}$. Then it is possible that the sequence of questions–answers was: "Is $x_{\text{secret}}$ even?" with answer YES and "Is $x_{\text{secret}}$ a multiple of 4?" with answer NO. But the sequence of questions–answers might also have been the following: "Is $x_{\text{secret}} = 1$?" with answer NO and "Is $x_{\text{secret}} \equiv 2 \bmod 4$?" with answer YES.

The actual listing of questions is irrelevant for the continuation of the game. What is important is that every number in $\Omega$ may be the secret and all the other numbers cannot. Hence, the *state of the game* at any stage is completely determined by $\Omega$, or equivalently, by its characteristic function

$$\omega(x) = \begin{cases} 1 & \text{if } x \in \Omega \\ 0 & \text{otherwise.} \end{cases} \tag{1}$$

The composition of two states represented by two characteristic functions $s_1(x)$ and $s_2(x)$ amounts to their infimum, i.e., the function defined by $(s_1 \wedge s_2)(x) = \min\{s_1(x), s_2(x)\}$. It is readily seen that whenever $\overline{a}_1$ and $\overline{a}_2$ are two sequences of answers represented by the states of knowledge $s_1$ and $s_2$, then the juxtaposition of $\overline{a}_1$ and $\overline{a}_2$ is represented by the pointwise infimum, $s_1 \wedge s_2$, of $s_1$ and $s_2$.

If a state $s_1$ is dominated by another state $s_2$, that is, if $s_1(x) \leq s_2(x)$ for all $x \in S$, then $s_1$ is *at least as informative as* $s_2$, in the sense that all numbers excluded by $s_2$ are also excluded by $s_1$. The constantly 0 state $\bar{0}$ is *the most informative one*, but it is too informative, because it corresponds to the inconsistent information. The constantly 1 state $\bar{1}$ is *the least informative one*, because it amounts to saying that the secret $x$ is a member of $S$—which is well known to both players from the very outset of the game.

The *meet of two states* $s_1$ and $s_2$ represents the minimal information which is obtained by joining the information contained in $s_1$ and the information contained in $s_2$. Remarkably, the *implication* $s_1 \to s_2$ *of two states* $s_1$ and $s_2$ is the state representing the smallest amount of information which added to $s_1$ yields a refinement of $s_2$. Hence, states may be combined by logical connectives. The resulting logic (that is, the set of formulas $\phi(p_1, \ldots, p_n)$ such that for any choice of states $s_1, \ldots, s_n$, $\phi(s_1, \ldots, s_n) = \bar{1}$) is classical logic. Indeed, there is an isomorphism between the Boolean algebra of subsets of $S$ and the algebra of states with join, meet, and negation, the last one defined by $\neg s = s \to \bar{0}$.

### 3.2 Ulam–Rényi game with a fixed maximum number of lies

Suppose now that Responder is allowed to lie. Clearly, it is impossible for Questioner to guess the secret if Responder can freely lie at any question. It is then stipulated that Responder can lie at most $e$ times, for some number $e \geq 0$ fixed in advance and known to both players.

It is clear that the strategies defined in the previous section do not provide enough information in this new variant of the game. As an example, let us assume $e = 1$ and the first question is: "Is $x_{\text{secret}}$ even?". If the answer is NO, Questioner cannot exclude the possibility that the secret is even, because the answer might be a lie. However, it is also clear that after this question the numbers 2 and 1 are in a different "state" with respect to their candidacy to be the secret. To see this, assume that the next question is: "Is $x_{\text{secret}} < 3$?". If the answer is NO, Questioner can now be sure that the secret is not 2. In fact, if the candidate was 2 Responder would have lied twice which is not allowed because of the lie-bound $e = 1$. On the other hand, it is still possible that the secret is 1.

A moment's reflection shows that Questioner can still effectively use a variant of Strategy 2, where each question is repeated three times. If for a question $Q$ Responder answers YES (resp. NO) at least twice, then Questioner can be sure that YES (resp. NO) is the correct answer to question $Q$ (for otherwise, Responder would have lied at least twice, which is not allowed).

This approach easily generalizes to any fixed number of possible lies. Assume that $e \geq 0$ and let Questioner's strategy be to ask all questions in Strategy 2 until either $e+1$ YES or $e+1$ NO answers have been received. Therefore, each question is asked at most $2e+1$ times. Hence, such a strategy guarantees that Questioner will be able to guess the secret with at most $(2e+1)\log N$ questions. Is this strategy best possible with respect to the maximum or expected number of questions asked by Questioner before identifying the secret? In other words, is there a better strategy ensuring that the secret is (expected to be) correctly identified with at most $q < (2e+1)\log N$ (if it is assumed that each number is the secret with probability $1/N$)?

Later we will provide answers to these and other related questions. For the moment, let us go back to the "logical machinery" underlying the evolution of the game.

### 3.3 Ulam–Rényi games and many-valued logics

In [31] Mundici proved that states (of knowledge) in Ulam–Rényi games obey the infinite-valued Łukasiewicz calculus. We will closely follow his original argument.

Like in the basic variant without lies considered in Section 3.1, at any point during the game, the "state" of knowledge of Questioner (hereafter simply referred to as *the state*) with respect to each number $x \in S$ is described by the number of answers falsified by $x$. A number $x$ is fully entitled to be the secret if it satisfies all the answers. In addition, $x$ cannot be the secret if it falsifies at least $e+1$ of Responder's answers. Besides these two 'classical' states, in the Ulam–Rényi game with $e$ lies, there is the possibility for a number to falsify $i$ questions for each $i = 1, \ldots, e$ and still be a candidate to be the secret. In other words, for each number $x$ there exist $e+1$ different states of "candidacy" to be the secret, and one not to be. It goes without saying that in a game with lies answers do not obey classical logic.

With pre-knowledge about the secret number, Questioner could attach to each answer a classical truth value and say that the answer is either *false* or *true*. However, during the game, and for the purpose of deciding his strategy this "God-given" truth-value is of little use (since it is unknown) to Questioner. The information acquired, determining his state of knowledge, is enough to assign to the proposition *the secret is* $x$ a truth-value given by the quantity $s(x)$, where the state function $s\colon x \in S \to s(x) \in \{0, \frac{1}{e+1}, \dots, \frac{e}{e+1}, 1\}$ is defined by:[2]

$$s(x) = \begin{cases} \frac{e+1-i}{e+1} & \text{iff } x \text{ falsifies exactly } i \text{ answers for some } 0 \leq i \leq e \\ 0 & \text{iff } x \text{ falsifies} \geq e+1 \text{ answers.} \end{cases} \tag{2}$$

We conclude that the logic of the game is *many-valued.* For instance, in the case of a game with only one lie allowed, besides the possibilities of being true or false, the proposition "$x$ is the secret" may be considered *true up to a lie*, i.e., $s(x) = 1/2$.

If a question is asked more than once, e.g. the question "Is the number even?" is asked twice, two opposite answers do not lead to *inconsistency*. For instance, the answers YES the secret number is even and NO the secret number is odd gives the Questioner the information that the number of possible lies is decreased by one. The connectives in the logic of the Ulam–Rényi game must be such that the *conjunction of two opposite answers need not express an unsatisfiable property* of numbers in the secret space.

Moreover, the conjunction of two answers, each saying "the secret number is even", can be more informative than a single answer. For instance, in a a game with at most one lie allowed, after the second answer, Questioner is sure that the number is even, since only one of the answers can be a lie. Technically speaking, the Ulam–Rényi game does not obey the contraction rule.

**States of knowledge are Łukasiewicz conjunctions of Post functions**

Given a set $S$, any function $f\colon S \to I_n = \{0, \frac{1}{n}, \frac{2}{n}, \dots, \frac{n-1}{n}, 1\}$, is a *Post function* (on S) *of order* $n$. Thus states of knowledge in the Ulam–Rényi game with $e$ lies are Post functions of order $e+1$.

Given a question $T$ in the Ulam–Rényi game with $e$ lies, and the corresponding answer $a$ to $T$, let us represent the information carried by the pair $\langle T, a\rangle$ by the function $f_{(T,a)}\colon S \to \{\frac{e}{e+1}, 1\}$ given by

$$\begin{array}{ll} f_{(T,a)}(x) = 1 & \text{if } (x \in T \text{ and } a = \text{yes}) \text{ or } (x \notin T \text{ and } a = \text{no}) \\ f_{(T,a)}(x) = \frac{e}{e+1} & \text{if } (x \notin T \text{ and } a = \text{yes}) \text{ or } (x \in T \text{ and } a = \text{no}). \end{array} \tag{3}$$

The Post function $f$ of Responder's answer $a$ to $T$ is given by $f = f_{(T,a)}$.

The Łukasiewicz conjunction $\odot$ and disjunction $\oplus$ are defined by

$$\begin{array}{ll} x \odot y = \max\{0, x+y-1\} & \text{for all } x, y \in [0,1] \\ x \oplus y = \min\{1, x+y\} & \text{for all } x, y \in [0,1]. \end{array}$$

The operations $\odot$ and $\oplus$ are canonically extended to functions by $(\psi \odot f)(x) = \psi(x) \odot f(x)$ and $(\psi \oplus f)(x) = \psi(x) \oplus f(x)$.

[2] The initial state is $s_{\mathtt{start}}$ such that $s_{\mathtt{start}}(x) = 1$, for each $x \in S$. A final state is any state $s$ such that $|\{x \mid s(x) > 0\}| \leq 1$. This definition allows for the situation in which no element of the secret space can be the secret, which can model the case where Responders lies more than it is allowed.

Let $s$ be the function representing the state immediately before Responder answers $a$ to the question $T$. Let $f = f_{(T,a)}$ be the Post function representing Responder's answer to $T$ and $s'$ be the state resulting from this answer.

Then

$$s' = s \odot f,$$

that is, for all $x \in S$, $s'(x) = s(x) \odot f(x)$.

We can conclude that: *The state after $n$ questions have been asked (and answered) is given by the Łukasiewicz conjunction of the Post functions of Responder's answers.*

**The Łukasiewicz sentential calculus**

For each $n = 2, 3, 4, \ldots$, the set of variables and connective symbols in the $n$-valued Łukasiewicz sentential calculus and the set of formulas are exactly the same as for the 2-valued (Boolean) calculus. Every formula $\eta$ with variables $x_1, x_2, \ldots, x_m$ represents a function $f_\eta\colon (I_n)^m \to I_n$ via the following inductive definition: Each variable $x_i$ represents the projection onto the $i$-th axis; if we know $f_\phi$ and $f_\psi$, then $f_{not\ \phi} = 1 - f_\phi$, $f_{\phi\ and\ \psi} = f_\phi \odot f_\psi$, and $f_{\phi\ or\ \psi} = f_\phi \oplus f_\psi$. Two formulas $\phi$ and $\psi$ are logically equivalent iff $f_\phi = f_\psi$. A *tautology* of the $(n+2)$-valued Łukasiewicz logic is a formula $\phi = \phi(x_1, \ldots, x_m)$, such that $f_\phi$ is constantly equal to 1 on $(I_n)^m$.

We consider $C(n,e) = \{\frac{1}{e+1}, \frac{e}{e+1}\}^n$ and $S = \{0, 1, \ldots, 2^n - 1\}$ and fix a bijection $\mu\colon S \to C(n,e)$.

The following proposition formalizes the Ulam–Rényi game with $e$ lies over the search space $S$ in the $(e+2)$-Łukasiewicz sentential calculus.

PROPOSITION 3.3.1 ([31]). *For each function $f\colon C(n,e) \to I_{e+2}$, there is a formula $\eta$ such that $f$ equals the restriction of $f_\eta$ to $C(n,e)$.*

*For each formula $\psi$, the restriction of $f_\psi$ to $C(n,e)$ is a Post function of order $e+2$ on $C(n,e)$.*

The *initial* state of knowledge corresponds to the function constantly equal to 1 over $C(n,e)$. Such a function is represented by any formula $\eta$, such that $f_\eta$ equals 1 on $C(n,e)$, for instance, a tautology. A *final* state of knowledge corresponds to a formula $\psi$ which is *uniquely satisfiable* in $C(n,e)$, that is, there exists exactly one $x \in C(n,e)$ such that $f_\psi(x) \neq 0$.

Let $T \subseteq S$ be a question. Let $T^\mu$ be the subset of $C(n,e)$ corresponding to $T$, via $\mu$. Responder's answer is in correspondence with a formula $\phi$, such that for the Post function $f_\phi$ it holds that $f_\phi(x) = 1$ for all $x \in T^\mu$ and $f_\phi(x) = \frac{e}{e+1}$, for all $x \in C(n,e) \setminus T^\mu$.

Suppose that the questions $T_1, \ldots, T_i$ have been asked and the answers $a_1, \ldots, a_i$ have been received. Since any pair $\langle T_j, a_j \rangle$ is represented by a formula $\alpha_j$ of the $(e+2)$-valued Łukasiewicz sentential calculus, there exists a formula $\phi$ in the Łukasiewicz sentential calculus, such that Questioner's state of knowledge corresponds to the restriction of the Post function $f_\phi$ to $C(n,e)$, where $\phi = \alpha_1$ and $\alpha_2 \ldots$ and $\alpha_i$.

The correspondence between Ulam–Rényi game and Łukasiewicz sentential calculus is summarized in Table 1.

| Ulam–Rényi game | Łukasiewicz logic |
|---|---|
| maximum number $e$ of lies | $e+2$ truth-values $0, \frac{1}{e+1}, \ldots, \frac{e}{e+1}, 1$ |
| search space $S = \{0,1\}^n$ | $C(n,e) = S^\mu = \{\frac{1}{e+1}, \frac{e}{e+1}\}^n$ |
| number $x \in S$ | point $\mu(x) \in C(n,e)$ |
| initial state of knowledge | tautology |
| final state of knowledge | formula uniquely satisfiable in $C(n,e)$ |
| current state of knowledge $\psi$ | formula $p$ such that $\psi^\mu = f_p$ in $C(n,e)$ |
| question $T$ | subset $T^\mu$ of $C(n,e)$ |
| positive answer $f_T$ to $T$ | formula $p$ with $(f_T)^\mu = f_p$ in $C(n,e)$ |
| state $\psi$ after answers to $T_1, T_2, \ldots, T_i$ | conjunction $p$ of corresponding formulas |
| set of excluded numbers $\psi^{-1}(0)$ | points of $C(n,e)$ falsifying $p$ |

Table 1. Ulam–Rényi game and its logical counterpart

# 4 The Combinatorics of Ulam–Rényi games

## 4.1 The fundamental bound on the strategy size

To discuss optimal strategies for the Ulam–Rényi game with lies, we will regard the current state of the game from a different perspective. Specifically, let us assume that some sequence $\psi$ of questions has been asked and their respective answers have been received.

Recall that $S$ denotes the search space. For each $i = 0, \ldots, e$, let $A_i \subseteq S$ be the set of numbers falsifying exactly $i$ answers. Clearly the $(e+1)$-tuple $\langle A_0, \ldots, A_e \rangle$ carries the same information as the truth function $s$ of Section 3.3. Therefore, we will also refer to $\langle A_0, \ldots, A_e \rangle$ as a state of the game.

THEOREM 4.1.1. *Let us suppose that Questioner, starting from a state $\sigma$ with $\sigma = \langle A_0, A_1, \ldots, A_e \rangle$, can guess the secret with $q$ questions. Then, we have*

$$q \geq \min\{i \mid \sum_{j=0}^{e} |A_j| \sum_{\ell=0}^{e-j} \binom{i}{\ell} \leq 2^i\}. \tag{4}$$

*Proof.* For each $j = 0, \ldots, e$, there are $|A_j|$ possibilities for the secret. For each $j = 0, \ldots, e$ and $x \in A_j$ Responder can choose the number $\ell \in \{0, \ldots, e-j\}$ of lies he will tell. Once $\ell$ is fixed, Responder can choose in $\binom{i}{\ell}$ ways the $\ell$ questions among the remaining $i$ to which he will answer with a lie.

Therefore, Responder has in total $\sum_{j=0}^{e} |A_j| \sum_{\ell=0}^{e-j} \binom{i}{\ell}$ many legal strategies (secret, number of lies, actual questions to which he will lie).

A strategy for Questioner can be represented by a binary tree, where each internal node represents a question and each one of the two edges stemming from an internal node $\nu$ represents a possible answer to the question asked in $\nu$. For a given strategy of Questioner, different strategies of Responder correspond to different root-to-leaf paths in the binary tree representing Questioner's strategy. Trivially, if each root-to-leaf path in a binary tree has length $\leq i$ then there are at most $2^i$ such paths. Therefore, if $i$ questions are enough to accommodate all of Responder's strategies it must hold that

$$\sum_{j=0}^{e} |A_j| \sum_{\ell=0}^{e-j} \binom{i}{\ell} \leq 2^i.$$

Since we are interested in the smallest possible number of questions that Questioner needs to ask, we have the desired result. □

Given a state $\sigma = \langle A_0, \ldots, A_e \rangle$, and an integer $q \geq 0$ we define the $q$th *weight* of $\sigma$ by $w_q(\sigma) = \sum_{j=0}^{e} |A_j| \sum_{\ell=0}^{e-j} \binom{i}{\ell}$. The *character* of $\sigma$, denoted $\mathrm{ch}(\sigma)$ is defined by

$$\mathrm{ch}(\sigma) = \min\{i \mid w_i(\sigma) \leq 2^i\}. \tag{5}$$

We have the following *conservation law*. Let $D$ be the question asked by Questioner when the state of the game is $\sigma = \langle A_0, \ldots, A_e \rangle$. Let $\sigma^{yes} = \langle A_0^{yes}, \ldots, A_e^{yes} \rangle$ and $\sigma^{no} = \langle A_0^{no}, \ldots, A_e^{no} \rangle$ be the states resulting from a *yes* or a *no* answer respectively. Then, for any integer $i \geq 1$ the following holds:

$$w_i(\sigma) = w_{i-1}(\sigma^{yes}) + w_{i-1}(\sigma^{no}). \tag{6}$$

By Theorem 4.1.1, the character of $\sigma$ yields a lower bound on the depth of any winning strategy for a game starting from $\sigma$. Since the character of a state only depends on the cardinality of the components of the state, in the following, we shall repeatedly use the notation $\mathrm{ch}(a_0, \ldots, a_e)$ with the understanding that this refers to the character of any state $\sigma = \langle A_0, \ldots, A_e \rangle$ such that for each $i = 0, \ldots, e$, it holds that $|A_i| = a_i$.

For all integers $M \geq 0$ and $e \geq 0$, let us define

$$N_{\min}(M, e) = \min\{q \mid M \sum_{i=0}^{e} \binom{q}{i} \leq 2^q\} \tag{7}$$

and let $N(M, e)$ denote the smallest number of questions that allow Questioner to identify the secret in a game played on the secret space of cardinality $M$ with e possible lies. Then, by Theorem 4.1.1 we immediately have

$$N(M, e) \geq N_{\min}(M, e). \tag{8}$$

Spencer [41] showed that for each $e \geq 0$ this inequality holds with equality for all sufficiently large $M$. We shall first present a stronger version of this result under the additional condition that the strategy must consist of two batches of non-adaptive questions. Then, we will consider a variant of the game with an additional constraint on the type of lies available to Responder. For this new game, Theorem 4.1.1 doesn't hold anymore. We shall show that even in this case it is possible to construct two-batch optimal strategies. Moreover, this variant in one of its possible generalizations will be used to provide semantics for the Hájek's BL logic.

### 4.2 The basic case — a two batch "optimal" strategy

In our analysis of the combinatorics of the Ulam–Rényi game, we shall focus on the case where the search space size is a power of two, i.e., there exists a natural number $m$ such that $M = 2^m$. Specifically, we assume that the secret belongs to the search space $S = \{0, 1, \ldots, 2^m - 1\}$. We shall prove that up to finitely many exceptional $m$, $N_{\min}(2^m, e)$ questions are sufficient. Moreover, we shall show that this is achievable even restricting our attention to strategies with questions divided into two non-adaptive batches: a first non-adaptive batch of $m$ questions and then, depending on the answers, a second batch of non-adaptive $N_{\min}(2^m, e) - m$ questions. The questions in each batch are non-adaptive in the sense that they are chosen independently of one another, and, for the questions in the second batch, only depending on the answers to the questions in the first batch.

**The first batch of questions.** For each $i = 1, 2, \ldots, m$ the $i$th question of the first batch $D_i^{(1)}$ is as follows "Is the $i$th binary digit of $x_{\text{secret}}$ equal to 1?".

Thus a number $y \in S$ belongs to $D_i^{(1)}$ iff the $i$th bit $y_i$ of its binary expansion $\vec{y} = y_1 \cdots y_m$ is equal to 1. Identifying a yes answer with the bit 1 and a no answer with the bit 0, let $b_i \in \{0, 1\}$ be the answer to question $D_i^{(1)}$. Let $\vec{b} = b_1 \cdots b_m$.

Therefore, it is not hard to see that the resulting state after the above first batch of questions is given by $\sigma^{\vec{b}} = \langle A_0, A_1, \ldots, A_e \rangle$, where $A_i$ is the set of numbers whose binary expansion differs from $\vec{b}$ in exactly $i$ bits. It follows that

$$|A_0| = 1, \quad |A_1| = m, \ldots, |A_e| = \binom{m}{e}.$$

We have $\operatorname{ch}(\sigma^{\vec{b}}) = N_{\min}(2^m, e) - m$. This can be easily shown by induction using (6) together with the fact that for each $i = 1, \ldots, m$, the question $D_i^{(1)}$ splits evenly the components of the intermediate state in which it is asked. Details can be found in [14].

Let $n = \operatorname{ch}(1, m, \binom{m}{2}, \ldots, \binom{m}{e})$. We will construct a non-adaptive strategy with $n$ questions, which is winning for the state $\sigma^{\vec{b}}$.

LEMMA 4.2.1. *Let $e \geq 1$, $m \geq 1$, and $n$ be integers such that $n = \operatorname{ch}(1, m, \ldots, \binom{m}{e})$. Then $m < \sqrt[e]{e!}\, 2^{\frac{n}{e}} + e$.*

*Proof.* Let us set $m^* = \sqrt[e]{e!}\, 2^{\frac{n}{e}} + e$. Observe that the function $\operatorname{ch}(1, m, \ldots, \binom{m}{e})$ is non-decreasing in $m$. By simple algebraic manipulation and the properties of binomial coefficients, we have

$$\sum_{j=0}^{e} \binom{m^*}{j} \sum_{\ell=0}^{e-j} \binom{n}{\ell} > \binom{m^*}{e} \frac{\left(\sqrt[e]{e!}\, 2^{n/e}\right)^e}{e!} = 2^n.$$

From this we obtain $\operatorname{ch}(1, m^*, \ldots, \binom{m^*}{e-1}) > n$. □

We now prove that for all sufficiently large $m$ there exists a second batch of $n = \operatorname{ch}(1, m, \binom{m}{2}, \ldots, \binom{m}{e})$ many non-adaptive questions allowing Questioner to infallibly guess Responder's secret number. We first need the following lemma.

LEMMA 4.2.2. *For any fixed* $e > 1$ *and all sufficiently large* $n$ *there exists* $\mathcal{C} \subseteq \{0,1\}^n$ *satisfying*

(i) $|\mathcal{C}| \geq 1 + m + \cdots + \binom{m}{e-1}$ *and*

(ii) *each two distinct sequences* $\vec{x}, \vec{y} \in \mathcal{C}$ *differ in at least* $2e$ *bits.*

*Proof.* From Lemma 4.2.1 together with the well known inequality $e! \leq \frac{(e+1)^e}{2^e}$, it follows that, for all sufficiently large $n$

$$\begin{aligned}\sum_{j=0}^{e-1}\binom{m}{j} &\leq e\binom{m}{e-1} < e(m)^{e-1} < e(\sqrt[e]{e!}\,2^{\frac{n}{e}} + e)^{e-1} \\ &\leq e(e\,2^{\frac{n}{e}})^{e-1} = e^e\,2^{n\frac{e-1}{e}} = \frac{2^n}{\frac{2^{\frac{n}{e}}}{e^e}} \leq \frac{2^n}{\sum_{j=0}^{2e}\binom{n}{j}}.\end{aligned}$$

The existence of the desired $\mathcal{C}$ is based on the following *ad hoc* variant of the well known Gilbert bound from the theory of error correcting codes [43].

Start by putting in $\mathcal{C}$ any arbitrary $\vec{x} \in \{0,1\}^n$, and then keep on adding to $\mathcal{C}$ binary sequences differing in at least $2e+1$ positions from all previously added ones. The process stops when, for each $\vec{y} \in \{0,1\}^n$, there is at least one $\vec{x} \in \mathcal{C}$ such that $\vec{y}$ differs from $\vec{x}$ in less than $2e+1$ bits. Thus the union of the set of sequences differing in at most $2e$ position from some $\vec{x} \in \mathcal{C}$ covers the whole space $\{0,1\}^n$. For each $\vec{x} \in \{0,1\}^n$, there are exactly $\sum_{j=0}^{2e}\binom{n}{j}$ elements of $\{0,1\}^n$ differing in at most $2e$ position from $\vec{x}$. It follows that $|\mathcal{C}|\,\sum_{j=0}^{2e}\binom{n}{j} \geq 2^n$, as required to complete the proof. □

Next let us fix a one-to-one a map $g_1\colon x \in \bigcup_{j=0}^{e} A_j \mapsto \vec{x} \in \mathcal{C}$, which is possible because of the cardinality of $\mathcal{C}$. Furthermore, let us fix a one-to-one map $g_2\colon x \in A_e \mapsto \vec{x} \in \{0,1\}^n$ such that for each $i = 0, \ldots, e$ $x \in A_e$ and $y \in A_i$ the two quantities $g_2(x)$ and $g_1(y)$ differ in at least $(e - i + 1)$ bits. For this we may argue as follows: for $i = 0, \ldots, e-1$ and for each $x \in A_i$ we remove from $\{0,1\}^n$ the sequence $\vec{x} = g_2(x)$ together with all the sequences which differ from it in up to $i$ bits. The remaining set is large enough to accommodate the range of $g_2$. In fact, from $\mathrm{ch}(A_0, \ldots, A_e) = n$ we get

$$|A_e| \leq 2^n - \sum_{i=0}^{e-1}|A_j|\sum_{j=0}^{i}\binom{n}{j}.$$

Finally, the (encoding) map $f\colon x \in \bigcup_{i=0}^{e} A_i \mapsto \vec{x} \in \{0,1\}^n$ is defined by stipulating that $f(x) = g_1(x)$ for each $i = 0, \ldots, e-1$ and $x \in A_i$, and $f(x) = g_2(x)$ for $x \in A_e$.

**The second batch of questions.** For each $i = 1, 2, \ldots, n$, the $i$th question $D_i^{(2)}$ is as follows: "Is the $i$th binary digit of $f(x_{\text{secret}})$ equal to 1?". Thus a number $y \in \bigcup_{i=0}^{e} A_i$ belongs to $D_i^{(2)}$ iff the $i$th bit $y_i$ of its encoding $\vec{y} = y_1 \cdots y_n = f(y)$ is equal to 1. Identifying a yes answer with the bit 1 and a no answer with the bit 0, let $b_i^{(2)} \in \{0,1\}$ be the answer to question $D_i^{(2)}$. Let $\vec{b}^{(2)} = b_1^{(2)} \cdots b_n^{(2)}$.

The secret number can now be identified by the following procedure:

(i) Let $i \in \{0, \ldots, e\}$ be such that $x_{\text{secret}} \in A_i$, then Responder can lie to at most $e - i$ questions of the second batch. Hence, $\vec{b}^{(2)}$ must differ in at most $e - i$ positions from $f(x)$.

(ii) For each $y \in A_i \setminus \{x_{\text{secret}}\}$, $f(y)$ differs in at least $2(e - i) + 1$ bits from $f(x_{\text{secret}})$—more precisely, if $i \neq e$ the difference is $2e + 1$ and if $i = e$, the difference is at least 1.

(iii) For each $j \neq i$ and $y \in A_j$, $f(y)$ differs in at least $2e - i - j + 1$ bits from $f(x_{\text{secret}})$—more precisely, if $i, j \neq e$ the difference is $2e + 1$ by the property of the range of $g_1$; and if $i = e$ (respectively $j = e$) the difference is at least $e - j + 1$ (resp. $e - i + 1$) by the property of the range of $g_2$.

As a consequence, for a unique index $i \in \{0, 1, \ldots, e\}$ there is $x \in A_i$ such that $f(x)$ and $\vec{b}^{(2)}$ differ in at most $e - i$ positions. This one $x$ necessarily coincides with $x_{\text{secret}}$.

### 4.3 Half lies and asymmetric channels

In [40], a variant of the Ulam–Rényi game was considered where a lie is only possible if the truthful answer to the question is YES. Therefore only Responder's NOs can be lies while all the YES answers are necessarily correct.

This variant is better explained if we momentarily change the setting of the game: Responder's answers are delivered to Questioner as bits (0 for NO and 1 for YES) via a noisy channel, which may alter up to $e$ of the bits transmitted. Responder is always sincere and mendacious answers are due to the effect of noise on the channel.[3]

The asymmetric variant of the game corresponds to the case where bits can be corrupted only by a $1 \mapsto 0$ transformation. This is typical of optical channels [37] where a bit 1 is encoded by the presence of a photon, and a bit 0 by the absence of a photon. Photons can be lost but spurious photons cannot be created. Therefore the only possible type of error in this channel occurs when answer 1 is received as 0.

This asymmetric type of channel is known as the *Z-channel* (see [29, 43]. Theorem 4.1.1 and its consequences (7) and (8) do not seem to have an immediate counterpart for the asymmetric variant. This fact makes optimization of encodings in asymmetric channels much more challenging than for symmetric coding. As a consequence, exact results for the Ulam–Rényi game with half lies are only known for the case $e = 1$ [11]. The techniques introduced in [11] have been subsequently generalized in [1, 2, 20], leading to some asymptotic characterization of perfect strategies for the Ulam–Rényi game.

### 4.4 The combinatorics of the half-lie Ulam–Rényi game

For the $Z$-channel with $e$ errors we have the following result [11, 20]: For all sufficiently large $q$ and for all sets $S$, there exists an encoding of length $q$ for $S$ iff

$$|S| \leq \frac{2^{q+e}}{\binom{q}{e}}. \tag{9}$$

[3] We are not discussing here the way questions can be efficiently and reliably delivered to Responder.

Although (7) and (8) may look similar to (9), the latter inequality does not yield a workable notion of a "weight function" on states. Recall that the weight function in the symmetric case is the basis of a conservation law and both of the lower bound and of the theoretical tools for the construction of the strategy. The difficulty in defining an appropriate weight function and a conservation law has been the main hurdle towards the characterization of optimal strategies for the asymmetric Ulam–Rényi game. Only very recently, besides the already mentioned [11], Dumitriu and Spencer [20] and Cicalese *et al.* [1], provided more comprehensive and general results. Here we will only outline the main ingredients necessary to the characterization of optimal strategies. We shall closely follow [11] and [20].

Let us first show that (9) provides a lower bound on the size of a strategy for the Ulam–Rényi game with $e$ half lies (equivalently, on the $Z$-channel with $e$ errors).

As in the proof of Theorem 4.1.1, a strategy with $q$ questions can be represented by a binary tree of height $q$ (nodes are questions and edges are answers) so that different strategies of Responder correspond to different root-to-leaf paths. In particular, for each $x \in S$ exactly one path is determined by the Responder's choice of $x$ as the secret if he decided never to lie. Let us refer to this one path as the *sincere path for* $x$. For each YES edge on this path, Responder may well choose to mendaciously answer NO. In this way we enter a new path (called a *one-lie path for* $x$), ending on a new leaf, in correspondence with Responder's decision to lie exactly once. Similarly, for each one-lie path for $x$ and YES edge in this path not belonging to the sincere path, a new path is determined if Responder decides to lie twice. Proceeding in this way for each $i \leq e$, we define *$i$-lie paths for* $x$. The number of all these paths depends on the number of YES edges in the sincere path for $x$ and on the number of YES edges on the one-lie paths for $x$ and so on.

Asymptotically with $q$, in a binary tree of height $q$, almost all paths contain an almost equal number of left and right branches. It follows that, for each $x \in S$ there are $\binom{q/2}{e}$ possible $e$-lie paths for $x$ in $\Sigma$, i.e., paths leading to a state of the form $\langle \emptyset, \dots, \emptyset, \{x\} \rangle$. Therefore a winning strategy $\Sigma$ must contain at least $|S|\binom{q/2}{e} \sim \frac{|S|}{2^e}\binom{q}{e}$ different paths. We have thus proved (9).

Let us now turn to the upper bound. We need to show that for any fixed $e$, arbitrarily large $q$, and all sets $S$ satisfying (9), Questioner has a winning strategy of size $q$ in a Ulam–Rényi game with $e$ half lies over the search space $S$. Equivalently, there exists an $e$-error correcting encoding for $S$ of length $q$ on the $Z$-channel (for a detailed account of the argument, we refer the reader to [1, 11, 20]).

In the basic (symmetric) Ulam–Rényi game a winning strategy achieving the lower bound was based on questions guaranteeing a strict decrease of the character of the states encountered during the game. In the present asymmetric variant, we will rely on a similar idea. We say that a question *splits evenly* a state $\langle S_0, S_1, \dots, S_e \rangle$ iff the two resulting states $\langle S_0^{yes}, S_1^{yes}, \dots, S_e^{yes} \rangle$ and $\langle S_0^{no}, S_1^{no}, \dots, S_e^{no} \rangle$, respectively arising from a YES or a NO answer, are termwise equal.

Let $\langle S_0, S_1, \dots, S_e \rangle$ be an arbitrary state of the game. It is easy to check that any feasible question $Q$ is even splitting whenever

$$|Q \cap S_i| = \sum_{j=0}^{i} \frac{|S_j|}{2^{i-j+1}} \tag{10}$$

In case $|S| = 2^m$ the initial state $\langle S_0^{(0)}, S_1^{(0)}, \ldots, S_e^{(0)} \rangle$ satisfies $|S_0^{(0)}| = 2^m$, and $|S_i^{(0)}| = 0$, for $i = 1, \ldots, e$. It is not hard to see that for each $j = 0, 1, \ldots, m - e - 1$, Questioner can choose his $(j+1)$th question $Q^{(j+1)}$ according to the following slightly modified version of (10):

$$\left| Q^{(j+1)} \cap S_i^{(j)} \right| = \min \left\{ \sum_{j=0}^{i} \frac{|S_j^{(j)}|}{2^{i-j+1}}, |S_i^{(j)}| \right\}, \tag{11}$$

where $\langle S_0^{(j)}, S_1^{(j)}, \ldots, S_e^{(j)} \rangle$ is Questioner's state resulting from the first $j$ answers.

Let $m' = m - e$. Then [1, Lemma 4] (see also [20, Lemma 3.7]) there are positive constants $c_0, c_1, \ldots, c_e$ such that, for all sufficiently large $q$, Questioner's state $\langle S_0^{(m')}, S_1^{(m')}, \ldots, S_e^{(m')} \rangle$ satisfies, for each $i = 0, \ldots, e$:

$$\left| S_i^{(m')} \right| \leq c_i \binom{m'}{i}$$

As shown in [1, 14, 20] this state has a Gilbert-like encoding, that can be used by Questioner to find the secret with $q$ questions. Specifically, the same strategy described above for the symmetric channel works well here. Since search on a symmetric channel is not easier than on the $Z$-channel, Questioner can successfully use Gilbert packing for this part of his strategy on the asymmetric channel.

# 5 Multichannel Ulam–Rényi games and Hájek Basic Logic: preliminaries

The transmission model using the $Z$-channel can be alternatively obtained by assuming that Responder can use two channels: an expensive Channel 1 which is noiseless and a cheap Channel 2 which is $e$-noisy, i.e., at most $e$ of the bits delivered can be corrupted. It is agreed that all NO answers must be transmitted via Channel 1, and all YES answers are to be sent using Channel 2. We assume that Questioner does not know which channel carried the bits/answers delivered to him. However, due to the characteristics of the channels and due to the above rules, Questioner is still aware that only NO answers (0-bits) can be lies. Therefore, this two channel game is perfectly equivalent to the Ulam–Rényi game with half lies (over the $Z$-channel).

In the multichannel variant of the Ulam-Rényi game considered in [13] Responder can use $k+1$ increasingly noisy channels. For each $i = 0, 1, \ldots, k$ channel $i$ can corrupt up to $e_i$ answers, and it holds that $0 \leq e_0 < e_1 < \cdots < e_k$. A *question* is a subset $Q$ of the search space together with an integer indicating to Responder which channel should be used for the answer. Therefore, we can represent an *answer* as a pair $\langle j, T \rangle$ where $j$ is the channel and $T = Q$ or $T = S \setminus Q$ according to whether the answer to the question is "YES the secret number belongs to $Q$" or "NO the secret number does not belong to Q", respectively.

Moreover, for any $i < j$, if an element $z \in S$ falsifies an answer $\langle i, T \rangle$, i.e., $z \notin T$, then any other information about $z$ delivered via channel $j$ has no value. Quoting [13], "negative information via channel $i$ pointwise supersedes all information given by the noisier channel $j$".

Given a record $\overline{a}$ of answers $\langle j_i, T_i \rangle$ for $i = 1, \ldots, q$, the state of the game is given by the truth-value function $s_{\overline{a}}$ assigning to each element $z \in S$ a truth-value which is a pair $\langle j, r \rangle$ defined as follows:

- If $z$ satisfies all the answers, i.e., $z \in T_i$ for each $i = 1, \ldots, q$, then $\langle j, r \rangle = \langle k, e_k + 1 \rangle$.
- Otherwise
  - $j$ is the smallest $c \in \{0, 1, \ldots, k\}$ such that $\langle c, T \rangle \in \overline{a}$ and $z \notin T$
  - $r = \max\{0, 1 - \dfrac{\text{fals}(z,c)}{e_c + 1}\}$ where $\text{fals}(z, c)$ denotes the number of answers $\langle c, T \rangle \in \overline{a}$ such that $z \notin T$.

Let $\overline{e} = \langle e_0, e_1, \ldots, e_k \rangle$ with $e_0 < e_1 < \ldots < e_k$. After taking isomorphic copies, we can assume that for $i \neq j$ and $i, j > 0$, $L_{e_0+1} \cap W_{e_i+1} = W_{e_i+1} \cap W_{e_j+1} = \{1\}$. Further, let $L^0_{e_0+1} = \{\langle k, e_k + 1 \rangle, \langle 0, 0 \rangle, \ldots \langle 0, e_0 \rangle\}$ and for $0 < i \leq k$, $W^i_{e_i+1} = \{\langle k, e_k + 1 \rangle, \langle i, 0 \rangle, \ldots \langle i, e_i \rangle\}$. Define a function from $\boldsymbol{B} = \boldsymbol{\text{Ł}}_{e_0+1} \oplus \boldsymbol{W}_{e_1+1} \oplus \ldots \oplus \boldsymbol{W}_{e_k+1}$ onto $K_{\overline{e}} = L^0_{e_0+1} \cup W^1_{e_1+1} \cup \ldots \cup W^k_{e_k+1}$ as follows:

$$\Phi(x) = \begin{cases} \langle k, e_k + 1 \rangle & \text{if} \quad x = 1 \\ \langle i, j \rangle & \text{if} \quad x = \frac{j}{e_i+1} \in W_{e_j+1} \setminus \{1\}. \end{cases}$$

Then $\Phi$ is a bijection, and hence it induces a structure $\boldsymbol{K}_{\overline{e}}$ on $K_{\overline{e}}$, which is isomorphic to $\boldsymbol{B}$ (for each binary operation $f$ on $\boldsymbol{B}$, define for $x, y \in K_{\overline{e}}$, $f^{\boldsymbol{K}_{\overline{e}}}(x, h) = \Phi^{-1}(f(\Phi(x), \Phi(y)))$. In the sequel we will identify the algebra $\boldsymbol{K}_{\overline{e}}$ of truth-values with the algebra $\boldsymbol{B} = \boldsymbol{\text{Ł}}_{e_0+1} \oplus \boldsymbol{W}_{e_1+1} \oplus \ldots \oplus \boldsymbol{W}_{e_k+1}$. In particular, the induced order of truth-values on $\boldsymbol{K}_{\overline{e}}$ is

$$\langle j, r \rangle < \langle j', r' \rangle \ \text{ iff } \ [\text{either } j < j' \ \text{ or } \ (j = j' \ \text{ and } \ r < r')]. \tag{12}$$

Hence, if $s \in S$ has truth-value $\langle j, r \rangle$, $s' \in S$ has truth-value $\langle j', r' \rangle$ and $\langle j, r \rangle < \langle j', r' \rangle$, then either $j < j'$, and then $s$ has been falsified by an answer on a more reliable channel, or $j = j'$ and $r < r'$, in which case $s$ has been falsified more times than $s'$ by answers on the same channel $j$. This definition formalizes the condition that negative information provided over cheap noisy channels is pointwise superseded by negative information provided over low-noise expensive channels.

Given any record $\overline{a}$, its truth function $s_{\overline{a}}$ belongs to $K^S_e$. The converse also holds:

LEMMA 5.0.1. *$K^S_{\overline{e}}$ is the set of all possible truth functions of all possible records (including the illegal ones).*

*Proof.* Clearly for every record $\overline{a}$, $s_{\overline{a}}$ is a map from $S$ into $K_{\overline{e}}$. Conversely, given a function $f$ from $S$ into $K_{\overline{e}}$, let $S' = \{z \in S \mid f(z) \neq (k, e_k + 1)\}$, and let for $z \in S'$, $f(z) = \langle i_z, h_z \rangle$. Consider the record $\overline{a}$ consisting, for all $z \in S'$ of $e_{i_z} + 1 - h_z$ questions of the form: "Is $x_{\text{secret}} = z$?" on channel $i_z$, with answer NO. An easy computation shows that $s_{\overline{a}} = f$. Notice that $\overline{a}$ might be illegal. This is the case if there is an $i$ such that for every $z \in S$, $f(z) = \langle i, 0 \rangle$. Indeed, $f$ is the truth function of a record $\overline{a}$ such that every $z \in S$ has been falsified by $e_i + 1$ answers on channel $i$. But this is only possible if the number of lies on channel $i$ is greater that $e_i$. □

The isomorphism $\Phi$ defined above induces a structure $\boldsymbol{K}_{\overline{e}}^{S}$ on $K_{\overline{e}}^{S}$, which makes it isomorphic to $\boldsymbol{B}^{S}$. Of course, the partial order of truth-values induces a partial order of truth functions: $s_{\overline{a}} \leq s_{\overline{b}}$ iff for all $z \in S$, $s_{\overline{a}}(z) \leq s_{\overline{b}}(z)$.

The *juxtaposition* $\overline{a} \odot \overline{b}$ of two records $\overline{a}$ and $\overline{b}$ is the record obtained by giving each answer $a_i$ a multiplicity equal to the sum of its multiplicity in $\overline{a}$ plus its multiplicity in $\overline{b}$.

It is easy to see that if $s_{\overline{a}}(x) = \langle i, k\rangle$ and $s_{\overline{b}}(x) = \langle j, h\rangle$, then

$$s_{\overline{a}\odot\overline{b}}(x) = \begin{cases} s_{\overline{a}}(x) & \text{if} \quad i < j \\ s_{\overline{b}}(x) & \text{if} \quad j < i \\ \langle i, \max\{0, e_i + 1 - k - h\}\rangle & \text{if} \quad i = j. \end{cases}$$

That is, up to isomorphism, $s_{\overline{a}\odot\overline{b}}$ is the pointwise product in $\boldsymbol{B}^{S} = \boldsymbol{K}_{\overline{e}}^{S}$ of $s_{\overline{a}}$ and $s_{\overline{b}}$. It follows that the truth function of the juxtaposition, $\overline{a} \odot \overline{b}$, of records $\overline{a}$ and $\overline{b}$, is the product $s_{\overline{a}} \cdot_{\boldsymbol{K}_{\overline{e}}^{S}} s_{\overline{b}}$ of the truth function of $\overline{a}$ and the truth function of $\overline{b}$ in $\boldsymbol{K}_{\overline{e}}^{S}$. As in any direct product of residuated lattices, the pointwise implication $\rightarrow_{\boldsymbol{K}_{\overline{e}}^{S}}$ in $\boldsymbol{K}_{\overline{e}}^{S}$ is then the residual of $\cdot_{\boldsymbol{K}_{\overline{e}}^{S}}$. Hence, the pointwise implication, $s_{\overline{a}} \rightarrow_{\boldsymbol{K}_{\overline{e}}^{S}} s_{\overline{b}}$ of two truth functions $s_{\overline{a}}$ and $s_{\overline{b}}$, is the maximum truth function $s_{\overline{c}}$ such that $s_{\overline{a}\odot\overline{c}} = s_{\overline{a}} \cdot_{\boldsymbol{K}_{\overline{e}}^{S}} s_{\overline{c}} \leq s_{\overline{b}}$. It represents the minimum information $\overline{c}$ that must be added to $\overline{a}$ in order to get a refinement of $\overline{b}$.

The representation of records can be given by means of equivalence classes. As in the error-free case, two records $\overline{a}$ and $\overline{b}$ are said to be *equivalent* iff they assign the same truth-value to each $z \in S$, in symbols,

$$\overline{a} \equiv \overline{b} \text{ iff } \forall z \in S \ s_{\overline{a}}(z) = s_{\overline{b}}(z). \tag{13}$$

We denote by $[\overline{a}]$ the equivalence class of $\overline{a}$ and by $[K_{\overline{e},S}]$ the set of equivalence classes of records modulo $\equiv$. Clearly, the map $s_{\overline{a}} \mapsto [\overline{a}]$ is a bijection from $K_{\overline{e}}^{S}$ onto $[K_{\overline{e},S}]$, and hence it induces a BL-algebra on $[K_{\overline{e},S}]$ isomorphic to $\boldsymbol{K}_{\overline{e}}^{S}$ and denoted by $[\boldsymbol{K}_{\overline{e},S}]$. The *initial* state of knowledge 1 is the equivalence class of the empty record (no answer received). It assigns truth value $\langle k, e_k + 1\rangle$ to each $z \in S$.

Given states $[\overline{a}]$ and $[\overline{b}]$ we write $[\overline{a}] \leq [\overline{b}]$ (read: "$[\overline{a}]$ is *more restrictive* than $[\overline{b}]$") iff $s_{\overline{a}}(z) \leq s_{\overline{b}}(z) \ \forall z \in S$.

From the isomorphism between $\boldsymbol{K}_{\overline{e}}^{S}$ and $[\boldsymbol{K}_{\overline{e},S}]$ we have the following.

PROPOSITION 5.0.2.

*(i) The binary relation $\leq$ is a partial order over $[K_{\overline{e},S}]$.*

*(ii) Juxtaposition $\odot$ of records canonically equips the set $[K_{\overline{e},S}]$ with an operation, also denoted $\odot$, given by*

$$[\overline{a}] \odot [\overline{b}] = [\overline{a} \odot \overline{b}].$$

*The $\odot$ operation is commutative, associative, and has the initial state 1 as its neutral element.*

*(iii) Given two states $[\overline{a}]$ and $[\overline{b}]$, among all $[\overline{t}] \in [K_{\overline{e},S}]$ such that $[\overline{a}] \odot [\overline{t}] \leq [\overline{b}]$ there is a least restrictive state, denoted $[\overline{a}] \rightarrow [\overline{b}]$.*

(iv) *The set* $[K_{\overline{e},S}]$, *equipped with the* $\odot$ *and* $\to$ *operations and the distinguished constants* 0 *(representing the equivalence class of an illegal record falsifying all numbers* $e_0 + 1$ *times on the safest channel* 0*) and* 1 *(representing the equivalence class of the empty record) is a* BL*-algebra.*

(v) *The partial order of* $[K_{\overline{e},S}]$ *is definable in* $[\boldsymbol{K}_{\overline{e},S}] = \langle K_{\overline{e},S}, \odot, \to, 0, 1\rangle$ *via the stipulation* $[\overline{a}] \leq [\overline{b}]$ iff $[\overline{a}] \to [\overline{b}] = 1$.

The following theorem was originally presented in [13]:

THEOREM 5.0.3. *Given* BL*-terms* $\sigma = \sigma(x_1, \ldots, x_n)$ *and* $\tau = \tau(x_1, \ldots, x_n)$, *the following conditions are equivalent for the equation* $\sigma = \tau$*:*

(i) *For every finite set* $S \neq \emptyset$ *and increasing* $k+1$*-tuple* $\overline{e} = \langle e_0 < e_1 < \ldots < e_k\rangle$ *of integers* $\geq 0$, *the equation is valid in the* BL*-algebra* $[\boldsymbol{K}_{\overline{e},S}]$ *of states in the* $k+1$*-channel Ulam–Rényi game over search space* $S$ *with* $\overline{e}$ *errors.*

(ii) *For every increasing* $k+1$*-tuple of positive integers* $\overline{e}$ *and singleton set* $\{s\}$, *the equation holds in the* BL*-algebra* $[\boldsymbol{K}_{\overline{e},\{s\}}] = \boldsymbol{K}_{\overline{e}}$.

(iii) *The equation holds in every finite ordinal sum of finite Łukasiewicz chains of increasing cardinalities.*

(iv) *The equation holds in every* BL*-algebra.*

*Proof.* (i) $\Leftrightarrow$ (ii) is trivial, because $[\boldsymbol{K}_{\overline{e},S}] = \boldsymbol{K}_{\overline{e}}^{S} = [\boldsymbol{K}_{\overline{e},\{s\}}]^S$ is a power of $[\boldsymbol{K}_{\overline{e},\{s\}}]$ and $[\boldsymbol{K}_{\overline{e},\{s\}}] = \boldsymbol{K}_{\overline{e}}$ is a quotient of $[\boldsymbol{K}_{\overline{e},S,}] = \boldsymbol{K}_{\overline{e}}^{S}$ through the projection into $s$. Hence, $[\boldsymbol{K}_{\overline{e},S}]$ and $\boldsymbol{K}_{\overline{e}}$ generate the same variety.

(ii) $\Leftrightarrow$ (iii) is also trivial because when $\overline{e}$ ranges over all increasing finite sequences of natural numbers, $\boldsymbol{K}_{\overline{e},\{s\}}$ covers all finite ordinal sums of increasingly large finite Łukasiewicz chains.

(iv) $\Rightarrow$ (iii) is trivial. The implication (iii) $\Rightarrow$ (iv) follows from Remark 2.0.6. □

# 6 A game semantics for Gödel logic

## 6.1 Gödel games

In this section we introduce a new class of multichannel Ulam–Rényi games, named Gödel games, their name referring to the fact that these games provide a sound and complete game semantics for Gödel logic G. A more detailed treatment of these games can also be found in [16].

A *Gödel game* $G(S, n)$ is defined by the search space $S = \{1, \ldots, N\}$ and a set of $n+1$ channels (for $n \geq 0$), which we will refer to by the natural numbers $0, \ldots, n$. As usual, Responder thinks of a number $x_{\text{secret}} \in S$ and Questioner has to guess it by means of YES–NO questions. In addition, Questioner can indicate the channel $i$ on which the answer will be sent. Channels have different accuracy and different costs. We assume that a probability $\alpha_i$ is associated to each channel $i$, so that each one of Responder's answers will travel on channel $i$ undistorted with probability $\alpha_i$, and with probability $1 - \alpha_i$ Responder's answers on channel $i$ may be distorted. Hence, with probability $\alpha_i$ all the answers received by Questioner on channel $i$ are reliable, and with probability

$1-\alpha_i$ there is no bound on the number of incorrect information that Questioner receives on this channel—this models the fact that conjunction in Gödel logic is idempotent.

We assume $1 = \alpha_0 > \alpha_1 > \cdots > \alpha_n = 1/2$. For instance, we can (and we will) take $\alpha_i = 1 - i/2n$. We also assume that if channel $i$ is reliable, then all channels $j$ with $0 \leq j \leq i$ are reliable. We can obtain the above model by the following procedure:

(i) Flip a fair coin. If the result is heads, then $n$ is reliable and all channels are reliable; otherwise, $n$ is not reliable.

(ii) By induction, assume that we have already established the reliability of channels $n, n-1, \ldots, i$, and $\alpha_j = 1 - \frac{j}{2n}$ for each $j = n, n-1, \ldots, i$. If $i$ is reliable, then each channel $h \leq i$ is also reliable. On the other hand if $i$ is not reliable, then we flip a biased coin and with probability $\frac{1}{i}$ we set $i$ to be reliable. Therefore, the probability that channel $i-1$ is reliable is given by the sum of the probability that $i$ is reliable and of the probability that $i-1$ is reliable and $i$ is not. Hence, the probability that $i-1$ is reliable is $\frac{2n-i}{2n} + \frac{i}{2n} \cdot \frac{1}{i} = \frac{2n-i+1}{2n}$, as desired.

We assume that asking a question on channel $i$ has a cost, which is maximum when $i = 0$ and minimum when $i = n$. The goal of Questioner is to guess the secret with minimum cost.

## 6.2 Some strategies

Of course, the best strategy for Questioner depends on the choice of the costs. Let us consider the following three different classes of strategies and their comparison according to some hypotheses about the costs.

Strategy 1. Guess the secret by means of $\log(N)$ questions on the safe channel 0.

Strategy 2. By means of $\log(N)$ questions on the cheapest channel $n$, guess a number $x$, which is the real secret with probability $\frac{1}{2}$. Then ask "Is $x_{\text{secret}} = x$?" on channel 0. If the answer is YES, you have guessed the secret; otherwise, you can guess the real secret by $\log(N)$ more questions on channel 0.

Strategy 2i. As in Strategy 2, but use channel $i$, with $0 < i < n$, instead of $n$, for the first $\log(N)$ questions.

Strategy 3. The start is as in Strategy 2: ask $\log(N)$ questions on channel $n$, so that if all answers are truthful you have guessed the secret. In this way, you guess a number $x$, which is the secret with probability $\frac{1}{2}$. Then ask "Is $x_{\text{secret}} = x$?" on channel 0. If the answer is YES, you have correctly guessed the secret, otherwise repeat the procedure with channel $n$ replaced by channel $n-1$, that is, by means of $\log(N)$ questions on $n-1$, guess a number $y$ and ask "Is $x_{\text{secret}} = y$?" on channel 0. If the answer is YES, you have guessed the secret, otherwise, repeat the procedure with $n-1$ replaced by $n-2$, etc.

In [13], the authors compared Strategies 1, 2, 2i and 3 in the following scenarios:

Case 1: The cost of any question on channel $i$ is equal to the probability $\alpha_i = \frac{2n-i}{2n}$ that $i$ is reliable.

Case 2: The cost of any question on Channel $i$ is $\alpha_i^2 = \left(\frac{2n-i}{2n}\right)^2$.

Case 3: The cost of any question on Channel $i$ is $2^{-i}$.

For simplicity, we assume $n$ to be large but negligible with respect to $\log(N)$ (this is an informal assumption that can be made precise). The following theorem records the comparison of the above strategies for Gödel games:

THEOREM 6.2.1 ([16]).

- *In Case 1, in terms of expected cost, Strategy 1 is the best one and Strategy 3 is the worst one.*
- *In Case 2, Strategy 2 is the best one and Strategy 3 is the worst one.*
- *In Case 3 Strategy 3 is the best one and Strategy 1 is the worst one. Moreover Strategy 2i is better than Strategy 2.*

### 6.3 Truth functions in Gödel games and completeness

Proceeding as in the case of the Ulam–Rényi game with finitely many lies, the information acquired by Responder after a sequence $\overline{a}$ of answers can be represented by a function $s_{\overline{a}}$ on $S$: Let $G_{n+2} = \{0, 1, \ldots, n, \top\}$, with $\top \notin \{0, 1, \ldots, n\}$, be ordered with the usual order among numbers and $\top$ as its maximum element. We obtain a Gödel chain with $n + 2$ elements, denoted by $\boldsymbol{G}_{n+2}$, where conjunction is interpreted as min and implication is interpreted as the residual of conjunction (hence, $x \to y = 1$ if $x \leq y$ and $x \to y = y$ otherwise). Then for all $x \in S$:

- If $x$ is not falsified by any answer in $\overline{a}$, set $s_{\overline{a}}(x) = \top$.
- Otherwise, let $i_x$ be the minimum natural number $i$ such that $x$ has been falsified by an answer in $\overline{a}$ on channel $i$. Then set $s_{\overline{a}}(x) = i_x$.

Functions from $S$ into $\{0, 1, \ldots, n, \top\}$ are called *truth functions*.

LEMMA 6.3.1. *For every state $f$, there is a record $\overline{a}$ such that $f = s_{\overline{a}}$.*

*Proof.* Let $Z = \{z \in S \mid f(z) < \top\}$. If $Z = \emptyset$, then let $\overline{a}$ be the empty record. Otherwise, let for all $z \in Z$, $a_z$ be the pair consisting of the question "Is $x_{\text{secret}} = z$?" and of the answer NO. Let $\overline{a}$ be the record consisting of all $a_z$ such that $z \in Z$. By an easy check, we see that $s_{\overline{a}} = f$.

Note that $\overline{a}$ may be legal iff $f$ is not identically equal to 0. Indeed, if $f$ is identically zero, then any record $\overline{a}$ such that $s_{\overline{a}} = f$ must falsify all numbers on the safe channel 0, thus violating the rules of the game. Conversely, if there is a $z_0$ such that $f(z_0) > 0$, then the record $\overline{a}$ defined above and such that $s_{\overline{a}} = f$ may be legal, provided that $z_0$ is the secret and only channel 0 is reliable. □

Truth functions are partially ordered by the pointwise order on $\boldsymbol{G}_{n+2}^S$, i.e., $s_{\overline{a}} \leq s_{\overline{b}}$ iff for all $x \in S$, $s_{\overline{a}}(x) \leq s_{\overline{b}}(x)$. Hence, if $s_{\overline{a}} \leq s_{\overline{b}}$, then every number which has been falsified by an answer in $\overline{b}$ through channel $i$ has been falsified by an answer in $\overline{a}$ through either channel $i$ or through a safer channel $j$. In other words, $s_{\overline{a}} \leq s_{\overline{b}}$ iff $\overline{a}$ is at least as informative as $\overline{b}$.

By an easy computation we see that the truth function $s_{\overline{a} \odot \overline{b}}$ of the juxtaposition $\overline{a} \odot \overline{b}$ of two records $\overline{a}$ and $\overline{b}$ is the pointwise product, $s_{\overline{a}} \cdot s_{\overline{b}}$, of $s_{\overline{a}}$ and $s_{\overline{b}}$ in $\boldsymbol{G}_{n+2}^S$. As in any direct product of residuated lattices, the pointwise implication of two truth functions $s_{\overline{a}}$ and $s_{\overline{b}}$ is the greatest truth function $s$ such that $s_{\overline{a}} \cdot s \leq s_{\overline{b}}$. By Lemma 6.3.1, there is a record $\overline{c}$ such that $s = s_{\overline{c}}$. Clearly, $\overline{c}$ is the least informative record which added to $\overline{a}$ gives a record at least as informative as $\overline{b}$. In this way, the algebra of truth functions, with the order $\leq$ and with the operations just defined, is isomorphic to $\boldsymbol{G}^S_{n+2}$. As usual, 1 is interpreted as $s_\emptyset$, the truth function corresponding to the empty record, and 0 is interpreted as the truth function corresponding to an inconsistent information falsifying all numbers on the safe channel 0.

REMARK 6.3.2. *Of course, we might also consider the set $G_{S,n}$ of all records in the Gödel game $G(S,n)$ modulo the equivalence $\overline{a} \equiv \overline{b}$ iff $s_{\overline{a}} = s_{\overline{b}}$. It is readily seen that through this representation we obtain an algebra isomorphic to $\boldsymbol{G}^S_{n+2}$.*

THEOREM 6.3.3. *The class of Gödel games constitutes a complete game semantics for Gödel logic* G. *More precisely, given Gödel terms $\sigma = \sigma(x_1, \ldots, x_n)$ and $\tau = \tau(x_1, \ldots, x_n)$, the following conditions are equivalent for the equation $\sigma = \tau$:*

*(i) For every finite set $S \neq \emptyset$ and for every $n$, the equation is valid in the Gödel algebra $\boldsymbol{G}^S_{n+2}$ of truth functions in the Gödel game $G(S,n)$.*

*(ii) For every natural number $n$ and singleton set $\{s\}$, the equation holds in the Gödel algebra $\boldsymbol{G}^{\{s\}}_{n+2}$ of truth functions in the Gödel game $G(\{s\},n)$.*

*(iii) The equation holds in every Gödel algebra.*

*Proof.* The equivalence between (i) and (ii) follows from the fact that $\boldsymbol{G}^S_{n+2}$ is a cartesian power of $\boldsymbol{G}^{\{s\}}_{n+2}$, and $\boldsymbol{G}^{\{s\}}_{n+2}$ is a quotient of $\boldsymbol{G}^S_{n+2}$. Hence, the two algebras generate the same variety.

That (iii) implies (ii) is trivial and that (ii) implies (iii) follows from the fact that the variety of Gödel algebras is generated by the class of all finite Gödel chains. For a proof of it is sufficient to note that Gödel algebras are locally finite (see [18]). □

# 7 A game semantics for product logic

## 7.1 The Product game

In this section, we investigate a game semantics for product logic, introduced first in [16]. The variant of the Ulam–Rényi game that we consider here is called Product game. It can be seen as an improvement of the game considered in [30], which also provides a complete semantics for Π but uses records that do not correspond to meaningful strategies of Questioner. The new game is based on the idea of combining Pelc's game (cf. [35]) with the Ulam–Rényi game with half lies investigated in [11].

Given a search space $S = \{1, \ldots, N\}$ (known to both players) and a real number $p$ with $\frac{1}{2} < p \leq 1$, the *Pelc game* $G_P(S,p)$ is defined as follows: as usual, Questioner has to find a secret number in $S$. Moreover, Questioner can ask a finite number of binary questions, and Responder's answers to each question will be truthful with probability $p$. We can imagine that Responder flips a coin, where heads has probability $p$ and tails has probability $1-p$. If the outcome is heads, then he answers truthfully, otherwise he gives the false answer. Unlike the case of Gödel game, where the reliability of the received answers is decided once and for all when the channels are fixed to be reliable or not, in the case of Pelc's game, for each question the decision whether to answer truthfully or not is taken independently.

Since it is not possible for Questioner to identify the secret with probability 1, the actual task is now: given a small positive real number $q$, devise an efficient strategy which guarantees to identify the secret with probability at least $1 - q$. It turns-out (see [35]) that a strategy with expected cost (number of questions) $\mathcal{O}(\log(n))^2$ exists if $p > 1/2$. Moreover, if in addition $p > 2/3$, there is also a better strategy with expected cost $\mathcal{O}(\log(n))$. In the sequel we will refer to the probability $p$ of a correct answer as the *reliability parameter.*

In perfect analogy with the case of the games analyzed before, a record $\overline{a}$ of answers can be represented by its truth-value function $s_{\overline{a}}$ mapping the search space $S$ into $[0, 1]$, which in this case is defined as follows: for all $x \in S$, $s_{\overline{a}}(x)$ expresses the conditional probability of (the answers contained in) the record $\overline{a}$ given that the secret is $x$.

Notice that for a Pelc's game $G_P(S, p)$ the truth-value function of a record $\overline{a}$ provides enough information to compute, for each $x \in S$, the conditional probability $Pr(x|\overline{a})$ that the secret is $x$ given the record $\overline{a}$. Indeed, by the laws of conditional probability we have

$$\Pr(x|\overline{a}) = \frac{\Pr(\overline{a} \mid x) \cdot \Pr(x)}{\sum_{y \in S} \Pr(\overline{a} \cap y)} = \frac{\frac{s_{\overline{a}}(x)}{N}}{\sum_{y \in S} \frac{s_{\overline{a}}(y)}{N}} = \frac{s_{\overline{a}}(x)}{\sum_{y \in S} s_{\overline{a}}(y)},$$

where $\overline{a} \cap y$ denotes the event: the current record is $\overline{a}$ and $x_{\text{secret}} = y$.

Thus truth-value functions are a complete record of the information about the probability of each number being the secret given the answers $\overline{a}$.

The truth value function $s_{\overline{a}}$ can be computed as follows: if $x$ is falsified exactly $k$ times in a record $\overline{a}$ of length $n$, then $s_{\overline{a}}(x) = (1-p)^k \cdot p^{n-k}$. Therefore, the truth-value function $s_{\overline{c}}$ associated to the juxtaposition $\overline{a} \odot \overline{b}$ of two records $\overline{a}$ and $\overline{b}$ is the pointwise product of $s_{\overline{a}}$ and $s_{\overline{b}}$.

One might now conjecture that the underlying logic of this variant of Pelc's game is product logic However, this is not true: first of all, because there does not exist any minimum truth-value function, and also because, up to a few uninteresting exceptions, the set of truth-value functions is not closed under product implication. For example, if $\overline{a}$ consists of one single question of the form "Is $x_{\text{secret}} \in X$?", with $\emptyset \subset X \subset S$, with answer YES and $\overline{b}$ consists of the same question but with answer NO, then assuming $1 - p < p$, the pointwise implication, $s_{\overline{a}} \to s_{\overline{b}}$ of $s_{\overline{a}}$ and $s_{\overline{b}}$, is the function $s$ defined by

$$s(x) = \begin{cases} 1 & \text{if } x \notin X \\ \frac{1-p}{p} & \text{otherwise.} \end{cases}$$

An easy argument now shows that the truth value function of a record $\overline{a}$ is either constantly equal to 1 (if $\overline{a} = \emptyset$) or different from 1 in all points. Hence, although the set of truth value functions is closed under product, it is not closed under residuation.

A key observation to explain the failure of the above attempt of using Pelc's game to model produce logics, is that product algebras have not only a multiplicative structure, representing records in Pelc's game, but also a Boolean structure, which represents records in the Ulam–Rényi game without lies.

To take care of both these structural features of product logic, we introduce a sort of Pelc's game with half lies, in which Responder must answer truthfully when the answer is NO and if the answer is YES, then he can answer YES with probability $p$ and NO with probability $1 - p$, where $p \in [0, 1]$, is chosen by Questioner, but, as in the case of the multichannel game, the cost $c(p)$ he has to pay for an answer with reliability $p$ depends on $p$. More precisely:

The *Product game* $\Pi(S)$, for a search space $S = \{1, \dots, N\}$, known to both players, is defined as follows: Responder chooses a secret number $x_{\text{secret}} \in S$. Then Questioner chooses a set $X \subseteq S$ and a number $p \in [0, 1]$, called *the reliability parameter*, pays $c(p)$ and asks: "Is $x_{\text{secret}} \in X$?". Responder has to answer NO if $x \notin X$. If $x \in X$, then Responder answers YES with probability $p$ and NO with probability $1 - p$. Then Questioner repeats the procedure until he can guess the secret number. Questioner's goal is to guess the number at the lowest cost possible.

Strictly speaking, there are many product games, one for each choice of the search space $S$. However, all product games have the same structure, and this is the reason for the choice of the name Product game instead of Product games.

### 7.2 Some strategies

According to the rules of the game, all YES answers must be truthful. Hence, a question with reliability parameter $\frac{1}{2}$ is not useless at all. Suppose for instance that Questioner repeats the questions "Is $x_{\text{secret}} \in X$?" and "Is $x_{\text{secret}} \in S \backslash X$?" until he gets answer YES. Then by the rules of the game, Questioner will know either that $x \in X$, or that $x \in S \setminus X$. Hence, with high probability (in fact, with probability 1), Questioner will eventually know the correct answer to the question "Is $x_{\text{secret}} \in X$?".

Before proving that the Product game constitutes a sound and complete semantics for $\Pi$, we outline two strategies for Questioner and we compare them according to two hypotheses on costs.

Strategy 1. Questioner guesses the secret by $\log(N)$ questions with reliability 1.

Strategy 2p. Consider a sequence of $\log(N)$ questions on sets $X_1, \dots, X_{\log(N)}$ such that it is possible to guess the secret when the answers to all questions are correct. Given $0 < p < 1$, for $i = 1, \dots, \log(N)$, Questioner asks both questions "Is $x_{\text{secret}} \in X$?" and "Is $x_{\text{secret}} \in S \backslash X$?" with reliability $p$ and repeats the questions until he gets an answer YES, which is necessarily truthful.

**Case 1:** *The cost of a question with reliability p is p.* Since the average number of questions to get a YES answer according to Strategy 2p is $\frac{2}{p}$, the average cost of Strategy 2p is $2\log(N)$. Since Strategy 1 has cost $\log(N)$, it is preferable to all Strategies 2p.

**Case 2:** *The cost of a question with reliability $p$ is $p^2$.* Using Strategy 2p, the expected number of questions needed is $\frac{2\log(N)}{p}$ and since each question has cost $p^2$, the average cost of all questions needed to guess the secret is $2p \cdot \log(N)$. Hence, for $p < 1/2$, Strategy 2p is better than Strategy 1. Moreover if $0 < q < p$, then Strategy 2q is better than Strategy 2p. Since with reliability 0 we cannot guess the secret because the answer is NO to every question, it follows that there is no best strategy: the total cost may be arbitrarily close to 0 but cannot be 0.

### 7.3 Truth functions in Product game and completeness

A *record* in the Product game is a sequence

$$\overline{a} = \langle (X_1, p_1, a_1), \ldots, (X_n, p_n, a_n) \rangle,$$

where each $X_i$ represents the question "Is $x_{\text{secret}} \in X_i$?", each $p_i$ represents the reliability parameter of the $i$th question, and $a_i \in \{\text{YES}, \text{NO}\}$ is Responder's answer. The *truth function*, $s_{\overline{a}}$, of a record $\overline{a}$ is defined, for all $x \in \overline{a}$, as the conditional probability of the record $\overline{a}$ given that the secret is $x$. Hence, $s_{\overline{a}}$ is defined by induction on the length, $\text{lth}(\overline{a})$, of $\overline{a}$. If $\text{lth}(\overline{a}) = 0$, then $s_{\overline{a}}(x) = 1$ for all $x$. Suppose $\text{lth}(\overline{a}) > 0$, and let $\overline{a}^-$ be $\overline{a}$ deprived of its last element. Suppose that the last question in $\overline{a}$ is "Is $x_{\text{secret}} \in X$?", with reliability parameter $p$. If the answer is YES and $x \notin X$, then $s_{\overline{a}}(x) = 0$; if the answer is YES and $x \in X$, then $s_{\overline{a}}(x) = s_{\overline{a}^-}(x) \cdot p$; if the answer is NO and $x \notin X$, then $s_{\overline{a}}(x) = s_{\overline{a}^-}(x)$; if the answer is NO and $x \in X$, then $s_{\overline{a}}(x) = s_{\overline{a}^-}(x) \cdot (1-p)$. In the sequel, the set of all truth functions, added with the constantly 0 function, corresponding to an illegal sequence of answers, will be denoted by $G_{\Pi,S}$.

LEMMA 7.3.1. $G_{\Pi,S} = [0,1]^S$.

*Proof.* Let $f \in [0,1]^S$, $Z_f = \{x \in S \mid f(x) = 0\}$, and $O_f = \{x \in \Omega \mid f(x) = 1\}$. Let $\overline{a}$ be the record consisting of the question "Is $x_{\text{secret}} \in S \setminus Z_f$?" with reliability 1 and answer YES. Then $s_{\overline{a}}$ is the characteristic function of $S \setminus Z_f$. Now for $x \in S \setminus (Z_f \cup O_f)$, consider the record $\overline{b}_x$ consisting of the question: "Is $x_{\text{secret}} = x$?", with reliability $1 - f(x)$ and answer NO. Then $s_{\overline{b}_x}(y) = 1$ if $y \neq x$ (because if the secret is not $x$ Responder has to answer NO) and $s_{\overline{b}_x}(x) = f(x)$ (because if the secret is $x$, then the answer NO is incorrect and it has probability $1 - (1 - f(x)) = f(x)$). Now let $\overline{c}$ be the juxtaposition of $\overline{a}$ and of all $\overline{b}_x$ such that $x \in S \setminus (Z_f \cup O_f)$, Then $f = s_{\overline{a}} \cdot \prod_{x \in S \setminus (Z_f \cup O_f)} s_{\overline{b}_x} = s_{\overline{c}}$. Hence, $f \in G_{\Pi,S}$, and by the arbitrariness of $f$, $G_{\Pi,S} = [0,1]^S$. □

We set $s_{\overline{a}} \preceq s_{\overline{b}}$ iff for all $x \in S$, $s_{\overline{a}}(x) \leq s_{\overline{b}}(x)$. The natural interpretation of $s_{\overline{a}} \preceq s_{\overline{b}}$ is that $\overline{a}$ is more informative than $\overline{b}$, in the sense that it contains more negative information than $\overline{b}$. It should be noted that this need not entail the that $\overline{a}$ is closer to guessing $x$ than $\overline{b}$. For instance, if $s_{\overline{a}}(x) = \varepsilon$ for all $x \in S$, where $\varepsilon$ is a small positive real number, then $\overline{a}$ is more informative than the empty sequence, but it is not more helpful to detect the secret than the empty information, as all elements of $S$ have the same chance to be the secret.

An easy check shows that the truth function $s_{\overline{a} \odot \overline{b}}$ of the juxtaposition $\overline{a} \odot \overline{b}$ of two records $\overline{a}$ and $\overline{b}$ is the pointwise product of $s_{\overline{a}}$ and $s_{\overline{b}}$. Hence, the pointwise implication of two truth functions $s_{\overline{a}}$ and $s_{\overline{b}}$ is the greatest truth function $s$ such that $s \cdot s_{\overline{a}} \leq s_{\overline{b}}$. By Lemma 7.3.1, there is a record $\overline{c}$ such that $s = s_{\overline{c}}$. By the definition of the order of truth functions, it follows that $\overline{c}$ is the less informative record that added to $\overline{a}$ gives a refinement of $\overline{b}$. Of course, $G_{\Pi,S}$, equipped with the order $\leq$, with the pointwise product and the pointwise implication (and with the constants 0 and 1) is a product algebra, denoted by $\boldsymbol{G}_{\Pi,S}$, which is isomorphic to $[0,1]_{\Pi}^{S}$.

THEOREM 7.3.2. *Given terms $\sigma = \sigma(x_1, \ldots, x_n)$ and $\tau = \tau(x_1, \ldots, x_n)$ of product algebras, the following conditions are equivalent for the equation $\sigma = \tau$:*

*(i) For every finite set $S \neq \emptyset$ and for every $n$, the equation is valid in the product algebra $\boldsymbol{G}_{\Pi,S}$ of truth functions in the product game $\Pi(S)$.*

*(ii) For every singleton set $\{s\}$, the equation holds in the product algebra $\boldsymbol{G}_{\Pi,\{s\}}$ of truth functions in the product game $\Pi(\{s\})$.*

*(iii) The equation holds in every product algebra.*

*Proof.* The equivalence between (i) and (ii) follows because $\boldsymbol{G}_{\Pi,S}$ is isomorphic to $\boldsymbol{G}_{\Pi,\{s\}}^{S}$ and $\boldsymbol{G}_{\Pi,\{s\}}$ is a quotient of $\boldsymbol{G}_{n+2}^{S}$. Hence, the two algebras generate the same variety.

That (iii) implies (ii) is trivial and that (ii) implies (iii) follows because the variety of product algebras is generated by $\boldsymbol{G}_{\Pi,\{s\}}$, since this algebra is isomorphic to $[0,1]_{\Pi}$. □

# 8 Game semantics for BL and for SBL

We now consider another game semantics for BL and for SBL, introduced first in [16] This semantics is similar, but not equivalent, to the one presented in [13] and outlined in Section 5. The main differences are: (1) we do not need the assumption on increasingly noisy channels, and hence we obtain Gödel games as special cases; (2) we consider the possibility of channels with half-lies, and hence we obtain the Product game as a special case.

## 8.1 BL games

As usual, Questioner has to guess a number $x_{\text{secret}}$ known to Responder and chosen in a search space $S = \{1, \ldots, N\}$, known to both players. Questioner can ask YES–NO questions using different channels $0, \ldots, n$. To each channel $i$, we associate a label $L_i$, which is either $\Pi$ or a non-negative natural number $e_i$, representing the maximum number of lies allowed. Moreover, to each channel $i$ it is also associated a probability $\alpha_i$, where $1 = \alpha_0 > \alpha_1 > \cdots > \alpha_n = \frac{1}{2}$ (as in the Gödel games, we may take $\alpha_i = 1 - \frac{2n-i}{2n}$).

The procedure in order to decide which channels are reliable and which are not is as in Gödel game. Once it is decided which channels are reliable (and this information is known to Responder but not to Questioner), Questioner asks YES–NO questions of the form "Is $x_{\text{secret}} \in X$?" and chooses a channel $i$. Moreover if $i$ is labelled by $\Pi$, then Questioner has also to chose a reliability parameter $p$.

If channel $i$ is not reliable, then Responder is allowed (but not forced) to lie through channel $i$ as many times as he wants.

If $i$ is reliable and is labelled by $\Pi$, then if the correct answer is NO, Responder is forced to answer NO, if the correct answer is YES, Responder answers YES with probability $p$ and NO with probability $1-p$, where $p$ is chosen by Questioner. If channel $i$ is reliable and is labelled by $e_i$, then Responder can lie on channel $i$ up to $e_i$ times, like in Ulam–Rényi game with $e_i$ lies.

DEFINITION 8.1.1. *A game of the form shown above is called* a BL-game. *If in addition channel* 0 *is labelled by* 0*, that is, no lies are allowed on it, then the game is said to be* an SBL-game.

*Let $S$ denote the search space and let $\overline{e} = \langle e_0, e_1, \ldots, e_n \rangle$ be the sequence of labels of the channels, where each $e_i$ is either a natural number representing the number of lies admitted, or $\Pi$. Then the game will be denoted by* $\mathrm{BL}(S, \overline{e})$.

Note that Gödel games arise as a special case when every channel is labelled by 0 (that is, if the channel is reliable then no lie is admitted through it), the Product game arises when there is only one channel labelled by $\Pi$, and Ulam–Rényi games arise when there is only one channel, and this is labelled with a natural number $e$. Finally, the Twenty Question game (the Ulam–Rényi game without lies) has precisely one channel, and this is labelled 0.

## 8.2 Truth functions and completeness for BL-games

Since there are really many structurally different BL-games, it is extremely difficult to investigate game strategies for all of them. Hence, we will only discuss representations of records and the completeness of BL (resp., SBL) with respect to the class of BL-games (resp., of SBL-games).

A record $\overline{a}$, consisting of a sequence of answers, possibly using different channels, is represented by a truth function $\sigma_{\overline{a}}$ as follows: let, for $i = 0, \ldots, n$, $\overline{a}_i$ denote the subsequence of $\overline{a}$ consisting of all answers through channel $i$ (possibly including the reliability parameter if $i$ is labelled by $\Pi$). Let $s_{\overline{a}_i}$ be the truth function corresponding to $\overline{a}_i$ in the Ulam–Rényi game with $e_i$ lies if $i$ is labelled by $e_i$ and in the Product game if $i$ is labelled by $\Pi$. Then define the truth function $\sigma_{\overline{a}}$ corresponding to $\overline{a}$ as follows: if no answer falsifies $x$, then set $\sigma_{\overline{a}}(x) = \top$ (where $0 < 1 < \cdots < n < \top$). Otherwise, let $i_x$ be the minimum $i$ such that $x$ has been falsified by an answer on channel $i$. Then set $\sigma_{\overline{a}_i}(x) = \langle i, s_{\overline{a}_i}(s) \rangle$.

Now consider a BL-game $\mathrm{BL}(S, \overline{e})$ with search space $S$ and $\overline{e} = \langle e_0, e_1, \ldots, e_n \rangle$ where for all $i$, $e_i$ is either a natural number or $\Pi$. Let for $i = 0, 1, \ldots, n$, $\boldsymbol{U}_{e_i} = [0,1]_\Pi$ if $e_i = \Pi$. Moreover let $\boldsymbol{U}_{e_0} = \boldsymbol{Ł}_{e_0+1}$ if $e_0 \in \mathsf{N}$ and let for $i > 0$, $\boldsymbol{U}_{e_i} = \boldsymbol{W}_{e_i+1}$ if $e_i \in \mathsf{N}$. After taking isomorphic copies, we can assume that for $i \neq j$, $U_{e_i} \cap U_{e_j} = \{1\}$ if $i \neq j$. Further, let $U^0_{e_i} = \{1, \langle i, 0 \rangle, \ldots \langle i, e_i \rangle\}$ if $e_i \in \mathsf{N}$ and $U^0_{e_i} = \{1\} \cup \{\langle i, x \rangle \mid x \in [0,1]\}$. Define the function $\Phi$ from $\boldsymbol{U} = \boldsymbol{U}_{e_0} \oplus \boldsymbol{U}_{e_1} \oplus \ldots \oplus \boldsymbol{U}_{e_n}$ onto $H_{\overline{e}} = U^0_{e_0} \cup U^0_{e_1} \cup \ldots \cup U^0_{e_n}$ as follows:

$$\Phi(x) = \begin{cases} \top & \text{if} \quad x = 1 \\ \langle i, x \rangle & \text{if} \quad x \in U_{e_i} \setminus \{1\} \qquad \text{and} \quad e_i = \Pi \\ \langle i, j \rangle & \text{if} \quad x = \frac{j}{e_i+1} \in U_{e_i} \setminus \{1\} \quad \text{and} \quad e_i \in \mathsf{N}. \end{cases}$$

Then $\Phi$ is a bijection, and hence it induces a structure $\boldsymbol{H}_{\overline{e}}$ on $H_{\overline{e}}$, which is isomorphic to $\boldsymbol{U}$ (for each binary operation $f$ on $\boldsymbol{U}$, define for $x, y \in H_{\overline{e}}$, $f^{\boldsymbol{H}_{\overline{e}}}(x,h) = \Phi^{-1}(f(\Phi(x),\Phi(y)))$. In the sequel we will identify the algebra $\boldsymbol{H}_{\overline{e}}$ of truth-values with the algebra $\boldsymbol{U} = \boldsymbol{U}_{e_0} \oplus \boldsymbol{U}_{e_1} \oplus \ldots \oplus \boldsymbol{U}_{e_m}$. In particular, the induced order of truth-values on $\boldsymbol{H}_{\overline{e}}$ is

$$\langle j,r\rangle < \langle j',r'\rangle \text{ iff } [\text{either } j<j' \text{ or } (j=j' \text{ and } r<r')].$$

Hence, if $s \in S$ has truth-value $\langle j,r\rangle$, $s' \in S$ has truth-value $\langle j',r'\rangle$ and $\langle j,r\rangle < \langle j',r'\rangle$, then either $j < j'$, and then $s$ has been falsified by an answer on a more reliable channel, or $j = j'$ and $r < r'$, in which case $s$ has received a stronger negative information than $s'$ by answers on the same channel $j$.

Truth functions are maps from $S$ into the truth-value algebra $\boldsymbol{H}_{\overline{e}} = \boldsymbol{U}$. Hence, the algebra of truth functions may be identified with $\boldsymbol{H}_{\overline{e}}^S = \boldsymbol{U}^S$.

LEMMA 8.2.1. *For every function $f \in H_{\overline{e}}^S$, there is a record $\overline{a}$ such that $f = \sigma_{\overline{a}}$.*

*Proof.* Clearly, for every record $\overline{a}$, $s_{\overline{a}}$ is a map from $S$ into $H_{\overline{e}}$. Conversely, given a function $f$ from $S$ into $H_{\overline{e}}$, let $S' = \{z \in S \mid f(z) \neq \top\}$, and let for $z \in S'$, $f(z) = \langle i_z, h_z\rangle$. Distinguish two cases: if $e_{i_z} = \Pi$, then let $\overline{a}_z$ be a record in the product game such that $s_{\overline{a}_z}(z) = h_z$, and consider the record $\overline{b}_z$ consisting of all answers in $\overline{a}_z$ through channel $i_z$. If $e_{i_z} \in \mathsf{N}$, then let $\overline{b}_z$ consist of $e_{i_z} + 1 - h_z$ questions of the form: "Is $x_{secret} = z$?" on channel $i_z$, with answer NO. Finally, let $\overline{a}$ the the juxtaposition of all $\overline{a}_z$ with $z \in S'$. An easy computation shows that the $s_{\overline{a}} = f$. □

REMARK 8.2.2. *In this game the only truth function corresponding to an illegal record is the truth function which is constantly equal to $\langle 0,0\rangle$. Indeed, if for some $z$, $f(z) = \langle i,x\rangle$ with $i > 0$, then it is possible that the secret is $z$ and that channel $i$ is unreliable. If for some $z$, $f(z) = \langle 0,x\rangle$ with $x > 0$, then it is possible that the secret is $z$.*

The partial order of truth-values induces a partial order of truth functions: $s_{\overline{a}} \leq s_{\overline{b}}$ iff for all $s \in S$, $s_{\overline{a}}(s) \leq s_{\overline{b}}(s)$.

The *juxtaposition* $\overline{a} \odot \overline{b}$ of two records $\overline{a}$ and $\overline{b}$ is the record obtained by giving each answer $a_i$ a multiplicity equal to the sum of its multiplicity in $\overline{a}$ plus its multiplicity in $\overline{b}$.

By an easy checking, we see that, if $s_{\overline{a}}(x) = \langle i,z\rangle$ and $s_{\overline{b}}(x) = \langle j,w\rangle$, then

$$s_{\overline{a}\odot\overline{b}}(x) = \begin{cases} \langle i,z\rangle & \text{if } \ i<j \\ \langle j,w\rangle & \text{if } \ j<i \\ \langle j,w\cdot z\rangle & \text{if } \ i=j \ \text{ and } \ e_i=\Pi \\ \langle j,\max\{w+z-e_i-1,0\}\rangle & \text{if } \ i=j \ \text{ and } \ e_i\in\mathsf{N}. \end{cases}$$

That is, up to isomorphism, $s_{\overline{a}\odot\overline{b}}$ is the pointwise product in $\boldsymbol{U}^S = \boldsymbol{H}_{\overline{e}}^S$ of $s_{\overline{a}}$ and $s_{\overline{b}}$. It follows that the truth function of the juxtaposition, $\overline{a} \odot \overline{b}$, of records $\overline{a}$ and $\overline{b}$, is the product $s_{\overline{a}} \cdot_{\boldsymbol{H}_{\overline{e}}^S} s_{\overline{b}}$ of the truth function of $\overline{a}$ and the truth function of $\overline{b}$ in $\boldsymbol{H}_{\overline{e}}^S$. Moreover the pointwise implication $\to_{\boldsymbol{H}_{\overline{e}}^S}$ in $\boldsymbol{H}_{\overline{e}}^S$ is the residual of $\cdot_{\boldsymbol{H}_{\overline{e}}^S}$. Hence, the pointwise implication, $s_{\overline{a}} \to_{\boldsymbol{H}_{\overline{e}}^S} s_{\overline{b}}$ of two truth functions $s_{\overline{a}}$ and $s_{\overline{b}}$, is defined as the maximum truth function $s_{\overline{c}}$ such that $s_{\overline{a}\odot\overline{c}} = s_{\overline{a}} \cdot_{\boldsymbol{K}_{\overline{e}}^S} s_{\overline{c}} \leq s_{\overline{b}}$. It represents the minimum information $\overline{c}$ that must be added to $\overline{a}$ in order to get a refinement of $\overline{b}$.

### 8.3 Completeness theorem

THEOREM 8.3.1. *Given* BL-*terms* $\sigma = \sigma(x_1, \ldots, x_n)$ *and* $\tau = \tau(x_1, \ldots, x_n)$, *the following conditions are equivalent for the equation* $\sigma = \tau$*:*

*(i) For every finite set* $S \neq \emptyset$ *and* $(k+1)$*-tuple* $\overline{e} = \langle e_0, e_1, \ldots, e_k \rangle$ *where each* $e_i$ *is either an integer* $\geq 0$ *or* $\Pi$, *the equation is valid in the* BL-*algebra* $\boldsymbol{H}^S_{\overline{e}}$ *of truth functions of records.*

*(ii) For every* $(k+1)$*-tuple as in (i) and singleton set* $\{s\}$, *the equation holds in the* BL-*algebra* $\boldsymbol{H}^{\{s\}}_{\overline{e}}$.

*(iii) The equation holds in every* BL-*algebra.*

*Proof.* (i) $\Leftrightarrow$ (ii) is simple: $\boldsymbol{H}^S_{\overline{e}} = (\boldsymbol{H}^{\{s\}}_{\overline{e}})^S$ is a power of $\boldsymbol{H}^{\{s\}}_{\overline{e}}$ and $\boldsymbol{H}^{\{s\}}_{\overline{e}}$ is a quotient of $\boldsymbol{H}^S_{\overline{e}}$ through the projection into $s$. Hence, $\boldsymbol{H}^{\{s\}}_{\overline{e}}$ and $\boldsymbol{H}^S_{\overline{e}}$ generate the same variety.

The implication (iii) $\Rightarrow$ (ii) is trivial and the implication (ii) $\Rightarrow$ (iii) follows from Remark 2.0.7. □

## 9 Historical remarks and further reading

The study of search problems with unreliable tests has a long history in computer science and discrete mathematics dating back to the mid of the last century. The name Ulam–Rényi game which is generally used for describing the model we studied here, originates from the following excerpts which we report verbatim.

In his autobiography *Adventures of a Mathematician* [42] Stanisłav Ulam raised the following question:

> Someone thinks of a number between one and one million (which is just less than $2^{20}$). Another person is allowed to ask up to twenty questions, to each of which the first person is supposed to answer only yes or no. Obviously the number can be guessed by asking first: Is the number in the first half-million? and then again reduce the reservoir of numbers in the next question by one-half, and so on. Finally the number is obtained in less than $\log_2(1,000,000)$. Now suppose one were allowed to lie once or twice, then how many questions would one need to get the right answer? One clearly needs more than $n$ questions for guessing one of the $2^n$ objects because one does not know when the lie was told. This problem is not solved in general.

Alfréd Rényi posed the very same problem in his half-fictitious book *A Diary on Information Theory* [39]

> ...I made up the following version, which I called "Bar-kochba with lies". Assume that the number of questions which can be asked to figure out the "something" being thought of is fixed and the one who answers is allowed to lie a certain number of times. The questioner, of course, doesn't know which answer is true and which is not. Moreover the one answering is not required to lie as many times as is allowed.

> For example, when only two things can be thought of and only one lie is allowed, then 3 questions are needed ... If there are four things to choose from and one lie is allowed, then five questions are needed. If two or more lies are allowed, then the calculation of the minimum number of questions is quite complicated ... It does seem to be a very profound problem ...

In these quotations Ulam and Rényi posed as an important open question what would be now called the solution to an instance of the Ulam–Rényi problem.

A search problem with erroneous tests had been already presented as a game in a 1961 paper of Rényi [38], where the author assumes that errors might be random:

> Two players are playing the game, let us call them $A$ and $B$. $A$ thinks of something and $B$ must guess it. $B$ can ask questions which can be answered by $yes$ or $no$ and he must find out what $A$ had thought from the answers. [...] it is better to suppose that a given percentage of the answers are wrong (because $A$ misunderstands the questions or does not know certain facts).

This is probably the first publication where searching with errors is formulated in terms of a game. More generally, problems of searching with lies had been also studied by Yaglom and Yaglom [44], who considered the following one: Determine which city, among A, B, C, you are in by asking yes/no questions to the people around. The only available information you have is that inhabitants of A always speak the truth; inhabitants of B always lie, and inhabitants of C alternate between one lie and one correct answer (but the nature of the first answer is not known). The same problem is also analyzed by Picard [36].

In [4], the game of the Twenty Questions with lies was used by Berlekamp to study error-correcting codes for the symmetric channel with feedback. In the same paper, Berlekamp introduced the concept of the *volume of a state* which in most of the later literature on the Ulam–Rényi game was renamed as the *weight of a state*—this is the term we used here, too. Theorem 4.1.1 is usually referred to as the Volume Bound. It also appeared first in [4]. Rivest et al. [40] showed a continuous variant of this bound for a variant of the game where the search space is the real interval $[0, 1]$. The proof we provided here for the discrete case is novel and induction-free.

Hill and Karim [26] were the first to give the solution for the Ulam–Rényi problem over a search space of cardinalities $2^{20}$ and $10^6$, for any number of lies $e$.

For $e = 1, 2, 3$, the exact value of $N(2^{20}, e)$—that is the case of the search space cardinality in Ulam's original formulation—had been computed by Pelc [34], Czyzowicz et al. [17] and Negro and Sereno [33], respectively. Specifically, these papers provide the exact value of $N(2^m, e)$ for all integers $m \geq 0$, for the cases $e = 1$, $e = 2$, $e = 3$, respectively. For $e = 1, 2, 3$, the exact value of $N(M, e)$ for all integers $M \geq 1$ was given by Pelc [34], Guzicki [24], and Deppe [19], respectively.

Spencer [41] showed that for each $e \geq 0$ and up to finitely many exceptional $M$, the minimum number of questions necessary and sufficient to identify the secret in a Ulam–Rényi game with $e$ lies over a search space of cardinality $M$ is given by the Volume Bound. It is also interesting to note that Spencer prefers to address the two players by the proper names Paul and Carole, with Paul being a reference to the great questioner Paul Erdős and Carole being her anagram: the Oracle.

Here we presented a stronger version of the Spencer's result under the additional condition that the strategy must consist of two batches of non-adaptive questions. This result originally appeared in [12, 14] and shows that two-batch strategies can be as powerful as the fully adaptive ones.

The binary game with asymmetric error was introduced by Rivest et al. in [40]. For the special case of only one error, $e = 1$, an exact estimate of the minimum number of questions necessary and sufficient to identify the secret was given by Cicalese and Mundici in [11]. In the same paper the authors introduced new analytic tools for the study of Ulam–Rényi games with asymmetric lies: a probabilistic/counting-based analysis of a strategy and a novel, weaker notion of weight conservation for states. These became the main ingredients in Dumitriu and Spencer's ingenious solution of a very general variant of the Ulam–Rényi game, where the question can be multiple-choice and the number of half-lies allowed to Responder is any constant $e \geq 0$ [20]. Subsequently, in [21] the results of [20] were also obtained with two-batch strategies.

Following [21], a new line of research started inspired by the relationship between Ulam–Rényi game and error correcting transmission. The main aim was to understand the power of adaptiveness with respect to the *combinatorial* nature of the noise acting on the channel used for Responder's answers. Games over $q$-ary symmetric channels and search spaces being of cardinality a power of $q$, i.e., $M = q^m$, were analyzed in [10]. The authors provided asymptotically tight bounds for some special combinatorial channels. This result was later generalized in [1] to the case of arbitrary search space dimension $M$ and more general models of asymmetric errors. The most general result, to date, which we presented in this chapter is due to Ahlswede et al. [2].

Many other variants of the Ulam–Rényi game have been investigated. Here we only focused on models also studied in the context of fuzzy logics. For a more comprehensive treatment of the combinatorial and algorithmic aspects of the Ulam–Rényi game as well as other search problems in the presence of unreliable information, the interested reader is referred to the recent book [9].

**Acknowledgements**

The authors would like to thank Daniele Mundici his for valuable comments on a draft version of this chapter.

Franco Montagna passed away unexpectedly before this manuscript was finished. The first author would like to dedicate this chapter to his memory, to the memory of the great scientist and the man, to his life spent in the search for truth, asking deep questions and giving simple answers.

# BIBLIOGRAPHY

[1] Rudolf Ahlswede, Ferdinando Cicalese, and Christian Deppe. Searching with lies under error cost constraints. *Discrete Applied Mathematics*, 156(9):1444–1460, 2008.

[2] Rudolf Ahlswede, Ferdinando Cicalese, Christian Deppe, and Ugo Vaccaro. Two batch search with lie cost. *IEEE Transactions on Information Theory*, 55(4):1433–1439, 2009.

[3] Libor Běhounek, Petr Cintula, and Petr Hájek. Introduction to mathematical fuzzy logic. In Petr Cintula, Petr Hájek, and Carles Noguera, editors, *Handbook of Mathematical Fuzzy Logic - Volume 1*, volume 37 of *Studies in Logic, Mathematical Logic and Foundations*, pages 1–101. College Publications, London, 2011.

[4] Elwyn R. Berlekamp. Block coding for the binary symmetric channel with noiseless, delayless feedback. In *Error-Correcting Codes*, pages 330–335. Wiley, New York, 1968.

[5] Willem J. Blok and Isabel Maria André Ferreirim. On the structure of hoops. *Algebra Universalis*, 43(2–3):233–257, 2000.

[6] Willem J. Blok and Don L. Pigozzi. *Algebraizable Logics*, volume 396 of *Memoirs of the American Mathematical Society*. American Mathematical Society, Providence, RI, 1989. Available at `http://orion.math.iastate.edu/dpigozzi/`.

[7] Manuela Busaniche and Franco Montagna. Hájek's logic BL and BL-algebras. In Petr Cintula, Petr Hájek, and Carles Noguera, editors, *Handbook of Mathematical Fuzzy Logic - Volume 1*, volume 37 of *Studies in Logic, Mathematical Logic and Foundations*, pages 355–447. College Publications, London, 2011.

[8] Agata Ciabattoni, Christian G. Fermüller, and George Metcalfe. Uniform rules and dialogue games for fuzzy logics. In *Proceedings of LPAR 2004*, volume 3452 of *Lecture Notes in Artificial Intelligence*, pages 496–510. Springer, 2005.

[9] Ferdinando Cicalese. *Fault Tolerant Search Algorithms*. Springer, Berlin, 2013.

[10] Ferdinando Cicalese, Christian Deppe, and Daniele Mundici. $q$-ary ulam-rényi game with weighted constrained lies. In *Proceedings of COCOON 2000*, volume 3106 of *Lecture Notes in Computer Science*, pages 82–91, 2004.

[11] Ferdinando Cicalese and Daniele Mundici. Optimal coding with one asymmetric error: Below the sphere packing bound. In *Proceedings of COCOON 2000*, volume 1858 of *Lecture Notes in Computer Science*, pages 159–169, 2000.

[12] Ferdinando Cicalese and Daniele Mundici. Perfect two-fault tolerant search with minimum adaptiveness. *Advances in Applied Mathematics*, 25(1):65–101, 2000.

[13] Ferdinando Cicalese and Daniele Mundici. Recent developments of feedback coding and its relations with many-valued logic. In Amithab Gupta, Rohit Parikh, and Johan van Benthem, editors, *Logic at the Crossroads: an Interdisciplinary View, Proceedings of the First Indian Congress on Logic and Applications*, pages 222–240, New Delhi, 2007. Allied Publishers Pvt.Ltd.

[14] Ferdinando Cicalese, Daniele Mundici, and Ugo Vaccaro. Least adaptive optimal search with unreliable tests. *Theoretical Computer Science*, 270(1–2):877–893, 2001.

[15] Petr Cintula and Ondrej Majer. Towards evaluation games for fuzzy logics. In Ondrej Majer, Ahti-Veikko Pietarinen, and Tero Tulenheimo, editors, *Games: Unifying Logic, Language, and Philosophy*, volume 15 of *Logic, Epistemology, and the Unity of Science*, pages 117–138. Springer, 2009.

[16] Esther Corsi and Franco Montagna. Ulam games and many-valued logic. To appear in *Fuzzy Sets and Systems*, doi: 10.1016/j.fss.2015.09.006.

[17] Jurek Czyzowicz, Andrzej Pelc, and Daniele Mundici. Ulam's searching game with lies. *Journal of Combinatorial Theory, Series A*, 52:62–76, 1989.

[18] Ottavio D'Antona and Vincenzo Marra. Computing coproducts of finitely presented Gödel algebras. *Annals of Pure and Applied Logic*, 142(1–3):202–211, 2006.

[19] Christian Deppe. Solution of Ulam's searching game with three lies or an optimal adaptive strategy for binary three-error-correcting-codes. Technical Report 98-036, University of Bielefeld, Faculty of Mathematics, 1998.

[20] Ioana Dumitriu and Joel Spencer. The liar game over an arbitrary channel. *Combinatorica*, 25(5): 537–559, 2005.

[21] Ioana Dumitriu and Joel Spencer. The two-batch liar game over an arbitrary channel. *SIAM Journal of Discrete Mathematics*, 19:1056–1064, 2006.

[22] Robin Giles. A non-classical logic for physics. *Studia Logica*, 33(4):399–417, 1974.

[23] Siegfried Gottwald. Axiomatizations of t-norm based logics - a survey. *Soft Computing*, 4:63–67, 2000.

[24] Wojciech Guzicki. Ulam's searching game with two lies. *Journal of Combinatorial Theory, Series A*, 54:1–19, 1990.

[25] Petr Hájek. *Metamathematics of Fuzzy Logic*, volume 4 of *Trends in Logic*. Kluwer, Dordrecht, 1998.

[26] Ray Hill and Jehangir P. Karim. Searching with lies: The Ulam problem. *Discrete Mathematics*, 106–7: 273–283, 1992.

[27] Jaakko Hintikka and Gabriel Sandu. Game-theoretical semantics. In Johan van Benthem and Alice ter Meulen, editors, *Handbook of Logic and Language*, pages 361–410. Elsevier and the MIT Press, Oxford/Shannon/Tokio/Cambridge (Massachusetts), 1997.

[28] Paul Lorenzen. Ein dialogisches Konstruktivitätskriterium. In *Infinitistic Methods, Proceedings of Symposium on Foundations of Mathematics, Warsaw, 1959*, pages 193–200. Pergamon Press, Oxford, 1961.

[29] Florence J. MacWilliams and Neil J. A. Sloane. *The Theory of Error Correcting Codes*, volume 16 of *North-Holland Mathematical Library*. North-Holland, Amsterdam, 1977.

[30] Franco Montagna, Claudio Marini, and Giulia Simi. Product logic and probabilistic Ulam games. *Fuzzy Sets and Systems*, 158(6):639–651, 2007.
[31] Daniele Mundici. The logic of Ulam's game with lies. In Cristina Bicchieri and M.L. Dalla Chiara, editors, *Knowledge, Belief, and Strategic Interaction (Castiglioncello, 1989)*, Cambridge Studies in Probability, Induction, and Decision Theory, pages 275–284. Cambridge University Press, Cambridge, 1992.
[32] Daniele Mundici. Ulam games, Łukasiewicz logic and AF $c^*$-algebras. *Fundamenta Informaticae*, 18:151–161, 1993.
[33] Alberto Negro and Matteo Sereno. Ulam's searching game with three lies. *Advances in Applied Mathematics*, 13:404–428, 1992.
[34] Andrzej Pelc. Solution of ulam's problem on searching with a lie. *Journal of Combinatorial Theory, Series A*, 44:129–142, 1987.
[35] Andrzej Pelc. Searching with known error probability. *Theoretical Computer Science*, 63:185–202, 1989.
[36] Claude Picard. *Theory of Questionnaires*. Gauthier-Villars, Paris, 1965.
[37] John R. Pierce. Optical channels: Practical limits with photon counting. *IEEE Transactions on Communications*, 26(12):1819–1821, 1978.
[38] Alfréd Rényi. On a problem in information theory. *Magyar Tud. Akad. Mat. Kutató Int. Közl.*, 6: 505–516, 1961. In Hungarian and Russian, English summary.
[39] Alfréd Rényi. *Napló az informácíóelméletről*. Gondolat, Budapest, 1976. English translation: A Diary on Information Theory, J.Wiley and Sons, New York, 1984.
[40] Ronald L. Rivest, Albert R. Meyer, Daniel J. Kleitman, Karl Winklmann, and Joel Spencer. Coping with errors in binary search procedures. *Journal of Computer and System Sciences*, 20(3):396–404, 1980.
[41] Joel Spencer. Ulam's searching game with a fixed number of lies. *Theoretical Computer Science*, 95:307–321, 1992.
[42] Stanislaw M. Ulam. *Adventures of a Mathematician*. Scribner's, New York, 1976.
[43] Jacobus Hendricus van Lint. *Introduction to Coding Theory*, volume 86 of *Graduate Texts in Mathematics*. Springer, Berlin, third edition, 1999.
[44] Akiva M. Yaglom and Isaak M. Yaglom. *Verojatnost' i Informacija*. Nakua, Moscow, 1957.

FERDINANDO CICALESE
Department of Computer Science
University of Verona
Strada Le Grazie 15
37134 Verona, Italy
Email: ferdinando.cicalese@univr.it

FRANCO MONTAGNA
Department of Information Engineering and Mathematics
University of Siena
Via Roma 56
53100 Siena, Italy

# Chapter XV:
# Fuzzy Logic with Evaluated Syntax

VILÉM NOVÁK

## 1 Introduction

This chapter[1] is devoted to a special kind of fuzzy logic which generalizes not only the classical two-valued semantics but also its syntax. It does this by relaxing the concept of axiom as an initially fully true formula. Namely, the axioms of formal theories can be only partially true and hence, form a fuzzy set. Consequently, we obtain a *fuzzy logic with evaluated syntax*, which we denote by $\mathrm{Ev}_{Ł}$.[2]

The concept of evaluated syntax means that each formula is, on a syntactical level, accompanied by a value representing the minimal degree in which the given formula can be true. We can thus express, e.g. that we cannot assume a given axiom or assumption to be fully true.[3] In fact, when dealing with the formula in a formal theory, it can attain various degrees. In such a context, the supremum of all of these degrees is called the *provability degree*, and we will demonstrate that it generalizes the classical concept of provability. As we will also see, the provability degree of a formula is infimum of its truth degrees in all models. This fact is the content of the completeness theorem, and we argue that it is a natural generalization of the classical completeness theorem.

It should be noted that traditional syntax can also be taken as evaluated, but the only evaluations of formulas are $\mathbf{1}$ or $\mathbf{0}$. This is an extreme situation and of course, it can hardly have any sense to write explicitly the evaluation $\mathbf{1}$ with each formula.

One of the problems of $\mathrm{Ev}_{Ł}$ discussed from the very beginning is the necessity to introduce *truth constants*, i.e., *names* of the truth values that belong to the syntax. First, note that truth constants are also contained in classical logic, namely, the constants $\bot, \top$ for falsity and truth. Thus, considering truth constants in fuzzy logic seems to be a direct generalization of this fact. The problem is how many constants are needed. In classical logic, we have truth constants for all (i.e., both) truth values. However, if the set of truth values is the whole interval $[0, 1]$ then introducing constants for all of them makes the language of $\mathrm{Ev}_{Ł}$ uncountable and this is undesirable.

[1] Editorial footnote: The notation of this chapter conforms the one used in the study of fuzzy logic with evaluated syntax and therefore differs from the rest of this Handbook.

[2] Since intermediary truth values occur both in its syntax as well as in its semantics it is sometimes argued that $\mathrm{Ev}_{Ł}$ is the "most genuine" fuzzy logic.

[3] For example, the formula stating that "if $n$ stones do not form a heap, then neither do $n + 1$ stones" cannot be fully true because by adding a stone to a collection of stones, we change it, even if by a very small amount.

Fortunately, this problem can be overcome in several ways. One possibility is to consider only formulas containing rational truth constants and use sets of the former to approximate formulas with irrational truth constants. Alternatively, we can also introduce only one truth constant $\bot$ and a special unary connective square root $\surd$ which enables us to get rid of the other truth constants. (This solution is only an indirect way how $\mathrm{Ev}_{\text{Ł}}$ can be freed of truth constants for irrational truth values.) There is also an interesting possibility is to represent $\mathrm{Ev}_{\text{Ł}}$ in Łukasiewicz logic.[4]

The solutions mentioned above seem not to be satisfactory from the point of view of a consistent development of $\mathrm{Ev}_{\text{Ł}}$. There is a direct way to address only a countable number of truth constants, and so have only a *countable language*. This is explained in this chapter.

The structure of this chapter is the following. In Section 2 we introduce the algebra of truth values, language and formulas. In Section 3, we analyze the difference between traditional and evaluated syntax, introduce classical and graded syntactic consequence and show that the latter generalizes the former. Then we define the semantics of $\mathrm{Ev}_{\text{Ł}}$, the concept of formal fuzzy theory and analyze conditions on the structure of truth values that must be satisfied if we want the logic to be complete. Section 4 is devoted to a detailed presentation of $\mathrm{Ev}_{\text{Ł}}$. In Section 5, we present a proof of the completeness theorem. Sections 6 and 7 are devoted to model theory and discuss selected additional problems. Finally, in Section 8, remarks on the history of $\mathrm{Ev}_{\text{Ł}}$ and hints for further reading are provided.

## 2 Language and truth values of $\mathrm{Ev}_{\text{Ł}}$

In this chapter, we denote formulas by capital Latin letters $A, B, \ldots$. The set of all well-formed formulas in the language $J$ is denoted by $F_J$. We suppose that the reader is familiar with the classical concepts of mathematical first-order logic, namely, those of axioms, inference rules, formal proof, provability, formal theory, model, etc. If $A$ is a formula, then $w_A$ is a formal proof of it. If $f$ is a function, then by $\mathrm{dom}(f)$ and $\mathrm{rng}(f)$ we denote its domain and range, respectively. The symbol $:=$ means "is defined as".

A fuzzy set $A$ is identified with its *membership function* $A\colon X \to L$, where $X$ is a universe and $L$ is a set of truth values (the support of a residuated lattice—see below). We often write $A \underset{\sim}{\subset} X$ to stress that $A$ is a fuzzy set in $X$. The set of all fuzzy sets on $X$ is denoted by $\mathcal{F}(X)$. Because fuzzy sets are functions, we order them classically, i.e., if $V, W \underset{\sim}{\subset} X$, then $V \leq W$ means $V(x) \leq W(x)$ for all $x \in X$. The *support* of a fuzzy set $A \underset{\sim}{\subset} X$ is the set $\mathrm{Supp}(A) = \{x \mid A(x) > \mathbf{0}\}$.

### 2.1 The algebra of truth values

When speaking about logic, we must first introduce an algebra of truth values $\boldsymbol{L}$. We will suppose that it is a residuated lattice, i.e., an integral, commutative, residuated, lattice ordered monoid:

$$\boldsymbol{L} = \langle L, \vee, \wedge, \otimes, \to, \mathbf{0}, \mathbf{1} \rangle, \tag{1}$$

[4] This representation was introduced by Hájek, who calls it Rational Pavelka logic (RPL).

where $\langle L, \vee, \wedge \rangle$ is a bounded lattice with the bottom element $\mathbf{0}$ and top element $\mathbf{1}$.The operations $\otimes$ and $\rightarrow$ are multiplication and residuum, respectively, joined by the adjunction condition

$$a \otimes b \leq c \iff a \leq b \rightarrow c, \qquad a, b, c \in L. \tag{2}$$

Finally, $\langle L, \otimes, \mathbf{1} \rangle$ is a commutative monoid.

The following operations on $L$ are, furthermore, defined:

$$\neg a = a \rightarrow \mathbf{0} \qquad \text{(negation)}$$

$$a \oplus b = \neg(\neg a \otimes \neg b) \qquad \text{(strong summation)}$$

$$a \leftrightarrow b = (a \rightarrow b) \wedge (b \rightarrow a) \qquad \text{(biresiduation)}$$

for all $a, b \in E$. We usually also require $\boldsymbol{L}$ to be *complete*, i.e., both supremum $\bigvee K$ and infimum $\bigwedge K$ exist for each $K \subseteq L$.

Later, we will demonstrate that $\boldsymbol{L}$ must be a complete MV-algebra, that is a residuated lattice fulfilling the following identities:

$$a \oplus b = b \oplus a \qquad a \otimes b = b \otimes a \tag{3}$$

$$a \oplus (b \oplus c) = (a \oplus b) \oplus c \qquad a \otimes (b \otimes c) = (a \otimes b) \otimes c \tag{4}$$

$$a \oplus \mathbf{0} = a \qquad a \otimes \mathbf{1} = a \tag{5}$$

$$a \oplus \mathbf{1} = \mathbf{1} \qquad a \otimes \mathbf{0} = \mathbf{0} \tag{6}$$

$$a \oplus \neg a = \mathbf{1} \qquad a \otimes \neg a = \mathbf{0} \tag{7}$$

$$\neg(a \oplus b) = \neg a \otimes \neg b \qquad \neg(a \otimes b) = \neg a \oplus \neg b \tag{8}$$

$$a = \neg\neg a \qquad \neg\mathbf{0} = \mathbf{1} \tag{9}$$

$$\neg(\neg a \oplus b) \oplus b = \neg(\neg b \oplus a) \oplus a. \tag{10}$$

Note that some of these identities are redundant.

We will also use the following notation:

$$a^n = \underbrace{a \otimes \ldots \otimes a}_{n\text{-times}} \qquad \text{(power)}$$

$$na = \underbrace{a \oplus \ldots \oplus a}_{n\text{-times}}, \qquad \text{(multiple)}$$

where $a \in L$. By $\operatorname{ord}(a)$ (order of $a$) we denote the smallest $n$ such that $na = \mathbf{1}$. An element $a < \mathbf{1}$ is *nilpotent* w.r.t. $\otimes$ if there is $n \in \mathsf{N}$ such that $a^n = \mathbf{0}$. A linearly ordered MV-algebra is called an MV*-chain*.

An important *representation theorem* says that any MV-algebra is isomorphic to a subdirect product of linearly ordered MV-algebras (see [4, 23, 36]). The following lemma is used below.

LEMMA 2.1.1. *Let* $\boldsymbol{L}$ *be an* MV*-algebra. If* $\operatorname{ord}(a) < \infty$ *holds for all* $a \in L \backslash \{\mathbf{0}\}$, *then* $\boldsymbol{L}$ *is* MV*-chain.*

*Proof.* This follows from [23, Propositions 4.2.9 and 4.2.11]. □

A reader who wants to learn more about the theory of MV-algebras is referred to Chapter VI of this Handbook.

A distinguished MV-algebra is the standard Łukasiewicz one

$$\boldsymbol{L}_Ł = \langle [0,1], \vee, \wedge, \otimes, \rightarrow, 0, 1\rangle, \tag{11}$$

where the operations are defined as follows: the operations $\vee, \wedge$ are maximum and minimum, respectively. Furthermore,

$$a \otimes b = 0 \vee (a + b - 1) \qquad \text{(Łukasiewicz conjunction)} \tag{12}$$

$$a \rightarrow b = 1 \wedge (1 - a + b), \qquad \text{(Łukasiewicz implication)} \tag{13}$$

where $a, b \in [0,1]$. Note that the negation in $\boldsymbol{L}_Ł$ is $\neg a = 1 - a$.

Let $L = \{\mathbf{0} = a_0 < a_1 < \ldots < a_n = \mathbf{1}\}$ be a finite chain. Then the finite MV-chain (finite MV-algebra)

$$\boldsymbol{L}_{Fin} = \langle L, \vee, \wedge, \otimes, \rightarrow, \mathbf{0}, \mathbf{1}\rangle \tag{14}$$

has the following operations:

$$a_p \vee a_q = a_{\max\{p,q\}}$$
$$a_p \wedge a_q = a_{\min\{p,q\}}$$
$$a_p \otimes a_q = a_{\max\{0,p+q-n\}}$$
$$a_p \rightarrow a_q = a_{\min\{1,n-p+q\}}.$$

In this chapter we will suppose that truth values form an MV-algebra and denote it by $\boldsymbol{L}$. Below, we will moreover give reasons for why it must be either a linearly ordered finite MV-algebra (14) or the standard Łukasiewicz MV-algebra (11).

## 2.2 Language and formulas

We will work with a first order language $J$ defined almost classically.

DEFINITION 2.2.1. *The language $J$ is a set containing the following symbols:*

*(i)* Object variables $x, y, \ldots$

*(ii)* Object constants $\mathbf{u} \in \mathrm{OC}(J)$

*(iii) A set of* functional symbols $f \in \mathrm{Func}(J)$ *of given arities*

*(iv) A set of* predicate symbols $P \in \mathrm{Pred}(J)$ *of given arities*

*(v) The binary* implication connective $\Rightarrow$

*(vi) The* general (or universal) quantifier $\forall$

*(vii) A set of* truth constants $\mathbf{a} \in \mathrm{TC}(J) = \{\mathbf{a} \mid a \in K\}$ *for some subset $K \subseteq L$ containing* $\mathbf{0}$ *and* $\mathbf{1}$ *(the truth constants for* $\mathbf{0}, \mathbf{1}$ *will be denoted (as usual) by* $\bot, \top$*, respectively)*

*(viii) Auxiliary symbols (brackets).*

Let us put $L_Q = [0,1] \cap Q$, where Q is the set of all rational numbers. If the algebra of truth values is finite, then we put $K = L$ so that $\mathrm{TC}(J)$ contains truth constants for all the truth values from $L$. If $L = [0,1]$, then we have two choices: either we put $K = L_Q$ and so the set of truth constants is $\mathrm{TC}_Q = \{\mathbf{a} \mid a \in L_Q\}$. The obtained countable language will be denoted by $J_{L_Q}$. Alternatively, we may put $K = L = [0,1]$, but then the obtained language $J$ is uncountable.

In this chapter we will usually assume that the language $J$ is countable. On some places, we also mention the situation when $K = L$. Then the corresponding language (possibly uncountable) will be denoted by $J_L$.

Terms are defined classically. The set of all terms of the language $J$ is denoted by $M_J$. The set of all terms without variables is denoted by $M_V$.

DEFINITION 2.2.2.

*(i) A truth constant $\mathbf{a}, a \in K$ is an (atomic) formula.*

*(ii) Let $P \in \mathrm{Pred}(J)$ be an $n$-ary predicate symbol and $t_1, \ldots, t_n$ be terms. Then the expression $P(t_1, \ldots, t_n)$ is an (atomic) formula.*

*(iii) If $A, B$ are formulas, then $A \Rightarrow B$ is a formula.*

*(iv) If $x$ is a variable and $A$ a formula, then $(\forall x)A$ is a formula.*

We will furthermore introduce the following abbreviations of formulas.

$$\begin{aligned}
\neg A &:= A \Rightarrow \bot && \text{(negation)}\\
A \vee B &:= (B \Rightarrow A) \Rightarrow A && \text{(disjunction)}\\
A \wedge B &:= \neg((B \Rightarrow A) \Rightarrow \neg B) && \text{(conjunction)}\\
A \mathbin{\&} B &:= \neg(A \Rightarrow \neg B) && \text{(strong conjunction)}\\
A \mathbin{\nabla} B &:= \neg(\neg A \mathbin{\&} \neg B) && \text{(strong disjunction)}\\
A \Leftrightarrow B &:= (A \Rightarrow B) \wedge (B \Rightarrow A) && \text{(equivalence)}\\
A^n &:= \underbrace{A \mathbin{\&} A \mathbin{\&} \cdots \mathbin{\&} A}_{n\text{-times}} && \text{($n$-fold strong conjunction)}\\
nA &:= \underbrace{A \mathbin{\nabla} A \mathbin{\nabla} \cdots \mathbin{\nabla} A}_{n\text{-times}} && \text{($n$-fold strong disjunction)}\\
(\exists x)A &:= \neg(\forall x)\neg A. && \text{(existential quantifier)}
\end{aligned}$$

The set of all formulas of the language $J$ is denoted by $F_J$. Note that truth constants can also be understood as nullary predicates. All the concepts of the scope of the quantifier, sets $\mathrm{FV}(t), \mathrm{FV}(A)$ of free variables, substitutable terms, and open and closed formula are defined classically. The symbol $A_x[t]$ denotes the formula in which all free occurrences of $x$ are replaced by the (substitutable) term $t$.

## 3 Traditional and evaluated syntax

In this section, we will recall some notions concerning the syntactic consequence operation and present in parallel the classical and graded approach. Our goal is to show that the evaluated syntax lead to a natural generalization of the classical concepts.

### 3.1 Classical syntactic consequence

Let us consider a logical system with a traditional syntax given by a language $J$ that contains, besides others, a truth constant $\perp$ and a set $R$ of $n$-ary inference rules.

A *consequence operation* is an operation on the power set of formulas being a *closure operation* $\mathcal{C}\colon P(F_J) \longrightarrow P(F_J)$, i.e., the conditions

(a) $X \subseteq \mathcal{C}(X)$,

(b) $X \subseteq Y$ implies $\mathcal{C}(X) \subseteq \mathcal{C}(Y)$,

(c) $\mathcal{C}(X) = \mathcal{C}(\mathcal{C}(X))$

hold for all $X, Y \subseteq F_J$.

Recall that in formal logical systems, we introduce two types of consequence operations: syntactic and semantic. Intuitively, the syntactic consequence operation assigns to a set of formulas all the formulas derivable from the former, i.e., those to which it is possible to find a proof. The semantic consequence operation assigns to a set of formulas all the formulas being true provided that the former formulas are true as well.

DEFINITION 3.1.1. *Let $V \subseteq F_J$ be a set of formulas and $r$ be an $n$-ary inference rule.*[5] *We say that $V$ is* closed *with respect to $r$ if $A_1, \dots, A_n \in V$ implies $r(A_1, \dots, A_n) \in V$ for all formulas $A_1, \dots, A_n \in \mathrm{dom}(r)$.*

Using the concept of sets closed with respect to inference rules, we define the syntactic consequence operation as follows.

DEFINITION 3.1.2. *The* syntactic consequence operation *is an operation $\mathcal{C}^{\mathrm{syn}}$ defined on $P(F_J)$ that assigns to every $X \subseteq F_J$ the set of formulas*

$$\mathcal{C}^{\mathrm{syn}}(X) = \bigcap\{V \subseteq F_J \mid X \subseteq V, V \text{ is closed w.r.t. all } r \in R\}.$$

It is easy to see that the operation $\mathcal{C}^{\mathrm{syn}}$ is a closure operation on $P(F_J)$. Recall that the intuitive idea behind the concept of the syntactic consequence operation is the requirement to find a proof to every formula being a consequence of formulas from $X$.

THEOREM 3.1.3. *Given a set $X \subseteq F_J$. Then $A \in \mathcal{C}^{\mathrm{syn}}(X)$ iff it is provable from $X$ (i.e., there is a proof of $A$ from $X$).*

*Proof.* Put

$$D = \{A \in F_J \mid \text{there is a formal proof of } A \text{ from } X\}.$$

Obviously, $X \subseteq D$, because if $A \in X$, then $w = A$ is a proof of $A$. We show that $D$ is closed with respect to all $r \in R$. Indeed, let $B_1, \dots, B_n \in D$ and consider $r \in R$. Then there are proofs $w_{B_1}, \dots, w_{B_n}$ of these formulas from $X$. Consequently,

$$w = w_{B_1}, \dots, w_{B_n}, r(B_1, \dots, B_n)$$

is a proof of the formula $A = r(B_1, \dots, B_n)$ from $X$ and we conclude that $A \in D$.

[5] Of course, the arity $n$ is uniquely given by $r$. We will silently assume this without special notice.

Now, let $A \in \mathcal{C}^{\mathrm{syn}}(X)$. Because $X \subseteq D$ and $D$ is closed with respect to all $r \in R$, we have $A \in D$, which means that there exists a formal proof of $w_A$ from $X$.

Vice-versa, we have to show that every provable formula, $A \in D$, belongs to $\mathcal{C}^{\mathrm{syn}}(X)$, i.e., that $D \subseteq V$ for all $V$, $X \subseteq V$, that are closed w.r.t. all $r \in R$. We will proceed by induction on the length of the (shortest) proof.

Let $w = A$, where $A \in X$. Obviously $A \in V$ by the definition of $V$. Let us now consider the proof $w = w_{B_1}, \dots, w_{B_n}, A$, where $A = r(B_1, \dots, B_n)$ and $w_{B_1}, \dots, w_{B_n}$ are proofs of some formulas $B_1, \dots, B_n$, respectively. By the inductive assumption, $B_1, \dots, B_n \in V$, but because $V$ is closed w.r.t. $r$, we have $A \in V$. □

## 3.2 Evaluated syntax and graded syntactic consequence

In this subsection, we will demonstrate that evaluated syntax is indeed a natural generalization of the traditional one. The truth values are considered to form a fixed MV-algebra $\boldsymbol{L}$.

DEFINITION 3.2.1. *Let $K \subseteq L$ be a set of truth values considered as a support for truth constants due to Definition 2.2.1. An* evaluated formula[6] *is a couple*

$$^{a}/A,$$

*where $A \in F_J$ is a formula and $a \in K$ is its syntactic evaluation.*

Any derivation in $\mathrm{Ev}_{\text{Ł}}$ proceeds with evaluated formulas. It should be emphasized that the syntactic evaluations of formulas *do not belong* to the language $J$.

DEFINITION 3.2.2. *An $n$-ary* inference rule *$r$ in evaluated syntax is an operation over evaluated formulas defined as follows:*

$$r\colon \frac{^{a_1}/A_1, \dots, ^{a_n}/A_n}{^{r^{\mathrm{evl}}(a_1,\dots,a_n)}/r^{\mathrm{syn}}(A_1,\dots,A_n)}. \tag{15}$$

*Using this definition, $^{r^{\mathrm{evl}}(a_1,\dots,a_n)}/r^{\mathrm{syn}}(A_1,\dots,A_n)$ is inferred from the evaluated formulas $^{a_i}/A_i, \dots, ^{a_n}/A_n$. The syntactic operation $r^{\mathrm{syn}}$ is a partial $n$-ary operation on $F_J$, and the evaluation operation $r^{\mathrm{evl}}$ is an $n$-ary lower semicontinuous operation on $L$ (i.e., it preserves arbitrary suprema in each variable).*

A natural generalization of Definition 3.1.1 is the following:

DEFINITION 3.2.3. *A fuzzy set $V \underset{\sim}{\subset} F_J$ of formulas is* closed *with respect to an $n$-ary inference rule $r$ (in the sense of Definition 3.2.2) if*

$$V(r^{\mathrm{syn}}(A_1,\dots,A_n)) \geq r^{\mathrm{evl}}(V(A_1),\dots,V(A_n)) \tag{16}$$

*holds for all formulas $A_1, \dots, A_n \in \mathrm{dom}(r^{\mathrm{syn}})$.*

---

[6] Some authors use also the term "graded formula" instead.

Thus, the membership degree of the formula $r^{\rm syn}(A_1, \ldots, A_n)$ in $V$ must be greater than the value $r^{\rm evl}(V(A_1), \ldots, V(A_n))$ as computed from the membership degrees $V(A_1), \ldots, V(A_n)$ using the operation $r^{\rm evl}$. When considering only two truth values, this definition becomes equivalent to Definition 3.1.1.

DEFINITION 3.2.4. *Let $R$ be a set of inference rules. Then the* fuzzy set of syntactic consequences *of the fuzzy set $X \lessapprox F_J$ is given by the membership function*

$$\mathcal{C}^{\rm syn}(X)(A) = \bigwedge\{V(A) \mid X \leq V \lessapprox F_J \text{ and } V \text{ is closed w.r.t. to all } r \in R\}. \quad (17)$$

Obviously, this is a *graded consequence operation* on $\mathcal{F}(F_J)$ assigning a fuzzy set $\mathcal{C}(X) \lessapprox F_J$ to a fuzzy set $X \lessapprox F_J$. Moreover, it is a graded closure operation, i.e., the following conditions are fulfilled:

(a) $X(A) \leq \mathcal{C}(X)(A)$,

(b) $X(A) \leq Y(A)$ implies $\mathcal{C}(X)(A) \leq \mathcal{C}(Y)(A)$,

(c) $\mathcal{C}(X)(A) = \mathcal{C}(\mathcal{C}(X))(A)$

for all fuzzy sets of formulas $X, Y \lessapprox F_J$ and all formulas $A \in F_J$.

A graded consequence operation $\mathcal{C}$ is called *compact* if to every $X \lessapprox F_J$ and $A \in F_J$ there is a finite set $G \subset F_J$ such that $\mathcal{C}(X)(A) = \mathcal{C}(X|G)(A)$, where

$$(X|G)(A) = \begin{cases} X(A) & \text{if } A \in G \\ \mathbf{0} & \text{otherwise.} \end{cases}$$

When considering a sequence of evaluated formulas, we naturally come to the concept of an evaluated formal proof.

DEFINITION 3.2.5. *An* evaluated formal proof *of a formula $A$ from a fuzzy set $X \lessapprox F_J$ is a finite sequence of evaluated formulas*

$$w = a_0/A_0,\ a_1/A_1, \ldots,\ a_n/A_n \quad (18)$$

*such that $A_n = A$, and for each $i \leq n$, either*

$$a_i/A_i = X(A_i)/A_i$$

*or there exists an $n$-ary inference rule $r$ such that*

$$a_i/A_i = r^{\rm evl}(a_{i_1}, \ldots, a_{i_n})/r^{\rm syn}(A_{i_1}, \ldots, A_{i_n}), \quad i_1, \ldots, i_n < i.$$

The evaluation $a_n$ of the last evaluated formula in (18) is the *value* of the evaluated proof $w$, and we write it as $\mathrm{Val}(w)$. If $w$ is an evaluated proof of $A$, then we will often write $w_A$. If it is clear from the context that we are dealing with an evaluated proof, we will usually omit the adjective "evaluated".

The relation between the concepts of evaluated proof and syntactic consequences is clarified in the following important theorem. It can also be seen as a generalization of Theorem 3.1.3.

THEOREM 3.2.6. *Let $X \subsetneq F_J$ and $R$ be a set of inference rules. Then*

$$\mathcal{C}^{\mathrm{syn}}(X)(A) = \bigvee\{\mathrm{Val}(w) \mid w \text{ is a proof of } A \text{ from } X\}. \tag{19}$$

*Proof.* Let $A \in F_J$. We denote by $W \subsetneq F_J$ the fuzzy set defined by the right-hand side of (19).

First, we show by induction on the length of a (shortest) proof $w_A$ that $\mathrm{Val}(w_A) \leq V(A)$ for every proof $w_A$ and every fuzzy set $V \subsetneq F_J$ such that $X \subseteq V$, i.e., that $W(A) \leq \mathcal{C}^{\mathrm{syn}}(X)(A)$.

Let the length of $w_A$ be 1. Then $A \in \mathrm{Supp}(X)$ and thus, by Definition 3.2.5 $\mathrm{Val}(w_A) = X(A) \leq V(A)$.

Let the length of $w_A$ be greater than 1 and $A = r^{\mathrm{syn}}(B_1, \ldots, B_n)$ for some already proved formulas $B_1, \ldots, B_n$ occurring in $w_A$. Then, for every $V$ from (17), we obtain

$$\mathrm{Val}(w_A) = r^{\mathrm{evl}}(\mathrm{Val}(w_{B_1}), \ldots, \mathrm{Val}(w_{B_n})) \leq r^{\mathrm{evl}}(V(B_1), \ldots, V(B_n)) \leq V(A)$$

by the inductive assumption and the fact that $V$ is closed with respect to $r$.

Vice-versa, because $w_A = X(A)/A$ is a proof with the value $\mathrm{Val}(w_A) = X(A)$, it holds that $X \leq W$. Therefore, by (17), it is sufficient to show that $W$ is closed with respect to all the inference rules from $R$.

Let $r \in R$ and $A = r^{\mathrm{syn}}(B_1, \ldots, B_n)$ for some formulas $B_1, \ldots, B_n$. First, realize that if $w_{B_1}, \ldots, w_{B_n}$ are proofs of the respective formulas, then

$$w_A = \mathrm{Val}(w_{B_1})/B_1, \ldots, \mathrm{Val}(w_{B_n})/B_n, r^{\mathrm{evl}}(\mathrm{Val}(w_{B_1}), \ldots, \mathrm{Val}(w_{B_n})/A$$

is a proof of $A$. Then

$$\begin{aligned} W(A) &= \bigvee\{\mathrm{Val}(w_A) \mid \text{all proofs } w_A\} \\ &\geq \bigvee\{r^{\mathrm{evl}}(\mathrm{Val}(w_{B_1}), \ldots, \mathrm{Val}(w_{B_n})) \mid \text{all proofs } w_{B_1}, \ldots, w_{B_n}\} \\ &= r^{\mathrm{evl}}\left(\bigvee\{\mathrm{Val}(w_{B_1}) \mid \text{all proofs } w_{B_1}\}, \ldots, \bigvee\{\mathrm{Val}(w_{B_n}) \mid \text{all proofs } w_{B_n}\}\right) \\ &= r^{\mathrm{evl}}(W(B_1), \ldots, W(B_n)) \end{aligned}$$

by the semicontinuity of $r^{\mathrm{evl}}$. □

This theorem provides a certain kind of finitistic characterization of the syntactic consequence operation. However, unlike the classical syntax, where finding one finite proof is sufficient, the situation here is not so nice. In general, a sequence (possibly infinite) of finite proofs must be considered because the maximum of their values may not exist, so we cannot confine ourselves to one proof only. Due to this deliberation, we rewrite (19) as follows.

COROLLARY 3.2.7.

$$\mathcal{C}^{\mathrm{syn}}(X)(A) = \bigvee\{\mathcal{C}^{\mathrm{syn}}(X|G)(A) \mid G \subset F_J \text{ and } G \text{ is finite}\}. \tag{20}$$

*Proof.* Let $G_{w_A}$ be a set of all formulas occurring in the proof $w_A$. Obviously, $G_{w_A}$ is finite. Using Theorem 3.2.6, we obtain

$$\begin{aligned}\mathcal{C}^{\mathrm{syn}}(X)(A) &= \bigvee\{\mathcal{C}^{\mathrm{syn}}(X|G_{w_A})(A) \mid w_A \text{ is a proof of } A\}\\ &\leq \bigvee\{\mathcal{C}^{\mathrm{syn}}(X|G)(A) \mid G \subset F_J, G \text{ is finite}\}\\ &\leq \mathcal{C}^{\mathrm{syn}}(X)(A).\end{aligned}$$

□

### 3.3 Semantics

The semantics of Ev$_Ł$ is defined as a generalization of the classical (Tarskian) semantics of predicate logic. The algebra of truth values $\boldsymbol{L}$ has the support $L$, and it is supposed to be a complete MV-algebra. A *structure* for the language $J$ is

$$\mathcal{M} = \langle M, \{P_M \mid P \in \mathrm{Pred}(J)\}, \{f_M \mid f \in \mathrm{Func}(J)\}, \{u \mid \mathbf{u} \in \mathrm{OC}(J)\}\rangle,$$

where $M$ is a set, each $P_M \subsetneqq M^n$ is an $n$-ary fuzzy relation assigned to the $n$-ary predicate symbol $P \in \mathrm{Pred}(J)$ ($n$ depends on $P$), each $f_M$ is an ordinary $n$-ary function on $M$ assigned to the $n$-ary functional symbol $f \in \mathrm{Func}(J)$, and each $u \in M$ is a designated element assigned to the object constant $\mathbf{u} \in \mathrm{OC}(J)$.

The connectives $\neg, \wedge, \vee, \&, \nabla, \Rightarrow, \Leftrightarrow$ are interpreted by the operations $\neg, \wedge, \vee, \otimes, \oplus, \rightarrow, \leftrightarrow$ on $L$, respectively. Terms are interpreted in the same way as in classical logic. If $t$ is a term interpreted by an element $m \in M$, then we will write $\mathcal{M}(t) = m$. Truth constants are interpreted by the truth value they represent, i.e., $\mathcal{M}(\mathbf{a}) = a$, $a \in L$ for all $\mathbf{a} \in \mathrm{TC}(J)$. For a precise definition of the interpretation of the other formulas, see [36].

A *truth valuation* of formulas from $F_J$ is a function $\mathcal{M}\colon F_J \to L$ that is a homomorphism w.r.t. truth constants, logical connectives and large suprema and infima. It fulfills the following conditions:

(i) If $\mathbf{a}$ is a truth constant for some $a \in L$, then $\mathcal{M}(\mathbf{a}) = a$.

(ii) If $\Box \in J$ is an $n$-ary connective interpreted by a function $h\colon L^n \to L$, then

$$\mathcal{M}(\Box(A_1, \dots, A_n)) = h(\mathcal{M}(A_1), \dots, \mathcal{M}(A_n)).$$

(iii) $\mathcal{M}((\forall x)A) = \bigwedge\{\mathcal{M}(A(x/v)) \mid v \in M\}$, where by $\mathcal{M}(A(x/v))$ we denote a truth value of the formula $A$ after replacing all free occurrences of $x$ by the element $v \in M$.[7]

DEFINITION 3.3.1. *Let $r$ be an inference rule (15). We say that it is* sound *if*

$$\mathcal{M}(r^{\mathrm{syn}}(A_1, \dots, A_n)) \geq r^{\mathrm{evl}}(\mathcal{M}(A_1), \dots, \mathcal{M}(A_n)) \tag{21}$$

*holds for all truth valuations $\mathcal{M}$ of formulas from $F_J$.*

By this definition, a sound inference rule cannot give a greater evaluation of its consequence than a truth value of the latter in any interpretation. This behavior ensures the soundness of the whole formal system (see below). Therefore, unless stated otherwise, we will always suppose that all the considered inference rules are sound.

[7] A pedantic definition requires first that the language $J$ be enriched by a collection of new constants to $J' = J \cup \{\mathbf{v} \mid v \in M\}$ and then that the interpretation $\mathcal{M}(A_x[\mathbf{v}])$ be constructed.

A typical principle considered in fuzzy set theory is the *maximality principle*: the set $\{{}^{a}/x \mid a \in K \subseteq L\}$ can be replaced by a singleton $\{(\bigvee_{a\in K} a)/x\}$. With this principle, each set of singletons is at the same time a fuzzy set. Due to the same principle, a set of evaluated formulas is the same as a fuzzy set of formulas. Thus, we can characterize $\mathrm{Ev}_{\text{Ł}}$, in general, as follows:

(i) There are no designated truth values; the *maximality principle* is considered instead.

(ii) $\mathrm{Ev}_{\text{Ł}}$ makes possible syntactical derivations concerning *any possible truth value.* Namely, each formula is characterized by its provability degree, that is, a bound for its truth degree in any model.

Note that the validity of the maximality principle in the evaluated syntax is ensured by Theorem 3.2.6.

DEFINITION 3.3.2. *Let $X \underset{\sim}{\subset} F$ be a fuzzy set of formulas. Then the fuzzy set of its* semantic consequences *is given by the membership function*

$$\mathcal{C}^{\mathrm{sem}}(X)(A) = \bigwedge\{\mathcal{M}(A) \mid \textit{for all truth valuations } \mathcal{M}\colon F_J \to L, X \leq \mathcal{M}\}. \quad (22)$$

Similarly as above, the semantic consequence operation $\mathcal{C}^{\mathrm{sem}}$ defined in (22) is a graded closure operation on $\mathcal{F}(F_J)$. Because all the truth values are equal in their importance, we may generalize the concept of tautology as follows.

DEFINITION 3.3.3. *We say that a formula $A$ is an $a$-*tautology *(*a tautology in the degree $a$*) if*

$$a = \mathcal{C}^{\mathrm{sem}}(\emptyset)(A) \quad (23)$$

*and write $\models_a A$. If $a = \mathbf{1}$, then we write simply $\models A$ and say that $A$ is a tautology.*

EXAMPLE 3.3.4. Consider as the algebra of truth values the standard Łukasiewicz MV-algebra $\boldsymbol{L}_{\text{Ł}}$. Then a formula of the form $A \vee \neg A$ is a 0.5-tautology. Then indeed, $\mathcal{M}(A \vee \neg A) = 0.5$ for $\mathcal{M}(A) = 0.5$ and $\mathcal{M}(A \vee \neg A) = \mathcal{M}(A) \vee \neg\mathcal{M}(A) \geq 0.5$ for any truth valuation $\mathcal{M}$.

Of course, due to the principle of equal importance of truth values, we also have $\mathbf{0}$-tautologies, though there is little sense in speaking about them.

### 3.4 Formal fuzzy theories

The main concept in evaluated syntax is that of the evaluated formula. Hence, it is natural to assume that axioms are evaluated formulas as well. However, because the evaluations can be interpreted as membership degrees in a fuzzy set, we conclude that axioms may form *fuzzy sets of formulas.* Similarly as in the classical case, we should distinguish logical axioms LAx and special axioms SAx. Axioms with a membership degree smaller than $\mathbf{1} \in L$, i.e., those having an initial syntactic evaluation different from $\mathbf{1}$, may be considered not fully true.

The fuzzy set of logical axioms should be a certain but fixed fuzzy set

$$\mathrm{LAx} \subset \{{}^{a}/A \mid A \text{ is an } a\text{-tautology}, a \in K\}.$$

We thus naturally arrive at the concept of formal fuzzy theory, which is a graded counterpart to the classical concept of the formal theory.

DEFINITION 3.4.1. *A formal fuzzy theory $T$ in the language $J$ is a fuzzy set of formulas from $F_J$ determined by the triple*

$$T = \langle \mathrm{LAx}, \mathrm{SAx}, R \rangle,$$

*where* $\mathrm{LAx} \lessapprox F_J$ *is a fuzzy set of logical axioms,* $\mathrm{SAx} \lessapprox F_J$ *is a fuzzy set of special axioms, and $R$ is a set of sound inference rules.*

A fuzzy theory can also be viewed as a fuzzy set $T = \mathcal{C}^{\mathrm{syn}}(\mathrm{LAx} \cup \mathrm{SAx}) \lessapprox F_J$. We will usually omit the adjectives "formal" and "fuzzy" when speaking about formal fuzzy theories with no danger of misunderstanding.

The language of a fuzzy theory $T$ will be denoted by $J(T)$. If $w$ is a proof in $T$ (i.e., the used axioms are those from LAx and SAx), then its value is denoted by $\mathrm{Val}_T(w)$.

DEFINITION 3.4.2. *Let $T$ be a fuzzy theory and $A \in F_J$ a formula.*

*(i) If* $\mathcal{C}^{\mathrm{syn}}(\mathrm{LAx} \cup \mathrm{SAx})(A) = a$*, then we say that the formula $A$ is a* theorem in the degree $a$ (provable in the degree $a$) *in the fuzzy theory $T$, and we write $T \vdash_a A$. By Theorem 3.2.6, we have*

$$T \vdash_a A \iff a = \bigvee\{\mathrm{Val}(w_A) \mid w_A \text{ is a proof from } \mathrm{LAx} \cup \mathrm{SAx}\}. \tag{24}$$

*If $a = \mathbf{1}$, then we write $T \vdash A$.*

*(ii) A truth valuation $\mathcal{M}$ is a* model of the fuzzy theory $T$, $\mathcal{M} \models T$ *whenever* $\mathrm{SAx}(A) \leq \mathcal{M}(A)$ *holds for all formulas $A \in F_J$.*

*(iii) If* $\mathcal{C}^{\mathrm{sem}}(\mathrm{LAx} \cup \mathrm{SAx})(A) = a$*, then we say that $A$ is* true in the degree $a$ *in the fuzzy theory $T$, and we write $T \models_a A$. Using (ii), we have*

$$a = \bigwedge\{\mathcal{M}(A) \mid \mathcal{M} \models T\}. \tag{25}$$

*If $a = \mathbf{1}$, then we write $T \models A$.*

It follows from (24) that finding a proof of a formula $A$ gives us information only about the lower bound of the degree in which it is a theorem in $T$. Let us emphasize that the provability (or truth) degree $a$ in $T \vdash_a A$ (or $T \models_a A$) can be an arbitrary element of $L$ (if $L = [0, 1]$, then it can even be irrational). The same holds for the truth values $\mathcal{M}(A)$ in any model $\mathcal{M} \models T$. However, the value $\mathrm{Val}_T(w_a)$ of any proof $w_A$ is always an element of $K \subseteq L$.

As a special case of Definition 3.4.2(i), if there is a proof $w_A$ of $A$ such that $\mathrm{Val}_T(w_A) = a$, then $A$ is *effectively provable in the degree $a$* in theory $T$. This means that $\mathcal{C}^{\mathrm{syn}}(\mathrm{LAx} \cup \mathrm{SAx})(A) = \mathcal{C}^{\mathrm{syn}}((\mathrm{LAx} \cup \mathrm{SAx})|G)(A)$ for some finite $G \subset F_J$.

Occasionally, SAx may be extended by some fuzzy set of formulas $\Gamma \lessapprox F_J$. Then by $T \cup \Gamma$ we understand an extended fuzzy theory

$$T \cup \Gamma = \langle \mathrm{LAx}, \mathrm{SAx} \cup \Gamma, R \rangle.$$

## 3.5 Soundness and completeness in evaluated syntax

Recall that the classical syntactic and semantic consequence operations are two different characterizations of true formulas. An important achievement of classical logic is the theorem stating that, even though each characterization is based on entirely different assumptions, both of them lead to the same result. We will show that such a conclusion also holds true in $\mathrm{Ev}_{\text{Ł}}$. We say that the evaluated syntax is *sound* if

$$\mathcal{C}^{\text{syn}}(X)(A) \leq \mathcal{C}^{\text{sem}}(X)(A) \tag{26}$$

holds for every $X \lessapprox F_J$ and any formula $A \in F_J$. When considering a formal fuzzy theory $T$, we may write (26) as

$$T \vdash_a A \quad \text{and} \quad T \models_b A \quad \text{implies} \quad a \leq b \tag{27}$$

holds for every theory $T$ and any formula $A \in F_J$.

Recall that in $\mathrm{Ev}_{\text{Ł}}$, all truth values are equal in their importance. Hence, we always have $T \vdash_a A$ as well as $T \models_b A$ for some $a \in L$ and $b \in L$. It is also possible that both degrees are equal to $\mathbf{0}$.

The evaluated syntax is *complete* if

$$\mathcal{C}^{\text{syn}}(X)(A) = \mathcal{C}^{\text{sem}}(X)(A) \tag{28}$$

holds for every $X \lessapprox F_J$ and a formula $A \in F_J$. Hence, a fuzzy theory $T$ is complete if

$$T \vdash_a A \iff T \models_a A \tag{29}$$

holds for every theory $T$ and every formula $A \in F_J$.

Completeness in the evaluated syntax, however, is not as straightforward as in the traditional syntax. The syntactic consequence operation in the traditional system means that some proof of a given formula is found (a single proof is sufficient). In the evaluated syntax, however, one proof may be sufficient only in a very special case. In general, we must find a set (possibly infinite in the case that $L = [0,1]$) of all the proofs and evaluate its supremum. Similarly, the semantic consequence operations means finding the infimum of all the possible truth evaluations. Hence, completeness here means a sort of limit coincidence of different truth characterizations.

On the other hand, we argue that the evaluated syntax is a natural generalization of the traditional one. The classical syntactic derivation means we start with axioms assumed to be hereditarily true, and using sound inference rules, we derive again true formulas. Hence, the syntax of classical logic is virtually evaluated as well, namely, by $\mathbf{1}$, because evaluation by $\mathbf{0}$ as the other possibility is uninteresting. Therefore, we do not need to speak about evaluated syntax in classical logic.

## 3.6 Restrictions when choosing the algebra of truth values

In this section, we will demonstrate that if we require equality (28) to hold true, then the algebra of truth values must have quite specific properties.

THEOREM 3.6.1. *If equality (28) holds and $\mathcal{C}^{\text{syn}}$ is compact, then the algebra of truth values $\mathbf{L}$ satisfies both the ascending and descending chain condition.*[8]

[8] An ordered set satisfies the as(des)cending chain condition if every such chain of its elements is finite.

*Proof.* Because equality (28) holds true, $\mathcal{C}^{\text{sem}}$ must be compact as well.

(a) Let $\boldsymbol{L}$ not fulfill the ascending chain condition. Let $a_0 < a_1 < \cdots < a_n < \cdots$ be an infinite ascending chain, and let $a = \bigvee\{a_n \mid n \in \mathsf{N}\}$. Let $B$ be a closed atomic formula, and define a fuzzy theory $T$ by the following fuzzy set of special axioms:

$$\text{SAx}(A) = \begin{cases} \mathbf{1} & \text{if } A = \mathbf{a}_n \Rightarrow B, n \in \mathsf{N} \\ \mathbf{0} & \text{otherwise.} \end{cases}$$

Then $T \models_a B$ because $a_n \leq \mathcal{M}(B)$ holds for each $n \in N$ in every model $\mathcal{M} \models T$. Let $G \subset \text{Supp}(\text{SAx})$ be a finite set, and let $T|G$ be fuzzy theory with the fuzzy set of special axioms $\text{SAx}|G$. Let $n_G \in \mathsf{N}$ be a maximal subscript among all the formulas $\mathbf{a}_n \Rightarrow B \in G$. Then $T|G \vdash_{a_{n_G}} B$, where $a_{n_G} < a$. Consequently, $\mathcal{C}^{\text{sem}}(\text{LAx} \cup \text{SAx})$ is not compact.

(b) Let $\boldsymbol{L}$ not fulfill the descending chain condition. Let $a_0 > a_1 > \cdots > a_n > \cdots$ be an infinite descending chain, and let $a = \bigwedge\{a_n \mid n \in \mathsf{N}\}$. Let $B$ be a closed atomic formula and define a fuzzy theory $T$ by the following fuzzy set of special axioms:

$$\text{SAx}(A) = \begin{cases} \mathbf{1} & \text{if } A = B \Rightarrow \mathbf{a}_n, n \in \mathsf{N} \\ \mathbf{0} & \text{otherwise.} \end{cases}$$

Then $T \models B \Rightarrow \mathbf{a}$ because $\mathcal{M}(B) \leq a_n$ holds for each $n \in N$ in every model $\mathcal{M} \models T$. Let $G \subset \text{Supp}(\text{SAx})$ be a finite set, and let $T|G$ be a fuzzy theory with the fuzzy set of special axioms $\text{SAx}|G$. Let $n_G \in \mathsf{N}$ be a maximal subscript among all the formulas $B \Rightarrow \mathbf{a}_n \in G$. Then $T|G \vdash_b B \Rightarrow \mathbf{a}_{n_G}$, where $b = a_{n_G} \to a < \mathbf{1}$. Consequently, $\mathcal{C}^{\text{sem}}(\text{LAx} \cup \text{SAx})$ is not compact. □

THEOREM 3.6.2. *Let the algebra of truth values* $\mathbf{L}$ *satisfy the ascending but not descending chain condition. Then equality (28) cannot hold.*

*Proof.* Let us assume that equality (28) holds. Because $\boldsymbol{L}$ satisfies the ascending condition, as a consequence of Theorem 3.2.6, the set of all proofs of a formula $A$ in (19) contains a maximal element. Then by the previous theorem, the consequence operation $\mathcal{C}^{\text{syn}}$ is compact—a contradiction. □

THEOREM 3.6.3. *Given a formal system of fuzzy logic with evaluated syntax with the implication connective* $\Rightarrow$ *interpreted by the residuation operation* $\to$ *on a complete residuated lattice (1) whenever* $\to$ *does not fulfill the equations*

$$\bigvee\{a \to b \mid b \in K\} = a \to \bigvee\{b \mid b \in K\} \tag{30}$$

$$\bigvee\{a \to b \mid a \in K\} = \bigwedge\{a \mid a \in K\} \to b \tag{31}$$

$$\bigwedge\{a \to b \mid b \in K\} = a \to \bigwedge\{b \mid b \in K\} \tag{32}$$

$$\bigwedge\{a \to b \mid a \in K\} = \bigvee\{a \mid a \in K\} \to b \tag{33}$$

*for an arbitrary subset* $K \subseteq L$*, then equality (28) does not hold.*

*Proof.* First, it may be verified that equalities (32) and (33) hold in every complete residuated lattice. Therefore, we have to check equalities (30) and (31).

Let us assume that equality (30) does not hold for a set $E \subseteq L$ and denote $e_1 = \bigvee\{a \to b \mid b \in E\}$ and $e_2 = a \to \bigvee\{b \mid b \in E\}$ for some $a \in L$. Then by the isotonicity of $\to$ in the second argument, we conclude that $e_1 < e_2$.

Let $B \in F_J$ and $T$ be a fuzzy theory given by the following fuzzy set of special axioms

$$\mathrm{SAx}(A) = \begin{cases} \mathbf{1} & \text{if } A = \mathbf{b} \Rightarrow B, b \in E \\ \mathbf{0} & \text{otherwise.} \end{cases}$$

Now, let $A = \mathbf{a} \Rightarrow B$ for the $a$ considered above. Then

$$\begin{aligned} \mathcal{C}^{\mathrm{sem}}(\mathrm{LAx} \cup \mathrm{SAx})(A) &= \bigwedge\{\mathcal{M}(A) \mid \mathcal{M} \models T\} = a \to \bigwedge_{\mathcal{M} \models T} \mathcal{M}(B) \\ &= a \to \bigvee_{i \in I_0} b_i = e_2 \end{aligned}$$

by equality (32) and because $\bigvee\{b \mid b \in E\} \leq \mathcal{M}(B)$ holds for every model $\mathcal{M} \models T$.

Because $\mathcal{C}^{\mathrm{syn}}$ fulfills equality (20), equality (28) implies that $\mathcal{C}^{\mathrm{sem}}$ must also fulfill (20). Let $G \subset F_J$ be a finite set. Then

$$\begin{aligned} \mathcal{C}^{\mathrm{syn}}((\mathrm{LAx} \cup \mathrm{SAx})|G)(A) &= \bigwedge\{\mathcal{M}(A) \mid ((\mathrm{LAx} \cup \mathrm{SAx})|G)(A) \leq \mathcal{M}(A), \mathcal{M} \models T\} \\ &= a \to e_G \end{aligned}$$

for some $e_G \leq \bigvee\{b \mid b \in E\}$. Using (20) we then obtain

$$\begin{aligned} \mathcal{C}^{\mathrm{syn}}(\mathrm{LAx} \cup \mathrm{SAx})(A) &= \bigvee\{\mathcal{C}^{\mathrm{syn}}((\mathrm{LAx} \cup \mathrm{SAx})|G)(A) \mid G \subset F_J, G \text{ finite}\} \\ &= \bigvee\{a \to e_G \mid G \subset F_J, G \text{ finite}\} \leq e_1 < e_2. \end{aligned}$$

Consequently, equality (28) is not satisfied.

If the second equality (31) does not hold, then we put

$$\mathrm{SAx}(A) = \begin{cases} \mathbf{1} & \text{if } A = B \Rightarrow \mathbf{b}, b \in E \\ \mathbf{0} & \text{otherwise,} \end{cases}$$

and we proceed analogously as above. □

The following lemma holds true for linearly ordered residuated lattices.

LEMMA 3.6.4. *Let $\mathbf{L}$ be a complete residuated chain. Let a topology $\tau$ be given by the open basis $B = \{\{x \in L \mid a < x < b\} \mid a, b \in L\}$. Then the residuation operation $\to$ is continuous with respect to $\tau$ iff it fulfills the equalities (30)–(33).*

*Proof.* All four conditions are equivalent to the upper as well as lower continuity of $\to$ in both arguments. Therefore, it is continuous. □

Theorems 3.6.1–3.6.3 have the following consequences for the choice of the algebra of truth values. Due to Theorem 3.6.3, the interpretation of the implication must satisfy equalities (30) and (31), which, however, are fulfilled in complete MV-algebras. Consequently, we are bound to assume that the structure of truth values $\boldsymbol{L}$ should form an MV-algebra.

A further question is whether $\boldsymbol{L}$ should be linearly ordered. By the representation theorem mentioned in Subsection 2.1, we may confine the scope to linear MV-algebras only. From the syntactical properties of $\mathrm{Ev}_{Ł}$ described below (see Lemma 4.5.9), it follows that a linear ordering of $\boldsymbol{L}$ is even necessary.

If $\boldsymbol{L}$ is a finite chain, then it satisfies both ascending and descending conditions, so we can construct a complete fuzzy logic with evaluated syntax (i.e., a logic for which equality (28) is satisfied). If $\boldsymbol{L}$ is a countably infinite residuated chain, then by Theorem 3.6.2 fuzzy logic completeness cannot be reached.

Because a countable chain is unacceptable, we should take $[0,1]$ as the set of truth values. However, due to Lemma 3.6.4, the residuation must be continuous. By [24, Proposition 2.3], any couple $\langle \otimes, \rightarrow \rangle$ of binary operations on $[0,1]$ adjoint by (2) where $\rightarrow$ is continuous are isomorphic with the couple of Łukasiewicz operations (12) and (13), respectively. Consequently, finite linearly ordered MV-algebra and the standard Łukasiewicz MV-algebra $\boldsymbol{L}_{Ł}$ are the only plausible choices for the structure of truth values for $\mathrm{Ev}_{Ł}$.

# 4 Syntax of $\mathrm{Ev}_{Ł}$

## 4.1 Logical axioms

Let us now define the fuzzy set LAx of logical axioms in $\mathrm{Ev}_{Ł}$.

DEFINITION 4.1.1. *The fuzzy set* LAx *of* logical axioms *is specified as follows:*

$$\mathrm{LAx}(A) = \begin{cases} a \in L & \text{if } A = \mathbf{a} \\ 1 & \text{if } A \text{ is one of (R1)–(R4), (B1), (T1)–(T2) specified below} \\ 0 & \text{otherwise.} \end{cases}$$

*(R1)* $A \Rightarrow (B \Rightarrow A)$

*(R2)* $(A \Rightarrow B) \Rightarrow ((B \Rightarrow C) \Rightarrow (A \Rightarrow C))$

*(R3)* $(\neg B \Rightarrow \neg A) \Rightarrow (A \Rightarrow B)$

*(R4)* $((A \Rightarrow B) \Rightarrow B) \Rightarrow ((B \Rightarrow A) \Rightarrow A)$

*(B1)* $(\mathbf{a} \Rightarrow \mathbf{b}) \Leftrightarrow (\overline{\mathbf{a} \rightarrow \mathbf{b}})$ — $\overline{\mathbf{a} \rightarrow \mathbf{b}}$ *denotes the logical constant (atomic formula) for the truth value* $a \rightarrow b$

*(T1)* $(\forall x)A \Rightarrow A_x[t]$ — *for any substitutable term* $t$

*(T2)* $(\forall x)(A \Rightarrow B) \Rightarrow (A \Rightarrow (\forall x)B)$ — *provided that* $x$ *is not free in* $A$.

Axioms (R1)–(R4) are those originally proposed by A. Rose and J.B. Roser for Łukasiewicz logic. Axiom (B1) is called *book-keeping*.[9] It ensures the correct behavior of truth constants with respect to the connectives. Axiom (T1) is the *substitution axiom.* Note that both (T1) and (T2) coincide with the corresponding classical logical axioms.

## 4.2 Inference rules and validity

The set $R$ of inference rules must contain, at least, the following three rules:

$$r_{\mathrm{MP}}\colon \frac{a/A,\, b/A \Rightarrow B}{(a \otimes b)/B} \qquad \text{(modus ponens)}$$

$$r_{\mathrm{G}}\colon \frac{a/A}{a/(\forall x)A} \qquad \text{(generalization)}$$

$$r_{\mathrm{LC}}\colon \frac{a/A}{(a \to a)/\mathbf{a} \Rightarrow A}. \qquad \text{(truth constant introduction)}$$

It is easy to prove that all these rules are valid and sound in the sense of Definitions 3.2.2 and 3.3.1. Note that the evaluation operation of $r_{\mathrm{LC}}$ is $r^{\mathrm{evl}}_{\mathrm{LC}}(x) = a \to x$.

LEMMA 4.2.1. *For every proof $w_A$ with $\mathrm{Val}(w_A) = a$, there is a proof $w_{\mathbf{b}\Rightarrow A}$ with the value $\mathrm{Val}(w_{\mathbf{b}\Rightarrow A}) = b \to a$.*

*Proof.*

$$\begin{aligned} w_{\mathbf{b}\Rightarrow A} = a/A \quad \{w_A\},\ 1/\mathbf{a} \Rightarrow A \quad \{r_{\mathrm{LC}}\},\ b \to a/\overline{\mathbf{b} \to \mathbf{a}} \quad \{\textit{logical axiom}\},\\ \ldots,\ 1/(\mathbf{b} \Rightarrow \mathbf{a}) \Rightarrow ((\mathbf{a} \Rightarrow A) \Rightarrow (\mathbf{b} \Rightarrow A)) \quad \{(R2)\}, \ldots,\\ b \to a/\mathbf{b} \Rightarrow A \quad \{r_{\mathrm{MP}}\}. \end{aligned}$$

□

THEOREM 4.2.2 (Validity theorem). *Let $T$ be a fuzzy theory. If $T \vdash_a A$ and $T \models_b A$, then $a \leq b$ holds for every formula $A$.*

*Proof.* The proof of this routine theorem can be given by checking that all axioms (R1)–(R4), (B1) and (T1), (T2) are tautologies in the degree 1. Moreover, $\mathcal{M}(\mathbf{a}) = a$ for all $a \in L$ holds in every model $\mathcal{M} \models T$. The theorem then follows from the soundness of all the inference rules $r_{\mathrm{MP}}, r_{\mathrm{G}}$ and $r_{\mathrm{LC}}$. □

From the validity theorem and the definition of the syntax and semantics of $\mathrm{Ev}_{Ł}$, we immediately obtain the following corollary.

COROLLARY 4.2.3. *In every fuzzy theory $T$, $T \vdash_a \mathbf{a}$ as well as $T \models_a \mathbf{a}$ holds for every truth constant $\mathbf{a}$, $a \in K$.*

The following theorem is a direct analogy of the corresponding classical one.

THEOREM 4.2.4 (Closure theorem). *Let $T$ be a fuzzy theory and $A \in F_{J(T)}$ be a formula whose free variables all occur among $x_1, \ldots, x_n$. Let $(\forall x_1)\cdots(\forall x_n)A$ be its universal closure. Then $T \vdash_a A \iff T \vdash_a (\forall x_1)\cdots(\forall x_n)A$.*

[9] This name has been proposed by P. Hájek.

*Proof.* Let $w_A$ be a proof of $A$. Using $r_G$, we obtain a proof $w_{(\forall x_1)\cdots(\forall x_n)A}$ of the universal closure of $A$ such that $\mathrm{Val}(w_A) = \mathrm{Val}(w_{(\forall x_1)\cdots(\forall x_n)A})$, and thus we have $T \vdash_b (\forall x_1)\cdots(\forall x_n)A$, $b \geq a$. The converse is proved classically using the substitution axiom (T1). □

### 4.3 Special provable tautologies and properties of the provability degree

In this section, we will introduce a few formal theorems needed in the proofs below. Note that all of them are theorems of Łukasiewicz logic and can be formally proved from the Rose–Roser's axioms introduced above.

LEMMA 4.3.1. *The following are schemes of effectively provable formal propositional theorems in* $\mathrm{Ev}_Ł$, *i.e., for each theorem there exists a proof with the degree equal to* 1.

*(P1)* $T \vdash (A \wedge B) \Rightarrow A$ $\qquad T \vdash (A \wedge B) \Rightarrow B$

*(P2)* $T \vdash (A \mathbin{\&} B) \Rightarrow A$ $\qquad T \vdash (A \mathbin{\&} B) \Rightarrow B$

*(P3)* $T \vdash A \Rightarrow (A \vee B)$ $\qquad T \vdash B \Rightarrow (A \vee B)$

*(P4)* $T \vdash (A \mathbin{\&} \neg A) \Rightarrow B$

*(P5)* $T \vdash (A \Rightarrow C) \Rightarrow ((B \Rightarrow C) \Rightarrow (A \vee B) \Rightarrow C)$

*(P6)* $T \vdash A \Rightarrow (B \Rightarrow (A \mathbin{\&} B))$

*(P7)* $T \vdash (A \Rightarrow B) \vee (B \Rightarrow A)$

*(P8)* $T \vdash (A \vee B)^n \Rightarrow (A^n \vee B^n)$ $\qquad n \in \mathsf{N}^+$

*(P9)* $T \vdash (A \Rightarrow (B \Rightarrow C)) \Leftrightarrow (B \Rightarrow (A \Rightarrow C))$

*(P10)* $T \vdash (A^m \Rightarrow (B \Rightarrow C)) \Rightarrow ((A^n \Rightarrow B) \Rightarrow (A^{m+n} \Rightarrow C))$ $\qquad m, n \in \mathsf{N}^+$

*(P11)* $T \vdash (A \mathbin{\&} B)^n \Rightarrow (A^n \Rightarrow B^n)$ $\qquad n \in \mathsf{N}^+$

*(P12)* $T \vdash (A \mathbin{\&} (B \vee C)) \Leftrightarrow ((A \mathbin{\&} B) \vee (A \mathbin{\&} C))$.

LEMMA 4.3.2. *The following are schemes of the effectively provable formal predicate theorems in* $\mathrm{Ev}_Ł$, *i.e., for each theorem there exists a proof with the degree equal to* 1.

*(Q1)* $T \vdash (\forall x)(A \Rightarrow B) \Rightarrow ((\forall x)A \Rightarrow (\forall x)B)$

*(Q2)* $T \vdash (\forall x)(A \mathbin{\&} B) \Leftrightarrow ((\forall x)A \mathbin{\&} B)$ $\qquad$ *provided that $x$ is not free in $B$*

*(Q3)* $T \vdash (\forall x)(A \Rightarrow B) \Leftrightarrow (A \Rightarrow (\forall x)B)$ $\qquad$ *provided that $x$ is not free in $A$*

*(Q4)* $T \vdash (\forall x)(A \Rightarrow B) \Leftrightarrow ((\exists x)A \Rightarrow B)$ $\qquad$ *provided that $x$ is not free in $B$*

*(Q5)* $T \vdash (\exists x)(A \Rightarrow B) \Leftrightarrow (A \Rightarrow (\exists x)B)$ $\qquad$ *provided that $x$ is not free in $A$*

*(Q6)* $T \vdash (\exists x)(A \Rightarrow B) \Leftrightarrow ((\forall x)A \Rightarrow B)$ $\qquad$ *provided that $x$ is not free in $B$*

*(Q7)* $T \vdash (\forall x)A \Leftrightarrow \neg(\exists x)\neg A$

*(Q8)* $T \vdash A_x[t] \Rightarrow (\exists x)A$ *for every term t*

*(Q9)* $T \vdash (\exists x)A^n \Leftrightarrow ((\exists x)A)^n$

*(Q10)* $T \vdash (\exists y)(A_x[y] \Rightarrow (\forall x)A)^n$.

The following theorem shows that the structure of the provability degrees is complex and does not, in general, mirror that of the truth degrees.

THEOREM 4.3.3. *Let $T \vdash_a A$ and $T \vdash_b B$. Then*

*(a) $T \vdash_c A \Rightarrow B$ implies $c \leq a \to b$.*

*(b) $T \vdash_c A \,\&\, B$ implies $c \geq a \otimes b$.*

*(c) If $T \vdash_c A \wedge B$, $T \vdash_a A$ and $T \vdash_b B$, then $c = a \wedge b$.*

*Proof.* (a) The case $a \leq b$ is trivial, so assume the opposite and let $c > a \to b$. Write down the proof

$$w = a'/A \quad \{\textit{some proof } w_A\},\ c'/A \Rightarrow B \quad \{\textit{some proof } w_{A \Rightarrow B}\},$$
$$a' \otimes c'/B \quad \{r_{\mathrm{MP}}\}$$

whence $T \vdash_d B$, $d \geq a \otimes c$. However, it follows from the assumption that $b < a \otimes c$ — a contradiction.

(b) Follows from (P6).

(c) From (P1) we obtain $a \wedge b \leq c$. Conversely, let $T \vdash_a A$ and $T \vdash_b B$. Then obviously, $T \vdash \mathbf{a} \Rightarrow A$ as well as $T \vdash \mathbf{b} \Rightarrow B$, and thus $T \vdash \mathbf{d} \Rightarrow A \wedge B$ for all $d \leq a \wedge b$. Therefore, $T \vdash_{a \wedge b} A \wedge B$. □

## 4.4 Consistency of fuzzy theories

Analogously to classical logic, we can also consider a formula $A$ and its negation $\neg A$ and ask whether both can be proved in a given fuzzy theory $T$. In classical logic, $T$ is immediately contradictory and collapses into a trivial theory in which everything is provable. One can thus be tempted to think that in fuzzy logic this can never happen because formulas can be provable in some degree, and so, when relaxing the simultaneous provability of $A$ and $\neg A$, the theory can remain consistent. The results below demonstrate that this is true only in a very limited sense.

DEFINITION 4.4.1. *A fuzzy theory $T$ is* contradictory *if there is a formula $A \in F_{J(T)}$ such that*

$$T \vdash_a A \quad \textit{and} \quad T \vdash_b \neg A \quad \textit{imply} \quad a \otimes b > 0.$$

*It is* consistent *otherwise.*

The following theorem states that a contradictory fuzzy theory collapses into a degenerated theory just as in classical logic.

THEOREM 4.4.2 (Contradiction). *A fuzzy theory $T$ is contradictory iff $T \vdash A$ holds for every formula $A \in F_{J(T)}$.*

*Proof.* Let us consider proofs $w_A$, $w_{\neg A}$, $\mathrm{Val}(w_A) = a$, $\mathrm{Val}(w_{\neg A}) = b$, $\mathbf{0} < a \otimes b < \mathbf{1}$ (the case $a \otimes b = \mathbf{1}$ is trivial). Write down the proof

$$\begin{aligned} w = {}& a/A \quad \{\textit{some proof } w_A\},\ b/\neg A \quad \{\textit{some proof } w_{\neg A}\},\\ & \mathbf{1}/A \Rightarrow (\neg A \Rightarrow (A \mathbin{\&} \neg A)) \quad \{\textit{instance of (P6)}\},\\ & \ldots,\ a \otimes b/A \mathbin{\&} \neg A \quad \{r_{\mathrm{MP}}\},\ \mathbf{1}/(A \mathbin{\&} \neg A) \Rightarrow \bot \quad \{\textit{instance of (P4)}\},\\ & a \otimes b/\bot \quad \{r_{\mathrm{MP}}\}, \ldots,\ \mathbf{1}/\overline{\mathbf{a} \otimes \mathbf{b}} \Rightarrow \bot \quad \{r_{\mathrm{LC}}\}, \ldots,\ \mathbf{1}/\overline{\mathbf{a} \otimes \mathbf{b} \to \mathbf{0}} \quad \{r_{\mathrm{MP}}\}, \end{aligned}$$

i.e., $T \vdash \mathbf{c}$ for some $c < \mathbf{1}$. Because $c$ is nilpotent with respect to $\otimes$, let us take $n \in \mathsf{N}$ such that $c^n = \mathbf{0}$. Then using (P6), we obtain a proof $w_{\mathbf{c}^n}$ with the value $\mathrm{Val}(w_{\mathbf{c}^n}) = \mathbf{1}$. Now, let $B \in F_{J(T)}$ be an arbitrary formula. Now, we may write the proof

$$\begin{aligned} w' = {}& \mathbf{1}/\mathbf{c}^n \quad \{w_{\mathbf{c}^n}\},\ \mathbf{0}/\bot,\ \mathbf{0} \to \mathbf{0} = \mathbf{1}/\mathbf{c}^n \Rightarrow \bot \quad \{\textit{Lemma 4.2.1}\},\ \mathbf{1}/\bot \quad \{r_{\mathrm{MP}}\},\\ & \mathrm{SAx}(B)/B \quad \{\textit{special axiom}\},\ \mathbf{0} \to \mathrm{SAx}(B) = \mathbf{1}/\bot \Rightarrow B \quad \{\textit{Lemma 4.2.1}\},\\ & \mathbf{1}/B \quad \{r_{\mathrm{MP}}\}, \end{aligned}$$

i.e., $T \vdash B$. Note that $\mathrm{SAx}(B)$ can also be $\mathbf{0}$. The converse implication is obvious. □

It follows from this theorem that contradiction is also disastrous when dealing with degrees of truth. Consequently, when applying fuzzy logic to model the vagueness phenomenon, this means that vague concepts may also lead to contradiction.

Using a technique similar to that used in the previous proof, we obtain the following corollary.

COROLLARY 4.4.3. *Let $T$ be a fuzzy theory. Then $T$ is contradictory iff any of the following holds:*

(a) *There is a formula $A$ and a proof $w_{A \& \neg A}$ such that* $\mathrm{Val}_T(w_{A \& \neg A}) > \mathbf{0}$.

(b) *There are $a < \mathbf{1}$ and a proof $w_{\mathbf{a}}$ such that* $\mathrm{Val}_T(w_{\mathbf{a}}) > a$.

(c) *There is a formula $A$ and $a > \mathbf{0}$ such that $T \vdash_a A \mathbin{\&} \neg A$.*

This corollary will often be used as a criterion for checking whether the theory of concern is consistent.

The following simple lemma shows that if a fuzzy theory has a model, then it is consistent. The converse, however, is nontrivial and will be proved below.

LEMMA 4.4.4. *Let a fuzzy theory $T$ have a model $\mathcal{M}$. Then it is consistent.*

*Proof.* Let $A$ be a formula and $\mathcal{M}(A) = a$. Then $\mathcal{M}(\neg A) = \neg a$. By the validity theorem, $T \vdash_b A$ and $T \vdash_c \neg A$, where $b \leq a$ and $c \leq \neg a$. Consequently, $\mathrm{Val}_T(w_A) \otimes \mathrm{Val}(w_{\neg A}) = \mathbf{0}$ for all proofs $w_A$ and $w_{\neg A}$. □

One may be tempted to consider a weaker concept of contradictory theory, namely, using the connective $\wedge$ instead of &. The following theorem shows that our possibilities are still quite limited.

THEOREM 4.4.5. *If for a fuzzy theory $T$ there is a formula $A$ and a proof $w_{A\wedge\neg A}$ of $A \wedge \neg A$ such that*

$$\mathrm{Val}_T(w_{A\wedge\neg A}) > \frac{1}{2},$$

*then the fuzzy theory $T$ is contradictory.*

*Proof.* Using the tautology $A \wedge B \Rightarrow A$ and the symmetry of $\wedge$, we obtain from a proof $w_{A\wedge\neg A}$ proofs $w_A$ and $w_{\neg A}$ such that

$$\mathrm{Val}_T(w_A) > \frac{1}{2} \text{ and } \mathrm{Val}_T(w_{\neg A}) > \frac{1}{2},$$

which immediately gives $\mathrm{Val}_T(w_A) \otimes \mathrm{Val}_T(w_{\neg A}) > 0$. Thus, $T$ is contradictory. □

## 4.5 Extension of fuzzy theories and deduction theorem

DEFINITION 4.5.1. *A language $J'$ is an* extension *of the language $J$ if $J \subseteq J'$. A fuzzy theory $T'$ is an* extension *of the fuzzy theory $T$ if $J(T) \subseteq J'(T')$ and $T \vdash_a A$ and $T' \vdash_b A$ implies $a \leq b$ for every formula $A \in F_{J(T)}$.*

*An extension $T'$ of $T$ is* conservative *if $T' \vdash_b A$ and $T \vdash_a A$ implies $a = b$ for every formula $A \in F_{J(T)}$. The extension $T'$ is a* simple extension *of $T$ if $J'(T') = J(T)$.*

Let $T_1 = \langle \mathrm{LAx}_1, \mathrm{SAx}_1, R\rangle$ and $T_2 = \langle \mathrm{LAx}_2, \mathrm{SAx}_2, R\rangle$ be fuzzy theories and $J(T_1) \subseteq J(T_2)$. If $\mathrm{SAx}_1 \subseteq \mathrm{SAx}_2$, then $T_2$ is an extension of $T_1$, where we understand that $\mathrm{SAx}_1(A) = \mathbf{0}$ for all $A \in F_{J(T_2)} \setminus F_{J(T_1)}$.

Let $T_1$ be a fuzzy theory and $\Gamma \subsetneqq F_{J(T_1)}$ be a fuzzy set of formulas. By $T_2 = T_1 \cup \Gamma$, we denote a fuzzy theory whose special axioms are $\mathrm{SAx}_2 = \mathrm{SAx}_1 \cup \Gamma$.

THEOREM 4.5.2 (Constants). *Let $T$ be a fuzzy theory and $V$ a set of constants not contained in $J(T)$. We put $\bar{J} = J(T) \cup V$ and define a fuzzy theory $\overline{T}$ by $\overline{\mathrm{SAx}}(A) = \mathrm{SAx}(A)$ for $A \in F_{J(T)}$ and $\overline{\mathrm{SAx}}(A) = \mathbf{0}$ otherwise. Then*

$$\overline{T} \vdash_a A_{x_1,\dots,x_n}[\mathbf{v}_1,\dots,\mathbf{v}_n] \iff T \vdash_a A$$

*holds for every formula $A \in F_{J(T)}$ and $\mathbf{v}_1,\dots,\mathbf{v}_n \in V$.*

*Proof.* Let $\overline{T} \vdash_a A_{x_1,\dots,x_n}[\mathbf{v}_1,\dots,\mathbf{v}_n]$. Because $\mathbf{v}_1,\dots,\mathbf{v}_n$ do not belong to $J(T)$, we can replace them in every proof by some variables $y_1,\dots,y_n$ not occurring in $A$. Consequently, $T \vdash_b A_{x_1,\dots,x_n}[y_1,\dots,y_n]$, $a \leq b$ and because $A_{x_1,\dots,x_n}[y_1,\dots,y_n] \in F_{J(T)}$, we obtain $T \vdash_b A$.

Conversely, from $T \vdash_c A$, we obtain $\overline{T} \vdash_d A_{x_1,\dots,x_n}[\mathbf{v}_1,\dots,\mathbf{v}_n]$, $c \leq d$, which implies the equivalence. □

Apparently, $\overline{T}$ is a conservative extension of $T$.

LEMMA 4.5.3. *Let $T$ be a fuzzy theory, $A$ be a closed formula and $T' = T \cup \{{}^{1}/A\}$.*

*(a) If $T \vdash_a A^n \Rightarrow B$ and $T' \vdash_b B$ for some $n$, then $a \leq b$.*

*(b) To every proof $w_B$ in $T'$ there are $n$ and a proof $w_{A^n \Rightarrow B}$ in $T$ such that*

$$\mathrm{Val}_{T'}(w_B) = \mathrm{Val}_T(w_{A^n \Rightarrow B}).$$

*Proof.* (a) follows from (P6) and modus ponens.

(b) is proved by induction on the length of $w_B$. The case $B = A$ is trivial. Let $w_B = {}^{b}/B \quad \{axiom\}$. Then

$$w_{A^n \Rightarrow B} = {}^{b}/B \quad \{axiom\},\ {}^{1}/B \Rightarrow (A^n \Rightarrow B) \quad \{log.\ axiom\ (R1)\},$$
$${}^{b}/A^n \Rightarrow B \quad \{r_{\mathrm{MP}}\}.$$

Let $B$ be obtained using the proof

$$w_B = {}^{c}/C \quad \{some\ proof\ w_C\},\ {}^{d}/C \Rightarrow B \quad \{some\ proof\ w_{C \Rightarrow B}\},$$
$${}^{c \otimes d}/B \quad \{r_{\mathrm{MP}}\}.$$

Then, by the inductive assumption,

$$w_{A^n \Rightarrow B} = {}^{c}/A^m \Rightarrow C \quad \{some\ proof\ w_{A^m \Rightarrow C}\},$$
$${}^{d}/A^p \Rightarrow (C \Rightarrow B) \quad \{some\ proof\ w_{A^p \Rightarrow (C \Rightarrow B)}\},$$
$${}^{1}/(A^p \Rightarrow (C \Rightarrow B)) \Rightarrow ((A^m \Rightarrow C) \Rightarrow (A^{m+p} \Rightarrow B)) \quad \{instance\ of\ (P10)\},$$
$$\ldots,\ {}^{c \otimes d}/A^n \Rightarrow B \quad \{r_{\mathrm{MP}}\},$$

where $n = m + p$.

Let $B = (\forall x)C$ and $w_B = {}^{c}/C \quad \{some\ proof\ w_C\},\ {}^{c}/(\forall x)C \quad \{r_{\mathrm{G}}\}$. Then, by the inductive assumption,

$$w_{A^n \Rightarrow B} = {}^{c}/A^n \Rightarrow C \quad \{some\ proof\ w_{A^n \Rightarrow C}\},\ {}^{c}/(\forall x)(A^n \Rightarrow C) \quad \{r_{\mathrm{G}}\},$$
$$\ldots,\ {}^{c}/A^n \Rightarrow (\forall x)C \quad \{r_{\mathrm{MP}}\}$$

using the axiom (T2).

If $B = \mathbf{c} \Rightarrow C$ and $w_B = {}^{c}/C \quad \{some\ proof\ w_C\},\ {}^{1}/\mathbf{c} \Rightarrow C \quad \{r_{\mathrm{LC}}\}$, then we use the inductive assumption and the tautology (P9). □

COROLLARY 4.5.4. *Let $A$ be a closed formula and $T' = T \cup \{{}^{a}/A\}$. Then, for every proof $w'_B$ in $T'$, there are an $n$ and a proof $w_{A^n \Rightarrow B}$ in $T$ such that* $\mathrm{Val}_{T'}(w') \leq \mathrm{Val}_T(w)$.

As seen, part (b) of the previous lemma is weaker than the opposite implication of part (a), so we do not obtain a sufficiently strong analogy of the classical deduction theorem. The problem is that part (b) assures the existence of some finite $n$, which is used in the formula $A^n \Rightarrow B$, to every proof of the latter. If $\boldsymbol{L}$ is infinite, then, because there may exist an infinite number of proofs, we cannot be sure that $n$ does not arbitrarily increase. It follows from the lemmas below, however, that this is impossible.

LEMMA 4.5.5. *If a fuzzy theory* $T' = T \cup \{{}^{1}/A\}$ *is contradictory, then, for every formula* $B \in J(T)$*, there is* $m$ *such that* $T \vdash A^m \Rightarrow B$.

*Proof.* By the proof of Theorem 4.4.2, there is a proof $w_B$ in $T'$ which has the value $\mathrm{Val}(w_B) = 1$. The lemma then follows from Lemma 4.5.3(b). □

LEMMA 4.5.6. *Let* $T$ *be a consistent fuzzy theory,* $T \vdash_a A$*,* $a < 1$ *and* $b > a$*. Then*

$$T' = T \cup \{{}^{1}/A \Rightarrow \mathbf{b}\}$$

*is a consistent extension of* $T$.

*Proof.* Let $T'$ be contradictory. Then $T' \vdash \bot$ by Theorem 4.4.2 and by Lemma 4.5.5, there is $m$ such that

$$T \vdash (A \Rightarrow \mathbf{b})^m \Rightarrow \bot. \tag{34}$$

At the same time, using (P7) and (P8), we have $T \vdash (A \Rightarrow \mathbf{b})^m \vee (\mathbf{b} \Rightarrow A)^m$, which gives

$$T \vdash (A \Rightarrow \mathbf{b})^m \vee (\mathbf{b} \Rightarrow A) \tag{35}$$

using (P2), (P3) and (P5). From (34) and (35), we obtain

$$T \vdash \neg(A \Rightarrow \mathbf{b})^m \,\&((A \Rightarrow \mathbf{b})^m \vee (\mathbf{b} \Rightarrow A)),$$

and using (P12), we finally derive that $T \vdash \mathbf{b} \Rightarrow A$.

Let us now write down the proof

$$w = \; {}^{b}/\mathbf{b} \quad \{\textit{logical axiom}\},\; {}^{u}/\mathbf{b} \Rightarrow A \quad \{\textit{some proof } w_{\mathbf{b}\Rightarrow A}\},\; {}^{b \otimes u}/A \quad \{r_{\mathrm{MP}}\},$$

where $\bigvee\{u = \mathrm{Val}\, w_{\mathbf{b}\Rightarrow A} \mid \text{all proofs } w_{\mathbf{b}\Rightarrow A}\} = 1$. Then $T \vdash_{b'} A$, $b' \geq b > a$ which is a contradiction with $T \vdash_a A$. □

Note that if in the previous lemma $a = 1$, then necessarily $b = 1$ as well, so Lemma 4.5.6 becomes trivial.

LEMMA 4.5.7. *Let* $T$ *be a consistent fuzzy theory and* $T \vdash_a A$*, where* $a < 1$.

(a) $T \vdash_b A^m$ *and* $b \leq a$ *holds for every* $m > 1$.

(b) *If* $p \geq q$*, then* $T \vdash A^p \Rightarrow A^q$.

(c) *There is* $n \geq 1$ *such that* $T \vdash_0 A^n$.

(d) *Let* $T \vdash_a A^m \Rightarrow B$*,* $a < 1$*. Then there is* $p$ *such that for every* $n \geq p$ *it holds that* $T \vdash A^n \Rightarrow B$.

*Proof.* (a) Let $b > a$, and let $w_{A^m}$ be a proof with the value $\mathrm{Val}_T(w_{A^m}) = b' \leq b$. Then, using (P2), we conclude that $T \vdash_c A$, where $c \geq b > a$—a contradiction.

(b) Follows from (a).

(c) Let $T \vdash_{b>0} A^n$ for all $n$, and let $\bar{b} > b$ be rational. By Lemma 4.5.6, the theory $T' = T \cup \{{}^1/A \Rightarrow \bar{\mathfrak{b}}\}$ is consistent. Let $m$ be such that $\bar{b}^m = \mathbf{0}$. Then $T' \vdash A^m \Rightarrow \bar{\mathfrak{b}}^m$, i.e., $T' \vdash A^m \Rightarrow \bot$. By the assumption, $T \vdash_{a>0} A^m$, and so $T' \vdash_{\bar{a}>0} A^m$, where $\bar{a} \geq a$. Using modus ponens, we obtain $T' \vdash_{\bar{a}} \bot$, i.e., $T'$ is contradictory by Corollary 4.4.3(b)—a contradiction.

(d) $T$ is consistent by the assumption. Let the proposition not hold. Then, for every $p$, there is $n \geq p$ such that $T \vdash_c A^n \Rightarrow B$ for some $c < \mathbf{1}$.

Choose $p_1, p_2$, and find $m \geq p_1$ and $n \geq p_2$ such that $T \vdash_{c<1} A^m \Rightarrow B$ and $T \vdash_{d<1} A^n \Rightarrow B$. Without losing generality, let $c < d$. Take (a rational) $e$ such that $d > e > c$, and consider a natural number $p \geq \max\{m, n\}$. By Lemma 4.5.6, the theory $T' = T \cup \{{}^1/(A^m \Rightarrow B) \Rightarrow \mathfrak{e}\}$ is consistent.

Let us now write a formal proof in $T'$:

$$w = {}^{d'}/A^n \Rightarrow B \quad \{\textit{some proof } w_{A^n \Rightarrow B}\},\ {}^1/A^p \Rightarrow A^n \quad \{\textit{by (b)}\},$$
$$ {}^{d'}/A^p \Rightarrow B \quad \{r_{\mathrm{MP}}, \textit{(R2)}\},\ {}^1/(A^p \Rightarrow B) \Rightarrow \mathfrak{e} \quad \{\textit{the new axiom, (b)}, r_{\mathrm{MP}}\},$$
$$ {}^{d'}/\mathfrak{e} \quad \{r_{\mathrm{MP}}\}.$$

Hence, $\bigvee\{\mathrm{Val}(w) \mid \text{all proofs } w_{A^n \Rightarrow B}\} = d$, so $T'_{d'' \geq d > e} \vdash \mathfrak{e}$. By Corollary 4.4.3(b), $T'$ is contradictory—a contradiction. □

THEOREM 4.5.8 (Deduction theorem). *Let $T$ be a fuzzy theory, $A$ be a closed formula and $T' = T \cup \{{}^1/A\}$. Then, for every formula $B \in F_{J(T)}$, there is an $n$ such that*

$$T \vdash_a A^n \Rightarrow B \iff T' \vdash_a B.$$

*Proof.* If $\boldsymbol{L}$ is finite, then the theorem follows immediately from Lemma 4.5.3. Otherwise, by Lemma 4.5.7(d), the exponent in $A^n \Rightarrow B$ has an upper bound. The theorem then again follows from Lemma 4.5.3. □

The deduction theorem has the following significant consequence.

LEMMA 4.5.9. *Let $T$ be a consistent fuzzy theory in the language $J_L$ containing truth constants* $\mathrm{TC}(J)$ *for all the truth values from L. Then the* MV*-algebra* $\boldsymbol{L}$ *of the truth values of any model $\mathcal{M}$ of $T$ must be linearly ordered.*

*Proof.* By Corollary 4.2.3, we have $T \vdash_a \mathfrak{a}$ for any $a \in L$. Let $a \in L \setminus \{\mathbf{0}\}$. By Corollary 4.4.3(b), a theory $T'$ in which $T' \vdash_b \mathfrak{a}$ for some $b > a$ is contradictory. Because the same holds for $T \vdash_c \neg\mathfrak{a}$ for some $c > \neg a$, the theory $T' = T \cup \{{}^1/\neg\mathfrak{a}\}$ is contradictory. This means that $T' \vdash \bot$.

By the deduction theorem, there is a natural number $n$ such that $T \vdash (\neg\mathfrak{a})^n \Rightarrow \bot$. Using the properties of $\mathrm{Ev}_{\text{Ł}}$, we obtain $T \vdash n\mathfrak{a}$. Let $\mathcal{M}$ be any model of $T$. Then $\mathcal{M}(\mathfrak{a}) = a \in L \setminus \{\mathbf{0}\}$ and there is $n \in \mathrm{N}$ such that $\mathcal{M}(n\mathfrak{a}) = na = \mathbf{1}$. Hence, $\mathrm{ord}(a) < \infty$ for any $a > \mathbf{0}$, so by Lemma 2.1.1, $\boldsymbol{L}$ is linearly ordered. □

The theorem below demonstrates that it is possible to denote an element that has a certain property represented by an existential formula by a new constant not contained in the language of the theory of concern. Similarly as in classical logic, it turns out that such an extension of the theory is conservative.

THEOREM 4.5.10. *Let $T$ be a consistent fuzzy theory, $T \vdash (\exists x)A(x)$ and $\mathbf{c} \notin J(T)$ be a new constant. Then the theory*

$$T' = T \cup \{^{\mathbf{1}}/A_x[\mathbf{c}]\}$$

*in the language $J(T) \cup \{\mathbf{c}\}$ is a conservative extension of the theory $T$.*

*Proof.* Let $\overline{T}$ be a conservative extension of $T$ by Theorem 4.5.2. Let $B \in F_{J(T)}$. By Theorems 4.5.2 and 4.5.8,

$$T' \vdash_b B \iff \overline{T} \vdash_b (A_x[\mathbf{c}])^n \Rightarrow B \iff T \vdash_b (A(x))^n \Rightarrow B$$

for some $n$. Furthermore, using (P6), we have $T \vdash ((\exists x)A)^n$. Then, using (Q9) and modus ponens, we obtain $T \vdash_b B$. □

## 4.6 Complete fuzzy theories

Complete fuzzy theories are the main tool by which we can syntactically prove the completeness theorem. In Ev$_Ł$, however, we may introduce complete theories in two slightly different ways depending on the chosen algebra $\boldsymbol{L}$ of the truth values, namely, whether the latter is finite or infinite. Moreover, if it is infinite, then we still may distinguish the case when $K = L$, or when $K$ consists of the set of rational numbers only.

### 4.6.1 Complete fuzzy theories for $K = L$

First, we consider the case when $K = L$ so that $\mathrm{TC}(J)$ contains truth constants for all the truth values from $L$ regardless of the cardinality of $L$. Because it is either equal to $[0, 1]$ or finite, the language $J_L$ can be uncountable. Note that for finite $L$, each formula is even effectively provable.

DEFINITION 4.6.1. *A fuzzy theory $T$ in the language $J_L$ is* complete *if it is consistent and*

$$T \vdash_a A \quad \textit{implies} \quad T \vdash A \Rightarrow \mathbf{a}$$

*holds for every closed formula $A$ and $a \in L$.*

LEMMA 4.6.2. *Let $T$ be a fuzzy theory in the language $J_L$.*

*(a) If $T$ is complete, then*

$$T \vdash_a A \iff T \vdash A \Leftrightarrow \mathbf{a}$$

*holds for every formula $A \in F_J$.*

*(b) Let $T$ have a model $\mathcal{M}$ such that for every closed formula $A \in F_J$ the following holds true:*

$$\mathcal{M}(A) = a \quad \textit{implies} \quad T \vdash_a A.$$

*Then $T$ is complete.*

*Proof.* (a) Immediate.

(b) By the definition of a model, $\mathcal{M}(A) = a$ iff $\mathcal{M}(A \Leftrightarrow \mathbf{a}) = 1$. Then $T \vdash A \Leftrightarrow \mathbf{a}$ by the assumption also $T \vdash A \Rightarrow \mathbf{a}$ which means $T$ is complete. □

THEOREM 4.6.3 (Completion). *Let $T$ be a consistent fuzzy theory. Then there exists a complete theory $\bar{T}$ that is a simple extension of $T$.*

*Proof.* In the same way as in the proof of Theorem 4.6.8, we construct a maximal theory $\bar{T}$. To show its completeness, let $\bar{T} \vdash_a A$, $b > a$ and put $\bar{T}' = \bar{T} \cup \{{}^{\mathbf{1}}/A \Rightarrow \mathbf{b}\}$. Because $\bar{T}'$ is consistent (by Lemma 4.5.6), we conclude by maximality of $\bar{T}$ that $\{{}^{\mathbf{1}}/A \Rightarrow \mathbf{b}\} \subseteq \overline{\text{SAx}}$. Hence, $\bar{T} \vdash A \Rightarrow \mathbf{b}$ for every $b > a$. Let $\bar{T} \vdash_c A \Rightarrow \mathbf{a}$, and write down the proof

$$w = {}^{\mathbf{1}}/A \Rightarrow \mathbf{b} \quad \{spec.\ axiom\},\ {}^{b \to a}/\mathbf{b} \Rightarrow \mathbf{a}) \quad \{log.\ axiom\}, \ldots,$$
$$\quad {}^{b \to a}/A \Rightarrow \mathbf{a} \quad \{r_{\text{MP}}\},$$

It then follows that $c \geq b \to a$ for every $b > a$, but then

$$\bigvee\{b \to a \mid b \in L, b > a\} = \bigwedge\{b \mid b \in L, b > a\} \to a = a \to a = \mathbf{1} \leq c.$$

We obtain that $\bar{T} \vdash A \Rightarrow \mathbf{a}$, i.e., $\bar{T}$ is complete. □

### 4.6.2 Complete fuzzy theories for $L = [0, 1]$

Let us consider $L = [0, 1]$ but a countable language $J_{L_Q}$. Then the definition of completeness must be modified because a formula can be provable in an irrational degree $a$ for which there is no truth constant in the language $J_{L_Q}$.

DEFINITION 4.6.4. *A fuzzy theory $T$ in the language $J_{L_Q}$ is* complete *if it is consistent and if the following holds for every closed formula $A$: If $T \vdash_a A$, then there is a set $H_A \subseteq L_Q$ such that $\bigwedge H_A = a$ and*

$$T \vdash A \Rightarrow \mathbf{b}$$

*holds for every $b \in H_A$.*

LEMMA 4.6.5. *Let $T$ be a complete fuzzy theory in the language $J_{L_Q}$, $T \vdash_b B$, and let $H_B \subseteq L_Q$ be such that $\bigwedge H_B = b$. Then*

$$T \vdash_{e(b')} \mathbf{b}' \Rightarrow B, \qquad b' \in H_B,$$

*where $e(b')$ is a provability degree (dependent on $b'$) such that $\bigvee\{e(b') \mid b' \in H_B\} = 1$.*

*Proof.* Let $w_B$ be a proof of $B$ and $b' \in H_B$. We construct a proof $w_{b'}$ of $\mathbf{b}' \Rightarrow B$ and then obtain

$$\bigvee\{\text{Val}(w_{b'}) \mid \text{all proofs } w_B\} =$$
$$\bigvee\{b' \to \text{Val}(w_B) \mid \text{all proofs } w_B\} = b' \to b \leq e(b')$$

for all $b' \in H_B$. It follows that

$$\bigvee_{b' \in H_B} (b' \to b) = \bigwedge_{b' \in H_B} b' \to b = \mathbf{1} \leq \bigvee_{b' \in H_B} e(b'). \qquad \square$$

LEMMA 4.6.6. *Let $T$ be a complete fuzzy theory in the language $J_{L_Q}$. Then, for every formula $A$, $T \vdash_a A$ implies the existence of a set $H_A \subseteq L_{\mathsf{Q}}$ such that*

$$T \vdash_{f(b)} A \Leftrightarrow \mathbf{b}, \qquad b \in H_A,$$

*where $f(b)$ is a provability degree (dependent on $b$) such that $\bigvee\{f(b) \mid b \in H_A\} = 1$.*

*Proof.* This follows from the completeness of $T$, Lemma 4.6.5, the definition of $A \Leftrightarrow \mathbf{b}$ and the tautology $(A \Rightarrow B) \Rightarrow ((A \Rightarrow C) \Rightarrow (A \Rightarrow (B \wedge C)))$. □

LEMMA 4.6.7. *Let $T$ be a complete fuzzy theory in the language $J_{L_Q}$, and let $B, C$ be formulas such that $T \vdash_b B$, $T \vdash_c C$. Then*

$$T \vdash_d B \Rightarrow C,$$

*where $d = b \to c$.*

*Proof.* Note that $d \le b \to c$ by the known properties of the provability degree (see [36, Theorem 4.10]). Furthermore, let $H_B, H_C \subseteq L_{\mathsf{Q}}$ be sets, $\bigwedge H_B = b$, $\bigwedge H_C = c$ existing on the basis of the completeness of $T$. Using Lemma 4.6.5 and the bookkeeping axiom, we can prove that

$$T \vdash_{f(b',c')} (\mathbf{b}' \Rightarrow \mathbf{c}') \Rightarrow (B \Rightarrow C), \tag{36}$$

where $f(b', c')$ is a provability degree (dependent on $b'$ and $c'$) such that $\bigvee\{f(b', c') \mid b' \in H_B, c' \in H_C\} = 1$. Let us, on the basis of (15), consider a proof $w_{b'c'}$ of $B \Rightarrow C$ with the value $b' \to c'$ (easily using the bookkeeping axiom and taking the truth constants $\mathbf{b}', \mathbf{c}'$ as axioms). Then

$$d = \bigvee\{\mathrm{Val}(w) \mid w \text{ is a proof of } B \Rightarrow C\} \ge$$
$$\ge \bigvee\{\mathrm{Val}(w_{b'c'}) \mid b' \in H_B, c' \in H_C\} = \bigvee\{b' \to c' \mid b' \in H_B, c' \in H_C\}.$$

However,

$$b \to c = \bigwedge_{b' \in H_B} b' \to \bigwedge_{c' \in H_C} c' \le \bigvee\{b' \to c' \mid b' \in H_B, c' \in H_C\}.$$

We have obtained $b \to c \le d \le b \to c$. □

THEOREM 4.6.8 (Completion). *Let $T$ be a consistent fuzzy theory in the language $J_{L_Q}$. Then there exists a complete fuzzy theory $\bar{T}$ which is a simple extension of $T$.*

*Proof.* Let $\langle E_i \mid i < q\rangle$ be a chain of fuzzy sets of formulas such that $T_0 = T$ and $T_{i+1} = T_i \cup E_i$ are consistent. Using Zorn's lemma, we construct a maximal fuzzy theory $\bar{T} = T \cup \bar{A}$, where $\bar{A}$ is a maximal fuzzy set, which is a simple consistent extension of $T$.

We will show that $\bar{T}$ is complete. Let $\bar{T} \vdash_a A$. If $a = 1$, then $\bar{T} \vdash A \Leftrightarrow \top$ and $H_A = \{1\}$. Let $a < 1$ and put $H_A = \{b \mid b \ge a, b \in L_{\mathsf{Q}}\}$. Then $H_A \subseteq L_{\mathsf{Q}}$ and $\bigwedge H_A = a$. Furthermore, for $b \in H_A$ put:

$$\bar{T}' = \bar{T} \cup \{1/A \Rightarrow \mathbf{b}\},$$

Then $\bar{T}'$ is a consistent extension of $\bar{T}$ by Lemma 4.5.6, and because $\bar{T}$ is maximal, we have

$$\bar{T} \vdash A \Rightarrow \mathbf{b},$$

that is, $\bar{T}$ is complete. □

The following lemma shows that if the language $J(T)$ contains truth constants for all the truth values, the above definition complies with Definition 4.6.1.

LEMMA 4.6.9. *Let $T$ be a complete fuzzy theory in the language $J_L$. Then*

$$T \vdash_a A \textit{ implies } T \vdash A \Rightarrow \mathbf{a}.$$

*Proof.* Let $T \vdash_a A$ and $H_A \subseteq L_Q$ be a set existing due to Definition 4.6.4, and take some $b \in H_A$. Let us consider a proof denoted by $w_a$ as follows:

(L1) $c_b / A \Rightarrow \mathbf{b}$ $\qquad c_b = \mathrm{Val}(w_b)$, ($w_b$ is a proof of $A \Rightarrow \mathbf{b}$)

(L2) $c_b \otimes (b \to a) / A \Rightarrow \mathbf{a}$ $\qquad$ (L1, bookkeeping, transitivity, $r_{MP}$)

Note that

$$\bigvee \{c_b \mid c_b = \mathrm{Val}(w_b), w_b \text{ is a proof of } A \Rightarrow \mathbf{b}\} = \mathbf{1}$$

by the completeness assumption. Then

$$\begin{aligned} &\bigvee \{\mathrm{Val}\, w \mid w \text{ is a proof of } A \Rightarrow \mathbf{a}\} \geq \\ &\quad \geq \bigvee \{c_b \otimes (b \to a) \mid c_b = \mathrm{Val}(w_b), w_b \text{ is a proof of } A \Rightarrow \mathbf{b}, b \in H_A\} = \\ &\qquad = \bigvee \{c_b \mid b \in H_A\} \otimes (\bigwedge H_A \to a) = \mathbf{1} \otimes (a \to a) = \mathbf{1}. \end{aligned}$$ □

It is possible to demonstrate that Lemma 4.6.2 can also be proved when considering Definition 4.6.4. Thus, Definition 4.6.1 is only a welcome simplification for the case when $K = L$.

The following example demonstrates that there exist incomplete fuzzy theories.

EXAMPLE 4.6.10. Let us consider the following fuzzy theory in the language $J_L$:

$$T = \{a_1/A_1, a_2/A_2, c_1/A_1 \Rightarrow B, c_2/A_2 \Rightarrow B\},$$

where $A_1$, $A_2$, and $B$ are closed formulas, $a_1 < a_2 < 1$, $c_1 < c_2 < 1$, and $0 < a_1 \otimes c_1 < a_2 \otimes c_2$. Then there is no model $\overline{\mathcal{M}}$ such that $\overline{\mathcal{M}}(A_1) = a_1$, $\overline{\mathcal{M}}(A_2) = a_2$, $\overline{\mathcal{M}}(A_1 \Rightarrow B) = c_1$, $\overline{\mathcal{M}}(A_2 \Rightarrow B) = c_2$. Indeed, the latter equalities follow that $\overline{\mathcal{M}}(B) = a_1 \otimes c_1 = a_2 \otimes c_2$ which is impossible.

On the other hand, let us consider three models $\mathcal{M}_1, \mathcal{M}_2, \mathcal{M}_3$ defined as follows (it is not difficult to find concrete numbers to fulfill all the conditions in items 1–3):

1. Set $\mathcal{M}_1(A_1) = a_1$, $\mathcal{M}_1(A_2) = a_2$, $\mathcal{M}_1(B) = 1$. One can see that $\mathcal{M}_1 \models T$.

2. Set $\mathcal{M}_2$ so that the following is fulilled: $a_1 \leq \mathcal{M}_2(A_1)$, $a_2 \leq \mathcal{M}_2(A_2)$, $c_1 = \mathcal{M}_2(A_1 \Rightarrow B)$ and $c_2 < \mathcal{M}_2(A_2 \Rightarrow B)$. One can see that $\mathcal{M}_2 \models T$. Moreover, $\mathcal{M}_2(B) = \mathcal{M}_2(A_1) \otimes c_1$ and $\mathcal{M}_2(A_2) \otimes c_2 < \mathcal{M}_2(A_1) \otimes c_1$.

3. Set $\mathcal{M}_3$ so that the following is fulilled: $a_1 \leq \mathcal{M}_3(A_1)$, $a_2 \leq \mathcal{M}_3(A_2)$, $c_2 = \mathcal{M}_3(A_2 \Rightarrow B)$ and $c_1 < \mathcal{M}_3(A_1 \Rightarrow B)$. One can see that $\mathcal{M}_3 \models T$. Moreover, $\mathcal{M}_3(B) = \mathcal{M}_3(A_2) \otimes c_2$ and $\mathcal{M}_3(A_1) \otimes c_1 < \mathcal{M}_2(A_2) \otimes c_2$.

Because $T$ has a model, we conclude that it is consistent. Furthermore, the definition of all three models implies that $T \vdash_{a_1} A_1$, $T \vdash_{a_2} A_2$, $T \vdash_{c_1} A_1 \Rightarrow B$ and $T \vdash_{c_2} A_2 \Rightarrow B$. Finally, $T \vdash_d (A_1 \Rightarrow B) \Rightarrow \mathbf{c_1}$ for some $d < 1$ due to 3. as well as $T \vdash_e (A_2 \Rightarrow B) \Rightarrow \mathbf{c_2}$ for some $e < 1$ due to 2. Hence, $T$ is incomplete.

### 4.7 Henkin fuzzy theories

Analogously as in classical logic, we can introduce a set of special constants and special axioms for all the formulas of the form $(\forall x)A$. The obtained fuzzy theory will be called Henkin. We prove that every consistent fuzzy theory can be conservatively extended into the Henkin one.

DEFINITION 4.7.1. *Given a formula $(\forall x)A(x)$, let $\mathbf{r}$ be a special constant for the formula. A fuzzy theory $T$ is called* Henkin *if for all formulas $A(x)$, the evaluated formulas*

$$1/A_x[\mathbf{r}] \Rightarrow (\forall x)A(x) \tag{37}$$

*are among special axioms of $T$.*

The evaluated formulas (37) are called Henkin axioms. Using contraposition, we can also derive the equivalent existential form

$$1/(\exists x)A(x) \Rightarrow A_x[\mathbf{r}]. \tag{38}$$

LEMMA 4.7.2. *Let $T$ be a consistent fuzzy theory and $\{E_0 \subseteq \cdots \subseteq E_\alpha \subseteq \cdots\}$ be a chain of fuzzy sets of formulas from $F_{J(T)}$, where $\alpha < \lambda$ for some ordinal $\lambda$. Furthermore, put $T_0 = T$, and let $T_{\alpha+1} = T_\alpha \cup E_\alpha$ be consistent for all $\alpha < \lambda$. Then*

$$\overline{T} = T \cup \bigcup_{\alpha<\lambda} E_\alpha$$

*is a consistent extension of $T$.*

*Proof.* Let $A \in F_{J(T)}$. By induction on the length of the proof and the semicontinuity of $r^{\text{evl}}$ in each variable, we prove that

$$\text{Val}_{\overline{T}}(w_A) = \bigvee\{\text{Val}_{T_\alpha}(w_A) \mid \alpha < \lambda\} \tag{39}$$

for every proof $w_A$ of the formula $A$. From the assumption, it follows that $\text{Val}_{T_\alpha}(w_A) \otimes \text{Val}_{T_\alpha}(w_{\neg A}) = \mathbf{0}$ for all $\alpha < \lambda$, which, together with (39), implies that $\overline{T}$ is consistent. □

THEOREM 4.7.3. *Let $T$ be a consistent fuzzy theory, let $K$ be a set of special constants for all the closed formulas $(\forall x)A$ and let $\mathrm{Ax}_H$ be a set of Henkin axioms (37) (a fuzzy set of Henkin axioms). Then the fuzzy theory $T_H = T \cup \mathrm{Ax}_H$ in the language $J(T_H) = J(T) \cup K$ is a conservative extension of the theory $T$.*

*Proof.* We proceed analogously to the classical proof for the Henkin extension. Namely, we construct sets of special constants $K_1, K_2, \ldots$ of the corresponding levels. Now, put $T_0 = T$ and

$$T_{i+1} = T_i \cup \{\mathbf{1}/A_x[\mathbf{r}] \Rightarrow (\forall x)A(x)\},$$

where $\mathbf{r} \in K_{i+1}$ is a special constant for $(\forall x)A$. Using formal theorem (Q9) of Lemma 4.3.2, we obtain $T_i \vdash (\exists y)(A_x[y] \Rightarrow (\forall x)A)$. Then, using Theorem 4.5.10 and formal theorem (Q10), we obtain that $T_{i+1}$ is a conservative extension of $T_i$ for every $i$. The theorem then follows from Lemma 4.7.2. □

The following lemma follows from the Henkin and substitution axioms.

LEMMA 4.7.4. *Let $T$ be a Henkin fuzzy theory and $\mathbf{r}$ a special constant for $(\forall x)A$. Then*

$$T \vdash_a (\forall x)A \iff T \vdash_a A_x[\mathbf{r}].$$

## 5 Completeness theorem

In this section, we will prove the completeness of $\mathrm{Ev}_{Ł}$. We formulate and prove two theorems analogous to the classical Gödel ones, so we will number them in the same way.

THEOREM 5.0.1 (Case $L = [0, 1]$). *A Henkin fuzzy theory $T$ in the language $J_{L_Q}$ is complete iff there is a model $\mathcal{M} \models T$ fulfilling the condition: $T \vdash_a A$ iff $\mathcal{M}(A) = a$ for every formula $A \in F_{J(T)}$.*

*Proof.* Let $T$ be complete. We define a canonical model

$$\mathcal{M}_0 = \langle M_0, P_{M_0}, \ldots, f_{M_0}, \ldots, u, \ldots \rangle$$

as follows:

$$\begin{aligned} M_0 &= M_V \\ \mathcal{M}_0(t) &= t && t \in M_V \\ f_{M_0}(t_1, \ldots, t_n) &= f(t_1, \ldots, t_n) \\ \mathcal{M}_0(\mathbf{a}) &= a && a \in L_Q \\ P_{M_0}(t_1, \ldots, t_n) = a &\iff T \vdash_a P(t_1, \ldots, t_n)). \end{aligned}$$

We show that this structure is a model of $T$ such that

$$T \vdash_a A \iff \mathcal{M}_0(A) = a \tag{40}$$

holds for every formula $A \in F_{J(T)}$.

Let $T'$ be a conservative Henkin extension of $T$ due to Theorem 4.7.3. By induction on the length of the formula, we show (40) for $T'$ and $\mathcal{M}_0$.

If $A$ is atomic, then (40) trivially holds by the definition of the canonical model. If $A = B \Rightarrow C$, then we obtain (40) using Lemma 4.6.7 and inductive assumption.

If $A = (\forall x)B$, then we have

$$T' \vdash_a A \iff T' \vdash_a B_x[\mathbf{r}] \iff \mathcal{M}_0(B_x[\mathbf{r}]) \iff \mathcal{M}_0((\forall x)B)$$

due to Lemma 4.7.4 and the inductive assumption. By the closure theorem, (40) holds for every formula $A \in F_{J(T)}$. Obviously, $\mathcal{M}_0$ is a model of $T'$, so it is also a model of $T$.

Conversely, let $\mathcal{M} \models T$ and $T \vdash_a A$. Put $H_A = \{b \mid b \geq a, b \in L_{\mathrm{Q}}\}$. Then $H_A \subseteq L_{\mathrm{Q}}$ and $\bigwedge H_A = a$. Moreover, $\mathcal{M}(A \Rightarrow \mathbf{b}) = \mathbf{1}$ for every $b \in H_A$, i.e., $T \vdash A \Rightarrow \mathbf{b}$, and so $T$ is complete. □

THEOREM 5.0.2 (Finite $L$). *A fuzzy theory $T$ is complete iff there is a model $\mathcal{M}$ fulfilling the condition: $T \vdash_a A$ iff $\mathcal{M}(A) = a$ for every $A$ and $a \in L$.*

*Proof.* This proof differs from the proof above only in the converse implication.

Let $\mathcal{M} \models T$ be a model and let $\mathcal{M}(A) = a$. Then $\mathcal{M}(A \Rightarrow \mathbf{a}) = \mathbf{1}$, i.e., $T \vdash A \Rightarrow \mathbf{a}$ which means that $T$ is complete. □

THEOREM 5.0.3 (Completeness theorem II). *A fuzzy theory $T$ is consistent iff it has a model.*

*Proof.* If $T$ has a model, then it is consistent by Lemma 4.4.4.

Conversely, using Theorems 4.7.3 and 4.6.8, we construct a conservative Henkin extension $T_H$ of the fuzzy theory $T$ and complete it to a theory $\bar{T}_H$. By Theorem 5.0.1, there is model $\bar{\mathcal{M}}_H \models \bar{T}_H$ which is a model of $T_H$. However, $T_H$ is an extension of $T$, and therefore, $\bar{\mathcal{M}}_H$ is a model of $T$. □

THEOREM 5.0.4 (Completeness theorem I). *Let $T$ be a fuzzy theory in the in the language $J_{L_Q}$. Then*

$$T \vdash_a A \iff T \models_a A$$

*holds for every formula $A \in F_{J(T)}$.*

*Proof.* Let $T \vdash_a A$ for some $a > 0$. Then $T \models_b A$ for some $b \geq a$ by the validity theorem. If $a = \mathbf{1}$ or $b = a$, then we are finished. Hence, let $b > a$. We have to show that for every such $b$, there is a model $\mathcal{M} \models T$ such that $\mathcal{M}(A) \leq b$.

By Lemma 4.5.6, $T' = T \cup \{^{\mathbf{1}}/A \Rightarrow \mathbf{b}\}$ is a consistent theory. By Theorem 5.0.3, it has a model $\mathcal{M} \models T'$. However, then $\mathcal{M} \models T$ and $\mathcal{M}(A \Rightarrow \mathbf{b}) = \mathbf{1}$, which gives $\mathcal{M}(A) \leq b$. □

Theorems 5.0.3 and 5.0.4 generalize the classical Gödel completeness theorems of classical first-order logic to the fuzzy one.[10] In this section, they are formulated and proved for $\mathrm{Ev}_{Ł}$ with rational truth constants only (i.e., countable number of them).

[10] These theorems can be extended to hold also for formulas containing new additional connectives—see Section 7 for more details.

# 6 Model theory

$\mathrm{Ev}_{Ł}$ is a rich and nontrivial mathematical theory that has many interesting properties. Though many of them have already been studied, there is still much more to be done.

In this section, we will focus on the model theory of $\mathrm{Ev}_{Ł}$. Note that in comparison with the classical model theory, in fuzzy logic, there are more options for how a generalization of classical concepts can be introduced. Let us also remark that most results obtained in model theory of general mathematical fuzzy logic (e.g., [5, 20]) are applicable to $\mathrm{Ev}_{Ł}$ as well.

DEFINITION 6.0.1. *Let $\mathcal{V}$, $\mathcal{W}$ be two structures for the language $J$, and let $\Gamma \subseteq F_J$.*

*(i)* *$\mathcal{V}$ is a* weak substructure *of $\mathcal{W}$ ($\mathcal{W}$ is a* weak extension *of $\mathcal{V}$), in symbols*

$$\mathcal{V} \subseteq \mathcal{W},$$

*if $V \subseteq W$, $f_\mathcal{V} = f_\mathcal{W}|V^n$ holds for every $n$-ary function assigned to a functional symbol $f \in \mathrm{Func}(J)$, $u_\mathcal{V} = u_\mathcal{W}$ for every object constant $\mathbf{u} \in J$ and*

$$P_\mathcal{V} \leq P_\mathcal{W}|V^n \tag{41}$$

*holds for the $n$-ary fuzzy relations $P_\mathcal{V}$ and $P_\mathcal{W}$ assigned to all the predicate symbols $P \in \mathrm{Pred}(J)$ in $\mathcal{V}$ and $\mathcal{W}$, respectively.*

*(ii)* *$\mathcal{V}$ is a* substructure *of $\mathcal{W}$ ($\mathcal{W}$ is an* extension *of $\mathcal{V}$), in symbols*

$$\mathcal{V} \subset \mathcal{W},$$

*if equality holds in (41) for all the predicate symbols $P \in \mathrm{Pred}(J)$.*

*(iii)* *$\mathcal{V}$ is a* strong $\Gamma$-substructure *of $\mathcal{W}$ ($\mathcal{W}$ is a* strong $\Gamma$-extension *of $\mathcal{V}$), in symbols*

$$\mathcal{V} \leq_\Gamma \mathcal{W},$$

*if $V \subseteq W$ and*

$$\mathcal{V}(A_{x_1,\dots,x_n}[\mathbf{v}_1,\dots,\mathbf{v}_n]) \leq \mathcal{W}(A_{x_1,\dots,x_n}[\mathbf{v}_1,\dots,\mathbf{v}_n]) \tag{42}$$

*holds for every formula $A \in \Gamma$ and $v_1,\dots,v_n \in V$.*

*If $\mathcal{V} \leq_\Gamma \mathcal{W}$ for $\Gamma = F_J$, then $\mathcal{V}$ is a* strong substructure *of $\mathcal{W}$ ($\mathcal{W}$ is a strong extension of $\mathcal{V}$), and we write $\mathcal{V} \leq \mathcal{W}$.*

*(iv)* *Let $V \subseteq W$. The* expanded structure *is*

$$\mathcal{W}_\mathcal{V} = \langle \mathcal{W}, \{\mathbf{v} \mid v \in V\} \rangle,$$

*where $\mathbf{v}$ are names for all the elements $v \in V$ taken in $\mathcal{W}_\mathcal{V}$ as new object constants, which, however, are interpreted by the same elements, i.e., $\mathcal{W}_\mathcal{V}(\mathbf{v}) = v$ for all $v \in V$.*

*(v) The* $\mathcal{V}$ *is an* elementary $\Gamma$-substructure *of* $\mathcal{W}$ *(*$\mathcal{W}$ *is an* elementary $\Gamma$-extension *of* $\mathcal{V}$*), in symbols* $\mathcal{V} \prec_\Gamma \mathcal{W}$ *whenever* $\mathcal{V} \subseteq \mathcal{W}$ *and*

$$\mathcal{V}(A_{x_1,\dots,x_n}[\mathbf{v}_1,\dots,\mathbf{v}_n]) = \mathcal{W}(A_{x_1,\dots,x_n}[\mathbf{v}_1,\dots,\mathbf{v}_n]) \qquad A(x_1,\dots,x_n) \in \Gamma \quad \textit{and} \quad v_1,\dots,v_n \in V. \tag{43}$$

*If* $\Gamma = F_J$*, then* $\mathcal{V}$ *is an* elementary substructure *of* $\mathcal{W}$ *(*$\mathcal{W}$ *is an* elementary extension *of* $\mathcal{V}$*) and write* $\mathcal{V} \prec \mathcal{W}$.

LEMMA 6.0.2. *Let* $\mathcal{V}_1$ *and* $\mathcal{V}_2$ *be two structures for a language* $J$.

*(a) If* $\mathcal{V}_1 \subseteq \mathcal{V}_2$ *(*$\mathcal{V}_1 \subset \mathcal{V}_2$*) and* $\mathcal{V}_2 \subseteq \mathcal{V}_1$ *(*$\mathcal{V}_2 \subset \mathcal{V}_1$*), then* $\mathcal{V}_1 = \mathcal{V}_2$.

*(b) If* $\mathcal{V}_1 \leq \mathcal{V}_2$*, then* $\mathcal{V}_1 \subseteq \mathcal{V}_2$.

*Proof.* This is an easy consequence of the previous definition. □

Note, however, that if $\mathcal{V}_1 \subseteq \mathcal{V}_2$, then there is no explicit relation between $\mathcal{V}_1(A)$ and $\mathcal{V}_2(A)$ for an arbitrary formula $A$.

DEFINITION 6.0.3. *Two structures,* $\mathcal{V}$ *and* $\mathcal{W}$ *are* isomorphic, $\mathcal{V} \cong \mathcal{W}$ *whenever there is a bijection* $g\colon V \to W$ *such that the following holds for all* $v_1,\dots,v_n \in V$*:*

*(i) For each couple of functions* $f_\mathcal{V}$ *in* $\mathcal{V}$ *and* $f_\mathcal{W}$ *in* $\mathcal{W}$ *assigned to a functional symbol* $f \in J$,

$$g(f_\mathcal{V}(v_1,\dots,v_n)) = f_\mathcal{W}(g(v_1),\dots,g(v_n)).$$

*(ii) For each predicate symbol* $P \in J$,

$$P_\mathcal{V}(v_1,\dots,v_n) = P_\mathcal{W}(g(v_1),\dots,g(v_n)). \tag{44}$$

*(iii) For each couple of constants* $u$ *in* $\mathcal{V}$ *and* $w$ *in* $\mathcal{W}$ *assigned to a constant symbol* $\mathbf{u} \in J$,

$$g(u) = w.$$

The concepts of isomorphic and elementary substructures were generalized so that they may hold only in some degree, see [36].

The $\Gamma$-*diagram* $D_\Gamma(\mathcal{V})$ of $\mathcal{V}$ is a fuzzy theory with the special axioms

$$\mathrm{SAx}_{D_\Gamma(\mathcal{V})} = \{{}^{a}/A_{x_1,\dots,x_n}[\mathbf{v}_1,\dots,\mathbf{v}_n] \mid A \in \Gamma, a = \mathcal{V}(A_{x_1,\dots,x_n}[\mathbf{v}_1,\dots,\mathbf{v}_n]), \; v_1,\dots,v_n \in V\}. \tag{45}$$

If $\Gamma = F_J$, then we speak about *diagram* of $\mathcal{V}$, and we write simply $D(\mathcal{V})$. Obviously, $\mathcal{V} \models D_\Gamma(\mathcal{V})$.

LEMMA 6.0.4. *Let* $\mathcal{V}$, $\mathcal{W}$ *be two structures for a language* $J$ *and* $V \subseteq W$. *Then* $\mathcal{V}$ *is a* $\Gamma$*-substructure of* $\mathcal{W}$ *iff* $\mathcal{W}_\mathcal{V} \models D_\Gamma(\mathcal{V})$.

*Proof.* Let $A \in \Gamma$ and $v_1, \ldots, v_n \in V$. Then

$$\mathcal{V}(A_{x_1,\ldots,x_n}[\mathbf{v}_1, \ldots, \mathbf{v}_n]) \leq \mathcal{W}(A_{x_1,\ldots,x_n}[\mathbf{v}_1, \ldots, \mathbf{v}_n]),$$

which means that $\mathcal{W}_\mathcal{V} \models D_\Gamma(\mathcal{V})$ because $\mathbf{v}_1, \ldots, \mathbf{v}_n$ are simultaneously object constants of $\mathcal{W}_\mathcal{V}$. The converse is obvious. □

THEOREM 6.0.5. *Let $T$ be a fuzzy theory of a language $J$, ND let $\mathcal{V}$ be a structure and $\Gamma \subseteq F_J$. Then there is a model $\mathcal{W} \models T$ that is an elementary $\Gamma$-extension of $\mathcal{V}$.*

*Proof.* We will consider a fuzzy theory $T'' = T' \cup D(\mathcal{V})$, where $T'$ is a conservative extension of $T$ by constants for the elements of $V$ and $D(\mathcal{V})$ is the diagram of $\mathcal{V}$.

The fuzzy theory $T''$ is consistent because otherwise there should exist formulas $A_1, \ldots, A_n$ and $m_1, \ldots, m_n \in \mathsf{N}^+$ such that $T' \vdash_d \neg((A'_1)^{m_1} \,\&\, \cdots \,\&\, (A'_n)^{m_n}))$ ($A'_i$ are $\mathcal{V}$-instances of $A_i$), so that

$$d > \neg(a_1^{m_1} \otimes \cdots \otimes a_n^{m_n}),$$

where $a_i = \mathcal{V}_\mathcal{V}(A'_i)$. Because $\mathcal{V}_\mathcal{V} \models T'$, we obtain

$$d \leq \mathcal{V}(\neg((A'_1)^{m_1} \,\&\, \cdots \,\&\, (A'_n)^{m_n}))) = \neg(a_1^{m_1} \otimes \cdots \otimes a_n^{m_n}) < d,$$

which is a contradiction. Hence, there is a model $\mathcal{W}'' \models T''$. We will replace $\mathcal{W}''$ by an isomorphic structure $\mathcal{W}$ such that $V \subset W$ holds for their supports. Then, by induction on the complexity of the formula, we will show that $\mathcal{W}(A) = \mathcal{V}(A)$ holds for every $\mathcal{V}$-instance of a formula $A$.

Let $A$ be a $\mathcal{V}$-instance of an atomic formula. Then $\mathcal{W}(A) \geq \mathcal{V}(A)$ as well as $\mathcal{W}(\neg A) \geq \mathcal{V}(\neg A)$ because $A$ and $\neg A$ are axioms of the theory $D(\mathcal{V})$. Consequently,

$$\mathcal{W}(A) = \mathcal{V}(A).$$

If $A = B \Rightarrow C$, then $\mathcal{W}(A) = \mathcal{V}(A)$ by the induction assumption.

Let $A = (\forall x)B$. Then

$$\mathcal{V}(A) \leq \bigwedge\{\mathcal{W}(B_x[\mathbf{w}]) \mid w \in W\},$$

which means that

$$\bigwedge\{\mathcal{W}(B_x[\mathbf{v}]) \mid v \in V\} \leq \mathcal{W}(B_x[\mathbf{w}]) \tag{46}$$

for every $w \in W - V$. By the inductive assumption, $\mathcal{W}(B_x[\mathbf{v}]) = \mathcal{V}(B_x[\mathbf{v}])$ for every $v \in V$. Consequently, by (46),

$$\begin{aligned}\mathcal{W}(A) &= \bigwedge\{\mathcal{W}(B_x[\mathbf{w}]) \mid w \in W\} \\ &= \bigwedge\{\mathcal{V}(B_x[\mathbf{v}]) \mid v \in V\} \wedge \bigwedge\{\mathcal{W}(B_x[\mathbf{w}]) \mid w \in W - V\} = \mathcal{V}(A).\end{aligned}$$

Thus, $\mathcal{W}$ is an elementary extension of $\mathcal{V}$. Finally, because $\Gamma \subseteq F_J$, $\mathcal{W}$ is an elementary $\Gamma$-extension. □

# 7 Some additional properties

## 7.1 Enriching the structure of truth values

Recall that the local character of logical connectives in fuzzy logic requires more interpretations of them depending on the given situation. From the formal point of view, we enrich the structure of truth values $\boldsymbol{L}$ being in the most general case of a residuated lattice by additional operations. A question arises of what properties should be fulfilled by them.

One of the most crucial properties is preservation of the logical equivalence. We thus come to the following definition.

DEFINITION 7.1.1. *Let* $\Box\colon L^n \to L$ *be an* $n$*-ary operation. We say that it is* logically fitting *if it satisfies the following condition: There are natural numbers* $k_1, \ldots, k_n > 0$ *such that*

$$(a_1 \leftrightarrow b_1)^{k_1} \otimes \cdots \otimes (a_n \leftrightarrow b_n)^{k_n} \leq \Box(a_1, \ldots, a_n) \leftrightarrow \Box(b_1, \ldots, b_n) \tag{47}$$

*holds for all* $a_1, \ldots, a_n, b_1, \ldots, b_n \in L$.

The following theorem demonstrates that all the considered operations are logically fitting.

THEOREM 7.1.2.

*(a) Let* $\boldsymbol{L}$ *be a residuated lattice. Then each basic operation* $\vee, \wedge, \otimes, \to$ *is logically fitting.*

*(b) Any composite operation obtained from logically fitting operations is logically fitting.*

*(c) An operation* $\Box\colon [0,1]^n \to [0,1]$ *in Łukasiewicz algebra* $\boldsymbol{L}_{Ł}$ *is logically fitting iff it is Lipschitz continuous.*

*Proof.* (a) can be proved in a straightforward way using the properties of the biresiduation $\leftrightarrow$.

(b) follows from the monotonicity of $\otimes$ after rewriting (47) for all the members of the composite operation.

(c) For simplicity, we assume $n = 2$. Then (47) holds iff there is some $k \in \mathsf{N}$ such that

$$((1 - |a_1 - b_1|) \otimes (1 - |a_2 - b_2|))^k \leq 1 - |\Box(a_1, a_2) - \Box(b_1, b_2)|$$

for all $a_1, a_2, b_1, b_2 \in [0, 1]$, i.e.,

$$|\Box(a_1, a_2) - \Box(b_1, b_2)| \leq k(|a_1 - b_1| + |a_2 - b_2|),$$

which is the condition for the Lipschitz continuity. □

DEFINITION 7.1.3. *Let* $\boldsymbol{L}$ *be a residuated lattice and* $\{\Box_i \mid i = 1, \ldots, k\}$ *be a set of additional* $n_i$*-ary operations on* $L$ *being logically fitting. The* enriched residuated lattice *is an algebra*

$$\boldsymbol{L}_e = \langle L, \vee, \wedge, \otimes, \to, \Box_1, \ldots, \Box_k, \mathbf{0}, \mathbf{1}\rangle. \tag{48}$$

Let $\boldsymbol{L}_e$ be an enriched residuated algebra. We will consider a language $J_e$ obtained from $J$ by adding a set $\{\Box_i \mid i = 1, \ldots, k\}$ of new $n$-ary connectives, each being interpreted by a logically fitting operation $\Box$ for the exponents $k_1, \ldots, k_n$. Furthermore, the fuzzy set of logical axioms LAx is extended by the congruence axiom corresponding to each new connective $\Box$:

$$1/(A_1 \Leftrightarrow B_1)^{k_1} \,\&\cdots\&(A_n \Leftrightarrow B_n)^{k_n} \Rightarrow \\ (\Box(A_1, \ldots, A_n) \Leftrightarrow \Box(B_1, \ldots, B_n)), \quad (49)$$

where $A_1, \ldots, A_n, B_1, \ldots, B_n \in F_{J_e}$.

As one can see, to be logically fitting, in fact, means that the additional connectives preserve logical equivalence. It is not surprising, then, that the completeness theorem is not harmed.

THEOREM 7.1.4 (Completeness in the enriched language). *Let $T$ be a consistent fuzzy theory in the language $J_e$. Then*

$$T \vdash_a A \iff T \models_a A$$

*holds for every formula $A \in F_{J_e}$.*

### 7.2 Sorites fuzzy theories

In this section, we will prove a theorem that formalizes the well-known Sorites paradox in $\mathrm{Ev}_{Ł}$ (with $L = [0, 1]$). First, we will recall the formulation of the paradox:

> *One grain of wheat does not make a heap, nor does two grains, or three, etc. Hence, there are no heaps.*

Let $\mathbf{Fe}(x)$ denote the proposition "the number $x$ does not make a heap". The problem lies in verification of the implication $\mathbf{Fe}(x) \Rightarrow \mathbf{Fe}(x+1)$. If the latter is taken classically (i.e., absolutely true), then starting from $\mathbf{Fe}(1)$, we necessarily arrive at the above paradox.

In general, verification that $x$ does not make a heap does not imply that we will be able to verify with the same effort that $x + 1$ also does not make a heap. For example, if we verify that 500 grains do not make a heap by counting them, then verification that 501 grains also do not make a heap means we must count one grain more, i.e., our effort to verify that we have 501 grains at disposal is slightly (imperceptibly) greater than that for 500 grains. Consequently, the above implication is not fully true. We see that the classical induction cannot be applied to the vague property of "being a heap".

$\mathrm{Ev}_{Ł}$ has means to express that the truth of the implication $\mathbf{Fe}(x) \Rightarrow \mathbf{Fe}(x+1)$ is smaller than 1 by setting its lower limit to $1 - \varepsilon$ for some $\varepsilon > 0$. As follows from the theorem below, the Sorites paradox disappears.

THEOREM 7.2.1. *Let $T_{PA}$ be a Peano theory whose fuzzy set of special axioms is the classical Peano axioms accepted in degree 1. Furthermore, let $0 < \varepsilon \leq 1$, and let $\mathbf{Fe} \notin J(T_{PA})$ be a new predicate. Then the fuzzy theory*

$$T_{Fe} = T_{PA} \cup \{1/\,\mathbf{Fe}(\bar{1}), 1 - \varepsilon/(\forall x)(\mathbf{Fe}(x) \Rightarrow \mathbf{Fe}(S(x))), 1/(\exists x)\neg\,\mathbf{Fe}(x)\} \quad (50)$$

*is a consistent conservative extension of $T_{PA}$.*

*Proof.* The proof is based on construction of the model $\mathcal{V} = \langle \mathsf{N}, +, \cdot, s, \leq, Fe, 1\rangle$, in which

$$Fe(n+1) = Fe(n) \otimes (1-\varepsilon), \qquad n \in \mathsf{N}, \tag{51}$$

where $\otimes$ is the Łukasiewicz conjunction (13) (the details can be found in [21, 36]). □

Note that the formulation of this theorem is analogous to the formulation provided by Parikh in [37], who showed that the contradiction does not occur when confining to proofs shorter than some limit. No such limitation is needed in $\mathrm{Ev}_{Ł}$.

COROLLARY 7.2.2. *For each* $n \in \mathsf{N}$,

$$T_{Fe} \vdash_{e(n)} \mathbf{Fe}(\bar{n}), \qquad e(n) = 0 \vee (1 - n\varepsilon).$$

*Proof.* Starting from ${}^{1}/\,\mathbf{Fe}(\bar{1})$, we construct a proof of $\mathbf{Fe}(\bar{n})$ with the value $e(n)$. On the other hand, by construction of the model $\mathcal{V} \models T_{Fe}$, we obtain $\mathcal{V}(\mathbf{Fe}(\bar{n})) = e(n)$. □

This solution to the Sorites paradox justifies well the transition from full truth to falsity without contradiction. Namely, it can be seen from Corollary 7.2.2 that $e(1) = 1$ and that there is a number $n_0$ such that $e(n_0) = 0$. The same holds in any model of $T_{Fe}$—there are numbers $m_0$ and $n_0$, $m_0 < n_0$ such that $Fe(m_0) = 1$ and $Fe(n_0) = 0$. The number $n_0$ stating that any greater or equal number of grains apparently makes a heap[11] is determined by the threshold $\varepsilon$ and it actually depends on the context. Thus, we can have, say, $n_0 = 10^{15}$ for computer memory, $n_0 = 10^5$ for the number of hairs on a human head and $n_0 = 100$ for a bus full of people. We argue that this is precisely in accordance with our intuition and observation.

### 7.3 $\mathrm{Ev}_{Ł}$ and Łukasiewicz logic

As the reader surely has noticed, the logical axioms (R1)–(R4), (T1) and (T2) are just axioms of Łukasiewicz first-order logic Ł∀. Thus, a natural question is raised: is $\mathrm{Ev}_{Ł}$ a new logic or it is just another form of Ł∀?

First, all the axioms above are taken as logical axioms in the degree **1**, so it is not difficult to see that Łukasiewicz logic is *included* in $\mathrm{Ev}_{Ł}$. Namely, all formulas provable in Ł∀ are effectively provable in the degree **1** in $\mathrm{Ev}_{Ł}$. Because the language of $\mathrm{Ev}_{Ł}$ is larger, it contains Ł∀ as its proper part.

However, there is also the opposite point of view. Namely, P. Hájek in [18] showed how $\mathrm{Ev}_{Ł}$ can be represented in Ł∀, which he calls *Rational Pavelka logic* (RPL). Being part of Ł∀, it has a classical notion of provability and classical inference rules of modus ponens and generalization. The provability degree is there introduced as an additional notion to the classical provability.

RPL is obtained from Ł∀ by extension of its language by truth constants $\mathfrak{a}$ for all $a \in L_{\mathsf{Q}}$ and the bookkeeping axiom (B1). The evaluated formula ${}^{a}/A$ is represented by the formula

$$\mathfrak{a} \Rightarrow A.$$

[11] Note that by no means can we say that $n_0$ is the first number making a heap; it is only a number that we surely know makes a heap.

Then the *provability degree* of a formula $A$ is defined as

$$|A|_T = \bigvee\{a \mid T \vdash \mathbf{a} \Rightarrow A\}.$$

Unfortunately, Hájek did not provide a sound justification of this notion and introduced it only as an additional and not really organic concept. RPL thus became a special extension of Łukasiewicz logic, for which it is unclear why it should be studied.

A very interesting result was published in [22]. The main theorem is the following.

THEOREM 7.3.1. *The first-order* RPL *is a conservative extension of* Ł∀, *i.e., for any formula A of the language of* Ł∀,

$$\vdash_{\mathrm{RPL}} A \iff \vdash_{\text{Ł}\forall} A.$$

The proof of this theorem is complicated, and thus we omit it. It uses algebraic arguments and assumes the validity of the completeness theorem for Łukasiewicz logic.

*Summarization:* The fuzzy logics introduced and studied by Hájek in [18] have a traditional syntax extended by several special axioms and many-valued (fuzzy) is only their semantics. Thus, they follow the idealistic assumption that all facts assumed in advance (i.e., axioms) are indisputably true. However, as everybody knows, this does not hold in reality. Many (if not most) facts are known with some certainty that need not be 100%, and so taking them as fully true is dubious. To bring logic closer to the real world, we must also consider axioms that are only partially true. However, this opens the door to evaluated syntax and to the provability degree concept. Because we strive for having a logic that is a very natural generalization of the classical logic, all concepts related to the evaluated syntax must not be something alien that has to be added (as is the case of RPL); rather, they must naturally arise from the pure character of the logic. We argue that $\mathrm{Ev}_{\text{Ł}}$ presented above fulfills this goal. Moreover, $\mathrm{Ev}_{\text{Ł}}$ enables us, as a side effect, to see classical logic from a different point of view and to understand more deeply why various studied concepts (including provability) work.

A very interesting fact is that when considering the interval $[0, 1]$ as the set of truth values, we are bound to just one (up to isomorphism) structure of truth values. An analogous situation is encountered in classical logic, where the Boolean algebra on $\{\mathbf{0}, \mathbf{1}\}$ is the only one. This is good for various kinds of applications because we do not face the delicate problem presented by the research in mathematical fuzzy logic: we have numerous different systems of fuzzy logic,[12] and it is unclear which of them should be chosen for our application. The uniqueness of $\mathrm{Ev}_{\text{Ł}}$ suggests a daring question: do we need any of them at all?

## 8 Historical remarks and further reading

The first paper containing analysis of the concept of fuzzy logic was published at the end of the 1960s by J.A. Goguen [13]. One of the essential ideas of this paper was to introduce two kinds of operations on the algebra of truth values that can be considered an interpretation of the logical conjunction. Consequently, there are two conjunctions

[12] For example, in [6], 66 different systems of fuzzy logic are analyzed.

in fuzzy logic—the first is interpreted by $\wedge$ and the second one by $\otimes$.[13] In classical logic, both conjunctions coincide. From the algebraic point of view, this idea gives rise to the concept of the integral, complete lattice-ordered monoid (i.e., our concept of a residuated lattice introduced in Section 2.1). Goguen gave arguments in favor of fuzzy logic developed on the basis of such an algebra of truth values.

Goguen's paper was followed by J. Pavelka, who in 1977 defended his PhD thesis in which he studied the theory of propositional fuzzy logic that generalizes classical logic in two aspects. Namely, it has not only many-valued semantics but also accepts degrees in its syntax. These degrees occur when assuming that axioms need not be fully true and so, they can form a fuzzy set. A further step is the introduction of many-valued inference rules and the concept of the provability degree. The thesis was published in 1979 in a sequence of 3 papers [38]. They contain an introduction to the concept of many-valued rules of inference, a detailed analysis of the metamathematical properties of his formal system of fuzzy logic and the proof of its completeness. Pavelka gave here also detailed arguments justifying necessity to consider complete MV-algebra as the algebra of truth values of $\mathrm{Ev}_{Ł}$ (these results were used in Subsection 3.6).

Pavelka's results were followed by the author of this chapter in his PhD thesis defended in 1988, where the propositional $\mathrm{Ev}_{Ł}$ was extended to a predicate first-order version. The thesis, published in [26], contains proof of the generalized Gödel completeness theorem of the predicate $\mathrm{Ev}_{Ł}$ as well as a form of the deduction theorem (Theorem 4.5.8). Analogous theorem was later proved also in various kinds of fuzzy logics with traditional syntax. $\mathrm{Ev}_{Ł}$ was further developed especially by the author of this chapter in many papers, e.g., [27–32, 34].

The most comprehensive elaboration of $\mathrm{Ev}_{Ł}$ including two different proofs of the completeness theorem, was provided in book [36]. In addition to many other results, this book contains a model of the approximate reasoning on the basis of $\mathrm{Ev}_{Ł}$. Two more books related to $\mathrm{Ev}_{Ł}$ were written by G. Gerla [11] and E. Turunen [40].

In the 1990s, $\mathrm{Ev}_{Ł}$ was overshadowed by fuzzy logics having *traditional syntax* and many-valued semantics. The turning point was the publication of the book [18] by P. Hájek. Let us remark that Hájek knew $\mathrm{Ev}_{Ł}$ because he was a reviewer of both Pavelka's and Novák's PhD theses. He appreciated these results but was unsatisfied with the idea of evaluated syntax and insisted on keeping the classical syntactical concepts. Therefore, he began to work on his concept of the theory of fuzzy logic. His effort was crowned by the cited book, where he introduced several systems of fuzzy logic based on variants of the residuated lattice of truth values, namely, the fuzzy logic BL and its three axiomatic extensions: the Łukasiewicz, product and Gödel logics. Hájek's work was followed by many other researchers, including F. Esteva, L. Godo, D. Mundici, F. Montagna, P. Cintula and others (see, e.g., [4, 6, 10, 15, 18, 35]).

Hájek also introduced in his book the Rational Pavelka Logic (RPL) as a representation of rational variant of $\mathrm{Ev}_{Ł}$ in Łukasiewicz logic. Let us emphasize that we by no means can identify $\mathrm{Ev}_{Ł}$ with RPL. He used the term "rational" because his RPL reduces the set of truth constants to rational ones over $L_{\mathbb{Q}}$. The reason is that considering truth constants over $L = [0, 1]$, i.e., having the language $J$ uncountable, makes the

[13] The $\wedge$-conjunction is, in general, greater than the $\otimes$-conjunction. In $\mathrm{Ev}_{Ł}$ the latter is even nilpotent. Therefore, it is often called *strong* or, in our case, Łukasiewicz one. In the literature, it is also called *fusion*.

logicians to question the value of such system, because they usually think that a computationally untractable logic cannot be used for real-life applications. However, it had already been shown before Hájek that we do not need truth constants for all $a \in [0, 1]$. One possibility was presented in [27, 36], where it was shown that we may confine the scope to formulas containing rational or dyadic truth constants only and approximate the other formulas by means of sets of the former. This solution, however, is an indirect approach to freeing $\mathrm{Ev}_{Ł}$ from truth constants for irrational truth values. The direct solution, also presented in this chapter, was published in [34].

A special problem was to find the minimal number of logical axioms (besides the book-keeping axiom and the truth constants, which are in $\mathrm{Ev}_{Ł}$ also logical axioms). Pavelka, in [38] considered more than thirty of them, adding a new axiom whenever he needed it. This number has been significantly reduced by Novák in [26]. As noted by Hájek in [16] and Gottwald in [14], the fuzzy set of propositional logical axioms can be reduced to the well-known Rose–Rosser axioms of Łukasiewicz logic.

Let us emphasize that there are many other interesting results in $\mathrm{Ev}_{Ł}$, for example, in model theory. Some of these results were obtained quite early by A. Di Nola and G. Gerla in [9] and by M. Ying in [42–44]. However, there are also newer results in model theory [33], on omitting types by V. Novák and P. Murinová in [25] and X. Caicedo and J. N. Iovino in [3], or those obtained by P. Dellunde in [7, 8].[14]

Notable also are results in the analysis of the arithmetical complexity of Łukasiewicz logic and $\mathrm{Ev}_{Ł}$. One of the first papers was written by Scarpellini [39], where the non-axiomatizability of first-order Łukasiewicz logic is proved. Hájek, in a sequence of papers [16, 17, 19], provided a detailed analysis of several kinds of fuzzy logics. He proved that if we limit $\mathrm{Ev}_{Ł}$ to rational truth constants then the complexity of the set of sentences provable in a rational degree is $\Pi_2$.

One interesting result was obtained by Bělohlávek in [1] (see also [2]), where he showed that it is possible to develop an equational fragment of fuzzy logic called *fuzzy equational logic*. This logic evaluates syntax in which the linearity of the algebra of truth values is not required and the completeness is proved for any complete residuated lattice as the structure of truth values. Various papers either mentioning or dealing directly with $\mathrm{Ev}_{Ł}$ appear from time to time, e.g., [12, 41].

**Acknowledgements**

The chapter has been supported by the project "LQ1602 IT4Innovations excellence in science".

# BIBLIOGRAPHY

[1] Radim Bělohlávek. *Fuzzy Relational Systems: Foundations and Principles*, volume 20 of *IFSR International Series on Systems Science and Engineering*. Kluwer and Plenum Press, New York, 2002.

[2] Radim Bělohlávek and Vilém Vychodil. *Fuzzy Equational Logic*, volume 186 of *Studies in Fuzziness and Soft Computing*. Springer, Berlin/Heidelberg, 2005.

[3] Xavier Caicedo and José N. Iovino. Omitting uncountable types and the strength of [0,1]-valued logics. *Annals of Pure and Applied Logic*, 165:1169–1200, 2014.

[14] To be precise, the authors do not explicitly mention $\mathrm{Ev}_{Ł}$ in these papers. Their results, however, can be applied also to it.

[4] Roberto Cignoli, Itala M.L. D'Ottaviano, and Daniele Mundici. *Algebraic Foundations of Many-Valued Reasoning*, volume 7 of *Trends in Logic*. Kluwer, Dordrecht, 1999.

[5] Petr Cintula, Francesc Esteva, Joan Gispert, Lluís Godo, Franco Montagna, and Carles Noguera. Distinguished algebraic semantics for t-norm based fuzzy logics: Methods and algebraic equivalencies. *Annals of Pure and Applied Logic*, 160(1):53–81, 2009.

[6] Petr Cintula, Petr Hájek, and Rostislav Horčík. Formal systems of fuzzy logic and their fragments. *Annals of Pure and Applied Logic*, 150(1-3):40–65, 2007.

[7] Pilar Dellunde. Preserving mappings in fuzzy predicate logics. *Journal of Logic and Computation*, 22(6):1367–1389, 2012.

[8] Pilar Dellunde. Revisiting ultraproducts in fuzzy predicate logics. *Multiple-Valued Logic*, 19:95–108, 2012.

[9] Antonio Di Nola and Giangiacomo Gerla. Fuzzy models of first-order languages. *Zeitschrift für Mathematische Logik und Grundlagen der Mathematik*, 32(19–24):331–340, 1986.

[10] Francesc Esteva and Lluís Godo. Monoidal t-norm based logic: Towards a logic for left-continuous t-norms. *Fuzzy Sets and Systems*, 124(3):271–288, 2001.

[11] Giangiacomo Gerla. *Fuzzy Logic—Mathematical Tool for Approximate Reasoning*, volume 11 of *Trends in Logic*. Kluwer and Plenum Press, New York, 2001.

[12] Giangiacomo Gerla. Effectiveness and multi-valued logics. *Journal of Symbolic Logic*, 71:137–162, 2006.

[13] Joseph Amadee Goguen. The logic of inexact concepts. *Synthese*, 19(3–4):325–373, 1969.

[14] Siegfried Gottwald. *A Treatise on Many-Valued Logics*, volume 9 of *Studies in Logic and Computation*. Research Studies Press, Baldock, 2001.

[15] Siegfried Gottwald. Mathematical fuzzy logics. *The Bulletin of Symbolic Logic*, 14(2):210–239, 2008.

[16] Petr Hájek. Fuzzy logic and arithmetical hierarchy. *Fuzzy Sets and Systems*, 73(3):359–363, 1995.

[17] Petr Hájek. Fuzzy logic and arithmetical hierarchy II. *Studia Logica*, 58(1):129–141, 1997.

[18] Petr Hájek. *Metamathematics of Fuzzy Logic*, volume 4 of *Trends in Logic*. Kluwer, Dordrecht, 1998.

[19] Petr Hájek. Fuzzy logic and arithmetical hierarchy III. *Studia Logica*, 68(1):129–142, 2001.

[20] Petr Hájek and Petr Cintula. On theories and models in fuzzy predicate logics. *Journal of Symbolic Logic*, 71(3):863–880, 2006.

[21] Petr Hájek and Vilém Novák. The sorites paradox and fuzzy logic. *International Journal of General Systems*, 32(4):373–383, 2003.

[22] Petr Hájek, Jeff Paris, and John C. Shepherdson. Rational Pavelka logic is a conservative extension of Łukasiewicz logic. *Journal of Symbolic Logic*, 65(2):669–682, 2000.

[23] Ioana Leuştean and Antonio Di Nola. Łukasiewicz logic and MV-algebras. In Petr Cintula, Petr Hájek, and Carles Noguera, editors, *Handbook of Mathematical Fuzzy Logic - Volume 2*, volume 38 of *Studies in Logic, Mathematical Logic and Foundations*, pages 469–583. College Publications, London, 2011.

[24] Jan Menu and Jan Pavelka. A note on tensor products on the unit interval. *Commentationes Mathematicae Universitatis Carolinae*, 17:71–83, 1976.

[25] Petra Murinová and Vilém Novák. Omitting types in fuzzy logic with evaluated syntax. *Mathematical Logic Quarterly*, 52(3):259–268, 2006.

[26] Vilém Novák. On the syntactico-semantical completeness of first-order fuzzy logic part I (syntax and semantic), part II (main results). *Kybernetika*, 26:47–66, 134–154, 1990.

[27] Vilém Novák. Fuzzy logic revisited. In *Proceedings of EUFIT 1994*, pages 496–499, Aachen, 1994. Verlag der Augustinus Buchhandlung.

[28] Vilém Novák. A new proof of completeness of fuzzy logic and some conclusions for approximate reasoning. In *Proceedings of FUZZ-IEEE/IFES 1995*, pages 1461–1468, Yokohama, 1995.

[29] Vilém Novák. Towards formalized integrated theory of fuzzy logic. In Zeungnam Bien and Kyung Chan Min, editors, *Fuzzy Logic and Its Applications to Engineering, Information Sciences, and Intelligent Systems*, pages 353–363. Kluwer, Dordrecht, 1995.

[30] Vilém Novák. Formal theories in fuzzy logic. In Didier Dubois, Erich Peter Klement, and Henri Prade, editors, *Fuzzy Sets, Logics, and Reasoning about Knowledge*, pages 213–235. Kluwer, Dordrecht, 1996.

[31] Vilém Novák. Paradigm, formal properties and limits of fuzzy logic. *International Journal of General Systems*, 24:377–405, 1996.

[32] Vilém Novák. Open theories, consistency and related results in fuzzy logic. *International Journal of Approximate Reasoning*, 18:191–200, 1998.

[33] Vilém Novák. Joint consistency of fuzzy theories. *Mathematical Logic Quarterly*, 48(4):563–573, 2002.

[34] Vilém Novák. Fuzzy logic with countable evaluated syntax revisited. *Fuzzy Sets and Systems*, 158: 929–936, 2007.

[35] Vilém Novák and Irina Perfilieva, editors. *Discovering the World With Fuzzy Logic*, volume 57 of *Studies in Fuzziness and Soft Computing*. Springer, Heidelberg, 2000.
[36] Vilém Novák, Irina Perfilieva, and Jiří Močkoř. *Mathematical Principles of Fuzzy Logic*. Kluwer, Dordrecht, 2000.
[37] Rohit Parikh. The problem of vague predicates. In Robert S. Cohen and Marx W. Wartofsky, editors, *Language, Logic and Method*, pages 241–261. D. Reidel, Dordrecht, 1983.
[38] Jan Pavelka. On fuzzy logic I, II, III. *Zeitschrift für Mathematische Logik und Grundlagen der Mathematik*, 25:45–52, 119–134, 447–464, 1979.
[39] Bruno Scarpellini. Die Nichtaxiomatisierbarkeit des unendlichwertigen Prädikatenkalküls von Łukasiewicz. *Journal of Symbolic Logic*, 27(2):159–170, 1962.
[40] Esko Turunen. *Mathematics Behind Fuzzy Logic*. Advances in Soft Computing. Springer, Heidelberg, 1999.
[41] Esko Turunen. Paraconsistent semantics for Pavelka style fuzzy sentential logic. *Fuzzy Sets and Systems*, 161:1926–1940, 2010.
[42] Mingsheng Ying. Deduction theorem for many-valued inference. *Zeitschrift für Mathematische Logik und Grundlagen der Mathematik*, 37:533–537, 1991.
[43] Mingsheng Ying. Compactness, the Löwenheim–Skolem property and the direct product of lattices of truth values. *Zeitschrift für Mathematische Logik und Grundlagen der Mathematik*, 38:521–524, 1992.
[44] Mingsheng Ying. The fundamental theorem of ultraproduct in Pavelka's logic. *Zeitschrift für Mathematische Logik und Grundlagen der Mathematik*, 38:197–201, 1992.

VILÉM NOVÁK
Institute for Research and Applications of Fuzzy Modeling
University of Ostrava & NSC IT4Innovations
30. dubna 22
701 03 Ostrava 1, Czech Republic
Email: Vilem.Novak@osu.cz

# Chapter XVI: Fuzzy Description Logics

FERNANDO BOBILLO, MARCO CERAMI, FRANCESC ESTEVA,
ÀNGEL GARCÍA-CERDAÑA, RAFAEL PEÑALOZA,
UMBERTO STRACCIA

## 1 Introduction

Description Logics (DLs) are a family of well established knowledge representation formalisms (see [4] for a complete overview on the subject). The languages for DLs are based on *concepts*, *roles* and *individuals*, that are, according to a well-defined Tarski style semantics, interpreted as sets (or unary predicates), binary relations (or binary predicates), and domain elements, respectively. From a suitable set of atomic concepts and roles, it is possible to build up complex concepts by means of a limited family of *concept constructors*. The different languages of DLs are identified by sets of concept constructors and they are more or less expressive, depending on the concept constructors that are either explicitly present in the language or implicitly definable from other constructors.

From concepts and roles it is possible to store structured knowledge under the form of *knowledge bases* that contain the relevant terminological knowledge about particular knowledge domains. Knowledge bases contain structured knowledge that gives an account of functional dependencies between complex concepts, or relations between individuals and either concepts or roles. In this sense, a knowledge base is usually compounded by a *terminological box*, or TBox, that contains inclusions between complex concepts, an *assertional box*, or ABox, that contains assertions of concepts or roles with respect to individuals or ordered couples of individuals respectively and, sometimes, by a *relational box*, or RBox, that contains inclusions between roles.

Since DLs are mainly oriented to applications, they need to exhibit good computational behavior. The reasoning services provided have to be decidable and, as much as possible, computable in a reasonable time. For this reason, historically the research on DLs has been characterized by the search of a fair trade-off between the expressivity of the languages considered and the computational complexity of the reasoning services.

There are two possible ways to present DL languages and their semantics. One way is to present them directly as representation languages as it is usually done. One example of this kind of presentation can be found in the introductory chapter [8] of the Description Logic Handbook [4]. Another way is to present basic DLs from the logical point of view as fragments of first order classical logic [111].

Fuzzy Description Logics (FDLs) were born in the nineties as a generalization of DLs to the fuzzy framework. The first generalizations consisted in having the same structure as DLs, but interpreting concepts and roles as fuzzy sets and fuzzy relations respectively (see the section on historical remarks at the end of this chapter). Nevertheless these first approaches had no clear logical counterpart. At the end of nineties, in the framework of Fuzzy Logic, Hájek [78] proposed many-valued residuated logical systems (studied in this handbook) as the kernel of Fuzzy Logic in narrow sense. This has been the starting point of that branch of Fuzzy Logic known as *Mathematical Fuzzy Logic* (MFL). The important development of this subject, of which the results provided in this handbook are a clear account, has established different consolidated logical systems. Within the framework of MFL, Hájek proposes in [80] (see also [82]) to define FDLs from the logical point of view in an analogous way as DLs could be defined from classical first order logic.

This chapter is structured as follows: In Section 2 we introduce the algebras of truth values defined by continuous t-norms (or divisible finite t-norms), their residua and the standard involutive negation. The semantics of our FDLs are defined as interpretations over these algebras. This is a technical section that could be omitted in a first reading. The reader can go back to it when she needs some notion or result concerning the algebras of truth values. In Section 3 we define FDLs taking as constructors the generalization of the classical ones. The semantics is defined by means of interpretations in the algebraic ordered structure defined on the real interval $[0, 1]$ by a continuous t-norm and its residuum, or in the one defined on a finite chain by a divisible finite t-norm and its residuum. We study the hierarchy of basic FDL languages, which is more complex than in the classical case due to the fact that some definabilities between certain constructors are not valid in the fuzzy framework. The axioms in knowledge bases in the fuzzy setting are graded. Reasoning tasks are also graded. We will be interested in stating satisfiability of a concept or subsumption between concepts in a certain degree. Section 4 is devoted to highlight the relationship among FDLs defined in the previous section and fuzzy predicate logics studied in the framework of MFL. First, we define logical systems underlying FDL languages from a semantical point of view as logics arising from particular algebras of truth values. Then we show that FDL languages defined in the previous sections can be seen as fragments of these first-order fuzzy logics, according to Hájek's proposal in [80]. The chapter follows with a section devoted to Fuzzy description languages with higher expressivity and a section dealing with different algorithms to decide the reasoning tasks. Section 7 is devoted to (un)decidability and complexity results. Finally, the last section contains the historical remarks that summarize the reported results from a historical point of view.

Let us remark that in this chapter we will pay special attention to the case of finite-valued FDLs, which seems to be the most interesting in practice since the basic reasoning tasks for quite expressive languages are decidable and the computational complexity of some of them is the same as the corresponding classical DLs.

Finally, let us stress that there are different issues that are treated in the fuzzy setting that are still not well modeled in the logical setting of MFL. Among the most important there are *fuzzy quantifiers*. Expressions like "the majority of ...", "a few ...", etc, that are crucial in the fuzzy approach are not yet considered in FDLs based in MFL. For an excellent report about the state of the art about fuzzy quantifiers in fuzzy logic, see [66], while for fuzzy quantification in FDLs see [109, 110]. Another topic not considered in this chapter is the treatment of uncertainty in FDLs (see [92] and the references therein for a report about the work done in this topic). Nevertheless, the framework of MFL still offers formal tools that can be used to try to give a mathematical logic treatment to this kind of notions. Undoubtedly, the framework of FDLs, above all in the MFL setting, can be a powerful source of motivation for undertaking this kind of research, as it was predicted by Hájek in the not so known and frequently visited second part of his seminal monograph.

# 2 Algebras of truth values defined by triangular norms

The fuzzy description logics we deal with in this chapter are based on interpretations on certain linearly ordered algebras of truth values called *canonical chains*. In the infinite case these algebras are those defined on the real interval $[0,1]$ by a continuous t-norm and its residuum, and by the involutive negation $N(x) = 1 - x$. In the finitely-valued case the canonical chains are the ones defined on the finite chain $\{0, \frac{1}{n-1}, \ldots, \frac{n-2}{n-1}, 1\}$ by a divisible finite t-norm, its residuum and the involutive negation $N$. In this preliminary section we recall the notions of continuous t-norm and divisible finite t-norm, and some notions and results about these canonical chains.

## 2.1 Triangular norms and conorms

For the notions and results presented in this section, a full reference is the book [91]. A *triangular norm* (or *t-norm*) is a binary operation defined on the real interval $[0,1]$ which is associative, commutative, non-decreasing in both arguments,[1] and has 1 as unit element. If the t-norm $*$ is continuous, we can define the operation $\Rightarrow_*$ satisfying, for all $x, y, z \in [0,1]$, the condition

$$x * z \leq y \quad \Leftrightarrow \quad z \leq x \Rightarrow_* y. \tag{1}$$

This operation is called the *residuum* of the t-norm and it can be defined by:

$$x \Rightarrow_* y := \max\{z \mid x * z \leq y\}. \tag{2}$$

PROPOSITION 2.1.1. *The following properties hold for any continuous t-norm $*$ and its residuum $\Rightarrow_*$:*

*a)* $x \Rightarrow_* y = 1$ *if and only if* $x \leq y$

*b)* $1 \Rightarrow_* x = x$

*c)* $\min\{x, y\} = x * (x \Rightarrow_* y)$ (divisibility)

*d)* $\max\{x, y\} = \min\{(x \Rightarrow_* y) \Rightarrow_* y, (y \Rightarrow_* x) \Rightarrow_* x\}$.

The reason for which property c) in the previous proposition is called *divisibility condition* is the fact that this is equivalent to the following one: for every $x, y \in [0,1]$ with $x > y$, there exists $z \in [0,1]$ such that $x * z = y$. Observe that this condition is

[1] A function $f\colon [0,1]^2 \to [0,1]$ is *non-decreasing* in both arguments if for each $x, y, z \in [0,1]$, $x \leq y$ implies $f(x,z) \leq f(y,z)$ and $f(z,x) \leq f(z,y)$.

| $*$ | Minimum (Gödel) | Product | Łukasiewicz |
|---|---|---|---|
| $x * y$ | $\min\{x, y\}$ | $x \cdot y$ | $\max\{0, x + y - 1\}$ |
| $x \Rightarrow_* y$ | $\begin{cases} 1 & \text{if } x \leq y \\ y & \text{otherwise} \end{cases}$ | $\begin{cases} 1 & \text{if } x \leq y \\ y/x & \text{otherwise} \end{cases}$ | $\min\{1, 1 - x + y\}$ |
| $\neg_*$ | $\begin{cases} 1 & \text{if } x = 0 \\ 0 & \text{otherwise} \end{cases}$ | $\begin{cases} 1 & \text{if } x = 0 \\ 0 & \text{otherwise} \end{cases}$ | $1 - x$ |

Table 1. The three main continuous t-norms

equivalent to the continuity of the functions defined by $f_y(x) = x * y$ for every $y \in [0,1]$ and also to the continuity of $*$ due to the fact that $*$ is commutative and non-decreasing in both arguments.

A *negation function* on $[0,1]$ is a unary operation $n\colon [0,1] \to [0,1]$ satisfying the following properties:

- $n(0) = 1$, $n(1) = 0$, and
- for all $x, y \in [0,1]$, $x \leq y \Rightarrow n(y) \leq n(x)$. (antimonotonicity)

We say that $n$ is *weak* if, for all $x \in [0,1]$, $x \leq n(n(x))$; and it is said to be *strong* or *involutive* if, for all $x \in [0,1]$, $n(n(x)) = x$. The main example of involutive negation is the so called *standard negation function* defined as $N(x) = 1 - x$. Each continuous t-norm $*$ has an associated negation function defined as $\neg_* x = x \Rightarrow_* 0$, which is always a weak negation function and it is involutive if and only if $*$ is the (finite or infinite) Łukasiewicz t-norm.

Table 1 shows the main continuous t-norms (Minimum, Product and Łukasiewicz) with their residua and the corresponding associated negations. The three continuous t-norms in Table 1 are the basic ones since any continuous t-norm can be expressed as an *ordinal sum* of copies of them [99]. Indeed, there are the following types of continuous t-norms that we will use in the chapter (see for instance [69]):

1. *Idempotent* t-norms: A t-norm is idempotent if for all $x \in [0,1]$, $x * x = x$.
2. *Strict*[2] t-norms: A t-norm is *strict* if for every $x, y \in [0,1]$ such that $x * y = 0$, then $x = 0$ or $y = 0$. This is equivalent to the following two conditions:
   (a) The associated negation is *Gödel negation*, i.e., $\neg_* 0 = 1$ and $\neg_* x = 0$, otherwise, and
   (b) for every $x \in [0,1]$, $\min\{x, \neg_* x\} = 0$.
3. *Nilpotent* t-norms: A t-norm is nilpotent if for each $x \in (0,1)$ there exists a natural $n$ such that $x^n = 0$. This is equivalent to the following conditions:
   (a) It is isomorphic to the Łukasiewicz t-norm and
   (b) the associated negation is involutive.

[2] Let us remark that the name *strict* is used in a different meaning than in [91], where *strict* means *strictly monotone*.

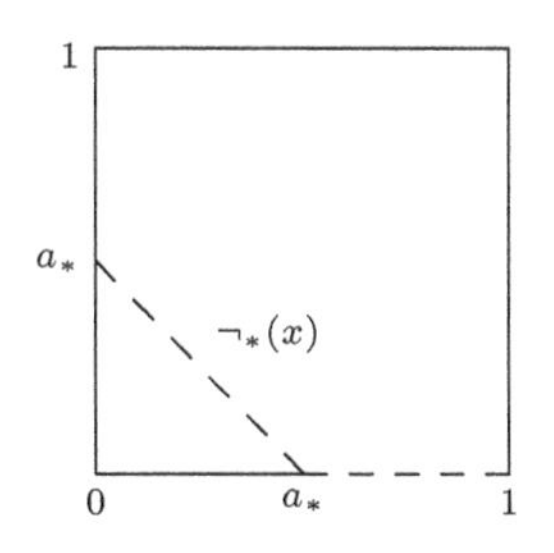


Figure 1. Graph of a residuated negation

The only idempotent continuous t-norm is the minimum and the only nilpotent continuous t-norm up to isomorphisms is Łukasiewicz. Strict continuous t-norms form a very large set containing minimum and product and any ordinal sum that has either no first component or whose first component is not Łukasiewicz t-norm. Finally, we have the set of non-strict continuous t-norms that contains Łukasiewicz t-norm plus any ordinal sum having Łukasiewicz as first component. Therefore, we can associate to each continuous t-norm $*$ an element, denoted $a_*$, defined as:

- 0 if $*$ is strict or
- the upper bound of the first component of the ordinal sum of $*$ if $*$ is non-strict.

Notice that $a_* = 1$ if and only if $*$ is Łukasiewicz. Let us remark that strict, nilpotent and non-strict continuous t-norms are characterized by their associated negation:

- The t-norm $*$ is strict if and only if $\neg_*$ is Gödel negation.
- The t-norm $*$ is nilpotent if and only if $\neg_*$ is involutive.
- The t-norm $*$ is non-strict if and only if $\neg_*$ is defined as $\neg_*(0) = 1$, $\neg_*(x) = a_* - x$ if $x \in (0, a_*]$ and $\neg_*(x) = 0$ otherwise.

Summarising, all residuated negation functions are definable in the form (see Figure 1):

$$\neg_*(x) = \begin{cases} 1 & \text{if } x = 0 \\ a_* - x & \text{if } x \in (0, a_*] \\ 0 & \text{otherwise.} \end{cases}$$

Observe that if $a_* = 0$, we obtain Gödel negation, and if $a_* = 1$, then $\neg_* = N$. Observe also that if $*$ is strict, then the double negation is given by $\neg_*\neg_*0 = 0$ and $\neg_*\neg_*x = 1$ for $x \neq 0$. Thus $\neg_*\neg_*$ can be seen as a crispification operator that preserves falsity.

The following result is used in Section 3.3 to state the relationship between both quantifier restrictions in our FDL languages.

PROPOSITION 2.1.2. *Let* $X \subseteq [0, 1]$. *If* $*$ *is a continuous t-norm, then*

*a)* $\neg_* \sup X = \inf\{\neg_* x \mid x \in X\}$.

*b)* $\neg_* \inf X \geq \sup\{\neg_* x \mid x \in X\}$; *the equality holds only if* $*$ *is the Łukasiewicz t-norm.*

| $\oplus$ | Maximum | Probabilistic sum | Łukasiewicz |
|---|---|---|---|
| $x \oplus y$ | $\max\{x, y\}$ | $(x + y) - (x \cdot y)$ | $\min\{1, x + y\}$ |

Table 2. The dual t-conorms corresponding to the three main continuous t-norms

*Proof.* By definition, $\neg_*$ is a decreasing continuous function over the interval $(0, 1]$. Therefore, if $\sup X \neq 0$, then $a)$ is obviously true. Analogously, if $\inf X \neq 0$, then $\neg_* \inf X = \sup\{\neg_* x \mid x \in X\}$. Moreover, if $\sup X = 0$, then $X$ can only contain the element 0 and thus $a)$ holds. Consider now the case where $\inf X = 0$. If $0 \in X$, then we have that $\neg_* \inf X = \sup\{\neg_* x \mid x \in X\}$, but if $0 \notin X$, then $\neg_* \inf X = 1$ and $\sup\{\neg_* x \mid x \in X\} = a_*$. In both cases, $\neg_* \inf X \geq \sup\{\neg_* x \mid x \in X\}$, in particular when $a = 1$, which is the Łukasiewicz case. □

The *standard chain relative to the continuous t-norm* $*$ is the structure

$$[0, 1]_* = \langle [0, 1], \max, \min, *, \Rightarrow_*, 0, 1 \rangle.$$

A *triangular co-norm* (or t-conorm) is a binary operation defined on $[0, 1]$ that is associative, commutative, non decreasing in both arguments, and having 0 as neutral element. We say that a t-norm $*$ and a t-conorm $\oplus$ are *dual* with respect to a negation $n$ if, for every $x, y \in [0, 1]$, the De Morgan laws hold:

$$n(x * y) = n(x) \oplus n(y) \qquad\qquad n(x \oplus y) = n(x) * n(y).$$

In Table 2 we show the t-conorms dual (with respect to the standard involutive negation function $N(x) = 1 - x$) to the three main continuous t-norms.

### 2.2 Divisible finite t-norms and t-conorms

The notion of t-norm can be extended to bounded finite chains with 0 and 1 as first and last element respectively (see [94, 95]). We will call this kind of operation *finite t-norm.* They fulfill the same properties as t-norms and since they are of finite range all of them have a residuum.

A finite t-norm satisfying the divisibility condition is called a *divisible finite t-norm.* Notice that such a t-norm also satisfies the properties stated for continuous t-norms in Proposition 2.1.1. In [94] it is shown that any divisible finite t-norm is either a finite Łukasiewicz t-norm or a finite minimum t-norm or a finite ordinal sum of copies of both kinds of such finite t-norms. In analogy with the case of t-norms we will consider the following types of divisible finite t-norms:

1. The only divisible finite *idempotent* t-norms are the finite minimum t-norms.
2. The divisible finite *strict* t-norms are the ordinal sums beginning with a finite minimum component.
3. The only divisible finite *nilpotent* t-norms are the Łukasiewicz finite t-norms.
4. The divisible finite *non strict* t-norms are the ordinal sums beginning with a finite Łukasiewicz component.

PROPOSITION 2.2.1. *Let $X \subseteq C_n$. If $*$ is a divisible finite t-norm, then*

$$\neg_* \sup X = \inf\{\neg_* x \mid x \in X\} \qquad \neg_* \inf X = \sup\{\neg_* x \mid x \in X\}.$$

*Proof.* It is an easy consequence of Proposition 2.1.2 taking into account that $X$ is finite and thus infima and suprema are in fact minima and maxima. □

Since all chains with $n$ elements are order-isomorphic, without loss of generality we can take as finite chain of $n$ elements the set:

$$C_n = \{0, \frac{1}{n-1}, \frac{2}{n-1}, \dots, \frac{n-2}{n-1}, 1\}.$$

Given a divisible finite t-norm $*$ over $C_n$, the structure

$$\boldsymbol{C}_* = \langle C_n, \max, \min, *, \Rightarrow_*, 0, 1\rangle$$

is called the *standard finite chain* relative to $*$. The standard chains of $n$ elements corresponding to the finite t-norms of Łukasiewicz and Minimum are restrictions to $C_n$ of the standard chains respectively defined on $[0,1]$ by Łukasiewicz and minimum t-norms and their residua.

The notion of triangular co-norm can also be extended to finite chains. A t-conorm $\oplus$ defined on a finite chain $C$ is *divisible* if for every $a, b \in C$ such that $b \geq a$, there exists $c \in C$ such that $b = a \oplus c$. We can define the notion of *duality* for finite t-norms and finite t-conorms in an analogous way to the case of t-norms and t-conorm.

## 2.3 Expanding the standard chain with an involutive negation

In this chapter we will deal with different fuzzy description languages provided with a complementation. In order to interpret complementation we need an involutive negation in our standard algebra of truth values. Consequently, when the negation $\neg_*$ is not involutive, we enrich the standard chain $[0,1]_*$ or $\boldsymbol{C}_*$ with the standard involutive negation $N(x) = 1 - x$. The structure obtained by enriching a standard chain either $[0,1]_*$ or $\boldsymbol{C}_*$ with $N$ will be called *canonical algebra relative to* $*$ and denoted by $\boldsymbol{T}_*$. We denote by $T$ the carrier of $\boldsymbol{T}_*$; thus $T$ means $[0,1]$ or $C_n$.

Having the standard involutive negation $N$ allows us to define in $\boldsymbol{T}_*$:

a) The dual $t$-conorm $x \oplus y := N(N(x) * N(y))$.

b) The operator known in the literature as *Monteiro–Baaz operator*, which is denoted by $\delta$ and defined on any chain as $\delta(1) = 1$ and $\delta(x) = 0$ if $x \neq 1$. This is definable in the following two cases:

   (a) When $*$ is a strict t-norm, by $\delta x = \neg_*(N(x))$

   (b) When $*$ is a divisible finite t-norm, by $\delta x = (\neg_*(N(x)))^n$.

Finally let us remark that in the canonical chain $\boldsymbol{T}_*$ defined over the product standard chain $\langle [0,1], \max, \min, *_\Pi, \Rightarrow_\Pi, 0, 1\rangle$, both the Łukasiewicz t-norm $*_Ł$ and its residuum $\Rightarrow_Ł$ are definable (see Section 4 of Chapter VIII of this Handbook and the references therein) as:

$$x *_Ł y = N(x *_\Pi N(x \Rightarrow_\Pi y)) \qquad x \Rightarrow_Ł y = x *_\Pi N(x \Rightarrow_\Pi N(y)).$$

Notice that the canonical chain relative to product t-norm embeds the canonical chain relative to Łukasiewicz t-norm.

### 2.4 Expanding the standard chain with other operators

There are several extra operators that are interesting in order to interpret some FDL contructors. Among them we highlight the following ones:

- The *Monteiro–Baaz delta operator* defined in the previous section. Observe that the operator $\delta$ is a *crispification* operator (the image is $\{0, 1\}$) and it is truth preserving ($\delta(x) = 1$ if and only if $x = 1$).
- *Truth stresser* and *truth depresser* functions:
  - A function $s\colon [0,1] \to [0,1]$ is a *truth stresser* if it is non-decreasing and *sub-diagonal* (for every $x \in [0,1], s(x) \leq x$) and $s(1) = 1$.
  - A function $d\colon [0,1] \to [0,1]$ is a *truth depresser* if it is non-decreasing and *super-diagonal* (for every $x \in [0,1], d(x) \geq x$) and $d(0) = 0$.

  These operators corresponds to *linguistic modifiers* (in the sense that composed with an evaluation we obtain a *modified* evaluation). For interested readers, see Chapter VIII of this Handbook or papers [71, 79]. Notice that the function $\delta$ is a particular case of truth stresser.
- *Aggregation operators* of arity $n$ defined as functions $g\colon [0,1]^n \to [0,1]$. The properties that these functions must satisfy depend on the type of aggregation operator we are defining. Typical aggregation operators are *weighted sum*, *ordered weighted averaging* (OWA) or more general integrals (see for instance [134]).

## 3 Introduction to Fuzzy Description Logics

Knowledge used in real applications is usually imperfect and has to address situations of uncertainty, imprecision and vagueness. From a real world viewpoint, vague concepts like "patient with a high fever" and "person living near Paris" have to be considered. Fuzzy Description Logics are natural generalizations of Description Logics to cope with vague concepts and roles (mainly graded concepts and roles). These generalizations interpret concepts and roles as fuzzy sets and fuzzy relations respectively. In this interpretation each individual is an instance of a concept in a certain degree, which is the membership degree in that such individual belongs to the corresponding fuzzy set. These degrees are values belonging to either the real unit interval $[0, 1]$ or a finite chain. In order to combine fuzzy concepts and roles in the basic FDL languages, we use the same basic constructors as in classical DLs but interpreted by means of the operations that are commonly used to combine fuzzy sets. Accordingly with this choice we will interpret the intersection of concepts as the intersection of fuzzy sets which is commonly point-wise defined by means of a continuous (or finite divisible) t-norm (for t-norms and related notions see Section 2). Given the continuous (or finite divisible) t-norm $*$ used to interpret concept intersection, and given its residuum $\Rightarrow_*$, the other basic constructors are interpreted as follows:

- Universal and empty concepts are interpreted as the constant functions 1 and 0 respectively.
- The complementation will be interpreted as the standard involutive negation function $N(x) = 1 - x$.
- The union will be interpreted as the dual t-conorm of $*$ with respect to $N$, denoted $\oplus$. That is, $x \oplus y = N(N(x) * N(y))$.
- The value restriction $\forall R.C$ is interpreted using the infimum. The degree $r$ in which an individual $a$ is an instance of $\forall R.C$ is computed as follows:
  - For each $b$ in the domain we compute the degree $\alpha$ in which $a$ is related to $b$ and the degree $\beta$ in which $b$ is an instance of $C$.
  - Then, $r$ is the infimum of all the values $\alpha \Rightarrow_* \beta$ computed for each $b$, where $\Rightarrow_*$ is the residuum of the t-norm $*$.
- The existential quantification $\exists R.C$ is interpreted using the supremum. The degree $r$ in which an individual $a$ is an instance of $\exists R.C$ is computed as follows:
  - For each $b$ in the domain we compute the degree $\alpha$ in which $a$ is related to $b$ and the degree $\beta$ in which $b$ is an instance of $C$.
  - Then, $r$ is the supremum of all the values $\alpha * \beta$ computed for each $b$.

Notice that the definitions of quantified concepts are generalizations of the ones in classical DL since if we reduce the set of truth values to $\{0, 1\}$, then we obtain the classical definitions.

Thus, in order to define an interpretation we need to first fix a chain of truth values ($[0, 1]$ or a finite chain) and a continuous or divisible finite t-norm over that chain. In fact, interpretations of the FDL languages can be seen as interpretations of the description language over a canonical chain defined by a t-norm $*$ and the involutive negation $N$.

If we want to have implication of concepts as a constructor in our basic language it must be included as a primitive constructor. The natural choice is to interpret the implication constructor by using the residuum $\Rightarrow_*$ of the t-norm (see [78, Chapter 2, 2.1.3] for a discussion about this choice in the fuzzy logic setting; see also Section 8.2 for a discussion about this choice in the framework of FDLs), and in general $x \Rightarrow_* y$ coincides neither with $\max\{N(x), y\}$ nor with $N(\min\{x, N(y)\})$ (see Example 3.3.1). Thus, we cannot define the implication constructor from union and complementation (or intersection and complementation). Moreover, having implication in the language, we can define:

- A weak complementation interpreted by the negation function $\neg_* x = x \Rightarrow_* 0$
- A weak intersection interpreted by the minimum which is definable as $\min(x, y) = x * (x \Rightarrow_* y)$
- A weak union interpreted by the maximum which is definable as $\max(x, y) = \min((x \Rightarrow_* y) \Rightarrow_* y, (y \Rightarrow_* x) \Rightarrow_* x)$.

Observe that definability of minimum is possible since the t-norm is continuous or finite divisible.

### 3.1 Basic concept constructors in FDLs

Let us define a *description signature* as a triple $\mathcal{D} = \langle N_I, N_A, N_R \rangle$, where $N_I$, $N_A$ and $N_R$ are pairwise disjoint sets of individual names, concept names (the *atomic concepts*), and role names (the *atomic roles*), respectively.

Our basic description languages for $\mathcal{D}$ will be propositional languages generated by the set $N_A$ of atomic concepts taken as set of propositional letters, and using as connectives some specific subset of the following set of *constructors*:

$$\mathfrak{C} = \{\sqcap, \sqcup, \rightarrow, \neg, \sim, \wedge, \vee\} \cup \{\bot, \top\} \cup \{\forall R. \mid R \in N_R\} \cup \{\exists R. \mid R \in N_R\}.$$

The set of *concepts* for each description language is inductively defined by the corresponding subset of the following syntactic rules (we use the symbols $C, C_1, C_2$ as metavariables for concepts):

$$\begin{array}{lrl}
C, C_1, C_2 \quad \rightsquigarrow & A \mid & \text{(atomic concept)} \\
& C_1 \sqcap C_2 \mid & \text{(concept strong intersection)} \\
& C_1 \sqcup C_2 \mid & \text{(concept strong union)} \\
& C_1 \rightarrow C_2 \mid & \text{(concept implication)} \\
& \neg C \mid & \text{(weak complementary concept)} \\
& \sim C \mid & \text{(strong complementary concept)} \\
& C_1 \wedge C_2 \mid & \text{(concept weak intersection)} \\
& C_1 \vee C_2 \mid & \text{(concept weak union)} \\
& \bot \mid & \text{(empty concept)} \\
& \top \mid & \text{(universal concept)} \\
& \forall R.C \mid & \text{(value restriction)} \\
& \exists R.C & \text{(existential quantification).}
\end{array}$$

EXAMPLE 3.1.1. Consider the atomic concepts $\mathsf{LovesCinema}$ and $\mathsf{LovesReading}$ and the atomic role $\mathsf{IsFriendOf}$. Then we can form the concepts:

$$\mathsf{LovesCinema} \sqcap \sim \mathsf{LovesReading}$$

"Person who loves cinema but does not love reading."

$$\mathsf{LovesReading} \sqcap \exists \mathsf{isFriendOf}. \sim \mathsf{LovesReading}$$

"Person who loves reading but has a friend who do not love reading."

$$\exists \mathsf{isFriendOf}.\mathsf{LovesReading} \rightarrow \mathsf{LovesReading}$$

"Person who loves reading if it has a friend who loves reading."

$$\forall \mathsf{isFriendOf}.(\mathsf{LovesCinema} \sqcup \mathsf{LovesReading})$$

"Person whose friends either love cinema or love reading."

## 3.2 Semantics for basic concept constructors

An *interpretation* for a description signature $\mathcal{D} = \langle N_I, N_A, N_R \rangle$ is a pair

$$\mathcal{I} = \langle \Delta^{\mathcal{I}}, \cdot^{\mathcal{I}} \rangle$$

consisting of a nonempty (crisp) set $\Delta^{\mathcal{I}}$ (called *domain*) and of a *fuzzy interpretation function* $\cdot^{\mathcal{I}}$ that assigns:

a) To each concept name $A \in N_A$ a fuzzy set, that is, a function

$$A^{\mathcal{I}} \colon \Delta^{\mathcal{I}} \to T,$$

b) To each role name $R \in N_R$ a fuzzy relation, that is, a function

$$R^{\mathcal{I}} \colon \Delta^{\mathcal{I}} \times \Delta^{\mathcal{I}} \to T,$$

c) To each individual name $a \in N_I$ an object $a^{\mathcal{I}} \in \Delta^{\mathcal{I}}$.

Let $T$ be the set of truth values, that is, either the real unit interval $[0,1]$ or a finite chain $C_n = \{0, \frac{1}{n-1}, \frac{2}{n-1}, \ldots, \frac{n-2}{n-1}, 1\}$ and fix a t-norm $*$ over $T$. Given $*$, and a description signature $\mathcal{D}$, and an $*$-interpretation is a pair

$$\mathcal{I}_* = \langle \mathcal{I}, * \rangle.$$

DEFINITION 3.2.1 (Semantics for basic languages). *Given an $*$-interpretation $\mathcal{I}_*$, the interpretation $\cdot^{\mathcal{I}}$ is extended to complex concepts by assigning to every complex concept $D$ a fuzzy set $D^{\mathcal{I}} \colon \Delta^{\mathcal{I}} \to T$ inductively defined as follows:*

$$D^{\mathcal{I}}(x) := \begin{cases} 0 & \textit{if } D = \bot \\ 1 & \textit{if } D = \top \\ A^{\mathcal{I}}(x) & \textit{if } D = A \in N_A \\ 1 - C^{\mathcal{I}}(x) & \textit{if } D = \sim C \\ C^{\mathcal{I}}(x) \Rightarrow_* \bot & \textit{if } D = \neg C \\ C_1^{\mathcal{I}}(x) * C_2^{\mathcal{I}}(x) & \textit{if } D = C_1 \sqcap C_2 \\ \min\{C_1^{\mathcal{I}}(x), C_2^{\mathcal{I}}(x)\} & \textit{if } D = C_1 \wedge C_2 \\ C_1^{\mathcal{I}}(x) \oplus C_2^{\mathcal{I}}(x) & \textit{if } D = C_1 \sqcup C_2 \\ \max\{C_1^{\mathcal{I}}(x), C_2^{\mathcal{I}}(x)\} & \textit{if } D = C_1 \vee C_2 \\ C_1^{\mathcal{I}}(x) \Rightarrow_* C_2^{\mathcal{I}}(x) & \textit{if } D = C_1 \to C_2 \\ \inf\{R^{\mathcal{I}}(x,y) \Rightarrow_* C^{\mathcal{I}}(y) \mid y \in \Delta^{\mathcal{I}}\} & \textit{if } D = \forall R.C \\ \sup\{R^{\mathcal{I}}(x,y) * C^{\mathcal{I}}(y) \mid y \in \Delta^{\mathcal{I}}\} & \textit{if } D = \exists R.C. \end{cases}$$

EXAMPLE 3.2.2. Consider the concepts $A_1 = \mathsf{LovesCinema}$ and $A_2 = \mathsf{LovesReading}$ and the atomic role $R = \mathsf{isFriendOf}$ from Example 3.1.1; we can form the concept $\forall R.(A_1 \sqcup A_2)$. Now take $T = \{0, \frac{1}{4}, \frac{2}{4}, \frac{3}{4}, 1\}$ and the minimum t-norm. Thus, now we have that $\oplus = \max$ and $\Rightarrow_*$ is the residuated implication corresponding to minimum (see Table 1). Consider now an interpretation $\mathcal{I}$ where $\Delta^{\mathcal{I}}$ is the a set of students of a postgraduate course in La Sorbone, Paris. Suppose that this set is formed by four students $\mathsf{marie}$ $(a)$, $\mathsf{jacques}$ $(b)$, $\mathsf{ivette}$ $(c)$, $\mathsf{pierre}$ $(d)$. The interpretations for the concepts $A_1, A_2$ and the role $R$ are the following fuzzy sets and fuzzy relation:

| | $a$ | $b$ | $c$ | $d$ |
|---|---|---|---|---|
| $A_1^{\mathcal{I}}$ | 1/2 | 1/2 | 1/4 | 1 |
| $A_2^{\mathcal{I}}$ | 3/4 | 1/4 | 3/4 | 0 |

| $R$ | $a$ | $b$ | $c$ | $d$ |
|---|---|---|---|---|
| $a$ | 1 | 3/4 | 1 | 1/4 |
| $b$ | 3/4 | 1 | 1/2 | 3/4 |
| $c$ | 3/4 | 1/2 | 1/2 | 1 |
| $d$ | 1/2 | 3/4 | 1 | 1/2 |

Let us compute the degree in which student $a$ is an instance of concept $\forall R.(A_1 \sqcup A_2)$:

| $y$ | $R^{\mathcal{I}}(a,y)$ | $A_1^{\mathcal{I}}(y)$ | $A_2^{\mathcal{I}}(y)$ | $R^{\mathcal{I}}(a,y) \Rightarrow_{\min} \max\{A_1^{\mathcal{I}}(y), A_2^{\mathcal{I}}(y)\}$ |
|---|---|---|---|---|
| $a$ | 1 | 1/2 | 3/4 | $1 \Rightarrow_{\min} \max\{1/2, 3/4\} = 3/4$ |
| $b$ | 3/4 | 1/2 | 1/4 | $3/4 \Rightarrow_{\min} \max\{1/2, 1/4\} = 1/2$ |
| $c$ | 1 | 1/4 | 3/4 | $1 \Rightarrow_{\min} \max\{1/4, 3/4\} = 3/4$ |
| $d$ | 1/4 | 1 | 0 | $1/4 \Rightarrow_{\min} \max\{1, 0\} = 1$ |

Thus we have:

$$(\forall R.(A_1 \sqcup A_2))^{\mathcal{I}}(a) = \inf\{R^{\mathcal{I}}(a,y) \Rightarrow_{\min} \max\{A_1^{\mathcal{I}}(y), A_2^{\mathcal{I}}(y)\} \mid y \in \Delta^{\mathcal{I}}\} = \frac{1}{2}.$$

In an analogous way we can compute that the degrees in which students $b, c, d$ are instances of concept $\forall R.(A_1 \sqcup A_2)$ are $\frac{1}{2}$, 1 and $\frac{1}{2}$, respectively.

REMARK 3.2.3. *Observe that constructors for classical DLs are the same that the ones previously defined for FDLs, but without the difference between strong and weak union, strong and weak intersection, and strong and weak complementation. Indeed, if in Definition 3.2.1, we substitute $\boldsymbol{T}_*$ by the Boolean algebra of two elements, we obtain the usual semantics for basic constructors of classical DLs. For the classical definition and semantics of basic languages as $\mathcal{ALC}$ see for instance [8].*

### 3.3 The hierarchy of fuzzy attributive languages

Next we will describe the hierarchy of FDLs languages that are built using the basic constructors introduced in previous section. In order to maintain as much as possible the similarity with classical DL notation, we propose the following fuzzy description languages:

- The language $\mathcal{EL}$, as in the classical case, contains
  - the universal concept $\top$,
  - the strong intersection $\sqcap$, and
  - the existential quantification $\exists R$,

  as concept constructors.
- The language $\mathfrak{N}\mathcal{EL}$ is built by adding the weak complementation $\neg$ to $\mathcal{EL}$.
- The language $\mathcal{ELC}$ is built by adding the strong complementation $\sim$ to $\mathcal{EL}$.
- The language $\mathcal{FL}_0$, as in the classical case, contains these concept constructors:
  - the universal concept $\top$,
  - the strong intersection $\sqcap$, and
  - the value restriction $\forall R$.
- The language $\mathcal{FL}^-$ is built by adding to the language $\mathcal{FL}_0$ the restricted existential quantification $\exists R.\top$.
- The language $\mathcal{AL}$ is built by adding to the language $\mathcal{FL}^-$ the strong complementation $\sim$ applied only to atomic concepts.
- The language $\mathcal{ALC}$ is built by adding the full strong complementation $\sim$ to the language $\mathcal{AL}$.

We will take as minimal languages of our hierarchy $\mathcal{EL}$ and $\mathcal{AL}$. Now we will construct the hierarchy of logical based FDL languages beginning with $\mathcal{AL}$. To this end, we need to introduce explicitly as primitive constructors the following ones: the existential quantifier ($\exists R.$), the complementation ($\sim$), the implication ($\rightarrow$), and the strong union ($\sqcup$). The presence of each one of these constructors in the language will be denoted by adding to the acronym of the language the symbols $\mathcal{E}$, $\mathcal{C}$, $\mathfrak{I}$,[3] and $\mathcal{U}$, respectively.

The hierarchy of languages that we obtain depends on the semantical choice, i.e., on the truth values set ($[0,1]$ or $C_n$), and the t-norm used to interpret strong conjunction. We can distinguish two different cases:

1. *The hierarchy obtained when strong conjunction is interpreted as (finite or infinite) Łukasiewicz t-norm.* This is similar to the classical case. Since in this case both complementations—weak and strong—coincide (see Section 2), other constructors are definable from intersection and complementation like in the classical case: (strong) union and existential restriction are definable by duality as $C \sqcup D \equiv \neg(\neg C \sqcap \neg D)$ and $\exists R.C \equiv \neg\forall R.\neg C$ respectively, and implication is definable using (strong) complementation and union (or intersection) by the equivalence $C \rightarrow D \equiv \neg C \sqcup D \equiv \neg(C \sqcap \neg D)$. Taking into account these definabilities, we have that $\mathcal{ALUEC} = \mathcal{ALC}$. Thus the graph of FDLs languages is the one given in Figure 2 that is the same as in the classical case.

[3] We have chosen $\mathfrak{I}$ because the calligraphic $\mathcal{I}$ is extensively used in the literature on DLs to denote the constructor for inverse roles.

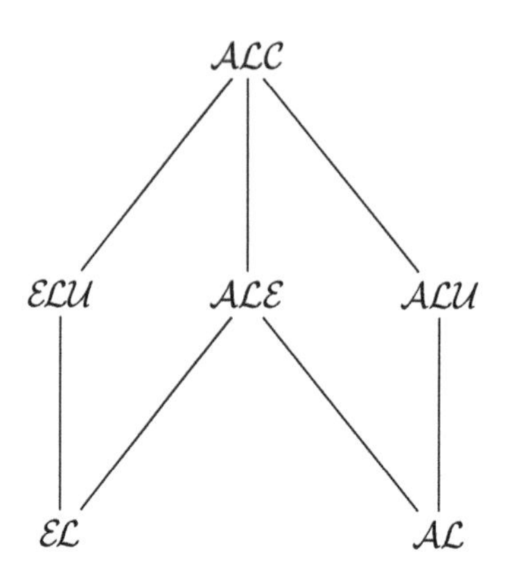


Figure 2. The hierarchy of basic languages under Łukasiewicz semantics

2. *The hierarchy obtained when strong conjunction is interpreted as a (finite or infinite) divisible t-norm different from Łukasiewicz.* In these cases, the hierarchy of basic languages obtained is more cumbersome because we do not have in general the same possibility of reducing languages as in the classical case. This is because in the semantics of our logics:

   - Although union is definable from conjunction by duality with respect to strong complementation by $C \sqcup D \equiv \sim(\sim C \sqcap \sim D)$, this is not true for the duality with respect to weak complementation. The reason is that the interpretation of $\sim$ is a continuous function, which is not the case for $\neg$.
   - When the t-norm is different from Łukasiewicz, weak and strong complementation do not coincide, and implication is not definable from intersection and complementation, neither the weak nor the strong one (as an illustration see Example 3.3.1).
   - The particular semantics of the universal constructor $\forall R.$ and the existential constructor $\exists R.$ does not allow to define one in terms of the other by duality (see Example 3.3.2 and Proposition 3.3.3).

   The new hierarchy of FDL languages over the logic of a continuous t-norm beginning with $\mathcal{AL}$ is represented in Figure 3, that shows the inclusions among the languages obtained by successively adding different basic operators. Observe that, due to the above defined semantics, in our framework the languages $\mathcal{ALE}$ and $\mathfrak{I}\mathcal{AL}$ are not strictly contained in $\mathcal{ALC}$.

EXAMPLE 3.3.1. Let $* = \min$, and take $x = 0.2$ and $y = 0.7$. Since $x < y$, we have $x \Rightarrow_{\min} y = 1$. However we have:

$$\max\{\neg_{\min}(x), y\} = \max\{0, 0.7\} = 0.7$$

$$\max\{N(x), y\} = \max\{0.8, 0.7\} = 0.8.$$

Observe that $N(0.2) = 0.8 \neq 0 = \neg_{\min}(0.2)$.

This example shows that Gödel implication is not definable using the standard operators of the Minimum t-norm. The following example shows that the duality of existential and value restrictions does not hold either.

EXAMPLE 3.3.2. Take a description signature with a unique atomic concept $C$ and a unique atomic role $R$. Take $T = \{0, \frac{1}{2}, 1\}$, and $* = \min$. Now consider a interpretation $\mathcal{I}$, where $\Delta^{\mathcal{I}} = \{a, b, c\}$, and $R$ and $C$ are interpreted as:

- $R^{\mathcal{I}}(a, b) = R^{\mathcal{I}}(b, c) = 1$; $R^{\mathcal{I}}(a, a) = R^{\mathcal{I}}(a, c) = \frac{1}{2}$; and $R^{\mathcal{I}}(x, y) = 0$ otherwise.
- $C^{\mathcal{I}}(b) = \frac{1}{2}$; and $C^{\mathcal{I}}(x) = 1$ otherwise.

Then, $(\exists R.C)^{\mathcal{I}}(a) = \sup\{\min\{R^{\mathcal{I}}(a, y), C^{\mathcal{I}}(y)\} \mid y \in \{a, b, c\}\} = \frac{1}{2}$ but

$$\begin{aligned}(\sim\forall R.\sim C)^{\mathcal{I}}(a) &= 1 - \inf\{R^{\mathcal{I}}(a, y) \Rightarrow_{\min} (1 - C^{\mathcal{I}}(y)) \mid y \in \{a, b, c\}\} \\ &= 1 - \inf\{\frac{1}{2} \Rightarrow_{\min} 0, 1 \Rightarrow_{\min} \frac{1}{2}, \frac{1}{2} \Rightarrow_{\min} 0\} \\ &= 1 - \inf\{0, \frac{1}{2}, 0\} = 1.\end{aligned}$$

PROPOSITION 3.3.3. *In any language containing value and existential restrictions and weak negation among its constructors, for each atomic role $R$ and concept $C$, it holds that $\neg\exists R.C = \forall R.\neg C$. However, in general, $\neg\forall R.C$ is not equal to $\exists R.\neg C$.*

*Proof.* Let $a \in \Delta^{\mathcal{I}}$. We have:

$$\begin{aligned}(\neg\exists R.C)^{\mathcal{I}}(a) &= (\exists R.C \to \bot)^{\mathcal{I}}(a) = (\exists R.C)^{\mathcal{I}}(a) \Rightarrow_* 0 \\ &= \sup\{R^{\mathcal{I}}(a, y) * C^{\mathcal{I}}(y) \mid y \in \Delta^{\mathcal{I}}\} \Rightarrow_* 0.\end{aligned}$$

And now, by applying Proposition 2.1.2 and the properties of residuation, we have:

$$\begin{aligned}(\neg\exists R.C)^{\mathcal{I}}(a) &= \sup\{R^{\mathcal{I}}(a, y) * C^{\mathcal{I}}(y) \mid y \in \Delta^{\mathcal{I}}\} \Rightarrow_* 0 \\ &= \inf\{(R^{\mathcal{I}}(a, y) * C^{\mathcal{I}}(y)) \Rightarrow_* 0 \mid y \in \Delta^{\mathcal{I}}\} \\ &= \inf\{R^{\mathcal{I}}(a, y) \Rightarrow_* (C^{\mathcal{I}}(y) \Rightarrow_* 0) \mid b \in \Delta^{\mathcal{I}}\} \\ &= \inf\{R^{\mathcal{I}}(a, y) \Rightarrow_* (\neg C)^{\mathcal{I}}(y) \mid y \in \Delta^{\mathcal{I}}\} = (\forall R.\neg C)^{\mathcal{I}}(a).\end{aligned}$$

Take now the language considered in Example 3.3.2, interpreting $R$ as the same fuzzy relation and $C$ as $C^{\mathcal{I}}(a) = \frac{1}{2}, C^{\mathcal{I}}(b) = 1, C^{\mathcal{I}}(c) = 0$. Then,

$$(\exists R.\neg C)^{\mathcal{I}}(a) = \sup\{\min\{R^{\mathcal{I}}(a, y), (\neg C)^{\mathcal{I}}(y)\} \mid y \in \{a, b, c\}\} = \frac{1}{2}$$

but

$$(\neg\forall R.C)^{\mathcal{I}}(a) = \neg_{\min} \inf\{R^{\mathcal{I}}(a, y) \Rightarrow_{\min} C^{\mathcal{I}}(y) \mid y \in \{a, b, c\}\} = 1. \qquad \square$$

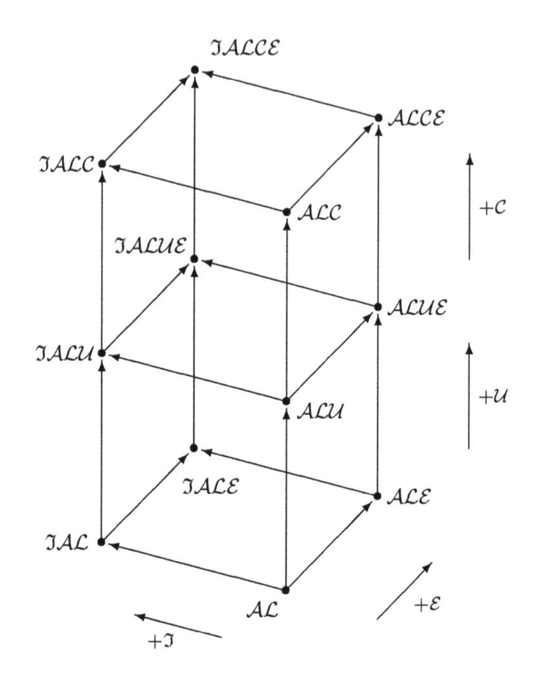


Figure 3. Hierarchy of languages under t-norm based logics different from Łukasiewicz

Since in languages with complementation $\sim$, the strong union $\sqcup$ is definable from the strong intersection $\sqcap$ and $\sim$, then the language $\mathcal{ALU}$ is strictly contained in the language $\mathcal{ALC} = \mathcal{ALCU}$. Following our convention to denote languages, the top of the hierarchy in Figure 3 (corresponding to $\mathcal{ALC}$ in the classical case) is called in our framework $\mathfrak{I}\mathcal{ALCE}$. In papers on FDLs (specially the initial ones) it is common the use of $\mathcal{ALC}$ for the top of the hierarchy which is acceptable in a broad sense, but if we want to make explicit which are the constructors, then we need to be more accurate in the cases having non-Łukasiewicz semantics.

An analogous hierarchy can be built from $\mathcal{EL}$, the main difference being that we need to change the role played by existential quantification for an analogous role for value restriction (universal quantification). Indeed, using $\mathcal{V}$ for universal quantification,[4] the language $\mathfrak{I}\mathcal{ELV}$ (obtained by adding $\bot$, universal quantification and implication to $\mathcal{EL}$) coincides with the language $\mathfrak{I}\mathcal{ALE}$ of Figure 3. Notice that $\mathfrak{N}\mathcal{EL}$ is an intermediate language between $\mathcal{EL}$ and $\mathfrak{I}\mathcal{EL}$.

[4] The '$\mathcal{V}$' in this case stands for 'value restriction.'

Finally, let us remark that some FDL languages have other constructors like modifiers, crispificaction, or aggregation (for definitions of these additional constructors see Section 5.1). In these cases, if we want to be accurate, we need to introduce some letter for each one of these constructors, for example $\mathcal{M}$, $\mathcal{D}$ or $\mathcal{G}$ respectively.

### 3.4 Knowledge bases

Like in classical DLs, knowledge bases are formed by two finite sets:

- A TBox (Terminological Box), which contains *graded inclusion axioms* stating that certain concept is subsumed by another at least in a given degree.
- An ABox (Assertional Box), which contains *graded assertion axioms* stating that, at least in a given degree, an individual is an instance of some concept, or a pair of individuals is an instance of some role.

A *knowledge base* is a pair $\mathcal{K} = \langle \mathcal{T}, \mathcal{A} \rangle$, where the first component is a TBox and the second one is an ABox. Let $C, D$ be concepts, $R$ be an atomic role, $a, b$ be individual names and $r \in T$.

A *graded concept inclusion axiom* is an expression of the form:

$$\langle C \sqsubseteq D, r \rangle.$$

A *graded concept assertion axiom* is an expression of the form:

$$\langle C(a), r \rangle.$$

A *graded role assertion axiom* is an expression of the form:

$$\langle R(a, b), r \rangle.$$

Fix a set of truth values $T$ and a t-norm $*$ defined on $T$, and let $\mathcal{I}$ be an interpretation.

- We say that $\mathcal{I}$ satisfies the *graded concept inclusion axiom* $\langle C \sqsubseteq D, r \rangle$ when

$$\inf\{C^{\mathcal{I}}(x) \Rightarrow_* D^{\mathcal{I}}(x) \mid x \in \Delta^{\mathcal{I}}\} \geq r.$$

- We say that $\mathcal{I}$ satisfies the *graded concept assertion axiom* $\langle C(a), r \rangle$ when

$$C^{\mathcal{I}}(a^{\mathcal{I}}) \geq r.$$

- We say that $\mathcal{I}$ satisfies the *graded concept inclusion axiom* $\langle R(a, b), r \rangle$ when

$$R^{\mathcal{I}}(a^{\mathcal{I}}, b^{\mathcal{I}}) \geq r.$$

- We say that a graded axiom is $*$-*satisfiable* if there exists an interpretation $\mathcal{I}$ which satisfies that axiom.

When $\mathcal{I}$ satisfies a graded axiom $\langle \alpha, r \rangle$ we say that $\mathcal{I}$ is an $*$-*model* (or simply a *model)* of $\langle \alpha, r \rangle$, and we write $\mathcal{I} \models_* \langle \alpha, r \rangle$ (or simply $\mathcal{I} \models \langle \alpha, r \rangle$ when the t-norm is clear from the context). We write $\mathcal{I} \not\models_* \langle \alpha, r \rangle$ otherwise. We will say that an interpretation $\mathcal{I}$ is a *model* of a knowledge base $\mathcal{K}$ if $\mathcal{I}$ is a model of all the axioms in $\mathcal{K}$.

For instance, in the setting of Example 3.2.2, the axioms:

$$\langle \mathsf{LovesCinema}(\mathsf{marie}), \frac{3}{4} \rangle, \langle \mathsf{LovesReading}(\mathsf{jacques}), \frac{1}{4} \rangle, \langle \mathsf{isFriendOf}(\mathsf{ivette}, \mathsf{marie}), \frac{1}{4} \rangle$$

are possible graded assertion axioms. Observe that for the interpretation $\mathcal{I}$ given in such example, we have:

$$\mathcal{I} \not\models \langle \mathsf{LovesCinema}(\mathsf{marie}), \frac{3}{4} \rangle \qquad \text{and} \qquad \mathcal{I} \models \langle \mathsf{LovesReading}(\mathsf{jacques}), \frac{1}{4} \rangle.$$

We also have that $\mathcal{I} \models \langle \mathsf{isFriendOf}(\mathsf{ivette}, \mathsf{marie}), \frac{1}{4} \rangle$. Therefore, $\mathcal{I}$ does not satisfy an ABox containing all these three graded assertion axioms. A possible graded inclusion axiom is $\langle \mathsf{LovesReading} \sqsubseteq \mathsf{LovesCinema}, \frac{1}{4} \rangle$. Observe that the interpretation $\mathcal{I}$ satisfies it since:

$$\inf\{\mathsf{LovesReading}^{\mathcal{I}}(x) \Rightarrow_* \mathsf{LovesCinema}^{\mathcal{I}}(x) \mid x \in \Delta^{\mathcal{I}}\} = \frac{1}{4}.$$

### 3.5 Reasoning tasks

Reasoning in FDLs involves the same kind of tasks as in the classical case but their results depend on the chosen truth values set $T$ and t-norm $*$. We consider graded notions of reasoning tasks. Given a knowledge base, $\mathcal{K}$, concepts $C$ and $D$, and a truth value $r \in T$, we define:

1. $\mathcal{K}$ is $*$-*consistent* if all the axioms in $\mathcal{K}$ are $*$-satisfiable, i.e., if there is a model of $\mathcal{K}$.

2. $C$ is $*$-*satisfiable to a degree greater than or equal to* $r$ with respect to $\mathcal{K}$ if there exists a model $\mathcal{I}$ of $\mathcal{K}$, and an element $u \in \Delta^{\mathcal{I}}$ such that $C^{\mathcal{I}}(u) \geq r$. In particular, $C$ is *positively* $*$-*satisfiable* with respect to $\mathcal{K}$ if it is $*$-satisfiable to a degree strictly greater than 0 with respect to $\mathcal{K}$.

3. $C$ is $*$-*subsumed* by $D$ *to a degree greater than or equal to* $r$ with respect to $\mathcal{K}$ if, for every model $\mathcal{I}$ of $\mathcal{K}$, $\inf\{C^{\mathcal{I}}(x) \Rightarrow_* D^{\mathcal{I}}(x) \mid x \in \Delta^{\mathcal{I}}\} \geq r$.[5]

4. An axiom is $*$-*entailed* by $\mathcal{K}$ if it is $*$-satisfied in every model of $\mathcal{K}$.

5. The *Best Entailment Degree* (BED) of $\tau \in \{C(a), R(a,b), C \sqsubseteq D\}$ with respect to $\mathcal{K}$ is the maximum degree $r$ such that $\langle \tau, r \rangle$ is $*$-*entailed* by $\mathcal{K}$.

6. The *Best Satisfiability Degree* (BSD) of $C$ with respect to $\mathcal{K}$ is the maximum degree $r$ such that $C$ is $*$-satisfiable to a degree greater than or equal to $r$.

In what follows, we will asume that $r > 0$ for $*$-satisfiability (consistency, subsumption) since otherwise the result of the reasoning test is trivially true.

[5] Observe that the logical notion of *validity* applied to a concept $C$ can be expressed in FDL as the fact that $C$ $*$-subsumes $\top$ to degree 1.

# 4 Fuzzy Description Logics from the point of view of MFL

In [80] (see also [82]) Hájek proposes to define FDLs from a logical point of view in an analogous way as DLs can be defined from classical first order logic. It is well known that classical $\mathcal{ALC}$ can be interpreted as a fragment of first order classical logic (see [111]). In a similar way, the fuzzy description logic $\mathfrak{I}\mathcal{ALCE}$ can be interpreted as a fragment of certain fuzzy first order logics. The fuzzy logics used in this interpretation are semantically defined as multi-valued logics that are complete with respect to interpretations on the algebraic structure defined on the chain either $[0,1]$ or $C_n$ by a divisible t-norm and its residuum and the involutive negation function $N$. It is worth to notice that the way we define these logical systems is different from the axiomatic approach mainly followed in the overall Handbook. Thus, we introduce them in next sections. We define the logics (propositional and first-order) where the description logic $\mathfrak{I}\mathcal{ALCE}$ is interpreted. In Section 4.3 we introduce the notion of *instance* of a description as a first-order open formula associated with it and we prove that the semantical interpretation of a concept coincides with the semantical interpretation of its instance in first-order logic.

## 4.1 A logical framework for FDLs: The propositional logics

Let us consider the language $\mathcal{L}$ inductively built from a countable set of propositional variables $\Phi = \{p_j \mid j \in J\}$ using the connectives $\vee$, $\wedge$, $\veebar$, $\&$, $\rightarrow$, $\neg$, $\sim$, and the constants $\bar{0}$ and and $\bar{1}$. Note that we have considered all possible connectives; in fact $\&$, $\rightarrow$, $\sim$, and $\bar{0}$ would be enough since the others are definable from them.

A *propositional evaluation* for the $\mathcal{L}$-formulas over the canonical chain $\boldsymbol{T}_*$ is a map $e\colon \Phi \rightarrow T$ which is extended to all formulas by setting:

$$
\begin{aligned}
e(\varphi \wedge \psi) &= \min\{e(\varphi), e(\psi)\} & e(\neg\varphi) &= e(\varphi) \rightarrow_* 0\\
e(\varphi \vee \psi) &= \max\{e(\varphi), e(\psi)\} & e(\sim\varphi) &= 1 - e(\varphi)\\
e(\varphi \,\&\, \psi) &= e(\varphi) * e(\psi) & e(\bar{0}) &= 0\\
e(\varphi \veebar \psi) &= e(\varphi) \oplus e(\psi) & e(\bar{1}) &= 1\\
e(\varphi \rightarrow \psi) &= e(\varphi) \rightarrow_* e(\psi).
\end{aligned}
$$

Given the propositional language $\mathcal{L}$, we define the logic $\Lambda^*$ semantically as the consequence relation associated with evaluations on the canonical chain $\boldsymbol{T}_*$ by putting, for all sets $\Gamma \cup \{\varphi\}$ of $\mathcal{L}$-formulas,

$$\Gamma \models_{\Lambda^*} \varphi \text{ iff for every evaluation } e \text{ over } \boldsymbol{T}_*, e(\gamma) = 1 \text{ for all } \gamma \in \Gamma \text{ implies } e(\varphi) = 1.$$

REMARK 4.1.1. *The first logic used in FDLs was the so-called Zadeh logic, semantically defined on the chain $\langle [0,1], \max, \min, \rightarrow, N, 0, 1\rangle$, where $\rightarrow$ is the so-called S-implication defined as $x \rightarrow y = \max(N(x), y)$, and $N$ is the standard negation $N(x) = 1 - x$. This logical system is not residuated and not considered in MFL (see the historical remarks for a discussion about the reasons for such decision). Nevertheless some earlier results on FDLs based on Zadeh logic are presented in this chapter since they are interesting for further results in the FDLs based on MFL.*

We could introduce some other connectives whose semantics is a function of the corresponding arity on the canonical chain. That is the case of the *Monteiro–Baaz* $\Delta$, that corresponds to crispification and whose semantics is defined by the function $\delta$, or *hedges*, which interpret linguistic modifiers and whose semantics are the unary functions called truth stressers or truth depressers. Another expansions can be obtained by adding *aggregation operators*, which are semantically defined as aggregation functions of the corresponding arity.

## 4.2 A logical framework for FDLs: The predicate logics

From the propositional logics defined above we can define their first order versions. As defined in Chapter I of this Handbook, a *predicate language* or *first order signature* is a triple $\Sigma = \langle \mathcal{P}, \mathcal{F}, \mathbf{ar} \rangle$, where $\mathcal{P} = \{P, Q, \ldots\}$ is a non empty set of *predicate symbols*, $\mathcal{F} = \{f, g, \ldots\}$ is a set (disjoint from $\mathcal{P}$) of *functional symbols*, and $\mathbf{ar}$ is a function assigning to each predicate and function symbol a natural number called the *arity* of the symbol. The functional symbols of arity 0 are called *object constants*. The predicate symbols of arity 0 are called *truth constants*. We will consider that the set of nullary predicates is parametrized by a subalgebra $\boldsymbol{S}$ of $\boldsymbol{T}_*$. Indeed this set will be denoted as $\{\bar{r} \mid r \in S\}$.

Let $V = \{x, y, \ldots\}$ be a countable set of objects whose elements are called *object variables*. The set of $\Sigma$-*terms* is the least set $X$ such that $V \subseteq X$, and if $f$ is a $k$-ary functional symbol and $t_1, \ldots, t_k \in X$, then $f(t_1, \ldots, t_k) \in X$. An *atomic* $\Sigma$-*formula* is an expression of the form $P(t_1, \ldots, t_k)$, where $P$ is a $k$-ary predicate symbol and $t_1, \ldots, t_k$ are $\Sigma$-terms. Atomic $\Sigma$-formulas and nullary logical connectives of $\mathcal{L}$ are called *atomic* $\langle \mathcal{L}, \Sigma \rangle$-*formulas*. The set of $\langle \mathcal{L}, \Sigma \rangle$-*formulas* is the least set such that:

- $X$ contains the atomic $\langle \mathcal{L}, \Sigma \rangle$-formulas,
- $X$ is closed under the logical connectives of $\mathcal{L}$, and
- if $\alpha \in X$ and $x$ is an object variable, then $(\forall x)\alpha, (\exists x)\alpha \in X$.

An $\langle \mathcal{L}, \Sigma \rangle$-*theory* is a set of $\langle \mathcal{L}, \Sigma \rangle$-formulas. For the sake of simplicity, since the propositional language is fixed, we will speak about $\Sigma$-formulas and $\Sigma$-theories. An occurrence of a variable $x$ in a formula $\varphi$ is *bound* if it is in the scope of some quantifier over $x$; otherwise it is called a *free* occurrence. A variable is free in a formula $\varphi$ if it has a free occurrence in $\varphi$. A term is *closed* if it contains no variables. A formula is *closed* if it has no free variables; closed formulas are also called *sentences*.

A $\boldsymbol{T}_*$-*structure* for the predicate language $\Sigma$ is a tuple

$$\mathbf{M} = \langle M, \{P^{\mathbf{M}} \mid P \in \mathcal{P}\}, \{f^{\mathbf{M}} \mid f \in \mathcal{F}\} \rangle, \text{ where}$$

1) $M$ is a non-empty set, called the *universe* of the structure.
2) For each $k$-ary predicate symbol $P \in \mathcal{P}$, with $k \geq 1$, $P^{\mathbf{M}}$ is a fuzzy $k$-ary relation defined on $M$, that is, a function $P^{\mathbf{M}} \colon M^k \to T$, and an element $r \in S \subseteq T$ if $P$ is the truth constant $\bar{r}$.
3) For each $k$-ary functional symbol $f \in \mathcal{F}$, with $k \geq 1$, $f^{\mathbf{M}}$ is a function $f^{\mathbf{M}} \colon M^k \to M$, and an element of $M$ if $f$ is an object constant.

Given a $\boldsymbol{T}_*$-structure $\mathbf{M}$, a map $v$ assigning an element $v(x) \in M$ to each variable $x$ is called an *assignment of the variables in* $\mathbf{M}$ (an $\mathbf{M}$-*assignment*). Given $\mathbf{M}$ and $v$, the *value of a term* $t$ in $\mathbf{M}$, denoted by $\|t\|_{\mathbf{M},v}$, is defined as $v(x)$ when $t$ is a variable $x$, and as $a^{\mathbf{M}}$ when $t$ is a constant $a$, and as $f_{\mathbf{M}}(\|t_1\|_{\mathbf{M},v}, \ldots, \|t_k\|_{\mathbf{M},v})$ when $t$ is of the form $f(t_1, \ldots, t_k)$. In order to emphasize that a formula $\alpha$ has its free variables in $\{x_1, \ldots, x_n\}$, we will denote it by $\alpha(x_1, \ldots, x_n)$. Let $v$ be an $\mathbf{M}$-assignation such that $v(x_1) = b_1, \ldots, v(x_n) = b_n$. The *truth value in* $\mathbf{M}$ *over the canonical chain* $\boldsymbol{T}_*$ *of the predicate formula* $\varphi(x_1, \ldots, x_n)$ *for the assignation* $v$, denoted by $\|\varphi\|^*_{\mathbf{M},v}$ or by $\|\varphi(b_1, \ldots, b_n)\|^*_{\mathbf{M}}$, is a value in the carrier $T$ of the canonical chain defined inductively as follows:

$$
\begin{array}{rll}
P^{\mathbf{M}}(\|t_1\|_{\mathbf{M},v}, \ldots, \|t_k\|_{\mathbf{M},v}) & \text{if} & \varphi = P(t_1, \ldots, t_k) \\
r & \text{if} & \varphi = \bar{r} \\
1 - \|\alpha\|^*_{\mathbf{M},v} & \text{if} & \varphi = \sim\alpha \\
\max\{\|\alpha\|^*_{\mathbf{M},v}, \|\beta\|^*_{\mathbf{M},v}\} & \text{if} & \varphi = \alpha \vee \beta \\
\min\{\|\alpha\|^*_{\mathbf{M},v}, \|\beta\|^*_{\mathbf{M},v}\} & \text{if} & \varphi = \alpha \wedge \beta \\
\|\alpha\|^*_{\mathbf{M},v} * \|\beta\|^*_{\mathbf{M},v} & \text{if} & \varphi = \alpha \& \beta \\
\|\alpha\|^*_{\mathbf{M},v} \Rightarrow_* \|\beta\|^*_{\mathbf{M},v} & \text{if} & \varphi = \alpha \to \beta \\
\inf\{\|\alpha(a, b_1, \ldots, b_n)\|^*_{\mathbf{M}} \mid a \in M\} & \text{if} & \varphi = (\forall x)\alpha(x, x_1, \ldots, x_n) \\
\sup\{\|\alpha(a, b_1, \ldots, b_n)\|^*_{\mathbf{M}} \mid a \in M\} & \text{if} & \varphi = (\exists x)\alpha(x, x_1, \ldots, x_n).
\end{array}
$$

We will write $\langle \boldsymbol{T}_*, \mathbf{M}\rangle \models \varphi$ if $\|\varphi\|^*_{\mathbf{M},v} = 1$ for every $\mathbf{M}$-assignation $v$. When $\boldsymbol{T}_*$ is known from the context, then we simply write $\mathbf{M} \models \varphi$. If $\langle \boldsymbol{T}_*, \mathbf{M}\rangle \models \varphi$ for each $\boldsymbol{T}_*$-structure $\mathbf{M}$, then we say that $\varphi$ is a $\boldsymbol{T}_*$-*tautology* and we write $\boldsymbol{T}_* \models \varphi$. A $\boldsymbol{T}_*$-structure $\mathbf{M}$ is a $\boldsymbol{T}_*$-*model*, or simply a *model of a first order theory*, of a theory $\Gamma$ if $\langle \boldsymbol{T}_*, \mathbf{M}\rangle \models \varphi$ for each $\varphi \in \Gamma$. If $\Gamma = \{\varphi\}$, we say that $\mathbf{M}$ is a model of $\varphi$. We will say that a formula $\varphi$ is 1-*satisfiable* if there is a $\boldsymbol{T}_*$-model of $\varphi$. We will say that a formula $\varphi$ is $s$-*satisfiable* if there exists a $\boldsymbol{T}_*$-structure $\mathbf{M}$, a value $s \in T$, and an assignation $v$ such that $\|\varphi\|^*_{\mathbf{M},v} = s$.

We define the logic $\Lambda^*\forall$ in the following way: Given a set $\Gamma \cup \{\varphi\}$ of $\langle \mathcal{L}, \Sigma\rangle$-formulas, we say that: $\varphi$ is *a consequence of the theory* $\Gamma$ *in* $\Lambda^*\forall$, and we write $\Gamma \models_* \varphi$ whenever every $\boldsymbol{T}_*$-model of $\Gamma$ is also a $\boldsymbol{T}_*$-model of $\varphi$.

Motivated by his studies on fuzzy description logics, Hájek introduced in [80] the notion of *witnessed models*. In what follows we present this notion because it is used in further sections of this chapter.

An interpretation $\mathbf{M}$ is *witnessed* if, for every formula $\varphi(x, y_1, \ldots, y_j)$ and any choice $w_1, \ldots, w_j \in M$ of values of $y_1, \ldots, y_j$, there exists $w \in M$ such that

$$\|(\forall x)\varphi(x, w_1, \ldots, w_j)\|^*_{\mathbf{M}} = \|\varphi(w, w_1, \ldots, w_j)\|^*_{\mathbf{M}},$$

and there exists $w' \in M$ such that

$$\|(\exists x)\varphi(x, w_1, \ldots, w_j)\|^*_{\mathbf{M}} = \|\varphi(w', w_1, \ldots, w_j)\|^*_{\mathbf{M}}.$$

REMARK 4.2.1. *For people acquainted with MFL it is interesting to notice that the* logic of a t-norm *as defined in MFL does not coincide with the logic* $\Lambda^*$. *The logic of a t-norm* $*$ *defined in the MFL setting (see Section 5 of Chapter V of this Handbook) is the fuzzy logic whose associated variety is the one generated by* $[0,1]_*$ *while the logic* $\Lambda^*$ *is the fuzzy logic semantically defined by evaluations on* $[0,1]_*$. *These two logics only coincide when the logic of a t-norm* $*$ *is standard complete. This is the case for both Gödel logic (the logic of minimum t-norm) and the logic of a finite divisible t-norm. The fact that, in general,* $\Lambda^*$ *does not coincide with the logic of a continuous t-norm is the main motivation to define the logic* $\Lambda^*$ *in this chapter. Notice also that the logic of a continuous t-norm (over* $[0,1]$ *or finite) is finitely axiomatizable even for its first-order version. This is not the case in general for logics* $\Lambda^*$ *and* $\Lambda^*\forall$. *Indeed, the logic* $\Lambda^*\forall$ *when* $*$ *is the Łukasiewicz t-norm is not recursively enumerable and when* $*$ *is the product t-norm is even not arithmetical (see Section 3 of Chapter XI of this Handbook).*

### 4.3 Basic fuzzy description logics as fragments of fuzzy first-order logics

In order to interpret our fuzzy description logics as fragments of a predicate logic $\Lambda^*\forall$, we associate any description signature:

$$\mathcal{D} = \langle N_I, N_A, N_R \rangle,$$

to the first-order signature $\Sigma_{\mathcal{D}} = \langle \mathcal{C}_{\mathcal{D}}, \mathcal{P}_{\mathcal{D}} \rangle$, being $\mathcal{C}_{\mathcal{D}} = N_I$ and $\mathcal{P}_{\mathcal{D}} = N_A \cup N_R$, in such a way that we interpret each individual name in $N_I$ as an object constant, each atomic concept in $N_A$ as a unary predicate symbol, and each atomic role in $N_R$ as a binary predicate symbol.

Now we introduce the notion of *instance of a description* as a first-order open formula associated with it. Then, we will prove that the semantical interpretation of a concept coincides with the semantical interpretation of its instance in first-order logic.

DEFINITION 4.3.1 (Instance of a description). Let $\mathcal{D}$ be a description signature. To each concept we assign a first order formula, its *instance*, in the following way: Given a term $t$ and a concept $D$, the *instance* $D(t)$ of $D$ is defined as follows:

$$\begin{array}{rll}
A(t) & \textit{if} & D \textit{ is an atomic concept } A\\
C_1(t)\&C_2(t) & \textit{if} & D = C_1 \sqcap C_2\\
C_1(t) \veebar C_2(t) & \textit{if} & D = C_1 \sqcup C_2\\
C_1(t) \to C_2(t) & \textit{if} & D = C_1 \to C_2\\
\neg C(t) & \textit{if} & D = \neg C\\
{\sim} C(t) & \textit{if} & D = {\sim} C\\
C_1(t) \wedge C_2(t) & \textit{if} & D = C_1 \wedge C_2\\
C_1(t) \vee C_2(t) & \textit{if} & D = C_1 \vee C_2\\
\bar{0} & \textit{if} & D = \bot\\
\bar{1} & \textit{if} & D = \top\\
(\forall y)(R(t,y) \to C(y)) & \textit{if} & D = \forall R.C\\
(\exists y)(R(t,y)\&C(y)) & \textit{if} & D = \exists R.C.
\end{array}$$

Finally, an *instance* of an atomic role $R$ is any atomic first order formula $R(t_1, t_2)$, where $t_1$ and $t_2$ are terms.

The next proposition states the relationship between interpretations of concepts in a description logic and interpretations of their instances in the corresponding first-order logic.

PROPOSITION 4.3.2. *Let $\mathcal{D} = \langle N_I, N_A, N_R\rangle$ be a description signature, and $*$ be a continuous or divisible finite t-norm.*

a) *Given an interpretation $\mathcal{I} = \langle \Delta^{\mathcal{I}}, \cdot^{\mathcal{I}}\rangle$ for the signature $\mathcal{D}$, extended to complex concepts by using $*$, we define the $\boldsymbol{T}_*$-interpretation:*

$$\mathbf{M}_{\mathcal{I}} = \langle \Delta^{\mathcal{I}}, \{A^{\mathcal{I}} \mid A \in N_A\} \cup \{R^{\mathcal{I}} \mid R \in N_R\}, \{a^{\mathcal{I}} \mid a \in N_I\}\rangle.$$

*Then, for every $x \in \Delta^{\mathcal{I}}$,*

$$||C(x)||^*_{\mathbf{M}_{\mathcal{I}}} = C^{\mathcal{I}}(x).$$

b) *Given a $\boldsymbol{T}_*$-interpretation for the signature $\mathcal{D}$,*

$$\mathbf{M} = \langle M, \{A^{\mathbf{M}} \mid A \in N_A\} \cup \{R^{\mathbf{M}} \mid R \in N_R\}, \{a^{\mathbf{M}} \mid a \in N_I\}\rangle,$$

*we define the interpretation: $\mathcal{I}_{\mathbf{M}} = \langle \Delta^{\mathcal{I}_{\mathbf{M}}}, \cdot^{\mathcal{I}_{\mathbf{M}}}\rangle$, where $\Delta^{\mathcal{I}_{\mathbf{M}}} = M$, $A^{\mathcal{I}_{\mathbf{M}}} = A^{\mathbf{M}}$, $R^{\mathcal{I}_{\mathbf{M}}} = R^{\mathbf{M}}$, and $a^{\mathcal{I}_{\mathbf{M}}} = a^{\mathbf{M}}$. Then, for every $w \in M$,*

$$C^{\mathcal{I}_{\mathbf{M}}}(w) = ||C(w)||^*_{\mathbf{M}}.$$

c) *For every interpretation $\mathcal{I}$ for the signature $\mathcal{D}$ extended to complex concepts by using $*$, the following holds:*

$$\mathcal{I}_{\mathbf{M}_{\mathcal{I}}} = \mathcal{I}.$$

d) *For every $\boldsymbol{T}_*$-interpretation $\mathbf{M}$ for the signature $\mathcal{D}$, the following holds:*

$$\mathbf{M}_{\mathcal{I}_{\mathbf{M}}} = \mathbf{M}.$$

*Proof.* Items $a)$ and $b)$ are proved by an easy induction. Items $c)$ and $d)$ are trivial. □

Therefore, the notion of instance allows us to understand description languages as fragments of predicate fuzzy logics. This means that we have a translation of our description languages into first-order fuzzy logic.[6] This translation can be extended to satisfiability of axioms of a knowledge base as stated in the following proposition.

PROPOSITION 4.3.3. *Given an interpretation $\mathcal{I} = \langle \Delta^{\mathcal{I}}, \cdot^{\mathcal{I}}\rangle$ for the signature $\mathcal{D}$ and its corresponding $\boldsymbol{T}_*$-interpretation $\mathbf{M}_{\mathcal{I}}$, then*

a) $\mathcal{I} \models_* \langle C \sqsubseteq D, r\rangle$ *if and only if* $\langle \boldsymbol{T}_*, \mathbf{M}_{\mathcal{I}}\rangle \models_* \bar{r} \to (\forall x)(C(x) \to D(x))$.

b) $\mathcal{I} \models_* \langle C(a), r\rangle$ *if and only if* $\langle \boldsymbol{T}_*, \mathbf{M}_{\mathcal{I}}\rangle \models_* \bar{r} \to C(a)$.

c) $\mathcal{I} \models_* \langle R(a,b), r\rangle$ *if and only if* $\langle \boldsymbol{T}_*, \mathbf{M}_{\mathcal{I}}\rangle \models_* \bar{r} \to R(a,b)$.

[6] In [54] direct translations of the considered description languages to the corresponding first order logics are given in a similar way as it is done in the classical case (see [30]).

*Proof.* We prove only $a$). The other items are proved analogously.

$$\begin{aligned}\mathcal{I} \models_* \langle C \sqsubseteq D, r\rangle &\Leftrightarrow \inf\{C^{\mathcal{I}}(x) \Rightarrow_* D^{\mathcal{I}}(x) \mid x \in \Delta^{\mathcal{I}}\} \geq r \\ &\Leftrightarrow \inf\{r \Rightarrow_* (C^{\mathcal{I}}(x) \Rightarrow_* D^{\mathcal{I}}(x)) \mid x \in \mathbf{M}_{\mathcal{I}}\} = 1 \\ &\Leftrightarrow \|\bar{r} \to (\forall x)(C(x) \to D(x))\|^*_{\mathbf{M}_{\mathcal{I}}} = 1 \\ &\Leftrightarrow \langle \boldsymbol{T}_*, \mathbf{M}_{\mathcal{I}}\rangle \models_* \bar{r} \to (\forall x)(C(x) \to D(x)). \qquad \square\end{aligned}$$

Since graded axioms are interpreted as first order sentences, we have that a knowledge base is interpreted as a first-order theory. Accordingly, we say that an interpretation $\mathbf{M}$ is a *model* of a knowledge base $\mathcal{K}$ if it satisfies all the first-order sentences corresponding to axioms of $\mathcal{K}$. As an easy consequence of the previous proposition we have that reasoning tasks can be fairly translated to first-order problems with respect to a theory.

Observe that the first order formulas corresponding to axioms of knowledge bases have the form of the so-called *evaluated formulas* in the setting of Mathematical Fuzzy Logic. Indeed, the truth constants appear only in the antecedent of implications in the formulas corresponding to the axioms.

Finally, it is interesting to remark the relationship between description and modal logics. It is well known that there exists a translation between the classical description logic $\mathcal{ALC}$ and the multi-modal logic $\mathbf{K}_m$ (see [112]). It is worth to say that there exists a similar translation between the FDL language $\mathfrak{I}\mathcal{ALCE}$ and a multi-modal many-valued logic (see [54, 57] for an extensive account on this subject).

## 5 Fuzzy Description Logic languages with higher expressivity

In Section 3.3 we have defined several FDL languages, the most expressive of them being $\mathfrak{I}\mathcal{ALCE}$. In classical DLs it is usual to consider even more expressive languages. In fact, the standard language for (crisp) ontology representation, OWL 2 [64], is based on the DL $\mathcal{SROIQ}(\mathbf{D})$, which is an extension of $\mathcal{ALC}$ (recall that in the classical case, $\mathfrak{I}\mathcal{ALCE}$ and $\mathcal{ALC}$ are equivalent). The purpose of this section is to discuss similar extensions of $\mathfrak{I}\mathcal{ALCE}$ in the fuzzy case.

FDLs described so far consider well-known logical connectives. However, the extension of more expressive DLs to the fuzzy case is not always unique. For instance, there could be different ways to extend the notion of cardinality in fuzzy logic. In this section, we will focus on the most common definitions in the FDL literature which often correspond to the most direct extensions [132].

The section is divided into five parts. The first four parts describe new concept constructors, role constructors, datatypes, and axioms, respectively. The fifth part summarizes some important expressive DL languages.

### 5.1 Fuzzy concept constructors

In this subsection we include fuzzy versions of the concept constructors corresponding to the DL $\mathcal{SROIQ}(\mathbf{D})$ together with some fuzzy concept constructors that do not have an equivalent in the classical case. More examples of new fuzzy concepts that do not have a crisp counterpart can be found in [23].

### 5.1.1 Nominals

The first extension that we will consider are *nominals*, which make it possible to give concept definitions by extension, using an enumeration of their instances. This constructor is denoted with the letter $\mathcal{O}$. Let $a$ be an individual name and let $r \in T$. The syntax of concepts is extended with a new case:

$$C \quad \rightsquigarrow \quad \{r/a\} \qquad \text{(nominal)}$$

and the semantics is the following:

$$\{r/a\}^{\mathcal{I}}(x) = \begin{cases} r & \text{if } x = a^{\mathcal{I}} \\ 0 & \text{otherwise.} \end{cases}$$

The idea is that if $x$ is interpreted as the same element of the domain than $a$, then the concept is evaluated as $r$. Otherwise, the concept is evaluated to 0. Although nominals can take an arbitrary truth degree $r$, in the literature they are usually restricted to the case where $r = 1$.

For example, the $\{1/\mathsf{germany}\} \sqcup \{1/\mathsf{austria}\} \sqcup \{0.67/\mathsf{switzerland}\}$ denotes the fuzzy concept of German-speaking country, to which germany and austria belong with degree 1 and switzerland, with 4 official languages, belongs with degree 0.67.

### 5.1.2 Unqualified cardinality restrictions

*Cardinality restrictions* (also called number restrictions) describe the number of $R$-successors that an element of the domain can have, for a given role $R$. Using these constructors, we can refer to those elements that have *at most* or *at least* a given number of successors through the role $R$. Logics allowing these concept constructors are denoted with the letter $\mathcal{N}$.

To allow for unqualified cardinality restrictions, the syntax of the language is extended to include the constructors

$$\begin{aligned} C \quad \rightsquigarrow \quad & \geq n\, R \mid && \text{(minimal cardinality restriction)} \\ & \leq n\, R && \text{(maximal cardinality restriction),} \end{aligned}$$

where $R \in N_R$ and $n \in \mathbb{N}$. These restrictions are called unqualified because we do not impose any other restriction on the objects of the relation. Following Example 3.1.1, $\geq 2$ isFriendOf denotes the fuzzy set of people having at least 2 friends.

It is worth to note that the semantics of this concept constructor is particularly debatable, since there exist a lot of definitions of cardinality of fuzzy sets. We will consider here the following translations to first order logic (although alternative definitions are possible):

$$(\geq n\, R) \mapsto (\exists y_1) \ldots (\exists y_n) \left( R(x, y_1) \wedge \cdots \wedge R(x, y_n) \wedge \bigwedge_{1 \leq i < j \leq n} y_i \neq y_j \right)$$

$$(\leq n\, R) \mapsto (\forall y_1) \ldots (\forall y_{n+1}) \left( R(x, y_1) \wedge \cdots \wedge R(x, y_{n+1}) \rightarrow \bigvee_{1 \leq j < k \leq n+1} y_j = y_k \right).$$

In fact, it is more common in the FDLs literature to interpret some of the conjunctions as strong conjunctions and not as weak ones. Recall that we are assuming that individual equality is a yes-no question, so the expressions $y_i \neq y_j$ and $y_j = y_k$ are evaluated in $\{0, 1\}$.

It is also possible to define *exact cardinality restrictions* $(= n\ R)$ combining a maximal and a minimal cardinality restriction: $(\geq n\ R) \sqcap (\leq n\ R)$. Moreover, functional roles can also be expressed in any logic that allows for unqualified number restrictions. Indeed, the axiom $\langle \top \sqsubseteq (\leq 1\ R), 1 \rangle$ restricts the role $R$ to be interpreted as a functional role (see Section 5.4.2 for a definition of functional roles).

### 5.1.3 Qualified cardinality restrictions

*Qualified cardinality restrictions* generalized the restrictions from Section 5.1.2 to consider only those $R$-successors that are elements of a specific concept $C$. The use of these constructors is denoted with the letter $\mathcal{Q}$. Languages with qualified number restrictions allow in their syntax the constructors

$$\begin{array}{lll} C \rightsquigarrow & \geq n\ R.C \mid & \text{(minimal qualified cardinality restriction)} \\ & \leq n\ R.C & \text{(maximal qualified cardinality restriction).} \end{array}$$

where $R \in N_R$, $n \in \mathbb{N}$, and $C$ is a concept from this language. For example, the concept $\geq 2\ R.Blond$ denotes the fuzzy set of people having at least 2 blond friends.

As in the previous case, an *exact qualified cardinality restriction* $= n\ R.C$ can be defined as $(\geq n\ R.C) \sqcap (\leq n\ R.C)$, and a possible semantics is obtained by translating these concepts to the following first order formulas:

$$(\geq n\, R.C) \mapsto (\exists y_1) \ldots (\exists y_n)(R(x, y_1) \wedge C(y_1) \wedge \cdots \wedge R(x, y_n) \wedge C(y_n) \wedge \bigwedge_{1 \leq i < j \leq n} y_i \neq y_j)$$

$$(\leq n\, R.C) \mapsto (\forall y_1) \ldots (\forall y_{n+1})(R(x, y_1) \wedge C(y_1) \wedge \cdots \wedge R(x, y_{n+1}) \wedge C(y_{n+1}) \rightarrow \bigvee_{1 \leq j < k \leq n+1} y_j = y_k).$$

Clearly, unqualified number restrictions are just special cases of qualified number restrictions; for instance, the concept $(\leq R)$ is equivalent to $(\leq R.\top)$.

Whenever we consider number restrictions, it is useful to allow for additional axioms for stating explicitly whether two individuals are equivalent (i.e., interpreted as the same element of the domain), or different. These axioms are introduced in Section 5.4.6.

### 5.1.4 Local reflexivity concepts

*Local reflexivity concepts* allow expressing that an individual is related to itself via some role $R$, without implying that the role is globally reflexive. To have them in the language, the syntax of concepts is extended as follows:

$$C \rightsquigarrow \quad \exists R.\mathtt{Self} \qquad \text{(local reflexivity concept)}$$

and the semantics is $(\exists R.\mathtt{Self})^{\mathcal{I}}(x) = R^{\mathcal{I}}(x, x)$. For example, $\exists\mathsf{likes}.\mathtt{Self}$ denotes narcisists, i.e., people who like themselves.

#### 5.1.5 Fuzzy modified concepts

In fuzzy logic it is usually desirable to consider language hedges (or fuzzy modifiers) interpreting linguistic modifiers. Given a hedge $h\colon T \to T$, it is possible to apply it to change the membership of a fuzzy concept, thus obtaining a *fuzzy modified concept* [128]. Given an hedge $h$, the syntax of concepts is extended as follows:

$$C \quad \rightsquigarrow \quad h(C) \qquad \text{(fuzzy modified concept)}$$

and the semantics is $(h(C))^{\mathcal{I}}(x) = h(C^{\mathcal{I}}(x))$. For example, we can apply the hedge $\mathsf{very}$ to the concept $\mathsf{Tall}$ to define $\mathsf{very}(\mathsf{Tall})$, denoting the fuzzy set of very tall people.

These concepts do not have an equivalent in classical DLs, since in classical logics no grading is possible. A particular case of fuzzy modifier is the constructor $\Delta$. Observe that the semantics is given by $(\Delta(C))^{\mathcal{I}}(x) = \delta(C^{\mathcal{I}}(x))$ (see Section 2.4).

#### 5.1.6 Aggregated concepts

Fuzzy aggregation operators $@\colon T^n \to T$ are used to aggregate $n$ values into a single one. In particular, they could be used to aggregate the evaluation of $n$ fuzzy concepts. The application of an aggregation operator to several concepts produce an *aggregated concept* [77]. Examples of these aggregation operators are the weighted sum, the Ordered Weighted Averaging (OWA) operator or more general fuzzy integrals. Given an aggregation operator @, the syntax of concepts is extended as follows:

$$C_1, \ldots, C_n \quad \rightsquigarrow \quad @(C_1, \ldots, C_n) \qquad \text{(aggregated concept)}$$

and the semantics is $(@(C_1, \ldots, C_n))^{\mathcal{I}}(x) = @(C_1^{\mathcal{I}}(x), \ldots, C_n^{\mathcal{I}}(x))$. This concept does not have an equivalent in classical DLs.

For example, given the fuzzy concepts $\mathsf{Tall}$ and $\mathsf{Blonde}$, we can define the fuzzy concept $\mathrm{WS}_{[0.67,0.33]}(\mathsf{Tall}, \mathsf{Blonde})$ which denotes the fuzzy set of tall blonde people, where the degree of being tall has double the importance than the degree of being blonde.

### 5.2 Fuzzy role constructors

In this subsection we include fuzzy versions of the role constructors in the DL $\mathcal{SROIQ}(\mathbf{D})$ together with some fuzzy roles without an equivalent in the classical case.

#### 5.2.1 Inverse roles

The inverse relations of the roles in the language are called *inverse roles*. This constructor is denoted by appending $\mathcal{I}$ to the language name. Up to now, we assumed that every role is atomic, that is, a member of $N_R$. To support inverse roles, the syntax need to be extended to allow roles of the form

$$R \quad \rightsquigarrow \quad R^- \qquad \text{(inverse role)},$$

where $R$ is an atomic role. For example, the roles $\mathsf{hasParent}$ and $\mathsf{hasChild}$ are inverse. To interpret these complex roles, we extend the interpretation function to define $(R^-)^{\mathcal{I}}(x, y) = R^{\mathcal{I}}(y, x)$. Note that $(R^-)^-$ is equivalent to $R$, and thus we can restrict without loss of generality to inverse of atomic roles.

### 5.2.2 Universal role

The *universal role* is similar to the top concept. It is denoted with the letter $U$ and is interpreted as $U^{\mathcal{I}}(x, y) = 1$ for all elements $x, y$ in the domain. Its use is often restricted, depending on the particular DL language (see [88] for details).

### 5.2.3 Fuzzy modified roles

Similarly to the case of fuzzy modified concepts, given an hedge $h$, the syntax of roles can be extended as follows:

$$R \quad \rightsquigarrow \quad h(R) \qquad \text{(fuzzy modified role).}$$

The semantics is $(h(R))^{\mathcal{I}}(x, y) = h(R^{\mathcal{I}}(x, y))$. For example, we can apply the hedge very to the role isFriendOf to define very(isFriendOf), denoting the fuzzy relation of very good friends. This role constructor does not have an equivalent in classical DLs.

## 5.3 Fuzzy datatypes

Up to now, the object of every relation is an individual of the world. The idea behind datatypes, also known as concrete domains, is to allow individuals having relations (e.g. "to have age") with certain kinds of data (e.g. the number "47") that do not belong to the represented domain, rather to a different domain that is already structured and whose structure is already known to the machine (for this reason it is "concrete"). These kinds of individuals, called *data values* can be numerical or textual, among many other possibilities. Clearly data values could be introduced as usual individuals with their attributes, but since their structure is already known for the machine (e.g. the machine already knows that $3 < 5$), in this way there is no need of introducing an ontology to fix the structure of datatypes. Possible datatypes include numbers (real, rational, integer, non-negative, etc.), strings, booleans, dates, times, or XML literals.

In the fuzzy case, in addition to relating an individual with some data value $dv$, it is possible to relate it with a fuzzy membership function $\mathbf{d}$. Typical examples of fuzzy datatypes, when the datatype domain is a dense total order, are the trapezoidal, triangular, left-shoulder or right-shoulder membership functions. For example, the datatype hasAgeAround$(n)$ can be defined using a triangular membership function with parameters $(\max\{0, n-3\}, n, n+3)$. This constructor is denoted by appending $(\mathbf{D})$ to the name of the language, e.g. $\mathcal{ALC}(\mathbf{D})$.

Datatypes can be used in complex concepts that use a role, namely in existential quantification, value restriction and, if allowed in the language, qualified cardinality restrictions. Hence, the syntax of the concept is extended as follows:

$$\begin{array}{lrl} C \rightsquigarrow & \forall R.\mathbf{d} \mid & \text{(data value restriction)} \\ & \exists R.\mathbf{d} \mid & \text{(data existential quantification)} \\ & \geq n\ R.\mathbf{d} \mid & \text{(minimal qualified data cardinality restriction)} \\ & \leq n\ R.\mathbf{d} & \text{(maximal qualified data cardinality restriction).} \end{array}$$

The semantics are similar to those of the respective cases $\forall R.C$, $\exists R.C$, $\geq n\ R.C$ and $\leq n\ R.C$ but replacing the interpretation (instantiation) of $C$ on an element $y \in M$ with the interpretation (instantiation) of $\mathbf{d}$ on a data value $dv$.

## 5.4 New axioms

We will consider in this subsection fuzzy versions of the axioms in the description logic $\mathcal{SROIQ}(\mathbf{D})$.

### 5.4.1 Transitive roles

The first extension regarding axioms that we are going to explore is the introduction of *transitive roles* in the language. For example, the role isSimilarTo can be considered as transitive. This element can be denoted by adding $\mathcal{R}_+$ to the language name. For example, the extension of $\mathcal{ALC}$ with transitive roles is denoted as $\mathcal{ALCR}_+$. In classical DLs, $\mathcal{ALCR}_+$ is usually shortened to $\mathcal{S}$ due to its equivalence with the modal logic **S4**. In FDLs, we prefer to use $\mathcal{S}$ to denote the extension of $\mathfrak{I}\mathcal{ALCE}$ with transitive roles.

In order to have transitive roles in the language, one must extend the syntax of the language to have another type of axioms of the form $\mathtt{trans}(R)$, where $R$ is a role. Its corresponding first order sentence is

$$(\forall x)(\forall y)(\forall z)(R(x,z) \mathbin{\&} R(z,y) \to R(x,y)).$$

Since the first order sentence should take value 1, an interpretation $\mathcal{I}$ satisfies an axiom $\mathtt{trans}(R)$ if and only if it verifies that for every $x, y \in M$,

$$R^{\mathcal{I}}(x,y) \geq \sup_{z \in M} R^{\mathcal{I}}(x,z) * R^{\mathcal{I}}(z,y).$$

### 5.4.2 Functional roles

In classical DLs, a role $R$ is called *functional* if any object of the domain can have at most one successor via the relation $R$. The use of this constructor in a DL is denoted with the letter $\mathcal{F}$. In the fuzzy case, we can assert that the concrete role livesInMainResidence is functional, because everybody has a unique main residence place, but this relation is graded to differentiate people who live always in their main residence from people who spend a lot of time out of their home.

Syntactically, functional roles are just a special kind of role. When defining an FDL that allows for functional roles, we simply assume that the set $N_R$ of role names is partitioned into two sets $N_{FR}$ and $N_{GR}$ of *functional roles* and *general roles*, respectively. Semantically, interpretations are required to treat the roles $R \in N_{FR}$ as fuzzy partial functions. While there are several possible ways in which partial functionality can be defined in the fuzzy setting, the most usual in the FDL literature is that an interpretation must satisfy the first order formula

$$(\forall y_1)(\forall y_2)\big((R(x,y_1) \wedge R(x,y_2)) \to (y_1 = y_2)\big),$$

for all functional roles $R \in N_{FR}$. Recall that $y_1 = y_2$ is evaluated in $\{0,1\}$.

### 5.4.3 Role inclusions

It is also natural to consider role inclusions (or role hierarchies) as part of the language in a similar way as it contains concept inclusions. For instance, the role hasSon is more specific than the role hasChild. This is denoted by appending $\mathcal{H}$ (as the initial letter of hierarchies) to the language name. In the fuzzy case, these inclusions can be graded, of course.

Let $R, S$ be two roles and let $r \in T$. A *graded role inclusion axiom* is an expression of one of the form

$$\langle R \sqsubseteq S, r \rangle,$$

whose corresponding first order sentence is

$$\bar{r} \to (\forall x)(\forall y)(R(x,y) \to S(x,y)).$$

### 5.4.4 More axioms involving roles

Traditionally, DLs have paid much more attention to the expressivity of concepts than to that of roles. Recently, several new axioms involving roles (and 5 other types of axioms) have been proposed [88]. These new axioms can be extended to the fuzzy case. Let $R, S$ be roles and $a, b$ be individuals. The syntax of axioms is extended as follows:

$$\begin{array}{rl} \langle \neg R(a,b), r \rangle \mid & \text{(weakly complemented role assertion)} \\ \langle \sim R(a,b), r \rangle \mid & \text{(strongly complemented role assertion)} \\ \langle R_1 \ldots R_m \sqsubseteq S, r \rangle \mid & \text{(role inclusion axiom)} \\ \texttt{dis}(R,S) \mid & \text{(disjoint roles)} \\ \texttt{ref}(R) & \text{(reflexive role).} \end{array}$$

Let us consider some examples. The assertion $\langle \neg \mathsf{isFriendOf}(\mathsf{alice}, \mathsf{bob}), 0.5 \rangle$ states that alice and bob cannot be considered friends with degree greater or equal than $0.5$. The axiom $\langle \mathsf{hasMotherhasSister} \sqsubseteq \mathsf{hasAunt}, 1 \rangle$ states the sister of one's mother is his/her aunt. The roles hasMother and hasFather are disjoint. Finally, the fuzzy role isCloseTo is reflexive.

The corresponding first order sentences are the following ones:

$$\begin{array}{rcl} \langle \neg R(a,b), r \rangle & \leadsto & r \to \neg R(a,b) \\ \langle \sim R(a,b), r \rangle & \leadsto & r \to \sim R(a,b) \\ \langle R_1 \ldots R_m \sqsubseteq S, r \rangle & \leadsto & r \to (\forall x_1) \cdots (\forall x_{m+1}) \big( (\exists x_2) \cdots (\exists x_m) \\ & & (R_1(x_1,x_2) \mathbin{\&} \cdots \mathbin{\&} R_m(x_m,x_{m+1})) \to S(x_1,x_{m+1}) \big) \\ \texttt{dis}(R,S) & \leadsto & (\forall x)(\forall y)(R(x,y) \wedge S(x,y) \to 0) \\ \texttt{ref}(R) & \leadsto & \bar{1} \to (\forall x) R(x,x). \end{array}$$

It is worth noticing that role inclusion axioms are strictly more expressive than role hierarchies. Moreover, irreflexive, symmetric, asymmetric and transitive roles can be simulated using the constructors and axioms introduced in this section. Let $R$ be a role, then $R$ is

$$\begin{array}{rll} \textit{irreflexive} & \text{if} & \langle \top \sqsubseteq \neg \exists R.\texttt{Self}, 1 \rangle \\ \textit{symmetric} & \text{if} & \langle R \sqsubseteq R^-, 1 \rangle \\ \textit{asymmetric} & \text{if} & \texttt{dis}(R, R^-) \\ \textit{transitive} & \text{if} & \langle RR \sqsubseteq R, 1 \rangle. \end{array}$$

There are some additional syntactic restrictions to guarantee decidability. The interested reader is referred to [26, 132].

#### 5.4.5 Crispification axioms

In some applications, it is important to restrict some concepts and roles to be *crisp*; that is, interpreted in such a way that all elements belong to them to a degree in $\{0,1\}$. We can achieve this by including axioms of the form $\mathrm{crisp}(C)$ and $\mathrm{crisp}(R)$ where $C$ and $R$ are (possibly complex) concepts and roles. An interpretation $\mathcal{I}$ satisfies such axioms if $(\mathrm{crisp}(C))^{\mathcal{I}}(x) \in \{0,1\}$ holds for all $x$, and $(\mathrm{crisp}(R))^{\mathcal{I}}(x,y) \in \{0,1\}$ holds for all $x, y$, respectively.

Notice that in FDL languages containing $\Delta$ these axioms can be formulated as $\Delta(C) = C$ and $\Delta(R) = R$ respectively.

#### 5.4.6 Individual restrictions

It is sometimes convenient to express that two individual names correspond to the same element of the domain, or that they must be interpreted as different elements. This can be achieved by allowing *individual restrictions* of the form $a = b$ and $a \neq b$, where $a, b$ are two individual names. The semantics of these axioms is the obvious one: the interpretation $\mathcal{I}$ satisfies $a = b$ if $a^{\mathcal{I}} = b^{\mathcal{I}}$; it satisfies $a \neq b$ if $a^{\mathcal{I}} \neq b^{\mathcal{I}}$.

### 5.5 Some important languages

To conclude this section, we will highlight some important languages whose classical versions are important from a practical point of view.

- The classical language $\mathcal{SHOIN}(\mathbf{D})$ is the logical counterpart behind the *Web Ontology Language* (OWL). $\mathcal{SHOIN}(\mathbf{D})$ is the extension of $\mathcal{ALC}$ with transitive roles, role hierarchies, nominals, inverse roles, cardinality restrictions, and datatypes. OWL was the standard ontology language until December 2012, when it was replaced by its successor OWL 2.
- Crisp $\mathcal{SROIQ}(\mathbf{D})$ is equivalent to the language OWL 2, the current standard language for (crisp) ontology representation. $\mathcal{SROIQ}(\mathbf{D})$ is the extension of $\mathcal{SHOIN}(\mathbf{D})$ with local reflexivity concepts, qualified cardinality restrictions, the universal role, complemented role assertions, role inclusion axioms, disjoint roles and reflexive roles.

The standardisation of OWL and OWL 2 has given rise to a massive increase of the development of crisp ontologies and tools,[7] such as reasoners, editors or APIs. In the fuzzy case, no standard language has been defined yet.

## 6 Reasoning algorithms

We start this section by discussing the equivalences between different reasoning tasks. This way, it is sufficient to provide a reasoning algorithm for only one of them. Then we provide an overview of the most important existing reasoning algorithms.

### 6.1 Equivalences between reasoning tasks in finitely-valued FDLs

In classical DLs, it is possible to show the inter-definability of the different reasoning tasks. In FDLs, this is not always the case, as depending on the fuzzy operators

[7] See `http://wiki.opensemanticframework.org/index.php/Ontology_Tools` for examples.

some of the reductions may not hold. In this subsection, we will restrict our attention to FDLs with the finitely-valued property and show the reduction of every reasoning task to $*$-consistency. To achieve this, we will assume that the language contains $*$, $\sim$, and $\rightarrow$.

Assuming a finite number of degrees makes it possible to reduce some tasks to a binary search over the degrees and consider the predecessor of every degree of truth. Furthermore, infinitely-valued DLs with GCIs are undecidable if $*$ is different from Gödel t-norm, as we will see in Section 7.

For every $r \in T, r > 0$, we use $\text{pred}(r)$ to denote its *predecessor*, and then we define $r^- = \text{pred}(r)$. This way, $\langle \sim C(a), r^- \rangle$ requires that the value of $C(a)$ is strictly smaller than $r$. Now, the following result can easily be shown:

THEOREM 6.1.1. *Let $\mathcal{K}$ be a knowledge base and $r \neq 0$. In $\mathfrak{I}\mathcal{ALCE}$, the following equivalences between reasoning tasks hold:*

- *$C$ is $*$-satisfiable to a degree greater than or equal to $r$ with respect to $\mathcal{K}$ if and only if $\mathcal{K} \cup \{\langle C(a), r \rangle\}$ is $*$-consistent, where $a$ is an individual not appearing in $\mathcal{K}$.*
- *$C$ is $*$-subsumed by $D$ to a degree greater than or equal to $r$ with respect to $\mathcal{K}$ if and only if $\mathcal{K} \cup \{\langle (\sim(C \rightarrow D))(a), r^- \rangle\}$ is not $*$-consistent, where $a$ is an individual not appearing in $\mathcal{K}$.*
- *$\mathcal{K}$ entails:*
  - *a) A graded concept assertion $\langle C(a), r \rangle$ if and only if $\mathcal{K} \cup \{\langle (\sim C)(a), r^- \rangle\}$ is not $*$-consistent.*
  - *b) A graded role assertion $\langle R(a, b), r \rangle$ if and only if*

    $$\mathcal{K} \cup \{\langle (\sim(\exists R.B))(a), r^- \rangle, \langle B(b), 1 \rangle\}$$

    *is not $*$-consistent, where $B$ is an atomic concept not appearing in $\mathcal{K}$.*
  - *c) A graded concept inclusion $\langle C \sqsubseteq D, r \rangle$ if and only if*

    $$\mathcal{K} \cup \{\langle (\sim(C \rightarrow D))(a), r^- \rangle\}$$

    *is not $*$-consistent, where $a$ is an individual not appearing in $\mathcal{K}$.*
- *The BED of an axiom $\tau$ can be computed as a binary search over the degrees in the finite chain of the maximum value $r \in T$ such that $\mathcal{K}$ entails $\langle \tau, r \rangle$, and entailment can be reduced to $*$-consistency.*
- *The BSD of a concept $C$ can be computed as a binary search over the degrees in the finite chain of the maximum value $r \in T$ such that $\mathcal{K} \cup \{\langle C(a), r \rangle\}$ is $*$-consistent, where $a$ is an individual not appearing in $\mathcal{K}$.*

Note the implicit assumptions of having existential restriction, concept implication, and involutive negation in the language. It is also worth to note that the BED and the BSD problems can be computed in a more efficient way by developing a specific reasoning algorithm, such as those described in Section 6.2.2, that requires to solve one single test instead of several ones.

## 6.2 Types of algorithms

The literature includes a substantial amount of work on proof methods for first order fuzzy logics. The interested reader is referred to [97] and, for the particularly interesting case of finite fuzzy logics, to [13]. Many specific algorithms to reason with FDLs have been proposed in the literature. They can be classified in the following categories:

- *Tableaux algorithms,* extending the tableaux algorithms for classical DLs to the fuzzy case.
- *Tableaux algorithms and optimization problems,* using a tableaux algorithm to reduce the reasoning to an optimization problem.
- *Automata-based algorithms,* adopting ideas similar to those used to prove complexity results in the classical case.
- *Reduction to classical DLs,* for which existing reasoning algorithms are known.
- *Reduction to propositional fuzzy logics,* for which reasoning has been studied.
- *Structural subsumption algorithms,* extending the structural subsumption algorithms for classical DLs with low expressivity to the fuzzy case.

In the rest of this section, we give a deeper look into each of these families of algorithms, discussing their advantages and limitations, and one salient example of algorithm of the family. For simplicity, we will consider algorithms for relatively simple logics. The most important results, involving more expressive logics, can be found in Section 8.

### 6.2.1 Tableau-based algorithms

Tableaux algorithms are the most popular solution to decide satisfiability or consistency in classical DLs [12]. Although these algorithms usually have a sub-optimal worst-case complexity, they typically behave well in practice.

Given a KB, tableau-based algorithms try to build a data structure (called a *tableau*) representing a finite description of a model of the KB and from which it is immediate to build a complete model. This structure is typically a graph where the nodes represent individuals of the domain (annotated with a list of concepts that the individual belongs to) and the arcs represent roles establishing relations between two individuals.

The graph is initialized using the assertions that appear in the KB. Then, tableau-expansion rules are used to decompose complex concepts into simpler ones. For instance, the conjunction rule decomposes a concept $C \sqcap D$ into both $C$ and $D$. Some of the rules are non-deterministic, making it sometimes necessary to try several possibilities. For instance, the disjunction rule decompose a concept $C \sqcup D$ by producing two possible tableaux, one where $C$ is added to the node and another one where $D$ is added. In most DLs, due to the presence of cycles in the axioms, it is also necessary to use a *blocking* mechanism to guarantee termination of the approach.

After all the possible rules have fired, the tableau is completely expanded and it only remains to check for obvious contradictions. For instance, if a node belongs to both a concept and its complement, then the result of the algorithm outputs as an answer that the input KB is inconsistent. Otherwise, the algorithm yields a positive answer to the consistency of the input KB.

**Algorithm 1** Consistency checking in $Z$-$\mathcal{ALC}$ without a TBox

**Require:** A fuzzy KB $\mathcal{K}$
**Ensure:** true if $\mathcal{K}$ is $*$-consistent; false otherwise

1: **for** each concept assertion $\langle C(a), r\rangle \in \mathcal{K}$ **do**
2: $\quad \mathcal{K} \leftarrow (\mathcal{K} \setminus \{\langle C(a), r\rangle\}) \cup \{\langle \mathsf{nnf}(C)(a), r\rangle\}$ {Negation Normal Form}
3: **end for**
{Initialization of the forest}
4: create an empty forest $\mathcal{F}$
5: **for** each concept assertion $\langle C(a), r\rangle \in \mathcal{K}$ **do**
6: $\quad$ create a node $v_a$ in $\mathcal{F}$ if it does not exist
7: $\quad \mathcal{L}(v_a) \leftarrow \mathcal{L}(v_a) \cup \{\langle C \geq r\rangle\}$
8: **end for**
9: **for** each role assertion $\langle R(a, b), r\rangle \in \mathcal{K}$ **do**
10: $\quad$ create nodes $v_a, v_b$ in $\mathcal{F}$ if they do not exist
11: $\quad$ create an edge $\langle v_a, v_b\rangle$ in $\mathcal{F}$ if it does not exist
12: $\quad \mathcal{L}(\langle v_a, v_b\rangle) = \mathcal{L}(\langle v_a, v_b\rangle) \cup \{\langle R \geq r\rangle\}$
13: **end for**
{Tableaux rules}
14: **for** each rule application to $\mathcal{F}$ **do**
15: $\quad \mathcal{F}' \leftarrow \mathcal{F}$
$\quad$ {build a completion graph of $\mathcal{F}'$}
16: $\quad$ **while** some tableaux rule is applicable to some node $v$ in $\mathcal{F}'$ **do**
17: $\quad\quad$ apply one of the rules in Table 3 to $v$
18: $\quad\quad$ mark the applied rule as not applicable to node $v$
$\quad\quad$ {Stop when a clash is found }
19: $\quad\quad$ **if** there is not some node of $\mathcal{F}'$ with a clash **then**
20: $\quad\quad\quad$ continue with the next iteration of the for loop
21: $\quad\quad$ **end if**
22: $\quad$ **end while**
$\quad$ {There is a clash-free completion graph }
23: $\quad$ **if** there is not any node in $\mathcal{F}'$ with a clash **then**
24: $\quad\quad$ **return** true
25: $\quad$ **end if**
26: **end for**
{There are no clash-free completion graphs }
27: **return** false

It is sometimes possible to extend these algorithms to work with FDLs. The main idea is that every node is annotated with restrictions of the form $\langle C \geq r\rangle$ or $\langle C \leq r\rangle$, where $C$ is a concept and $r \in T$, stating that the individual nodes belong to the concept $C$ at least with degree $r$. The rules are updated to take into account the semantics of the concepts in the fuzzy case. For instance, $\langle C \wedge D \geq r\rangle$ is decomposed into both $\langle C \geq r\rangle$ and $\langle D \geq r\rangle$. The notion of contradiction is updated as well since now it is usually possible that a node belongs to some degree to both a concept and its complement.

| | preconditions | actions |
|---|---|---|
| $(\neg_+)$ | $\langle \neg C \geq t_i \rangle \in \mathcal{L}(v)$ | $\mathcal{L}(v) = \mathcal{L}(v) \cup \{\langle C \leq t_{n-i} \rangle\}$ |
| $(\neg_-)$ | $\langle \neg C \leq t_i \rangle \in \mathcal{L}(v)$ | $\mathcal{L}(v) = \mathcal{L}(v) \cup \{\langle C \geq t_{n-i} \rangle\}$ |
| $(\sqcap_+)$ | $\langle C \sqcap D \geq t_i \rangle \in \mathcal{L}(v)$ | $\mathcal{L}(v) = \mathcal{L}(v) \cup \{\langle C \geq t_i \rangle, \langle D \geq t_i \rangle\}$ |
| $(\sqcap_-)$ | $\langle C \sqcap D \leq t_i \rangle \in \mathcal{L}(v)$ | $\mathcal{L}(v) = \mathcal{L}(v) \cup \{X\}$ for some $X \in \{\langle C \leq t_i \rangle, \langle D \leq t_i \rangle\}$ |
| $(\sqcup_+)$ | $\langle C \sqcup D \geq t_i \rangle \in \mathcal{L}(v)$ | $\mathcal{L}(v) = \mathcal{L}(v) \cup \{X\}$ for some $X \in \{\langle C \geq t_i \rangle, \langle D \geq t_i \rangle\}$ |
| $(\sqcup_-)$ | $\langle C \sqcup D \leq t_i \rangle \in \mathcal{L}(v)$ | $\mathcal{L}(v) = \mathcal{L}(v) \cup \{\langle C \leq t_i \rangle, \langle D \leq t_i \rangle\}$ |
| $(\exists_+)$ | $\langle \exists R.C \geq t_i \rangle \in \mathcal{L}(v)$ $\nexists y$ with $\langle R \geq t_j \rangle \in \mathcal{L}(\langle v, y \rangle)$ $\langle C \geq t_k \rangle \in \mathcal{L}(y), j, k \geq i$ | create a new node $w$ $\mathcal{L}(\langle v, w \rangle) = \{\langle R \geq t_i \rangle\}$ $\mathcal{L}(w) = \{\langle C \geq t_i \rangle\}$ |
| $(\exists_-)$ | $\langle \exists R.C \leq t_i \rangle \in \mathcal{L}(v)$ $\langle R \geq t_j \rangle \in \mathcal{L}(\langle v, w \rangle, j > i$ | $\mathcal{L}(w) = \mathcal{L}(w) \cup \{\langle C \leq t_i \rangle\}$ |
| $(\forall_+)$ | $\langle \forall R.C \geq t_i \rangle \in \mathcal{L}(v)$ $\langle R \geq t_j \rangle \in \mathcal{L}(\langle v, w \rangle, j > n - i$ | $\mathcal{L}(w) = \mathcal{L}(w) \cup \{\langle C \geq t_i \rangle\}$ |
| $(\forall_-)$ | $\langle \forall R.C \leq t_i \rangle \in \mathcal{L}(v)$ $\nexists y$ with $\langle R \geq t_{n-j} \rangle \in \mathcal{L}(\langle v, y \rangle)$ $\langle C \leq t_k \rangle \in \mathcal{L}(y), j, k \leq i$ | create a new node $w$ $\mathcal{L}(\langle v, w \rangle) = \{\langle R \geq t_{n-i} \rangle\}$ $\mathcal{L}(w) = \{\langle C \geq t_i \rangle\}$ |

Table 3. Rules of the tableaux algorithm

In the case of FDLs the obtained tableau has a contradiction if a node has two pairs $\langle C \geq r_1 \rangle$ and $\langle C \leq r_2 \rangle$ with $r_1 < r_2$.

This family of algorithms can be used to decide $*$-consistency in Zadeh FDLs or if the degrees of truth is taken from a residuated De Morgan lattice, as we will describe below. Indeed, in all these FDLs the weak and the strong conjunction coincide, and then its semantics is the minimum and thus $\langle C \wedge D \geq r \rangle$ can be decomposed into both $\langle C \geq r \rangle$ and $\langle D \geq r \rangle$. In more general cases, tableaux algorithms are combined with an optimization problem, as we will see in the next subsection.

Let us see in detail Algorithm 1 to decide the consistency of the fuzzy knowledge base $\mathcal{K}$ in $Z$-$\mathcal{ALC}$ with an empty TBox. For simplicity, we will restrict to a chain $T = \{t_0, t_1, \ldots, t_n\}$ of degrees of truth, but the extension to $[0, 1]$ is easy [125].

Like most of the tableaux algorithms, the algorithm works on *completion-forests* since an ABox might contain several individuals with arbitrary roles connecting them. A completion-forest $\mathcal{F}$ for $\mathcal{K}$ is a collection of nodes arbitrarily connected by edges.

- Each node $v$ is labeled with a set $\mathcal{L}(v)$ of expressions of the form $\langle C \geq t_i \rangle$ and $\langle D \leq t_j \rangle$, meaning that $v$ is an instance of $C$ to degree greater than or equal to $t_i$ and an instance of $D$ to degree less or equal than $t_j$.
- Each edge $\langle v, w \rangle$ is labeled with a set $\mathcal{L}(\langle v, w \rangle)$ of expressions of the form $\langle R \geq t_i \rangle$ and $\langle R \leq t_j \rangle$.

$\mathcal{F}$ is then expanded by repeatedly applying the tableaux rules in Table 3. Note that there are two non-deterministic rules, namely $(\sqcup_+)$ and $(\sqcap_-)$. In classical DLs, the conjunction does not generate a non-deterministic rule. Similarly, the rules $(\exists_+)$ and $(\forall_-)$ generate new individuals; while in classical DLs only existential restrictions do.

If the rules can be applied in such a way that they yield a complete and clash-free completion-forest, $\mathcal{K}$ is $*$-consistent. Otherwise, $\mathcal{K}$ is not $*$-consistent. Since there are non-deterministic rules, there are many possible ways to apply the rules, so the algorithm must consider each possible rule application until a clash-free completion-forest is found. This can be computed by using a search with backtracking. A node $v$ contains a clash in any of the following situations:

- $\langle \bot \geq t_i \rangle \in \mathcal{L}(v), i > 0$,
- $\langle \top \leq t_i \rangle \in \mathcal{L}(v), i < n$, or
- $\langle C \geq t_i \rangle \in \mathcal{L}(v), \langle C \leq t_j \rangle \in \mathcal{L}(v), j < i$.

Under Zadeh fuzzy logic, we can assume w.l.o.g. that concept are in *Negation Normal Form* (NNF) where the negation only appears before an atomic concept. In fact, a fuzzy concept $\neg C$ can be transformed into NNF by recursively applying the function $\mathsf{nnf}$ defined as follows: $\mathsf{nnf}(\neg\top) = \bot$, $\mathsf{nnf}(\neg\bot) = \top$, $\mathsf{nnf}(\neg A) = \neg A$, $\mathsf{nnf}(\neg\neg C) = C$, $\mathsf{nnf}(\neg(C \sqcap D)) = \neg C \sqcup \neg D$, $\mathsf{nnf}(\neg(C \sqcup D)) = \neg C \sqcap \neg D$, $\mathsf{nnf}(\neg\exists R.C) = \forall R.\neg C$, and $\mathsf{nnf}(\neg\forall R.C) = \exists R.\neg C$.

EXAMPLE 6.2.1. Consider the chain of 11 degrees of truth and the fuzzy KB

$$\{\langle(\forall R.C)(a), 0.7\rangle, \langle R(a,b), 0.6\rangle, \langle(\neg C)(b), 0.8\rangle\}.$$

The completion graph after the initialization and the application of the rules $(\neg)$ and $(\forall)$ can be depicted as:

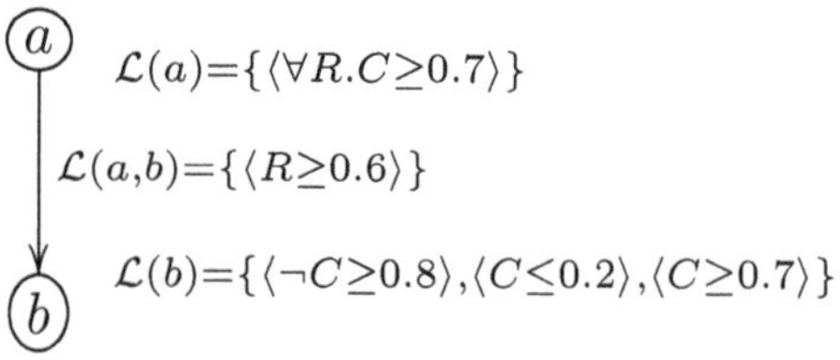

$a$ $\quad \mathcal{L}(a)=\{\langle\forall R.C\geq 0.7\rangle\}$

$\mathcal{L}(a,b)=\{\langle R\geq 0.6\rangle\}$

$b$ $\quad \mathcal{L}(b)=\{\langle\neg C\geq 0.8\rangle,\langle C\leq 0.2\rangle,\langle C\geq 0.7\rangle\}$

It can be seen that the KB is $*$-inconsistent because there is a clash: node $b$ contains both $\langle C \leq 0.2\rangle$ and $\langle C \geq 0.7\rangle$.

### 6.2.2 Tableaux algorithms and optimization problems

In general, the tableaux algorithms already described are not appropriate for other t-norms different from Gödel or to manage fuzzy datatypes or fuzzy concepts without a counterpart in crisp DLs, such as fuzzy modified concepts or aggregated concepts. Intuitively, given $\langle C \sqcap D \geq r\rangle$ we cannot infer both $\langle C \geq r\rangle$ and $\langle D \geq r\rangle$, but only $\langle C \geq r_1\rangle$ and $\langle D \geq r_2\rangle$ for $r_1, r_2 \in T$ such that $r_1 * r_2 \geq r$. Hence, the tableau rules not only decompose complex concept expressions into simpler ones but also generate a system of constraints. For example, in the case of the Łukasiewicz t-norm, the restriction $r_1 *_Ł r_2 \geq r$ can be encoded using the set of linear constraints $\{y \leq 1 - r, r_1 + r_2 - 1 \geq r - y, y \in \{0,1\}\}$. In this case, the tableau rules are deterministic and only one optimization problem is obtained. Indeed, in the previous example the two possibilities $y = 0$ and $y = 1$ encode the non-deterministic choice.

After the tableau is completely expanded, an optimization problem must be solved before obtaining the final solution. The optimization problem has a solution iff the fuzzy KB is consistent. Note that we are using the reduction from entailment to inconsistency provided by Theorem 6.1.1. In general, the algorithm produces a bounded Mixed Integer Non Linear Programming (bMINLP) optimization problem. In some particular cases, an easier problem is obtained, such as a bounded Mixed Integer Linear Programming (bMILP) problem in Łukasiewicz DLs, or a bounded Mixed Integer Quadratically Constrained Programming (MIQCP) problem in Product DLs.

Now we discuss in detail Algorithm 2 to compute the BED of $C(a)$ with respect to $\mathcal{K}$ in Ł-$\mathcal{ALC}$ with an empty TBox [60]. In order to determine the BED problem of $C(a)$ with respect to $\mathcal{K}$, we consider an expression of the form $\langle(\neg C)(a), 1 - x\rangle$ (informally, $\langle C(a) \leq x\rangle$), where $x$ is a $[0,1]$-valued variable. Then we construct a tableaux for $\mathcal{K}' = \mathcal{K} \cup \{\langle(\neg C)(a), 1 - x\rangle\}$ in which the application of satisfiability preserving rules generates new fuzzy assertion axioms together with *inequations* over $[0,1]$-valued variables. These inequations have to hold in order to respect the semantics of the DL constructors. Hence, we *minimize* the original variable $x$ such that all constraints are satisfied.

Like most of the tableaux algorithms, our algorithm works on *completion-forests* since an ABox might contain several individuals with arbitrary roles connecting them.

A completion-forest $\mathcal{F}$ for $\mathcal{K}$ is a collection of trees whose distinguished roots are arbitrarily connected by edges.

- Each node $v$ is labeled with a set $\mathcal{L}(v)$ of concepts $C$. If $C \in \mathcal{L}(v)$ then we consider a variable $x_{C(v)}$. The intuition is that $v$ is an instance of $C$ to degree equal or greater than the value of the variable $x_{C(v)}$.
- Each edge $\langle v, w\rangle$ is labeled with a set $\mathcal{L}(\langle v, w\rangle)$ of roles $R$ and if $R \in \mathcal{L}(\langle v, w\rangle)$ then we consider a variable $x_{R(v,w)}$ representing the degree of being $\langle v, w\rangle$ and instance of $R$.

The forest has associated a set $\mathcal{C}_{\mathcal{F}}$ of constraints of the form $l \leq l'$, $l = l'$, $x_i \in [0,1], y_i \in \{0,1\}$, where $l, l'$ are linear expressions using the variables occurring in the forest. $\mathcal{F}$ is then expanded by repeatedly applying the tableaux rules in Table 4. We note that $x_1 \to x_2 \geq z$ can be encoded as $1 - x_1 \oplus x_2 \geq z$ and that $x_1 \otimes x_2 \geq z$, $x_1 \otimes x_2 \leq z$ and $x_1 \oplus x_2 \geq z$, $(x_1, x_2, z \in [0,1])$ can be encoded in MILP as follows:

- $x_1 \otimes x_2 \geq z \mapsto \{y \leq 1 - z, x_1 + x_2 - 1 \geq z - y, y \in \{0,1\}\}$, where $y$ is a new variable.
- $x_1 \otimes x_2 \leq z \mapsto \{x_1 + x_2 - 1 \leq z\}$.
- $x_1 \oplus x_2 \geq z \mapsto \{x_1 + x_2 \geq z\}$.

**Algorithm 2** Computation of the BED of $C(a)$ in Ł$\mathcal{ALC}$ without a TBox

**Require:** A concept assertion $C(a)$, a fuzzy KB $\mathcal{K}$
**Ensure:** BED of $C(a)$ with respect to $\mathcal{K}$

1: $\mathcal{K} \leftarrow \mathcal{K} \cup \{\langle \mathsf{nnf}(\neg C)(a), 1 - x \rangle\}$
   {Negation Normal Form}
2: **for** each concept assertion $\langle C(a), r \rangle \in \mathcal{K}$ **do**
3: $\quad \mathcal{K} \leftarrow (\mathcal{K} \setminus \{\langle C(a), r \rangle\}) \cup \{\langle \mathsf{nnf}(C)(a), r \rangle\}$
4: **end for**
   {Initialization of the forest}
5: create an empty forest $\mathcal{F}$
6: $\mathcal{C}_{\mathcal{F}} \leftarrow \emptyset$
7: **for** each concept assertion $\langle C(a), r \rangle \in \mathcal{K}$ **do**
8: $\quad$ create a node $v_a$ in $\mathcal{F}$ if it does not exist
9: $\quad \mathcal{L}(v_a) \leftarrow \mathcal{L}(v_a) \cup \{C\}$
10: $\quad \mathcal{C}_{\mathcal{F}} \leftarrow \mathcal{C}_{\mathcal{F}} \cup \{x_{C(v_a)} \geq r\}$
11: **end for**
12: **for** each role assertion $\langle R(a,b), r \rangle \in \mathcal{K}$ **do**
13: $\quad$ create nodes $v_a, v_b$ in $\mathcal{F}$ if they do not exist
14: $\quad$ create an edge $\langle v_a, v_b \rangle$ in $\mathcal{F}$ if it does not exist
15: $\quad \mathcal{L}(\langle v_a, v_b \rangle) \leftarrow \mathcal{L}(\langle v_a, v_b \rangle) \cup \{R\}$
16: $\quad \mathcal{C}_{\mathcal{F}} \leftarrow \mathcal{C}_{\mathcal{F}} \cup \{x_{R(v_a, v_b)} \geq r\}$
17: **end for**
   {Tableaux rules}
18: **while** some tableaux rule is applicable to some node in $\mathcal{F}$ **do**
19: $\quad$ apply one of the rules in Table 4 to a node $v$
20: $\quad$ mark the applied rule as not applicable to node $v$
21: **end while**
   {Solve the optimization problem}
22: **for** each variable $x$ of the forms $x_{C(v)}$ or $x_{R(v,w)} \in \mathcal{K}$ **do**
23: $\quad \mathcal{C}_{\mathcal{F}} = \mathcal{C}_{\mathcal{F}} \cup \{x \in [0,1]\}$
24: **end for**
25: **if** $\mathcal{C}_{\mathcal{F}}$ has a solution **then**
26: $\quad$ solve the optimization problem
27: $\quad$ **return** x
28: **else**
29: $\quad$ **return** 1 {$\mathcal{K}$ is inconsistent}
30: **end if**

Taking into account that the residuated negation in Łukasiewicz is involutive, we can use the recursive definition of $\mathsf{nnf}$ to transform any concept into Negation Normal Form, where the negation only appears before an atomic concept (see pag. 1140).

Another important observation is that since there is not a TBox, there is a model of $\mathcal{K}$ iff there is a witnessed model of $\mathcal{K}$ [80].

| | preconditions | actions |
|---|---|---|
| $(\bot)$ | $\bot \in \mathcal{L}(v)$ | $\mathcal{C}_{\mathcal{F}} = \mathcal{C}_{\mathcal{F}} \cup \{x_{v:\bot} = 0\}$ |
| $(\top)$ | $\top \in \mathcal{L}(v)$ | $\mathcal{C}_{\mathcal{F}} = \mathcal{C}_{\mathcal{F}} \cup \{x_{v:\top} = 1\}$ |
| $(\neg)$ | $\neg A \in \mathcal{L}(v)$ | $\mathcal{C}_{\mathcal{F}} = \mathcal{C}_{\mathcal{F}} \cup \{x_{v:\neg C} = 1 - x_{v:C}\}$ |
| $(\sqcap)$ | $C \sqcap D \in \mathcal{L}(v)$ | $\mathcal{L}(v) = \mathcal{L}(v) \cup \{C, D\}$ |
| | | $\mathcal{C}_{\mathcal{F}} = \mathcal{C}_{\mathcal{F}} \cup \{x_{v:C} \otimes x_{v:D} \geq x_{v:C \sqcap D}\}$ |
| $(\sqcup)$ | $C \sqcup D \in \mathcal{L}(v)$ | $\mathcal{L}(v) = \mathcal{L}(v) \cup \{C, D\}$ |
| | | $\mathcal{C}_{\mathcal{F}} = \mathcal{C}_{\mathcal{F}} \cup \{x_{v:C} \oplus x_{v:D} \geq x_{v:C \sqcup D}\}$ |
| $(\exists)$ | $\exists R.C \in \mathcal{L}(v)$ | create a new node $w$ |
| | | $\mathcal{L}(\langle v, w\rangle) = \mathcal{L}(\langle v, w\rangle) \cup \{R\}$, and |
| | | $\mathcal{L}(w) = \mathcal{L}(w) \cup \{C\}$, and |
| | | $\mathcal{C}_{\mathcal{F}} = \mathcal{C}_{\mathcal{F}} \cup \{x_{(v,w):R} \otimes x_{w:C} \geq x_{v:\exists R.C}\}$ |
| $(\forall)$ | $\forall R.C \in \mathcal{L}(v)$ | $\mathcal{L}(w) = \mathcal{L}(w) \cup \{C\}$ |
| | $R \in \mathcal{L}(\langle v, w\rangle$ | $\mathcal{C}_{\mathcal{F}} = \mathcal{C}_{\mathcal{F}} \cup \{x_{w:C} \geq x_{v:\forall R.C} \otimes x_{(v,w):R}\}$ |

Table 4. Rules of the tableaux algorithm combined with an optimization problem

EXAMPLE 6.2.2. Consider the fuzzy KB

$$\{\langle(\forall R.C)(a), 0.7\rangle, \langle R(a,b), 0.6\rangle, \langle\neg C(b), 0.8\rangle\}.$$

The completion graph after the initialization and the application of the rules $(\neg)$ and $(\forall)$ can be depicted as:

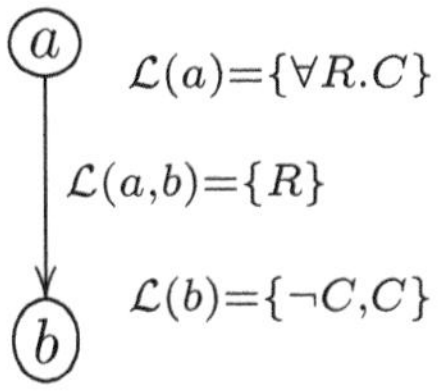

$$a \quad \mathcal{L}(a)=\{\forall R.C\}$$
$$\mathcal{L}(a,b)=\{R\}$$
$$b \quad \mathcal{L}(b)=\{\neg C, C\}$$

where $\mathcal{C}_{\mathcal{F}} = \{x_{(\forall R.C)(a)} \in [0,1], x_{R(a,b)} \in [0,1], x_{(\neg C)(b)} \in [0,1], x_{C(b)} \in [0,1], x_{(\forall R.C)(a)} \geq 0.7, x_{R(a,b)} \geq 0.6, x_{(\neg C)(b)} \geq 0.8, x_{(\neg C)(b)} = 1 - x_{C(b)}, x_{C(b)} \geq x_{(\forall R.C)(a)} \otimes x_{R(a,b)}\}$.

It can be seen that this system of constraints has no solution since

$$x_{C(b)} \geq x_{(\forall R.C)(a)} \otimes x_{R(a,b)} \geq 0.7 \otimes 0.6 = 0.3 \text{ and}$$

$$x_{C(b)} = 1 - x_{(\neg C)(b)} \leq 1 - 0.8 = 0.2.$$

Hence, the KB is inconsistent.

Although the complexity of these algorithms has not been deeply studied, bMINLP and bMILP problems are NP-HARD problems, so these families have a high worst-case computational complexity.

### 6.2.3 Automata-based algorithms

Another important approach for reasoning in classical DLs is to reduce the reasoning problem to decide whether the language accepted by an automaton is empty or not. An advantage of this approach is that automata can handle non-determinism and infinite structures in a simple and elegant way. Thus, automata-based methods do not need complex blocking conditions or backtracking to ensure that a model exists. The reduction to automata has been successfully used to prove tight complexity bounds for a large family of classical DLs [1, 5]. Recently, it has been also explored for reasoning in FDLs. The main idea is to exploit the fact that many of these logics enjoy the *tree-model property*: a KB has a model iff it has a well-structured tree-shaped model. In order to decide $*$-consistency, one only needs to decide the existence of one such model.

The typical approach is to abstract even further and construct an automaton whose runs describe so-called *Hintikka trees*. Hintikka trees can be seen as an abstract representation of the tree-shaped models, in which the membership degree of all relevant concepts is explicitly stated. More precisely, a Hintikka tree is a labeled infinite tree in which every node is labeled with a partial function, called the Hintikka function, from the set of all concepts appearing in the KB to the set of membership degrees, and every edge is labeled with a membership degree. The function at each node needs to be locally consistent; that is, the membership degree of e.g. the concept $C \sqcap D$ needs to be consistent with the membership degrees of $C$ and $D$ at each node of the tree. Likewise, it must be consistent with every GCI in the KB. This ensures that the semantics are locally preserved at every node. To handle existential and value restrictions, the functions associated at each node and its successors are also restricted in an appropriate manner; this restriction is called the Hintikka condition. Each such Hintikka tree represents a set of infinite models, that have as domain all the nodes in the tree, and the concept names and role names are interpreted according to their respective labels in the tree.

Since all the conditions are local, it is easy to build an automaton that receives as input an unlabeled infinite tree, and labels it through its runs in a way that successful runs correspond exactly to the set of Hintikka trees: the states are the locally consistent Hintikka functions, and the transitions are those tuples satisfying the Hintikka condition. Thus, the input KB is consistent iff this automaton has a successful run. The latter problem is known to be decidable in polynomial time on the size of the automaton.

Automata-based algorithms to decide witnessed $*$-consistency for the logics between fuzzy $\mathcal{ALC}$ and $\mathcal{SHI}$ with GCIs for any finite residuated De Morgan lattice with a t-norm $*$ are summarized in [42]. Using more advanced automata-based techniques, it can be shown that the complexity in these logics is preserved.

Using an automata-based method to decide witnessed G-consistency for $\mathfrak{I}\mathcal{ALCE}$ in exponential time has been discussed in [35]. In this case, a model cannot be built directly, since that would require an automaton with infinitely many states. Instead, a novel idea is used, based on the defining properties of the minimum t-norm. The main insight is that the operators of this logic are invariant w.r.t. the specific membership degrees used, but depend only on the order between them. Thus, in this case it is possible to use preorders between the concepts, rather than explicit membership degrees, to label the nodes of the Hintikka tree. Each such Hintikka tree potentially represents uncountably many models, namely all those that share the exact same ordering among the relevant concepts.

While automata-based methods are useful for proving tight complexity bounds, they are not used in practical applications. The reason for this is that their worst-case behavior, used for providing complexity upper bounds, matches its best-case behavior. Before any reasoning step can be done, an automaton that is typically of exponential size needs to be constructed. Thus, even for simple input KBs, an exponential runtime cannot be avoided.

EXAMPLE 6.2.3. Consider the fuzzy KB

$$\{\langle \forall R.C(a), 0.8\rangle, \langle \top \sqsubseteq C(b), 0.7\rangle\}.$$

Three possible Hintikka functions are

$$\begin{aligned} H_0 &:= \forall r.C \mapsto 0.8,\ C \mapsto 1,\ \rho \mapsto 1, \\ H_1 &:= C \mapsto 0.7,\ \rho \mapsto 0.9, \\ H_2 &:= \rho \mapsto 0. \end{aligned}$$

Hintikka trees for this KB have arity 1, since only one quantified concept exists. One possible such tree can be depicted as:

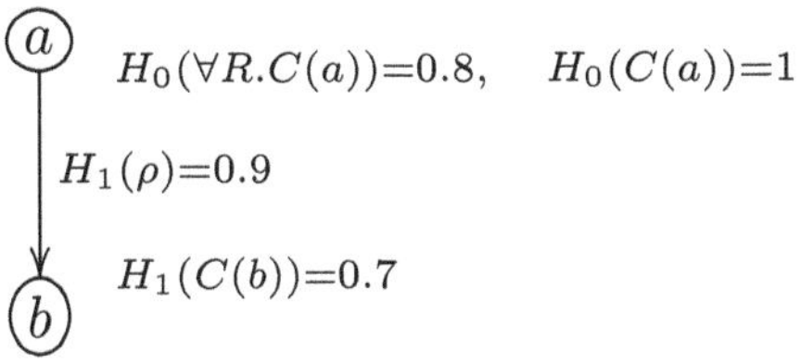

$(a)$ $H_0(\forall R.C(a))=0.8$, $H_0(C(a))=1$

$H_1(\rho)=0.9$

$H_1(C(b))=0.7$ $(b)$

Recall that Hintikka trees are infinite by definition; in this case, the node $b$ has a successor labeled with the Hintikka function $H_2$, which means that it is a successor to degree 0. Thus, it is not depicted in the figure. The Hintikka automaton for this very small KB contains over 16 states.

### 6.2.4 Reduction to classical DLs

The idea of this family of algorithms is to reduce fuzzy reasoning to finitely many crisp DL reasoning tasks. If the number of degrees is finite, it is possible to simulate fuzzy concepts and fuzzy roles using several $\alpha$-cuts of the fuzzy sets and the fuzzy properties. The $\alpha$-cut of a fuzzy set $X$ is the crisp set that contains all the elements which belong to $X$ with a degree greater than or equal to $\alpha$. In this way, every axiom in the fuzzy ontology can be represented using these $\alpha$-cuts. It can be shown that a fuzzy KB is $*$-consistent if and only if its equivalent crisp KB is; other reasoning tasks can be reduced to $*$- consistency.

In general, the translation of the equivalent crisp KBs grows exponentially with the size of the fuzzy KB. This approach is theoretically interesting to prove the decidability of FDLs. From an implementation point of view, it makes it possible to reuse existing crisp reasoners but since the translation to the crisp case is exponential it does not provide a tight complexity bound.

**Algorithm 3** Reduction of a finitely-valued fuzzy $\mathcal{ALCH}$ KB into a crisp KB

**Require:** A fuzzy KB $\mathcal{K}$
**Ensure:** true if $\mathcal{K}$ is $*$-consistent; false otherwise

1: crisp($\mathcal{K}$) $\leftarrow \emptyset$
{Axioms to keep the semantics of the new crisp concepts and roles}
2: **for** $i := 1$ to $n-1$ **do**
3: **for** each $A \in N_A$ **do**
4: crisp($\mathcal{K}$) $\leftarrow$ crisp($\mathcal{K}$) $\cup \{A_{\geq t_{i+1}} \sqsubseteq A_{\geq t_i}\}$
5: **end for**
6: **for** each $R \in N_R$ **do**
7: crisp($\mathcal{K}$) $\leftarrow$ crisp($\mathcal{K}$) $\cup \{R_{\geq t_{i+1}} \sqsubseteq R_{\geq t_i}\}$
8: **end for**
9: **end for**
{Crisp reduction of the axioms in the fuzzy KB}
10: **for** each axiom of the form $\langle \tau, r \rangle \in \mathcal{K}$ **do**
11: crisp($\mathcal{K}$) $\leftarrow$ crisp($\mathcal{K}$) $\cup\, \kappa(\langle \tau, r \rangle)$
12: **end for**
13: **if** crisp($\mathcal{K}$) is consistent **then**
14: **return** true
15: **else**
16: **return** false
17: **end if**

To illustrate this family of algorithms, Algorithm 3 shows the process of reducing a $\mathcal{ALCH}$ KB in finite Gödel extended with an involutive negation to an equivalent crisp KB. It is an adaptation of the more general algorithm proposed in [18].

It is convenient to recall that $N_A$ is the set of atomic concepts and $N_R$ the set of atomic roles in a fuzzy knowledge base $\mathcal{K}$, that we assume a finite chain of degrees of truth $T = \{t_0, t_1, \ldots, t_n\}$ with $t_0 = 0$ and $t_n = 1$, and that axioms of the form $\langle \tau, t_0 \rangle$ are not allowed as they are tautologies. We will use $T^+$ to denote $T \setminus \{t_0\}$.

First, some new crisp concepts and roles representing $\alpha$-cuts of the fuzzy concepts and roles are introduced. Having assumed a finite number of degrees of truth makes it possible to add $\alpha$-cuts for every possible degree. The semantics of these newly introduced atomic concepts and roles is preserved by some new axioms (Lines 2–9). The idea is that an $\alpha$-cut is subsumed by a $\beta$-cut if $\beta \leq \alpha$. Then, a mapping $\kappa$ represents every axiom in the fuzzy KB using a crisp axiom that relies on these new crisp elements. Formally, the definition of $\kappa$ is the following:

$$\begin{aligned}
\kappa(\langle C(a), r \rangle) &= \{\rho(C \geq r)(a)\} \\
\kappa(\langle R(a,b), r \rangle) &= \{\rho(R \geq r)(a,b)\} \\
\kappa(\langle C \sqsubseteq D, r \rangle) &= \textstyle\bigcup_{t_j \in T^+, t_j \leq r} \{\rho(C \geq t_j) \sqsubseteq \rho(D \geq t_j)\} \\
\kappa(\langle R \sqsubseteq S, r \rangle) &= \textstyle\bigcup_{t_j \in T^+, t_j \leq r} \{\rho(R \geq t_j) \sqsubseteq \rho(S \geq t_j)\}.
\end{aligned}$$

$$
\begin{aligned}
\rho(\top \geq t_i) &= \top \\
\rho(\top \leq t_i) &= \bot \\
\rho(\bot \geq t_i) &= \bot \\
\rho(\bot \leq t_i) &= \top \\
\rho(A \geq t_i) &= A_{\geq t_i} \\
\rho(A \leq t_i) &= \neg A_{\geq t_{i+1}} \\
\rho(\neg C \geq t_i) &= \rho(C \leq t_{n-i}) \\
\rho(\neg C \leq t_i) &= \rho(C \geq t_{n-i}) \\
\rho(\sim C \geq t_i) &= \rho(C \leq t_0) \\
\rho(\sim C \leq t_i) &= \rho(C \geq t_1) \\
\rho(C \sqcap D \geq t_i) &= \rho(C \geq t_i) \sqcap \rho(D \geq t_i) \\
\rho(C \sqcap D \leq t_i) &= \rho(C \leq t_i) \sqcup \rho(D \leq t_i) \\
\rho(C \sqcup D \geq t_i) &= \rho(C \geq t_i) \sqcup \rho(D \geq t_i) \\
\rho(C \sqcup D \leq t_i) &= \rho(C \leq t_i) \sqcap \rho(D \leq t_i) \\
\rho(C \rightarrow D \geq t_i) &= \sqcap_{t_j \in T^+, t_j \leq t_i} \left(\neg\rho(C \geq t_j) \sqcup \rho(D \geq t_j)\right) \\
\rho(C \rightarrow D \leq t_i) &= \sqcup_{t_j \in T, t_j \leq t_i} \left(\rho(C \geq t_{j+1}) \sqcap \rho(D \leq t_j)\right) \\
\rho(\forall R.C \geq t_i) &= \sqcap_{t_j \in T^+, t_j \leq t_i} \left(\forall\rho(R \geq t_j).\rho(C \geq t_j)\right) \\
\rho(\forall R.C \leq t_i) &= \sqcup_{t_j \in T, t_j \leq t_i} \left(\exists\rho(R \geq t_{j+1}).\rho(C \leq t_j)\right) \\
\rho(\exists R.C \geq t_i) &= \exists\rho(R \geq t_i).\rho(C \geq t_i) \\
\rho(\exists R.C \leq t_i) &= \forall\rho(R \geq t_{i+1}).\rho(C \leq t_i) \\
\rho(R \geq t_i) &= R_{\geq t_i}
\end{aligned}
$$

Table 5. Mapping of concept and role expressions

$\kappa$ needs an auxiliary mapping $\rho$. Given a fuzzy concept $C$, $\rho(C \geq t_i)$ is a crisp set containing all the elements which belong to $C$ with a degree greater than or equal to $t_i$. $\rho(R \geq t_i)$ is defined in a similar way for fuzzy roles, and $\rho(C \leq t_i)$ constains individuals belonging to $C$ with a degree less or equal than $t_i$. Formally, the definition of $\rho$ can be found in Table 5.

EXAMPLE 6.2.4. Assume a set of degrees of truth $\{0, 1/3, 2/3, 1\}$. In Gödel fuzzy logic, the fuzzy knowledge base $\mathcal{F} = \{\langle C(a), 1/3\rangle, \langle C(b), 2/3\rangle\}$ is equivalent to the crisp knowledge base

$$\mathcal{K} = \{C_{1/3}(a), C_{2/3}(b), C_{2/3} \sqsubseteq C_{1/3}, C_1 \sqsubseteq C_{2/3}\},$$

where $C_{1/3}, C_{2/3}, C_1$ are crisp concepts.

**Algorithm 4** Reduction into a crisp KB for t-norms without zero divisors

**Require:** A fuzzy KB $\mathcal{K}$
**Ensure:** true if $\mathcal{K}$ is $*$-consistent; false otherwise

1: $\text{crisp}(\mathcal{K}) \leftarrow \emptyset$
2: **for** each axiom of the form $\langle \tau, r \rangle \in \mathcal{K}$ **do**
3: $\quad \text{crisp}(\mathcal{K}) \leftarrow \text{crisp}(\mathcal{K}) \cup \{\tau\}$
4: **end for**
5: **if** $\text{crisp}(\mathcal{K})$ is consistent **then**
6: $\quad$ **return** true
7: **else**
8: $\quad$ **return** false
9: **end if**

Another alternative is to ignore fuzziness by just discarding the degree of truth in the fuzzy axioms (the details of this process are provided in Lemma 7.3.1). In some cases, it can be shown that KB $*$-consistency is trivially reducible to crisp reasoning. Hence, KB $*$-consistency and crisp consistency have the same computational complexity. It is not possible if the language contains an involutive negation or has upper bounds in the axioms, i.e., axioms of the form $\langle \tau \leq r \rangle$. If this is not the case and we consider a strict t-norm $*$ (that is, if $*$ is an ordinal sum that does not start with a Łukasiewicz component), a fuzzy KB in $\mathcal{SHOI}$ is $*$-consistent if and only if it has a crisp model [34]. Algorithm 4 illustrates this result. The same idea can be applied to the concept $*$-satisfiability but not to $*$-subsumption. In the less expressive FDL $\mathcal{EL}$, the same result does hold for $*$-subsumption [43].

EXAMPLE 6.2.5. In Gödel fuzzy logic, the fuzzy knowledge base

$$\mathcal{F} = \{\langle C(a), 1/3 \rangle, \langle C(b), 2/3 \rangle\}$$

is equivalent to the crisp base $\mathcal{K} = \{C(a), C(b)\}$, where $C$ is a crisp concept.

### 6.2.5 Reduction to propositional fuzzy logics

P. Hájek proposed a reduction of FDL concepts into propositional fuzzy formulas [80]. In particular, he considered the FDL $\mathcal{IALCUE}$ under arbitrary continuous t-norms and defined a reasoning algorithm for witnessed $*$- satisfiability and $*$-validity of concepts based on a reduction to BL. Please note that the involutive negation is not part the language when the semantics is not based on Łukasiewicz t-norm. The strong disjunction was also not considered in the original formulation, but, since the algorithm presented in [80] is based on a reduction to the respective propositional calculus, it can be easily expanded in order to cope with the strong disjunction too. The same author extended the results by introducing rational truth degrees, and thus making it possible to have ABoxes [81].

This approach also has some limitations: the translation is more than exponential in the size of the fuzzy concept (see [54] for a detailed proof), it only works for witnessed concept satisfiability, and it does not consider TBoxes in the language.

Let us now describe in detail Hájek's algorithm [80]. To decide the satisfiability of a concept $C$, we start with an instance of the form $C(a)$ for a new individual $a$. Next, a finite *witnessing theory* $T(C(a))$ is computed from $C(a)$. $C(a)$ can be seen as a propositional combination of instances of non-quantified concepts and instances of quantified concepts. The idea is that for every instance of a quantified concept, a new witness individual and some instances keeping the semantics of the quantified concept are created. The new instances are quantifier-free formulas. This is similar to what tableaux algorithms do. However, instead of looking for a clash-free completion forest, all the instances are reduced to a propositional fuzzy logic formula.

Firstly, we will show how to compute a finite theory $T(C(a))$ from an instance $C(a)$. Then, we will show how to compute a propositional formula $\text{prop}(C(a))$ for every instance $C(a)$. $T(C(a))$ can be computed as shown in Algorithm 5. It uses an iterative process that is repeated $n$ times, where $n$ is the degree of nesting of quantifiers in $C$, denoted $\text{nest}(C)$ and defined as follows:

$$\begin{aligned} \text{nest}(A) &= 0 \\ \text{nest}(\sim C) &= \text{nest}(C) \\ \text{nest}(C \sqcap D) = \text{nest}(C \vee D) = \text{nest}(C \wedge D) &= \max\{\text{nest}(C), \text{nest}(D)\} \\ \text{nest}(\forall R.C) = \text{nest}(\exists R.C) &= 1 + \text{nest}(C). \end{aligned}$$

Every step $i$ creates a set of new individuals (denoted $I_i$), adds new axioms to the theory (denoted $T(C(a))$), and introduces new instances to be processed by the next step (denoted $S_i$). The $i$-th step considers the set of instances $S_i$ introduced in the previous step and produces a new set with a degree of nesting $n - i$. The algorithm starts with the initial instance $C(a)$. Instances $\alpha$ of the forms $(\exists R.C)(b)$ and $(\forall R.C)(b)$ can generate new individuals $d_\alpha$ which are witnesses of the existential and universal restrictions. All the individuals created in the $i$-th step are collected in a set $I_i$ and will be revisited to add some new axioms, keeping the semantics of existential and universal restrictions. The axiom $x \equiv y$ is used as a shorthand for the pair of formulas $x \to y$ and $y \to x$.

Now, it remains to show how to compute $\text{prop}(C(a))$. These formulas are defined by induction, where the base case are atomic concepts, existential quantifications and value restrictions, which are assigned a propositional variable $p_{C,a}$. In particular, $\text{prop}(C(a))$ is defined as follows:

$$\begin{aligned} \text{prop}(A(a)) &= p_{A,a} \\ \text{prop}(\exists R.C(a)) &= p_{\exists R.C,a} \\ \text{prop}(\forall R.C(a)) &= p_{\forall R.C,a} \\ \text{prop}((\bot)(a)) &= \bar{0} \\ \text{prop}(C \sqcap D(a)) &= \text{prop}(C(a)) \;\&\; \text{prop}(D(a)) \\ \text{prop}(C \wedge D(a)) &= \text{prop}(C)(a) \;\&\; (\text{prop}(C(a)) \to \text{prop}(D(a))) \\ \text{prop}(C \vee D(a)) &= \text{prop}((C \to D) \to D(a)) \;\&\; \\ &\quad (\text{prop}((C \to D) \to D(a)) \to \text{prop}((D \to C) \to C(a))) \\ \text{prop}((C \to D(a)) &= \text{prop}(C(a)) \to \text{prop}(D(a)) \\ \text{prop}(\neg C(a)) &= \text{prop}(C(a)) \to \bar{0}. \end{aligned}$$

**Algorithm 5** Hájek's reduction to propositional fuzzy logics

**Require:** $C$
**Ensure:** true if $C$ is $*$-satisfiable; false otherwise

```
1: a ← some constant
2: S_0 ← {C(a)}
   {Compute T(C(a))}
3: for i := 1 to nest(C) do
4:    I_i ← ∅
5:    T(C(a)) ← ∅
6:    S_i ← ∅
      {Create witnesses and add witnessing axioms}
7:    for each α ∈ S_{i−1} do
8:       if α of the form ∀R.D(b) then
9:          create a new individual d_α
10:         I_i ← I_i ∪ {d_α}
11:         T(C(a)) ← T(C(a)) ∪ {∀R.D(b) ≡ (R(b, d_α) → D(d_α))}
12:         S_i ← S_i ∪ {D(d_α)}
13:      else if α is of the form ∃R.D(b) then
14:         create a new individual d_α
15:         I_i ← I_i ∪ {d_α}
16:         T(C(a)) ← T(C(a)) ∪ {∃R.D(b) ≡ (R(b, d_α) & D(d_α))}
17:         S_i ← S_i ∪ {D(d_α)}
18:      end if
19:   end for
      {Apply restrictions to each individual generated in this step using the same role}
20:   for each α ∈ S_{i−1} of the form (∀R.D)(b) and each d_β ∈ I_i with α ≠ β do
21:      if β has the form ∃R.X(b) or ∀R.X(b) and α ≠ β then
22:         T(C(a)) ← T(C(a)) ∪ {∀R.D(b) → (R(b, d_β) → D(d_β))}
23:         S_i ← S_i ∪ {C(d_β)}
24:      end if
25:   end for
26:   for each α ∈ S_{i−1} of the form ∃R.D(b) and each d_β ∈ I_i with α ≠ β do
27:      if β has the form ∃R.X(b) or ∀R.X(b) and α ≠ β then
28:         T(C(a)) ← T(C(a)) ∪ {(R(b, d_β) & D(d_β)) → ∃R.D(b)}
29:         S_i ← S_i ∪ {C(d_β)}
30:      end if
31:   end for
32: end for
33: return T(C(a))
    {Solve the propositionally equivalent formula}
34: if prop(C(a)) ∪ prop(T(C(a))) is ∗- satisfiable then
35:    return true
36: else
37:    return false
38: end if
```

EXAMPLE 6.2.6. Consider the concept

$$\forall R.C \sqcap \exists R.\neg C.$$

According to Algorithm 5, the list of propositional formulas produced from the above KB at the end of the process is the following:

1. $p_{\forall R.C} \,\&\, p_{\exists R.\neg C}$
2. $p_{\forall R.C(d)} \equiv p_{R(d,d_{\forall R.C(b)})} \rightarrow p_{C(d_{\forall R.C(b)})}$
3. $p_{\exists R.\neg C(d)} \equiv p_{R(d,d_{\exists R.\neg C(b)})} \,\&\, \neg p_{C(d_{\exists R.\neg C(b)})}$
4. $p_{\forall R.C(d)} \rightarrow (p_{R(d,d_{\exists R.\neg C(b)})} \rightarrow p_{C(d_{\exists R.\neg C(b)})})$
5. $(p_{R(d,d_{\forall R.C(b)})} \,\&\, \neg p_{C(d_{\forall R.C(b)})}) \rightarrow p_{\exists R.\neg C(d)}$.

We show that the propositional theory above is not 1-satisfiable; suppose that there exists a propositional evaluation $e$ such that $e(p_{\forall R.C} \,\&\, p_{\exists R.\neg C}) = 1$, then

- by 1. both $e(p_{\forall R.C}) = 1$ and $e(p_{\exists R.\neg C}) = 1$, hence
- by 3. both $e(p_{R(d,d_{\exists R.\neg C(b)})}) = 1$ and $e(p_{C(d_{\exists R.\neg C(b)})}) = 0$, therefore
- by 4. $e(p_{\forall R.C(d)}) = 0$, a contradiction.

### 6.2.6 Structural subsumption algorithms

A structural subsumption algorithm is a relatively simple algorithm suited for solving subsumption with respect to empty knowledge bases in very basic FDLs. Both in the case of classical DLs and of FDLs it has been historically the first algorithm to be designed (see Section 8 for more historical details). Its problem is that it turns out to be incomplete for more expressive languages.

Let us now describe in detail the structural subsumption algorithm for subsumption with respect to empty knowledge bases in language $\mathcal{FL}^-$, from [61]. To decide if a concept $C$ is subsumed by a concept $D$, we recursively build a set of matrices $\mathbf{E}_{C,D}$. These matrices contain a row for each value restriction $\forall R.F$ which is a conjunct of $D$ and a column for each value restriction $\forall R.E$ which is a conjunct of $C$. In particular, if there are multiple occurrences of the same expression then there are multiple rows for them. The same observation applies for the columns in $\mathbf{E}_{C,D}$. The only non-deterministic problem is to decide whether every value restriction $\forall R.F$ which is a conjunct of a given subconcept $D'$ of $D$ is subsumed by a different value restriction $\forall R.E$ which is a conjunct of a given subconcept $C'$ of $C$. But, as in [108], instead of checking out this fact for different non-deterministic guesses, a suitable procedure for the bipartite matching problem (see [86] for example) can give an answer in polynomial time.

**Algorithm 6** Computing structural subsumption for $\text{Ł}_n$-$\mathcal{FL}^-$

**Require:** $C, D$

**Ensure:** true if $C$ is 1-subsumed by $D$; false otherwise

  **if** there is an occurrence of an atomic or existential conjunct $A$ of $D$ that is not in $C$ where concept $A$ appears in $C$ strictly less $n-1$ times **then**
    **return** false
  **else**
    $\mathbf{E}_{C,D} := \emptyset$
    **for** value restriction $\forall R.F$ which is a conjunct of $D$ **do**
      **for** value restriction $\forall R.E$ which is a conjunct of $C$ **do**
        $\mathbf{E}_{C,D}(\forall R.F, \forall R.E) := \text{Ł}_n\text{-SUBS}(1, F, E)$
      **end for**
    **end for**
    **if** there is a maximal bipartite matching for $\mathbf{E}_{C,D}$ **then**
      **return** true
    **else**
      **return** false
    **end if**
  **end if**

EXAMPLE 6.2.7. Consider the concepts

$$\forall R.(\forall P.(A \sqcap C) \sqcap \forall P.C) \sqcap \forall R.(C \sqcap D) \quad \text{and} \quad \forall R.(\forall P.A \sqcap \forall P.B) \sqcap \forall R.C.$$

According to Algorithm 6, if we want to see whether the first concept is subsumed by the second, the following matrices will be created:

| | $\forall P.A$ | $\forall P.B$ |
|---|---|---|
| $\forall P.(A \sqcap C)$ | $\times$ | |
| $\forall P.C$ | | |

| | $\forall R.(\forall P.A \sqcap \forall P.B)$ | $\forall R.C$ |
|---|---|---|
| $\forall R.(\forall P.(A \sqcap C) \sqcap \forall P.C)$ | | |
| $\forall R.(C \sqcap D)$ | | $\times$ |

It can be seen that in the first matrix the subconcept $\forall P.C$ is not subsumed by any subconcept of $\forall R.(\forall P.A \sqcap \forall P.B) \sqcap \forall R.C$. Therefore, recursively, the concept $\forall R.(\forall P.(A \sqcap C) \sqcap \forall P.C) \sqcap \forall R.(C \sqcap D)$ is not subsumed by $\forall R.(\forall P.A \sqcap \forall P.B) \sqcap \forall R.C$.

# 7 Decidability and complexity issues

In this section, we present the known results on the decidability and complexity of reasoning in FDLs. We first prove that KB consistency is undecidable for the DL $\mathfrak{I}\mathcal{ALCE}$ if the semantics is based on the product t-norm. The proof presented is intended only as a prototype for a general method of proving undecidability that has been applied to a large class of description logic languages. These further results are summarized next, with a general idea on how the framework can be applied to obtain them. Afterwards, we provide an overview of the complexity of reasoning in the known decidable cases.

## 7.1 Undecidability of consistency in $\mathfrak{I}\mathcal{ALCE}$

The consistency problem has been recently shown to be undecidable for several FDLs, where general concept inclusion axioms are allowed [36] (see Section 8 for more historical details). While these proofs of undecidability differ in their details, they are all based on the same general idea of modeling instances of the *Post correspondence problem* through structured tree-shaped models of a KB. In this section, we provide a prototypical proof of undecidability by showing that $\sqcap$-consistency of $\mathfrak{I}\mathcal{ALCE}$ knowledge bases (w.r.t. witnessed models) is undecidable. This proof is intended to highlight the core steps followed by most other undecidability proofs for FDLs, and can thus be used as a basic model for understanding these other proofs.

For this whole section, we fix the semantics of $\mathfrak{I}\mathcal{ALCE}$ to consider only the product t-norm and its associated operators plus the standard involutive negation for interpreting the different constructors. This is an FDL language defined on the first-order version of the logic $\text{Ł}\Pi\frac{1}{2}$ (see Section 5.4 of Chapter VIII of this Handbook). The proof of undecidability is based on a reduction from a slight variant of the Post correspondence problem, which is well-known to be undecidable [106].

DEFINITION 7.1.1. *Let* $\langle v_1, w_1\rangle, \dots, \langle v_m, w_m\rangle$*be a finite list of pairs of words over a finite alphabet* $\Sigma = \{1, \dots, s\}, s > 1$. *The* Post correspondence problem *(PCP) consists of deciding whether there is a sequence* $i_1, i_2, \dots, i_k$, $1 \leq i_j \leq m$, *such that* $v_1 v_{i_1} v_{i_2} \cdots v_{i_k} = w_1 w_{i_1} w_{i_2} \cdots w_{i_k}$. *If such a sequence exists, the word* $i_1 i_2 \cdots i_k$ *is called a* solution *of the problem.*

Given a word $\mu = i_1 i_2 \cdots i_k \in \{1, \dots, m\}^*$, we will use $v_\mu$ and $w_\mu$ to denote the words $v_1 v_{i_1} v_{i_2} \cdots v_{i_k}$ and $w_1 w_{i_1} w_{i_2} \cdots w_{i_k}$, respectively.

Intuitively, we can see every instance of the PCP as an $m$-ary infinite tree such that (i) the root node is labeled with $\langle v_\varepsilon, w_\varepsilon\rangle := \langle v_1, w_1\rangle$, and (ii) if a node is labeled with the pair $\langle v, w\rangle$, then its $i$-th successor is labeled with the pair $\langle vv_i, ww_i\rangle$ (see Figure 4).[8] The PCP then consists in deciding whether there is a node in this tree whose label $\langle v, w\rangle$ is such that $v = w$.

Recall from Definition 7.1.1 that the alphabet $\Sigma$ over which an instance of the PCP is described consists of the first $s$ positive integers. We can thus view every word in $\Sigma^*$ as a natural number represented in base $s + 1$ in which 0 never occurs. Slightly abusing this intuition, we will express the empty word as the number 0. To reduce the PCP, we need to be able to express words not as integers, but rather as real numbers in the interval $[0, 1]$. To achieve this, we encode each word $u$ in $\Sigma^*$ through the number $2^{-u} \in [0, 1]$.

The main idea of the reduction is to construct a knowledge base $\mathcal{K}_\mathcal{P} = \langle \mathcal{T}_\mathcal{P}, \mathcal{A}_\mathcal{P}\rangle$ whose models encode an instance of the PCP, viewed as a tree as in Figure 4. In other words, every model contains, for each possible solution $\mu \in \{1, \dots, m\}^*$, a domain element at which the interpretation of two designated concept names $A$ and $B$ will correspond to the words $v_\mu, w_\mu$, respectively. More formally, for a given instance $\mathcal{P} = \langle\langle v_1, w_1\rangle, \dots, \langle v_m, w_m\rangle\rangle$ of the PCP, we define an ABox $\mathcal{A}_\mathcal{P}$ and a TBox $\mathcal{T}_\mathcal{P}$ such that for every witnessed model $\mathcal{I}$ of $\mathcal{A}_\mathcal{P}$ and $\mathcal{T}_\mathcal{P}$ and every $\mu \in \{1, \dots, m\}^*$

[8] Notice that the first word concatenated is always $v_1$ and $w_1$.

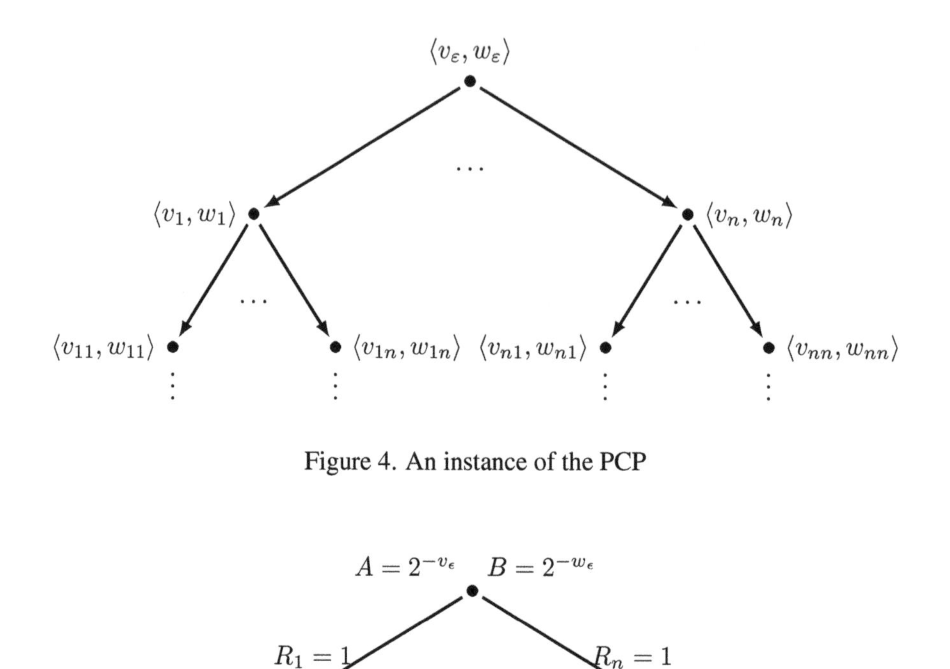


Figure 4. An instance of the PCP

Figure 5. The interpretation $\mathcal{I}_\mathcal{P}$

there is $\delta_\mu \in \Delta^\mathcal{I}$ with $A^\mathcal{I}(\delta_\mu) = 2^{-v_\mu}$ and $B^\mathcal{I}(\delta_\mu) = 2^{-w_\mu}$. Additionally, we will show that there is a witnessed model $\mathcal{I}_\mathcal{P}$ of $\mathcal{A}_\mathcal{P}$ and $\mathcal{T}_\mathcal{P}$ whose domain has only these elements (see Figure 5). Then, the instance $\mathcal{P}$ has a solution iff for every witnessed model $\mathcal{I}$ of the constructed knowledge base $\mathcal{K}_\mathcal{P}$ there exists a $\delta \in \Delta^\mathcal{I}$ such that $A^\mathcal{I}(\delta) = B^\mathcal{I}(\delta)$.

We first introduce some abbreviations that will be helpful in simplifying the description of the construction. We use the expression $C^n$ to denote the $n$-ary conjunction of a concept description $C$ with itself; formally, $C^0 := \top$ and $C^{n+1} := C \sqcap C^n$ for every $n \geq 0$. Given two concept descriptions $C, D$ and a role name $R$, we use the expression $\langle C \overset{R}{\leadsto} D \rangle$ to denote the two axioms $\langle C \sqsubseteq \forall R.D, 1 \rangle$ and $\langle \exists R.D \sqsubseteq C, 1 \rangle$. These axioms are used to transfer specific values along role connections of degree 1, as described in the following lemma.

$$\mathcal{A}_\mathcal{P} := \{\langle A(a), 2^{-v_1}\rangle, \langle B(a), 2^{-w_1}\rangle, \langle \sim A(a), 1-2^{-v_1}\rangle, \langle \sim B(a), 1-2^{-w_1}\rangle\} \cup \{\langle V_i(a), 2^{-v_i}\rangle, \langle W_i(a), 2^{-w_i}\rangle, \langle \sim V_i(a), 1-2^{-v_i}\rangle, \langle \sim W_i(a), 1-2^{-w_i}\rangle \mid 1 \le i \le m\}$$

$$\mathcal{T}^0_\mathcal{P} := \{\langle V_j \overset{R_i}{\rightsquigarrow} V_j\rangle, \langle W_j \overset{R_i}{\rightsquigarrow} W_j\rangle \mid 1 \le i, j \le m\}$$
$$\mathcal{T}^i_\mathcal{P} := \{\langle \top \sqsubseteq \exists R_i.\top, 1\rangle,\ \langle (V_i \sqcap A^{(s+1)^{|v_i|}}) \overset{R_i}{\rightsquigarrow} A\rangle,\ \langle (W_i \sqcap B^{(s+1)^{|w_i|}}) \overset{R_i}{\rightsquigarrow} B\rangle\}$$
$$\mathcal{T}_\mathcal{P} := \bigcup_{i=0}^m \mathcal{T}^i_\mathcal{P}$$
$$\mathcal{T}'_\mathcal{P} := \mathcal{T}_\mathcal{P} \cup \{\langle \top \sqsubseteq \sim((A \to B) \sqcap (B \to A)), 0.5\rangle\}$$

$$\mathcal{K}_\mathcal{P} := \langle \mathcal{T}_\mathcal{P}, \mathcal{A}_\mathcal{P}\rangle$$
$$\mathcal{K}'_\mathcal{P} := \langle \mathcal{T}'_\mathcal{P}, \mathcal{A}_\mathcal{P}\rangle$$

Figure 6. The knowledge bases $\mathcal{K}_\mathcal{P}$ and $\mathcal{K}'_\mathcal{P}$

LEMMA 7.1.2. *For every interpretation $\mathcal{I}$ and all $x, y \in \Delta^\mathcal{I}$, we have*

- $(C^n)^\mathcal{I}(x) = (C^\mathcal{I}(x))^n$.
- *If $\mathcal{I}$ satisfies $\langle C \overset{R}{\rightsquigarrow} D\rangle$ and $R^\mathcal{I}(x, y) = 1$, then $C^\mathcal{I}(x) = D^\mathcal{I}(y)$.*

*Proof.* The first equality obviously follows from the definition of $C^n$ and the fact that we are using the product t-norm. For the second equality, since $\mathcal{I}$ satisfies the GCI $\langle C \sqsubseteq \forall R.D, 1\rangle$, we have that $C^\mathcal{I}(x) \le (\forall R.D)^\mathcal{I}(x)$, and thus

$$C^\mathcal{I}(x) \quad \le \quad \inf_{\delta \in \Delta^\mathcal{I}} (R^\mathcal{I}(x, \delta) \Rightarrow_\sqcap D^\mathcal{I}(\delta)) \le R^\mathcal{I}(x, y) \Rightarrow_\sqcap D^\mathcal{I}(y) = D^\mathcal{I}(y).$$

Dually, since $\mathcal{I}$ satisfies $\langle \exists R.D \sqsubseteq C, 1\rangle$, we obtain $(\exists R.D)^\mathcal{I}(x) \le C^\mathcal{I}(x)$, and hence

$$C^\mathcal{I}(x) \quad \ge \quad \sup_{\delta \in \Delta^\mathcal{I}} (R^\mathcal{I}(x, \delta) * D^\mathcal{I}(\delta)) \ge R^\mathcal{I}(x, y) * D^\mathcal{I}(y) = D^\mathcal{I}(y).$$

These two facts together imply that $C^\mathcal{I}(x) = D^\mathcal{I}(y)$, as claimed. □

We are now ready to construct a knowledge base that encodes the search space for a solution of $\mathcal{P}$ as depicted in Figure 5. Given an instance $\mathcal{P}$, we construct the knowledge base $\mathcal{K}_\mathcal{P}$ depicted in Figure 6. To aid in the understanding of this construction, we now develop it step-wise, explaining the use of each axiom in the reduction. We first show how to enforce the tree-like structure into every model. Afterwards, we add new axioms that allow us to detect whether a solution exists or not.

### 7.1.1 Constructing the successor nodes

We start by assuming that we have already constructed a node $\delta \in \Delta^\mathcal{I}$ in some interpretation that encodes $v, w \in \Sigma^*$; that is, $A^\mathcal{I}(\delta) = 2^{-v}$ and $B^\mathcal{I}(\delta) = 2^{-w}$. We show how to generate the first such node in Section 7.1.2; our goal now is to ensure, for each $i, 1 \le i \le m$, the existence of an $R_i$-successor of $\delta$ that encodes the concatenation of $v, w$ with the $i$-th pair from $\mathcal{P}$, i.e., $vv_i$ and $ww_i$. We assume for now that there are concept names $V_i, W_i$ encoding $v_i$ and $w_i$; more precisely, we know that $V_i^\mathcal{I}(\delta) = 2^{-v_i}$ and $W_i^\mathcal{I}(\delta) = 2^{-w_i}$ for each $i, 1 \le i \le m$. We construct the TBoxes:

$$\mathcal{T}^i_\mathcal{P} := \{\langle \top \sqsubseteq \exists R_i.\top, 1\rangle,\ \langle (V_i \sqcap A^{(s+1)^{|v_i|}}) \overset{R_i}{\rightsquigarrow} A\rangle,\ \langle (W_i \sqcap B^{(s+1)^{|w_i|}}) \overset{R_i}{\rightsquigarrow} B\rangle\}.$$

Recall that we are viewing each word in $\Sigma^*$ as a natural number in base $s+1$. Therefore, under this view, the concatenation of two words $u, u'$ corresponds to the operation $u \cdot (s+1)^{|u'|} + u'$ on natural numbers. Consider the concept $V_i \sqcap A^{(s+1)^{|v_i|}}$, the interpretation $\mathcal{I}$ satisfies

$$(V_i \sqcap A^{(s+1)^{|v_i|}})^{\mathcal{I}}(\delta) = V_i^{\mathcal{I}}(\delta) \cdot (A^{\mathcal{I}}(\delta))^{(s+1)^{|v_i|}} = 2^{-(v(s+1)^{|v_i|}+v_i)} = 2^{-vv_i}.$$

In other words, this concept encodes the word $vv_i$, which is the word that should appear in the $i$-th successor of $\delta$. If $\mathcal{I}$ is a witnessed model of $\mathcal{T}_{\mathcal{P}}^i$, then from the first axiom it follows that there must exist an element $\gamma \in \Delta^{\mathcal{I}}$ with $R_i^{\mathcal{I}}(\delta, \gamma) = 1$. As argued above, the last two axioms then ensure that $A^{\mathcal{I}}(\gamma) = 2^{-vv_i}$ and $B^{\mathcal{I}}(\gamma) = 2^{-ww_i}$; thus, the concept names $A$ and $B$ encode, at node $\gamma$, the words $vv_i$ and $ww_i$, as desired.

The idea is to use this construction recursively so that, starting from an appropriate root node, we can guarantee the existence of the full infinite tree that describes the instance $\mathcal{P}$. To achieve this, we need to guarantee that at this new node $\gamma$ we also have that $V_j^{\mathcal{I}}(\gamma) = 2^{-v_j}$ and $W_j^{\mathcal{I}}(\gamma) = 2^{-w_j}$ hold for every $j, 1 \leq j \leq m$, since this was a necessary condition for the correctness of the construction. This can be done by including the axioms

$$\mathcal{T}_{\mathcal{P}}^0 \quad := \quad \{\langle V_j \overset{R_i}{\rightsquigarrow} V_j\rangle,\ \langle W_j \overset{R_i}{\rightsquigarrow} W_j\rangle \mid 1 \leq i, j \leq m\}$$

to the TBox.

The union of the TBoxes $\mathcal{T}_{\mathcal{P}}^i$ for $0 \leq i \leq m$ guarantees the construction of the search tree for $\mathcal{P}$, provided that we can build a node that encodes the appropriate restrictions at the root of this tree.

### 7.1.2 Constructing the root node

We now need to ensure that there exists a node $\delta_\varepsilon \in \Delta^{\mathcal{I}}$ such that $A^{\mathcal{I}}(\delta_\varepsilon) = 2^{-v_1}$ and $B^{\mathcal{I}}(\delta_\varepsilon) = 2^{-w_1}$; that is, where $A$ and $B$ encode $v_\varepsilon$ and $w_\varepsilon$, respectively. Additionally, at this node $\delta_\varepsilon$, $V_j^{\mathcal{I}}(\delta_\varepsilon) = 2^{-v_j}$, $W_j^{\mathcal{I}}(\delta_\varepsilon) = 2^{-w_j}$ must hold for every $j$, $1 \leq j \leq m$. To enforce these conditions, we use the ABox:

$$\mathcal{A}_{\mathcal{P}} := \{\langle A(a), 2^{-v_1}\rangle, \langle B(a), 2^{-w_1}\rangle, \langle \sim A(a), 1-2^{-v_1}\rangle, \langle \sim B(a), 1-2^{-w_1}\rangle\} \cup$$

$$\cup \{\langle V_i(a), 2^{-v_i}\rangle, \langle W_i(a), 2^{-w_i}\rangle, \langle \sim V_i(a), 1-2^{-v_i}\rangle, \langle \sim W_i(a), 1-2^{-w_i}\rangle \mid 1 \leq i \leq m\}.$$

Recall that $\sim$ stands for the strong complementation constructor. Any interpretation satisfying the two axioms $\langle A(a), 2^{-v_1}\rangle$ and $\langle \sim A(a), 1-2^{-v_1}\rangle$ contains an element $\delta_\varepsilon := a^{\mathcal{I}}$ such that $A^{\mathcal{I}}(\delta_\varepsilon) \geq 2^{-v_1}$ and $1 - A^{\mathcal{I}}(\delta_\varepsilon) = (\sim A)^{\mathcal{I}}(\delta_\varepsilon) \geq 1 - 2^{-v_1}$ and hence $A^{\mathcal{I}}(\delta_\varepsilon) = 2^{-v_1}$. A similar argument shows that in every model of $\mathcal{A}_{\mathcal{P}}$, $\delta_\varepsilon$ is interpreted to the desired membership degrees of the concepts $B$ and $V_j, W_j$ for each $j, 1 \leq j \leq m$. Thus, this element can be used to represent the root of the tree.

### 7.1.3 The canonical model

We now show that every model of the axioms defined so far encodes the instance $\mathcal{P}$ of the PCP. Let $\mathcal{T}_{\mathcal{P}} := \bigcup_{i=0}^m \mathcal{T}_{\mathcal{P}}^i$ and $\mathcal{K}_{\mathcal{P}} = \langle \mathcal{T}_{\mathcal{P}}, \mathcal{A}_{\mathcal{P}}\rangle$. We define the *canonical interpretation* as the interpretation $\mathcal{I}_{\mathcal{P}} := \langle \Delta^{\mathcal{I}_{\mathcal{P}}}, \cdot^{\mathcal{I}_{\mathcal{P}}}\rangle$ whose domain is the set

$\Delta^{\mathcal{I}_\mathcal{P}} = \{1, \ldots, m\}^*$ of all finite words over $\{1, \ldots, m\}$ and the interpretation function $\cdot^{\mathcal{I}_\mathcal{P}}$ is such that: (i) the individual name $a$ is mapped to $a^{\mathcal{I}_\mathcal{P}} = \varepsilon$, (ii) for every $\mu \in \Delta^{\mathcal{I}_\mathcal{P}}$, $A^{\mathcal{I}_\mathcal{P}}(\mu) = 2^{-v_\mu}$, and $B^{\mathcal{I}_\mathcal{P}}(\mu) = 2^{-w_\mu}$, and (iii) for all $j, 1 \leq j \leq m$,

- $V_j^{\mathcal{I}_\mathcal{P}}(\mu) = 2^{-v_j}$ and $W_j^{\mathcal{I}_\mathcal{P}}(\mu) = 2^{-w_j}$.
- $R_j^{\mathcal{I}_\mathcal{P}}(\mu, \mu j) = 1$ and $R_j^{\mathcal{I}_\mathcal{P}}(\mu, \mu') = 0$ if $\mu' \neq \mu j$.

This interpretation is precisely the one depicted in Figure 5. It is easy to see that $\mathcal{I}_\mathcal{P}$ is a model of $\mathcal{K}_\mathcal{P}$, and is in fact witnessed, since every node has exactly one $R_i$-successor with degree greater than 0, for every $i, 1 \leq i \leq m$. More interesting, however, is that for every witnessed model $\mathcal{I}$ of $\mathcal{K}_\mathcal{P}$, there is a homomorphism from $\mathcal{I}_\mathcal{P}$ to $\mathcal{I}$ as described in the following lemma.

LEMMA 7.1.3. *Let $\mathcal{I}$ be a witnessed model of $\mathcal{K}_\mathcal{P}$. There is a function $f\colon \Delta^{\mathcal{I}_\mathcal{P}} \to \Delta^\mathcal{I}$ such that, for every $\mu \in \Delta^{\mathcal{I}_\mathcal{P}}$ and every concept name $C$ appearing in $\mathcal{A}_\mathcal{P}$ or in $\mathcal{T}_\mathcal{P}$, it holds that $C^{\mathcal{I}_\mathcal{P}}(\mu) = C^\mathcal{I}(f(\mu))$.*

*Proof.* The function $f$ is built inductively on the length of $\mu$. For $\mu = \epsilon$, $\mathcal{A}_\mathcal{P}$ fixes the interpretation of all relevant concept names at $a^\mathcal{I}$ and hence defining $f(\varepsilon) := a^\mathcal{I}$ satisfies the condition of the lemma.

Let now $\mu$ be such that $f(\mu)$ is already defined. By induction, assume that $A^\mathcal{I}(f(\mu)) = 2^{-v_\mu}$, $B^\mathcal{I}(f(\mu)) = 2^{-w_\mu}$, and that for every $j$, where $1 \leq j \leq m$, $V_j^\mathcal{I}(f(\mu)) = 2^{-v_j}$, $W_j^\mathcal{I}(f(\mu)) = 2^{-w_j}$. Since $\mathcal{I}$ is a witnessed model of the axiom $\langle \top \sqsubseteq \exists R_i.\top, 1\rangle$, for all $i, 1 \leq i \leq m$ there exists a $\gamma \in \Delta^\mathcal{I}$ with $r^\mathcal{I}(f(\mu), \gamma) = 1$. As $\mathcal{I}$ satisfies all the axioms of the form $\langle C \overset{R}{\rightsquigarrow} D\rangle \in \mathcal{T}_\mathcal{P}$, it follows that

$$A^\mathcal{I}(\gamma) = 2^{-v_\mu v_i} = 2^{-v_{\mu i}}, \quad B^\mathcal{I}(\gamma) = 2^{-w_\mu w_i} = 2^{-w_{\mu i}},$$

and for all $j, 1 \leq j \leq m$, $V_j^\mathcal{I}(\gamma) = 2^{-v_j}, W_j^\mathcal{I}(\gamma) = 2^{-w_j}$. Setting $f(\mu i) := \gamma$ thus satisfies the required property. □

From this lemma it then follows that if the instance $\mathcal{P}$ of the PCP has a solution $\mu \in \{1, \ldots, m\}^*$, then every witnessed model $\mathcal{I}$ of $\mathcal{K}_\mathcal{P}$ contains a node $\delta = f(\mu)$ such that $A^\mathcal{I}(\delta) = B^\mathcal{I}(\delta)$; that is, where $A$ and $B$ encode the same word. Conversely, if every witnessed model contains such a node, then in particular $\mathcal{I}_\mathcal{P}$ does, and thus $\mathcal{P}$ has a solution. The question now is how to detect whether a node with this characteristics exists in every model.

### 7.1.4 Finding a solution

To detect whether there must exist a domain element that belongs to the concepts $A$ and $B$ to the same degree, we extend $\mathcal{T}_\mathcal{P}$ with axioms that further restrict $\mathcal{I}_\mathcal{P}$, and hence indirectly all models, to satisfy $A^{\mathcal{I}_\mathcal{P}}(\mu) \neq B^{\mathcal{I}_\mathcal{P}}(\mu)$ for every $\mu \in \{1, \ldots, m\}^*$. This will ensure that the extended ontology will have a model iff $\mathcal{P}$ has no solution.

Suppose for now that, for some $\mu \in \{1, \ldots, m\}^*$, it holds that

$$2^{-v_\mu} = A^{\mathcal{I}_\mathcal{P}}(\mu) > B^{\mathcal{I}_\mathcal{P}}(\mu) = 2^{-w_\mu}.$$

We then have that $v_\mu < w_\mu$ and hence $w_\mu - v_\mu \geq 1$. It thus follows that

$$A^{\mathcal{I}_\mathcal{P}}(\mu) \Rightarrow_\sqcap B^{\mathcal{I}_\mathcal{P}}(\mu) = 2^{-w_\mu}/2^{-v_\mu} = 2^{-(w_\mu - v_\mu)} \leq 2^{-1} = 0.5.$$

Likewise, if $A^{\mathcal{I}_\mathcal{P}}(\mu) < B^{\mathcal{I}_\mathcal{P}}(\mu)$, we get $B^{\mathcal{I}_\mathcal{P}}(\mu) \Rightarrow_\sqcap A^{\mathcal{I}_\mathcal{P}}(\mu) \leq 0.5$. Additionally, if we have $A^{\mathcal{I}_\mathcal{P}}(\mu) = B^{\mathcal{I}_\mathcal{P}}(\mu)$, then it is easy to verify that

$$A^{\mathcal{I}_\mathcal{P}}(\mu) \Rightarrow_\sqcap B^{\mathcal{I}_\mathcal{P}}(\mu) = B^{\mathcal{I}_\mathcal{P}}(\mu) \Rightarrow_\sqcap A^{\mathcal{I}_\mathcal{P}}(\mu) = 1.$$

From all this it follows that, for every $\mu \in \{1, \dots, m\}^*$,

$$A^{\mathcal{I}_\mathcal{P}}(\mu) \neq B^{\mathcal{I}_\mathcal{P}}(\mu) \quad \text{iff} \quad \begin{array}{ll} \text{either} & A^{\mathcal{I}_\mathcal{P}}(\mu) \Rightarrow_\sqcap B^{\mathcal{I}_\mathcal{P}}(\mu) \leq 0.5 \\ \text{or} & B^{\mathcal{I}_\mathcal{P}}(\mu) \Rightarrow_\sqcap A^{\mathcal{I}_\mathcal{P}}(\mu) \leq 0.5. \end{array} \tag{3}$$

Thus, the instance $\mathcal{P}$ has no solution iff for every $\mu \in \{1, \dots, m\}^*$ one of the two restrictions $A^{\mathcal{I}_\mathcal{P}}(\mu) \Rightarrow_\sqcap B^{\mathcal{I}_\mathcal{P}}(\mu) \leq 0.5$ or $B^{\mathcal{I}_\mathcal{P}}(\mu) \Rightarrow_\sqcap A^{\mathcal{I}_\mathcal{P}}(\mu) \leq 0.5$ is satisfied. We now define the TBox $\mathcal{T}'_\mathcal{P}$ that ensures this behavior in every model. Let

$$\mathcal{T}'_\mathcal{P} := \mathcal{T}_\mathcal{P} \cup \{\langle \top \sqsubseteq {\sim}((A \to B) \sqcap (B \to A)), 0.5\rangle\} \qquad \mathcal{K}'_\mathcal{P} = \langle \mathcal{T}'_\mathcal{P}, \mathcal{A}_\mathcal{P}\rangle$$

This KB can be used to decide whether $\mathcal{P}$ has a solution:

LEMMA 7.1.4. *The instance $\mathcal{P}$ of the PCP has a solution iff the knowledge base $\mathcal{K}'_\mathcal{P}$ is not $\sqcap$-consistent w.r.t. witnessed models.*

*Proof.* Assume that $\mathcal{P}$ has a solution $\mu = i_1 \cdots i_k$ and let $u = v_\mu = w_\mu$. Suppose there is a witnessed model $\mathcal{I}$ of $\mathcal{K}'_\mathcal{P}$. Since $\mathcal{T}_\mathcal{P} \subseteq \mathcal{T}'_\mathcal{P}$, $\mathcal{I}$ is also a (witnessed) model of $\mathcal{K}_\mathcal{P}$. From Lemma 7.1.3 it follows that there is a node $\delta \in \Delta^\mathcal{I}$ where

$$A^\mathcal{I}(\delta) = A^{\mathcal{I}_\mathcal{P}}(\mu) = B^{\mathcal{I}_\mathcal{P}}(\mu) = B^\mathcal{I}(\delta).$$

Then, $A^\mathcal{I}(\delta) \Rightarrow_\sqcap B^\mathcal{I}(\delta) = 1$ and $B^\mathcal{I}(\delta) \Rightarrow_\sqcap A^\mathcal{I}(\delta) = 1$. This then implies that

$$({\sim}((A \to B) \sqcap (B \to A)))^\mathcal{I}(\delta) = 0,$$

which violates the axiom $\langle \top \sqsubseteq {\sim}((A \to B) \sqcap (B \to A)), 0.5\rangle$ from $\mathcal{T}'_\mathcal{P}$. Thus $\mathcal{I}$ cannot be a model of $\mathcal{T}'_\mathcal{P}$, nor of $\mathcal{K}'_\mathcal{P}$.

For the converse, assume that $\mathcal{K}'_\mathcal{P}$ is not witnessed $\sqcap$-consistent. Then $\mathcal{I}_\mathcal{P}$ cannot be a model of $\mathcal{K}'_\mathcal{P}$. Since it is a model of $\mathcal{K}_\mathcal{P}$, $\mathcal{I}_\mathcal{P}$ must violate the only axiom in $\mathcal{T}'_\mathcal{P} \setminus \mathcal{T}_\mathcal{P}$. Thus, there is a node $\mu \in \{1, \dots, m\}^*$ where $({\sim}((A \to B) \sqcap (B \to A)))^{\mathcal{I}_\mathcal{P}}(\mu) < 0.5$ or, equivalently, where $(A^{\mathcal{I}_\mathcal{P}}(\mu) \Rightarrow_\sqcap B^{\mathcal{I}_\mathcal{P}}(\mu)) \cdot (B^{\mathcal{I}_\mathcal{P}}(\mu) \Rightarrow_\sqcap A^{\mathcal{I}_\mathcal{P}}(\mu)) > 0.5$. But this means that each of these residua is strictly greater than 0.5, and hence $\mathcal{P}$ has a solution. □

Lemma 7.1.4 implies that we can reduce the PCP to $\sqcap$-consistency in the FDL $\mathfrak{I}\mathcal{ALCE}$, and in particular, that the latter problem is undecidable.

THEOREM 7.1.5. *Witnessed $\sqcap$-consistency of knowledge bases in $\mathfrak{I}\mathcal{ALCE}$ is undecidable.*

Observe that with obvious changes the proof of Theorem 7.1.5 is also valid for any FDL language of $\mathfrak{I}\mathcal{ALCE}$ type when taking as semantic of the complementation any involutive negation on $[0,1]$ (different from the standard one). Moreover we cannot drop the condition of models being witnessed. Take the concept $B := \forall R.C \& \neg\forall R.(C \sqcap C)$, given by Hájek in [80], which is 1-satisfiable but not witnessed 1-satisfiable. In the same way the concept $\neg\Delta B$ is a tautology for witnessed models but not for all $[0,1]$-models.

## 7.2 Further undecidability results

Using similar techniques to those presented in Section 7.1, it is possible to prove undecidability of other reasoning tasks and for FDLs using other continuous t-norms for their semantics. The undecidability of $*$-consistency holds even for less expressive FDLs. In fact, it was shown in [41, 60] that for every continuous t-norm $*$ that is not idempotent (i.e., not the minimum t-norm), $*$-consistency is undecidable already in $\mathcal{ELC}$.

As explained before, the proof of this claim follows very closely the construction presented in Section 7.1 for reducing an instance of the PCP to consistency of a knowledge base. However, some additional technicalities are needed to remove any use of the constructors $\forall$ and $\to$ from the axioms used. As it can be seen from Figure 6, value restrictions are only used in the definition of the axioms $\langle C \overset{R}{\rightsquigarrow} D \rangle$. These axioms are used to transfer the membership degree of a concept to a given $R$-successor. We can obtain the same result using existential restrictions and strong negation, instead of value restrictions. To do this, simply redefine the expression $\langle C \overset{R}{\rightsquigarrow} D \rangle$ to stand for the two axioms $\langle \exists R. \sim D \sqsubseteq \sim C, 1 \rangle$ and $\langle \exists R.D \sqsubseteq C, 1 \rangle$. It is easy to see that Lemma 7.1.2 still holds under this definition.

The implication constructor is only used in the axiom

$$\langle \top \sqsubseteq \sim((A \to B) \sqcap (B \to A)), 0.5 \rangle$$

added to the knowledge base $\mathcal{K}_{\mathcal{P}}$ to verify the existence of a solution for $\mathcal{P}$. Recall that this axiom is intended to enforce that each element of the tree-model belongs to the concepts $A$ and $B$ to a different membership degree, and hence cannot be a solution. It is possible to remove the constructor $\to$ with the help of auxiliary concept names that verify that $A$ and $B$ never receive the same interpretation. Since this construction is quite technical, we do not reproduce it here and refer the interested reader to the proofs appearing in [41]. Additionally, if the t-norm used is not $\Pi$, then the words from the instance $\mathcal{P}$ need to be encoded in a different manner. Overall, we obtain the following theorem.

THEOREM 7.2.1. *Witnessed $*$-consistency of $\mathcal{ELC}$ knowledge bases is undecidable if $*$ is a non-idempotent continuous t-norm.*

Notice that strong negation is fundamental for the constructions presented so far. In fact, even for initializing the recursive construction of the tree, we need to ensure the existence of a domain element that belongs to the concepts $A$ and $B$ with specific degrees. Since concept assertions only provide lower bounds for the membership degrees, we require strong negation to provide a matching upper bound. It is well-known that for the Łukasiewicz t-norm, strong and weak negations coincide. Thus, a trivial consequence of Theorem 7.2.1 is that Ł-consistency of $\mathfrak{N}\mathcal{EL}$ knowledge bases is also undecidable. Observe that in fact, in Łukasiewicz, $\mathfrak{N}\mathcal{EL}$ is equivalent to $\mathcal{ALC}$.

Let now $*$ be any non-strict continuous t-norm. It is well-known that such t-norms and their residua behave as the Łukasiewicz operators, scaled down to an initial subinterval from $[0,1]$. More precisely, if $*$ is a non-strict continuous t-norm, then there exists a $b \in (0,1]$ such that for all $x, y \in [0,b]$:

$$\begin{aligned} x * y &= b \cdot (\frac{x}{b} *_{Ł} \frac{y}{b}) \\ x \rightarrow_* y &= b \cdot (\frac{x}{b} \rightarrow_{Ł} \frac{y}{b}). \end{aligned}$$

Using this property, it is possible to scale the reduction of the PCP to Ł-consistency down to the interval $[0,b]$, thus obtaining undecidability also for the nilpotent t-norm $*$. Clearly, in this scaling, we cannot consider strong negation, as there is no guarantee on how operators behave beyond the value $b$. Thus, this result can only hold if we allow weak negation, which is defined through the residuum and the lower bound 0.

THEOREM 7.2.2. *Witnessed $*$-consistency of $\mathfrak{NEL}$ knowledge bases is undecidable if $*$ is a non-strict continuous t-norm.*

We now briefly describe how to apply the same ideas described in this section to also show undecidability of subsumption and concept satisfiability. Recall that for the reduction presented in Section 7.1, we encoded each natural number $v$ as the number $2^{-v} \in (0,1]$. Notice that in this case, the base 2 is arbitrary, and the reduction would still apply, with minor modifications, if the encoded had the form $p^v$ for any $p \in (0,1)$. Moreover, the ABox $\mathcal{A}_{\mathcal{P}}$ is only used to initialize the root node to specific powers of $\frac{1}{2}$.

Suppose that we can guarantee the existence of an element $\delta$ that belongs to some concept name $X$ to a degree $p \in (0,1)$, then the TBox

$$\begin{aligned} \mathcal{T}_X := {} & \{\langle \top \sqsubseteq X, 0.1\rangle, \langle \top \sqsubseteq \exists S.\top\rangle, \langle X^{v_1} \overset{S}{\rightsquigarrow} A\rangle, \langle X^{w_1} \overset{S}{\rightsquigarrow} B\rangle\} \cup \\ & \{\langle X^{v_j} \overset{S}{\rightsquigarrow} V_j\rangle, \langle X^{w_j} \overset{S}{\rightsquigarrow} W_j\rangle \mid 1 \leq i, j \leq m\} \end{aligned}$$

ensures the existence of an element $\gamma$ such that $A^{\mathcal{I}}(\gamma) = p^{v_1}$, $B^{\mathcal{I}}(\gamma) = p^{w_1}$, and $V_j^{\mathcal{I}}(\gamma) = p^{v_j}$ and $W_j^{\mathcal{I}}(\gamma) = p^{w_j}$ holds for every $j, 1 \leq j \leq m$; that is, this $\gamma$ behaves like the root node for the search tree in the reduction of $\mathcal{P}$. Additionally, it guarantees that $X$ is always interpreted to a degree greater than or equal to $0.1$; in particular, at the element $\delta$ we have that $X^{\mathcal{I}}(\delta) > 0$. Using the arguments from Section 7.1, the following lemma can be easily proven.

LEMMA 7.2.3. *The instance $\mathcal{P}$ of the PCP has a solution if and only if any of the following holds:*

1. *$X$ is $\sqcap$-satisfiable to degree smaller than or equal to 0.9 w.r.t. $\mathcal{T}'_{\mathcal{P}} \cup \mathcal{T}_X$.*
2. *$\top$ is* not *$\sqcap$-subsumed by $X$ to degree 1 w.r.t. $\mathcal{T}'_{\mathcal{P}} \cup \mathcal{T}_X$.*

A simple consequence of this lemma is that $\sqcap$-satisfiability and $\sqcap$-subsumption are undecidable in $\mathfrak{IALCE}$. As before, these undecidability results can be strengthened to less expressive logics, and to a larger class of t-norms. We thus obtain the following theorem, whose detailed proof can be found in [2, 36].

THEOREM 7.2.4. *Witnessed $*$-satisfiability and witnessed $*$-subsumption w.r.t. TBoxes are undecidable*

- *in $\mathcal{ELC}$ if $*$ is a non-idempotent continuous t-norm and*
- *in $\mathfrak{NEL}$ if $*$ is a non-strict continuous t-norm.*

Notice that all these undecidability results hold only for witnessed models. In fact, the only step in which the properties of witnessed interpretations are used is to guarantee that there are $R_i$-successors with degree 1, using the axiom $\langle \top \sqsubseteq \exists R_i.\top \rangle$. Thus, we can weaken this restriction to consider only so-called *weakly-witnessed* interpretations. An interpretation is weakly-witnessed if, for every $\delta \in \Delta^{\mathcal{I}}$ such that $(\exists R.\top)^{\mathcal{I}}(\delta) = 1$, there is a $\gamma \in \Delta^{\mathcal{I}}$ with $R^{\mathcal{I}}(\delta, \gamma) = 1$.

When considering general models, the constructions presented so far do not work for encoding instances of the PCP. The reason for this is that the existential restriction can be satisfied through an infinite sequence of successors that get a degree arbitrarily close to 1, but this limit is never reached. In such a situation, the preconditions of Lemma 7.1.2 do not hold, and hence we cannot guarantee that the desired degrees are transferred to a specific successor.

To overcome this problem, it is proposed in [39] to encode words using intervals, rather than just numbers from $[0, 1]$. The idea is that such an interval allows for a margin of error in the transfer of values through successors with degrees close to, but not precisely equal to 1. While the main idea is the same, the additional handling of this error-margin makes the construction rather technical. Moreover, this idea has only been applied to non-strict t-norms, yielding the following result.

THEOREM 7.2.5. *For every non-strict t-norm $*$, $*$-consistency, $*$-satisfiability, and $*$-subsumption are undecidable in $\mathfrak{NEL}$ w.r.t. general models.*

### 7.3 Complexity bounds

We now study the complexity of reasoning in FDLs. For a historical overview of the known decidability results for these logics, we refer the interested reader to Section 8.

In order to study the complexity of reasoning in FDLs, recall first that in the classical case, reasoning in $\mathcal{FL}_0$ w.r.t. general TBoxes is already exponential [3, 133]; moreover, under the classical semantics, all the logics between $\mathfrak{NEL}$ and $\mathcal{ELC}$ from below, and $\mathfrak{IALCE}$ from above are equivalent to the DL $\mathcal{ALC}$, for which the standard reasoning problems are all ExpTime-complete. This implies that reasoning in the fuzzy variants of all these logics is already ExpTime-hard, regardless of the t-norm chosen for their semantics. As shown earlier in this section, these lower complexity bounds are not necessarily tight, since in many cases these logics are undecidable. We now prove that in all other cases, KB consistency and concept satisfiability w.r.t. witnessed models are decidable in exponential time. Afterwards, we discuss the issues with general models, and the problem of subsumption.

Let $*$ be a strict continuous t-norm; in other words, a continuous t-norm with Gödel negation as its corresponding residuated negation. We want to show that $*$-consistency of $\mathfrak{NEL}$ knowledge bases is in ExpTime. We will show a stronger result, proving this complexity upper bound for the more expressive $\mathfrak{IALUE}$.

Since $*$ is a strict t-norm, we know that its weak negation corresponds to the Gödel negation that maps 0 to 1 and all other values to 0. This means that its double negation $\neg\neg$ maps every positive value to 1 and the least element 0 to itself. Moreover, [62] proves that the double negation operator is an automorphism in any BL-chain. This result can be extended to existential and value restrictions, provided that the latter are witnessed.

LEMMA 7.3.1. *Let $*$ be a strict continuous t-norm. For all non-empty sets $X \subseteq [0,1]$ it holds that*

1. $\neg\neg(\sup\{x \mid x \in X\}) = \sup\{\neg\neg x \mid x \in X\}$.
2. *If* $\min\{x \mid x \in X\}$ *exists, then* $\neg\neg(\min\{x \mid x \in X\}) = \min\{\neg\neg x \mid x \in X\}$.

*Proof.* To prove (1), observe that $\sup X = 0$ iff $X = \{0\}$, which yields

$$\begin{aligned}\neg\neg(\sup X) = 0 \quad \text{iff} \quad \sup X = 0 \quad &\text{iff} \quad X = \{0\} \quad \text{iff} \quad \{\neg\neg x \mid x \in X\} = \{0\}\\ &\text{iff} \quad \sup\{\neg\neg x \mid x \in X\} = 0.\end{aligned}$$

Assume now that $\min X = x_{\min}$ exists. To prove (2) observe that:

$$\begin{aligned}\neg\neg(\min X) = 0 \quad \text{iff} \quad x_{\min} = 0 \quad &\text{iff} \quad 0 \in \{\neg\neg x \mid x \in X\}\\ &\text{iff} \quad \min\{\neg\neg x \mid x \in X\} = 0.\end{aligned}$$

□

In a natural way, the $\neg\neg$ operator induces, for each interpretation $\mathcal{I}$, the crisp interpretation $\mathcal{I}^\star$ defined by $A^{\mathcal{I}^\star}(\delta) = \neg\neg(A^{\mathcal{I}}(\delta))$ and $R^{\mathcal{I}^\star}(\delta,\gamma) = \neg\neg(R^{\mathcal{I}}(\delta,\gamma))$ for all $\delta,\gamma \in \Delta^{\mathcal{I}}$, concept name $A$ and role name $R$. Let $\mathcal{K}$ be a knowledge base. From $\mathcal{K}$ we can construct a new (crisp) knowledge base $\mathcal{K}^\star$. The axioms in $\mathcal{K}^\star$ are:

1. A concept assertion $\langle C(a), \neg\neg r\rangle$ for every concept assertion $\langle C(a), r\rangle$ in $\mathcal{K}$
2. A role assertion $\langle R(a,b), \neg\neg r\rangle$ for every role assertion $\langle R(a,b), r\rangle$ in $\mathcal{K}$
3. A GCI $\langle C \sqsubseteq D, \neg\neg r\rangle$ for every GCI $\langle C \sqsubseteq D, r\rangle$ in $\mathcal{K}$.

THEOREM 7.3.2. *A witnessed interpretation $\mathcal{I}$ is a model of the $\mathcal{K}$ if and only if $\mathcal{I}^\star$ is a model of $\mathcal{K}^\star$.*

Since $*$ is strict, it follows that $\mathcal{I}^\star$ only uses the extreme membership degrees 0 and 1; that is, it is a crisp interpretation. Likewise, $\mathcal{K}^\star$ can be seen as a classical KB, since it is not graded. From this it follows that $*$-consistency of $\mathfrak{I}\mathcal{ALUE}$ knowledge bases can be reduced in linear time to classical consistency in the DL $\mathcal{ALC}$, yielding the following result [34].

THEOREM 7.3.3. *If $*$ is a strict continuous t-norm then $*$-consistency in $\mathfrak{I}\mathcal{ALUE}$ is in* ExpTime.

The only remaining case is for logics that allow strong negation as a constructor. As stated in Theorem 7.2.1, in any logic that extends $\mathcal{EL}$ with strong negation, $*$-consistency is undecidable if $*$ is not the idempotent t-norm. Thus, we only need to study the case when the semantics is based on the minimum t-norm. It has been recently shown, using an automata-based method, that G-consistency of $\mathfrak{I}\mathcal{ALCE}$ can also be decided in

| | $\mathcal{NEL}$ | $\mathfrak{IELU}$ | $\mathfrak{IALUE}$ | $\mathcal{ELC}$ | $\mathfrak{IALCE}$ |
|---|---|---|---|---|---|
| classical | Ł⊕ | Ł⊕ | Ł⊕ | $\overline{\mathsf{G}}$ | $\overline{\mathsf{G}}$ |
| general | Ł⊕ | Ł⊕ | Ł⊕ | $\overline{\mathsf{G}}$ | $\overline{\mathsf{G}}$ |

Table 6. Undecidability of consistency in fuzzy description logics

exponential time [35]. The main insight in this approach is that the precise membership degrees are irrelevant for deciding whether a model exists; it is only necessary to know the *order* between this degrees. Since only a finite number of such orderings are possible, one can construct an automaton that accepts abstract representations of the tree-shaped models of the KB. For the full details of this construction see [35].

THEOREM 7.3.4. *G-consistency in $\mathfrak{IALCE}$ is in* EXPTIME.

An upper bound in the complexity of $*$-consistency also provides a complexity upper bound for the problem of $*$-satisfiability of concepts to a degree greater than or equal to some $r$. Indeed, it is easy to see that a concept $C$ is $*$-satisfiable to degree at least $r$ w.r.t. the knowledge base $\langle \mathcal{T}, \mathcal{A} \rangle$ if and only if the knowledge base $\langle \mathcal{T}, \{\langle C(a), r \rangle\}\rangle$ is $*$-consistent. If strong negation is allowed, then $*$-satisfiability to a degree smaller than or equal to $r$ and $*$-subsumption can also be reduced to $*$-consistency in the obvious manner. This yields the following result.

THEOREM 7.3.5. *G-satisfiability and G-subsumption in $\mathfrak{IALCE}$ are in* EXPTIME. *Moreover, if $*$ is a strict continuous t-norm, then $*$-consistency to a degree greater than or equal to $r$ in $\mathfrak{IALUE}$ is also in* EXPTIME.

Unfortunately, this reduction does not work for deciding satisfiability to a degree smaller or equal to some $r$, or subsumption. It is important to notice that the proof of decidability of consistency in $\mathfrak{IALUE}$ constructs an equi-consistent classical KB from which consistency can be decided. However, this new knowledge base $\mathcal{K}^\star$ is not necessarily equivalent to $\mathcal{K}$; the latter can have many more models than the former. The translation $\mathcal{K} \to \mathcal{K}^\star$ guarantees that if there is any model, then there is also a model that uses only the degrees 0 and 1. However, nothing further is known of the existence of a model that gives a degree lower than some upper bound (for satisfiability), nor of all models (for subsumption). It is still an open problem whether these problems are even decidable for the $\mathfrak{IALUE}$ description language.

The undecidability and complexity results for $*$-consistency are summarized in Table 6. Each cell of the table describes the class of t-norms $*$ for which $*$-consistency in the corresponding FDL is known to be undecidable, w.r.t. witnessed and general models. In this case, Ł⊕ denotes the class of non-strict continuous t-norms; that is, those t-norms starting with the Łukasiewicz t-norm, and $\overline{\mathsf{G}}$ is the class of non-idempotent continuous t-norms, i.e., all continuous t-norms except the minimum t-norm. Cells with a gray background express that for all t-norms that are *not* in the class, $*$-consistency is known to be decidable and EXPTIME-complete. Notice that there are two white cells for the case of general models. In these cases, it is unknown whether the problem is decidable for any t-norm that is outside the respective class.

### 7.4 Reasoning with finite chains

We finish this section with a brief observation on the complexity of reasoning in FDLs when the semantics is based on a finite chain. It is easy to see, using standard techniques from model theory, that when the space of truth degrees is finite, then $\mathfrak{I}\mathcal{ALCE}$ has the bounded-model property; that is, if a knowledge base $\mathcal{K}$ has a model, then it also has a finite model with size bounded by a function on the size of $\mathcal{K}$. This immediately implies decidability of all the standard reasoning tasks in this logic.

We can further improve this positive complexity result. In fact, using a technique known as *unraveling*, it is also possible to prove that if the knowledge base $\mathcal{K}$ has a model, then it also accepts a Hintikka tree [35, 42, 44]. Abstract representations of these Hintikka trees can be accepted by an automaton that uses as states all the types for the relevant concepts from $\mathcal{K}$, and whose transition relation verifies that the semantics of the existential and value restrictions are satisfied. Since the number of types is exponential on the size of $\mathcal{K}$, and deciding whether an automaton accepts at least one tree is polynomial on its number of states, this yields the following result.

THEOREM 7.4.1. *Standard reasoning in $\mathfrak{I}\mathcal{ALCE}$ over finite chains is* EXPTIME-*complete.*

The result in Theorem 7.4.1 matches the same complexity result for classical $\mathcal{ALC}$. The same matching has been proved for different languages over (not necessarily linear) finite truth algebras. Nevertheless, in [33] has been proved that the complexity of KB consistency for language $\mathcal{EL}$ under the finite Łukasiewicz $t$-norm jumps from PTIME-complete in the classical case to EXPTIME-complete in the finite-valued case.

## 8 Some historical remarks and further reading

We now provide a brief overview of the history of fuzzy description logics.

### 8.1 Classical Description Logics in short

*Description Logics* (DLs), previously called *Terminological Logics* (e.g. [100]) and *conceptual languages* (e.g. [114]), are usually considered as an evolution of *frame-based systems* (e.g. [84, 98]) and *Inheritance Networks* (e.g. [46, 107, 135, 139]), the latter two lacking a formal (logic based) semantics. DLs have been introduced to overcome to this deficiency. In short,[9] the history of DLs can be split into four phases:

**Phase 1.** During the '80s, DLs' development has been characterised by so-called *structural subsumption algorithms* and implementation of first DL systems such as KL-ONE [51, 140], KL-TWO [138], KRYPTON [48], KANDOR [104], LOOM [93], BACK [101], NIKL [89], LASSIE [67], K_REP [96] and CLASSIC [47]. Some of these algorithms have been found incomplete and the subsumption problem may be undecidable (e.g. [105, 113]), but most of the cases these systems have reduced expressivity and computational complexity of the subsumption problem (e.g. [49, 115]).

[9] For more on classical DLs' history see [4].

**Phase 2.** The early '90s have been characterized by the introduction of complete *tableaux based reasoning methods* [68, 85, 114], first complexity results, first investigation on optimization methods and systems such as KRIS [6], KSAT [74], CRACK [72], whose computational complexity of the subsumption problem is in PSPACE.

**Phase 3.** From the mid '90s, tableaux methods and systems have been developed with good practical performance on very expressive DLs with high worst case complexity, essentially in EXPTIME, and the study of relationships to other fragments of First-Order Logic [30, 65, 112]. Example systems of this period include FACT [87], DLP, RACER [76], CEL [7] and KAON2.[10]

**Phase 4.** This latest phase, is characterised by the development and application of mature, highly optimised implementations such as PELLET [117], HERMIT [116], FACT++ [137], RACERPRO [75], ELK [90], MASTRO STUDIO [63], used as the reasoning engines for knowledge representation languages in the realm of the Semantic Web, such as OWL 2 [102] and their profiles [103].

### 8.2 Fuzzy Description Logics

Initially, the work on *Fuzzy* DLs (FDLs) addressed the problem to equip DLs with a fuzzy semantics. A first step was to generalize the semantics of atomic concepts and roles from crisp sets and relations to fuzzy sets and fuzzy relations, respectively, and the semantics of subsumption to the inclusion between fuzzy sets. Nevertheless this does not mean there is a wide agreement on how to generalize the semantics of complex concepts and, thus, it is not surprising that various proposals have been made.

The first attempt in this direction is [141], which proposes a generalization of [49]. The language defined in this work, called $\mathcal{FTSL}^-$, uses concept constructors conjunction (: $\texttt{and}\, C_1, \ldots, C_n$), value restriction (: $\texttt{all}\, R\, C$), restricted existential quantification (: $\texttt{some}\, R\, \top$), modifiers (: `NOT`, : `VERY`, : `SLIGHTLY`, etc.) and an early version of concrete domains. The semantics was called *test score semantics* (see [142]) as *scores* (what we nowadays call "truth values") are assigned to concepts. However, the interesting point are the truth functions used to calculate the truth values of complex concepts. Specifically,

- It is suggested to use the min function to compute the value of a conjunction of concepts. Besides this suggestion, the author not only recognizes that any other t-norm can be used as the semantics of conjunction, but also that both lower and upper bounds for conjunctions can be computed considering min and Łukasiewicz t-norms as upper and lower bounds respectively.
- The semantics of value restriction (: $\texttt{all}\, R\, C$) is defined in two alternative ways. Either through a fuzzy implication operator, as it is done nowadays, or through the notion of conditional necessity from possibility theory. The author, however, adopts the second option.

[10] However, it should be noted that KAON2 maps ontologies into disjunctive logic programs; `http://kaon2.semanticweb.org/#introduction`.

- The semantics of number restriction is defined just by means of a function that normalizes the sum of the relation degree of a given individual with respect to each other individual in the domain. As an example we report the definition of the *at-least number restriction*:

$$\mu_{(\texttt{:at-least}\ n\ R)}(x) = \mu_{\textit{at-least-n}}\Big(\sum_y \mu_R(x,y)\Big),$$

where $\mu_X$ is the characteristic function of the set (relation) $X$ and $\mu_{\textit{at-least-n}}$ is defined in the following way:

$$\mu_{\textit{at-least-n}}(z) = \begin{cases} 0 & \text{if } z \leq n-1 \\ z-n+1 & \text{if } n-1 \leq z \leq n \\ 1 & \text{if } z \geq n, \end{cases}$$

that is quite different from the solution adopted in the latest works on FDL.

The semantics defined in [141] was general enough to leave open the adoption of a truth function for conjunction. Nevertheless, this work was inspired by practical purposes and its goal was to provide a more refined tool for knowledge representation.

Later on, in [124, 136] a more theoretic approach was considered. The evolution of the notation towards a logical-like abstraction, that was experimented by the classical DL community, influenced these works, which utilizes the same modern notation reported in Section 3.1. The language studied in [136] was called $\mathcal{ALC}_{F_M}$ (the subindex $F_M$ stands for infinitely many truth values). It presents, as concept constructors, conjunction $\sqcap$, disjunction $\sqcup$, value restriction $\forall R.C$, existential quantifier $\exists R.C$ and *manipulators*[11] $M_i C$ [136], together with a many-valued logic style semantics [77]. The choice of the truth functions for the logical connectives falls on min and max for conjunction and disjunction, respectively. The semantics for the existential quantifier is the one provided in Definition 3.2.1 and it is the first place where it has been defined in this way. This work is also the first (together with [124]) in defining the semantics for value restriction $\forall R.C$ by means of the so-called *Kleene–Dienes implication*, defined on $[0,1]$ as:

$$r \rightarrow s := \max\{1-r, s\}$$

which is a straightforward generalization of the classical one. In particular, if $\mathcal{I}$ is an FDL interpretation, the semantics of value restriction $\forall R.C$, based on Kleene–Dienes implication, is defined in the following way:

$$(\forall R.C)^{\mathcal{I}}(v) = \inf_{w \in M}\{\max\{1 - R^{\mathcal{I}}(v,w), C^{\mathcal{I}}(w)\}\}.$$

Finally, for the semantics of manipulators $M_i C$, unary functions on $[0,1]$ were used, in the same way as is done in the framework of Mathematical Fuzzy Logic when the semantics of *fuzzy hedges* is defined (see Chapter VIII of this Handbook and therein references for details).

[11] The notion of manipulators corresponds to the logical notion of *hedges* modeling *linguistic modifiers* used in this chapter.

In [124], the language studied is called (and, indeed, it is) $\mathcal{ALC}$ and the semantics adopted is the same as the one used in [136], plus a unary function that gives the semantics to concept complementation defined as:

$$\neg r := 1 - r.$$

The set of operations that includes $\min\{r, s\}, \max\{r, s\}, \max\{1 - r, s\}$ and $1 - r$ on the real unit interval is commonly denoted with the name of *Zadeh's semantics*. The strength of [124] and of its journal version [125], is that they set up a clear syntax and semantics, very close to the classical ones and relate each other without the intermediate step of many-valued first order logic, like in [136]. In this way Fuzzy Description Logic is set up as an autonomous discipline with a clearly defined syntax and semantics as the one provided in [125].

Moreover, in [124] for the first time fuzzy axioms have been defined in this way, since in previous papers the notion of fuzzy axioms was not considered. In [124], just non-strict lower bound axioms are considered. In [125] axioms stating a non-strict upper bound have been introduced as well, but strict bound axioms are not considered. Since then, some works on FDL consider strict bound axioms, like [26, 122, 126, 131] and some others do not consider strict bound axioms, like [9, 15, 22, 38, 56, 58, 127, 129]. Later, in [127], the more general framework of semantics based on lattices has been considered that are supposed to be not necessarily chains, too.

These works, indeed, opened the door to the possibility of expanding the language in order to cover the advances that had been done in the classical framework. A fuzzy semantics for concrete domains was introduced in [130]. A semantics for unqualified number restriction, role hierarchies, inverse and transitive roles was introduced in [122]. A semantics for nominals was introduced in [121]. A semantics for qualified number restriction was introduced in [16, 17] and later on discussed and improved in [24, 27].

As it is known in the framework of MFL, the presence of Kleene–Dienes implication leads to some counter-intuitive facts due to its behavior with respect to the order in $[0, 1]$. Indeed if $x \to y = (1 - x) \vee y$ then the implication is true iff either $x = 0$ or $y = 1$ that is not intuitive. Moreover if $x \geq a$ and $x \to y \geq b$ nothing could be deduced about the value of $y$. This means that we have no Modus Ponens rule of any type. This fact, in the setting of FDLs based on Zadeh semantics, has been pointed out by Hájek in [80], where the example of the assertion "all hotels near to the main square are expensive" is presented in order to highlight the consequences of using Kleene–Dienes implication in the semantics of value restriction. Such assertion can be formally expressed as

$$\langle \forall \mathtt{hasNear}.\mathtt{Expensive}(\mathtt{MainSquare}) \geq 1 \rangle. \tag{4}$$

Here we will further develop this example, as done in [54]. Consider the following fuzzy ABox HOTELS:

$$\langle \mathtt{hasNear}(\mathtt{MainSquare}, \mathtt{Hotel}_1) = 0.9 \rangle \quad \langle \mathtt{Expensive}(\mathtt{Hotel}_1) = 0.9 \rangle$$
$$\langle \mathtt{hasNear}(\mathtt{MainSquare}, \mathtt{Hotel}_2) = 0.5 \rangle \quad \langle \mathtt{Expensive}(\mathtt{Hotel}_2) = 0.5 \rangle$$
$$\langle \mathtt{hasNear}(\mathtt{MainSquare}, \mathtt{Hotel}_3) = 0.1 \rangle \quad \langle \mathtt{Expensive}(\mathtt{Hotel}_3) = 0.1 \rangle.$$

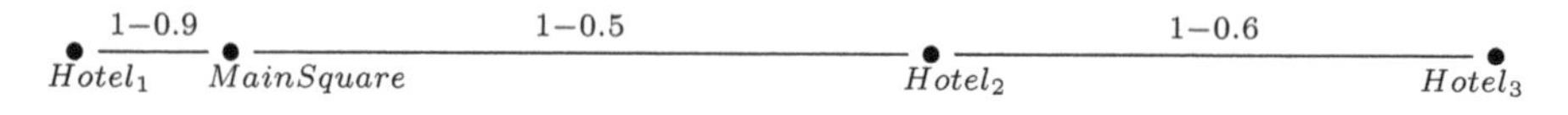


Figure 7. Interpretation satisfying HOTELS

The HOTELS ABox indeed depicts the ideal situation, where "for each hotel the degree of its being near to the main square equals the degree of its being expensive" and where "there is at least one hotel which is near to the main square in degree 0.5". In Figure 7 we report an interpretation that satisfies ABox HOTELS, where the distance between MainSquare and $\texttt{Hotel}_\texttt{x}$ is calculated as

$$1 - \texttt{hasNear}(\texttt{MainSquare}, \texttt{Hotel}_\texttt{x}).$$

In this ideal situation the truth value of assertion (4) should be 1, because hotels are at least as expensive as they lie near the main square. Now, if the truth value of (4) is calculated using the truth function of any residuated implication, its value is indeed 1, which means that the ABox HOTELS $\cup$ $\{(4)\}$ is consistent. Nevertheless, using the truth function of Kleene–Dienes implication, the result is different. In fact, in every interpretation $\mathcal{I}$ that is a model of HOTELS, we have that:

$$\begin{aligned}
& (\forall \texttt{hasNear}.\texttt{Expensive}(\texttt{MainSquare}))^{\mathcal{I}} = \\
= \ & \inf_{v \in \Delta^{\mathcal{I}}} \{\texttt{Near}^{\mathcal{I}}(\texttt{MainSquare}^{\mathcal{I}}, v) \Rightarrow_* \texttt{Expensive}^{\mathcal{I}}(v)\} \leq \\
\leq \ & \inf\{\max\{1-0.9, 0.9\}, \max\{1-0.5, 0.5\}, \max\{1-0.1, 0.1\}\} = \\
= \ & \inf\{0.9, 0.5, 0.9\} = \\
= \ & 0.5.
\end{aligned}$$

Hence, the truth value of assertion (4), using Kleene–Dienes implication, is at most 0.5 in every model of HOTELS, against the intuition, reflected in HOTELS, that its truth value should be 1. Hence, the ABox HOTELS $\cup$ $\{(4)\}$ is inconsistent.

For these reasons, many-valued logics studied in the framework of MFL use residuated implications (satisfying $x \leq y$ iff $x \to y = 1$ and a Modus Ponens rule saying that if $x \geq a$ and $x \to y \geq b$ then $y \geq a * b$). Based on MFL, Hájek in [80] defines logical based FDLs as fragments of first order residuated many-valued logics in an analogous way as DLs are fragments of first order classical logic. In this new framework, not only the semantics of conjunction is a t-norm, but it is also recovered the idea, firstly proposed in [124, 136], of a tight relation between FDLs and first order fuzzy logic. As we have seen, with the only exception of [141], the operation min is the only function adopted as a semantics for the conjunction concept constructor until [80]. Hájek's proposal defines a family of FDLs depending on the t-norm used to interpret strong conjunction. In the same year, Straccia defined and studied in [128–130] some FDLs where conjunction is interpreted as a t-norm different from the minimum as well.

The new framework inspired several successive works on FDL such as, [10, 18, 22, 24, 25, 27, 55, 131]. Among the ones that deepen the relationships between FDL and MFL we can find [56, 58, 59, 73].

The new framework of FDLs based on t-norms supposed also a re-thinking about the notation used in FDLs. Indeed, the use of the same notation of DLs for FDLs has been based on the fact that, in order to generalize DLs to the multi-valued framework, it seemed enough to generalize the semantics of concepts and roles to fuzzy sets and fuzzy relations. With this idea it is obvious that the same concept constructors (and, with them, the same formal languages) could be maintained in a multi-valued framework. This formalization worked indeed well when the semantics adopted as underlying truth value algebra was the Zadeh's semantics. But, since [80, 128–130], some works on FDLs began to consider the use of residuated implications as formalized in the framework of MFL. However, adopting a framework based on MFL and maintaining the same notation as in the classical case, could produce a slight confusion. This is due to several reasons related to differences between the classical and the many-valued framework. Commonly, with some exceptions, such differences include the following items:

1. Two kinds of conjunctions, with different mathematical properties, can be considered in the many-valued framework and the same holds for disjunction.
2. Implication is, in general, not definable from other connectives.
3. The quantifiers are not definable from each other by means of the duality through negation $\exists R.C \equiv \neg\forall R.\neg C$.
4. The disjunction is not definable from the residuated negation $\neg C := C \to \bot$ and the conjunction $\sqcap$.

All these items must be taken into account both when choosing the symbols denoting the constructors of our description languages and when building the hierarchy of fuzzy description languages, as we have already seen in Section 3.3. As an example recall that, in classical DLs, $\mathcal{ALE}$ is strictly contained in $\mathcal{ALC}$, while within many FDLs, by item 3 above, this is not the case.

In particular, we find worth discussing the case of implication. In classical DLs, the implication is not usually a primitive concept constructor, even though implication is often implicitly used. This is due to the fact that the implication is definable from conjunction and negation. Nevertheless, in the logic BL and many of its extensions, implication is in general not definable from other connectives. The first time that the concept constructor $\to$ for the implication is included in the definition of the language as a primitive connective has been in [80].

The introduction of a new symbol for implication allows to utilize a concept constructor that is not otherwise definable. Moreover, it is useful in order to define, in BL and its extensions, other concept constructors like weak conjunction, weak disjunction and residuated negation.

### 8.3 Computational issues in a historical perspective

Like for classical DLs, also in the framework of t-norm based FDL the computational issues have marked the recent history and the development of the subject. Computational issues have been already extensively addressed in Sections 6 and 7. Here we want to summarize them in a historical perspective. For this reason, in this section we omit the technical details, the definitions of the notions cited and any explanation of how they work. For a deeper understanding of these notions the reader is addressed to Sections 6 and 7.

Decidability has been a central matter since the beginning of the research in FDL. As in the classical case, the first problem to be dealt with has been concept subsumption. In order to deal with this problem a structural subsumption based procedure has been employed. In [49] a structural subsumption algorithm SUBS?$[a,b]$ for deciding classical concept subsumption in $\mathcal{FL}^-$ with respect to an empty KB is presented. This kind of algorithms perform a comparison in the syntactic structure of two given concept descriptions after having transformed them in a suitable normal form. The fact that SUBS?$[a,b]$ calculates subsumption in polynomial time relies on the fact that every $\mathcal{FL}^-$ concept $C$ is equivalent to a $\mathcal{FL}^-$ concept $C^*$ where each value restriction $\forall R.C$ appears at most once for each nesting level. That is, for example concepts $\forall R.(C \sqcap D)$ and $\forall R.C \sqcap \forall R.D$ are equivalent. Moreover, the possibility of applying structural algorithms to a given calculus is due to the fact that concept conjunctions can be considered as sets of concepts. The problem of generalizing structural subsumption algorithms to the many-valued framework has been addressed in [141] and [136]. In [141] it is proven that the concept subsumption problem, with respect to empty knowledge bases, of a language denoted as $\mathcal{FTSL}^-$ is decidable. The main result in [136] is the decidability proof for the concept subsumption problem, with respect to empty knowledge bases, of a language denoted as $\mathcal{ALC}_{FM}$. The structural subsumption algorithm SUBS?$[a,b]$ can be indeed consistently used in order to decide 1-subsumption[12] for $\mathbf{G_n}$-$\mathcal{FL}^-$. This is due to the fact that the Gödel $t$-norm $\wedge$ works well with its residuum $\Rightarrow_{\mathbf{G_n}}$. That is, for every $x,y,z \in \mathbf{G_n}$ it holds that $x \Rightarrow_{\mathbf{G_n}} (y \wedge z) = (x \Rightarrow_{\mathbf{G_n}} y) \wedge (x \Rightarrow_{\mathbf{G_n}} z)$. Unfortunately, the same result does not hold for non-idempotent t-norms. In this case, there may be $x,y,z \in T$ such that $x \Rightarrow_* (y * z) \neq (x \Rightarrow_* y) * (x \Rightarrow_* z)$. In [61] the problem has been studied for the 1-subsumption problem with empty KB under the finite Łukasiewicz $t$-norm $*_{Ł_n}$. Since Łukasiewicz conjunction is not idempotent, complex concepts where just $\sqcap$ appears as concept constructor can not be seen as sets of atomic concepts. Nevertheless, as shown in [61], complex concepts in $\mathbf{Ł}_n$-$\mathcal{FL}^-$ can be seen as *multisets* of simpler concepts, that is, different *occurrences* of atomic concepts are now seen as different elements of a given complex concept. This gives us the possibility of still defining structural subsumption algorithms for $\mathbf{Ł}_n$-$\mathcal{FL}^-$.

In [124] and [125] the decidability of the entailment problem with respect to empty TBoxes of language $\mathcal{ALC}$ under Zadeh's semantics is proven. The procedure used is a tableaux algorithm based on the recursive production of a set of constraints until either a clash is produced or the set of constraint is complete, in the sense that no further rule can be applied to the existing set of constraints. Some years later, more general tableau algorithms were presented, managing GCIs and the more expressive FDLs $\mathcal{SHIN}$ [123] and $\mathcal{SHOIQ}$ [120] under Zadeh's semantics. The reasoner `Fire` implements a tableaux algorithm to solve several reasoning tasks in Zadeh $\mathcal{SHIN}$ [118]. If the TBox is empty, these algorithms have the same computational complexity as the tableaux algorithms

[12] Note that subsumption between two concepts in $\mathbf{G_n}$-$\mathcal{FL}^-$ always takes either value 0 or value 1. Therefore, speaking about $(\geq r)$- or $(= r)$-subsumption in $\mathbf{G_n}$-$\mathcal{FL}^-$ does not make sense.

for their corresponding crisp languages. If there are GCIs, it is conjectured that the algorithms do not provide a tight complexity bound. Tableau algorithms can also be used for finite t-norms if the degrees of truth are taken from a residuated De Morgan lattice. The idea here is that all the tableau rules can be applied in a non-deterministic way by producing all the combinations of degrees of truth that satisfy a restriction. An algorithm for $*$-consistency in $\mathcal{SHI}$ without GCIs has been presented in [44]. Since there are exponentially many rule applications and exponentially many ways to apply a single rule, it is conjectured that this algorithm does not provide a tight complexity bound.

As explained in Section 6.2.4, the decidability of the reasoning with several FDLs have been proved by a reduction to reasoning with classical DLs. In [126] KB consistency of the language $\mathcal{ALCH}$ with role hierarchies over Zadeh's semantics is reduced to reasoning with classical $\mathcal{ALCH}$. An algorithm for Z-consistency in the more expressive $\mathcal{SROIQ}(\mathbf{D})$ has also been presented in [18]. If one restricts his attention to finite chains, there are algorithms for G-consistency [20] and Ł-consistency [27] in $\mathcal{SROIQ}$. These results have been extended to $*$-consistency for every finite t-norm [21, 29] (where it is also possible to combine operators belonging to different finite fuzzy logics). The `DeLorean` reasoner implements a reasoning algorithm for fuzzy extensions of OWL 2 using Zadeh and Gödel fuzzy logics [19].

In [128] a procedure based on a reduction to the Mixed Integer Linear Programming problem is proposed in order to prove decidability of language $\mathcal{ALC}$ extended with concrete domains and fuzzy modifiers under Zadeh's semantics or Łukasiewicz semantics for *acyclic TBoxes*. This procedure has been explained in Section 6.2.1. This family of algorithms based on a combination of tableaux rules and optimization problems has been used to solve the BED problem in more expressive logics. An algorithm to solve the witnessed BED problem in $\mathfrak{I}\mathcal{ALC}$ without TBoxes for any left-continuous t-norm was proposed in [25]. This result was later extended to consider acyclic TBoxes [60]. Similarly, an algorithm checking the $*$-consistency in $\mathcal{SI}$ without TBoxes for any continuous t-norm was proposed in [119]. These algorithms can be extended to support fuzzy aggregation operators as shown in [28]. `fuzzyDL` [23] reasoner uses one of these algorithms to solve several reasoning tasks for Zadeh and Łukasiewicz $\mathcal{SHIF}(\mathbf{D})$.

Until [80], the algebra of truth values considered is mainly the real unit interval [0,1] with operations max, min, $1 - x$ and Kleene–Dienes implication (also known as the Zadeh semantics). In [80] and [128–130], for the first time a t-norm based semantics is considered. Specifically, the main result proved in [80] is the decidability of the satisfiability and subsumption problems, with respect to empty knowledge bases, for language $\mathcal{ALC}$ over a standard algebra restricted to witnessed models. The result is achieved by means of a recursive reduction of the concept satisfiability and subsumption to satisfiability and logical consequence in the corresponding propositional calculus. This procedure has been explained in Section 6.2.5. Other works where the same research line is applied are [55] and [45]. The first one tried to prove decidability of concept satisfiability in $\Pi$-$\mathfrak{I}\mathcal{ALCUE}$ without knowledge bases and with respect to so-called *quasi-witnessed models*. In recent times an error was found in the proof, but a correct proof can be found in [53]. The second one proposes an implementation in PSPACE of Algorithm 5 in the case of any finite t-norm.

Further works on the decidability of FDLs are the following. In [21] a quite expressive FDL over a set of operations that join minimum t-norm with Zadeh's operations is presented. The decidability of the KB consistency problem is proved through a recursive reduction to classical DL when the set of truth values is fixed in advance and finite. A similar result can be obtained by considering degrees of truth taken from a residuated De Morgan lattice [44]. In [34] it is proved that KB consistency for an FDL with role constructors and nominals over any t-norm with Gödel negation is linearly reducible to classical DL. In [39] it is proved undecidability of KB consistency with respect to unrestricted interpretations for a really basic FDL over Łukasiewicz t-norm. In [40] different results are provided for quite expressive FDLs over complete residuated De Morgan lattices, including some decidability results involving finite lattices. An interesting feature of this paper is the use of tableaux algorithms for the decidability results.

In the last years there have been several works dealing with undecidability of FDLs over a BL-chain determined by a continuous t-norm. Early works on FDLs based on non-idempotent t-norms, are spotted by examples of languages that do not enjoy the *finite model property* (FMP), that is a fundamental property to easily prove decidability of a calculus. In [80] it is proven that FMP fails with general models for the concept satisfiability problem without TBox in $[0,1]_\Pi$-$\mathfrak{I}\mathcal{ALE}$. In [25] the lack of FMP is proved for knowledge base consistency in $[0,1]_{Ł\Pi}$-$\mathcal{ALC}$. At that time a proof of undecidability was still not taken into account, because the lack of FMP was not considered to be a serious problem. On the other hand, in [25] the problem is restricted to acyclic knowledge bases and witnessed models in order to easily obtain decidability. Some years later, it was proven that the languages $[0,1]_Ł$-$\mathcal{ALC}$ and $[0,1]_\Pi$-$\mathfrak{I}\mathcal{ALCE}$ do not enjoy the finite model property when general TBoxes are considered, even if we consider witnessed models only [15]. Although this work does not give any undecidability result, it casts doubts on the decidability of those languages when general TBoxes are considered. The first actual undecidability result is given in [9], where undecidability of KB consistency for language $[0,1]_\Pi$-$\mathfrak{I}\mathcal{ALCE}$ is proved. The method used in this work as well as in the subsequent works on undecidability is a reduction to PCP similar to the one explained in Section 7. However, in order to prove undecidability, this work requires the use of inclusion axioms with a strict lower bound; but the authors were not able to strengthen the construction to prove undecidability of the same problem without this kind of axioms. Nevertheless the full undecidability of KB consistency for language $[0,1]_\Pi$-$\mathfrak{I}\mathcal{ALCE}$ was proved later in [10] and [11]. Note that in these works the undecidability result is proved for the cited problem when restricted both to witnessed interpretations and to strongly witnessed interpretations. In [38] the undecidability of concept satisfiability with respect to a KB for the language $\mathcal{ALC}$ over a De Morgan chain containing the standard Łukasiewicz chain as a subalgebra is proved. In the same paper [38] it is also shown that if the De Morgan lattice considered as underlying algebra of truth values is finite, the same problem is indeed decidable. The result is achieved through a recursive reduction to a decidable problem from automata theory. Along the same line, in [60] similar results under Łukasiewicz semantics for $\mathcal{ALC}$ are proven. In [41] a systematic study of undecidability in FDL is undertaken yielding sufficient conditions for proving undecidability in a wide class of FDLs. A quite complete overview of the decidability and undecidability results known for DLs can be found in [2, 36].

Computational complexity is a problem that has been addressed only in recent years. In [31] it is proved that concept witnessed $r$-satisfiability with respect to a general TBox for language $\mathcal{ALC}$ over (not necessarily linear) finite De Morgan lattices is EXPTIME-complete. The result is proved by means of a reduction to automata theory. Notice that in [31] the truth function considered for concept constructors are min and max for conjunction and disjunction respectively, a unary function that satisfies the De Morgan laws for the negation and Kleene–Dienes implication. This implies that the implication constructor and the existential quantifier are definable like in Zadeh's semantics. In [38] the result of [31] is enhanced by adding a finite t-norm to the operations of the De Morgan lattice and using its residuum in the semantics of the value restriction and of inclusion axioms. With such semantics, the existential quantifier is no more definable as in the De Morgan lattices considered in [31]. So, the language considered in [38] is $\mathcal{ALCE}$. Nevertheless, concept satisfiability with respect to general knowledge bases is proved to be EXPTIME-complete. Again, the result is proved by means of a reduction to automata theory. The semantics considered in [37] is the same as in [38] and the language is $\mathcal{ALCE}$ with inverse roles. By means of a reduction to automata theory it is proved that entailment of lower bound inclusion axioms by general knowledge bases is EXPTIME-complete. The same problem is proved to be PSPACE-complete if the TBox is acyclic. Moreover, concept satisfiability with respect to acyclic TBoxes is proved to be PSPACE-complete as well. Again, the proof is based on a recursive reduction to automata theory. The result in [37] generalizes the one in [38], not only because the language considered is more expressive, but also because the concept subsumption problem is considered, that was not considered in [38]. Moreover the problem of concept satisfiability with respect to acyclic TBoxes is considered.

The result in Theorem 7.4.1 matches the same complexity result for classical $\mathcal{ALC}$. The same matching has been proved for different languages, e.g., for $\mathcal{ALC}$ over (not necessarily linear) finite De Morgan lattices [31], for $\mathcal{ALCE}$ over finite De Morgan lattices with an added t-norm [38], for $\mathcal{ALCE}$ with inverse roles over finite De Morgan lattices [37], for $\mathfrak{I}\mathcal{ALCE}$ with the empty KB over finite chains [45] and for $\mathcal{FL}^-$ with an empty KB under the finite Łukasiewicz $t$-norm [61]. These are all languages whose complexity of the respective reasoning tasks has been proved to match the classical results. Nevertheless, in [33] has been proved that the complexity of KB consistency for language $\mathcal{EL}$ under the finite Łukasiewicz $t$-norm does not match with the complexity of the same problem for classical $\mathcal{EL}$. Indeed, KB consistency jumps from PTIME-complete in the latter case to EXPTIME-complete in the former case.

Some complementary related work about FDLs can be found in [32, 54, 132].

**Acknowledgements**

We want to thank Brunella Gerla and Ulrike Sattler for their reading of the first version of the chapter and their helpful comments and corrections.

Fernando Bobillo acknowledges support by two Spanish CICYT projects: TIN2012-30939 and TIN2013-46238-C4-4-R.

Marco Cerami acknowledges support by the european project POST UP II, number CZ.1.07/2.3.00/30.0041. The present work has been written while this author was working as a postdoctoral researcher at University Palacký in Olomouc, under this project.

Francesc Esteva acknowledges support by Spanish MICINN project EdeTRI (TIN 2012-39348-C02-01) and grant 2014SGR-118 from the Generalitat de Catalunya.

Àngel García-Cerdaña acknowledges support by Spanish MICINN projects EdeTRI (TIN2012-39348-C02-01) and MTM 201125747, the CSIC project 201450E045 and grant 2014SGR-788 from the Generalitat de Catalunya.

Rafael Peñaloza has been partially supported by the German Research Foundation (DFG) under grant BA 1122/17-1 (FuzzyDL) and within the Cluster of Excellence 'Center for Advancing Electronics Dresden' (cfAED). Most of the work presented here was developed while he was still affiliated to TU Dresden and the Center for Advancing Electronics Dresden, Germany.

# BIBLIOGRAPHY

[1] Franz Baader. Using automata theory for characterizing the semantics of terminological cycles. *Annals of Mathematics and Artificial Intelligence*, 18(2–4):175–219, 1996.

[2] Franz Baader, Stefan Borgwardt, and Rafael Peñaloza. On the decidability status of fuzzy $\mathcal{ALC}$ with general concept inclusions. *Journal of Philosophical Logic*, 44(2):117–146, 2015.

[3] Franz Baader, Sebastian Brandt, and Carsten Lutz. Pushing the $\mathcal{EL}$ envelope. In Leslie Pack Kaelbling and Alessandro Saffiotti, editors, *Proceedings of the 19th International Joint Conference on Artificial Intelligence (IJCAI 2005)*, pages 364–369. Morgan Kaufmann, 2005.

[4] Franz Baader, Diego Calvanese, Deborah L. McGuinness, Daniele Nardi, and Peter F. Patel-Schneider, editors. *The Description Logic Handbook: Theory, Implementation, and Applications*. Cambridge University Press, New York, NY, USA, 2003.

[5] Franz Baader, Jan Hladik, and Rafael Peñaloza. Automata can show PSPACE results for description logics. *Information and Computation*, 206(9–10):1045–1056, 2008.

[6] Franz Baader and Bernhard Hollunder. KRIS: Knowledge representation and inference system, system description. *ACM SIGART Bulletin*, 2:8–14, 1991.

[7] Franz Baader, Carsten Lutz, and Boontawee Suntisrivaraporn. CEL—A polynomial-time reasoner for life science ontologies. In Ulrich Furbach and Natarajan Shankar, editors, *Automated Reasoning. Proceedings of the 3rd International Joint Conference on Automated Reasoning (IJCAR 2006)*, volume 4130 of *Lecture Notes in Artificial Intelligence*, pages 287–291. Springer, 2006.

[8] Franz Baader and Werner Nutt. Basic description logics. In Franz Baader, Diego Calvanese, Deborah L. McGuinness, Danielle Nardi, and Peter F. Patel-Schneider, editors, *The Description Logic Handbook: Theory, Implementation, and Applications*, pages 47–100. Cambridge University Press, 2003.

[9] Franz Baader and Rafael Peñaloza. Are fuzzy description logics with general concept inclusion axioms decidable? In *Proceedings of the 20th IEEE International Conference on Fuzzy Systems (FUZZ-IEEE 2011)*, pages 1735–1742. IEEE Press, 2011.

[10] Franz Baader and Rafael Peñaloza. GCIs make reasoning in fuzzy DL with the product t-norm undecidable. In Riccardo Rosati, Sebastian Rudolph, and Michael Zakharyaschev, editors, *Proceedings of the 24th International Workshop on Description Logics (DL 2011)*, volume 745, pages 37–47. CEUR Workshop Proceedings, 2011.

[11] Franz Baader and Rafael Peñaloza. On the undecidability of fuzzy description logics with gcis and product $t$-norm. In Cesare Tinelli and Viorica Sofronie-Stokkermans, editors, *Proceedings of the 8th International Symposium Frontiers of Combining Systems (FroCos 2011)*, number 6989 in Lecture Notes in Artificial Intelligence, pages 55–70. Springer, 2011.

[12] Franz Baader and Ulrike Sattler. An overview of tableau algorithms for description logics. *Studia Logica*, 69:5–40, 2001.

[13] Matthias Baaz, Christian G. Fermüller, Gernot Salzer, and Richard Zach. Labeled calculi and finite-valued logics. *Studia Logica*, 61(1):7–33, 1998.

[14] Libor Běhounek, Petr Cintula, and Petr Hájek. Introduction to mathematical fuzzy logic. In Petr Cintula, Petr Hájek, and Carles Noguera, editors, *Handbook of Mathematical Fuzzy Logic - Volume 1*, volume 37 of *Studies in Logic, Mathematical Logic and Foundations*, pages 1–101. College Publications, London, 2011.

[15] Fernando Bobillo, Félix Bou, and Umberto Straccia. On the failure of the finite model property in some fuzzy description logics. *Fuzzy Sets and Systems*, 172(1):1–12, 2011.

[16] Fernando Bobillo, Miguel Delgado, and Juan Gómez-Romero. A crisp representation for fuzzy $\mathcal{SHOIN}$ with fuzzy nominals and general concept inclusions with fuzzy nominals and general concept inclusions. In Paulo Cesar G. da Costa, Kathryn B. Laskey, Kenneth J. Laskey, Francis Fung, and Michael Pool, editors, *Proceedings of the 2nd International Workshop on Uncertainty Reasoning for the Semantic Web (URSW 2006)*, volume 218. CEUR Workshop Proceedings, 2006.

[17] Fernando Bobillo, Miguel Delgado, and Juan Gómez-Romero. Optimizing the crisp representation of the fuzzy description logic $\mathcal{SROIQ}$. In Fernando Bobillo, Paulo Cesar G. da Costa, Claudia d'Amato, Nicola Fanizzi, Francis Fung, Thomas Lukasiewicz, Trevor Martin, Matthias Nickles, Yun Peng, Michael Pool, Pavel Smrž, and Peter Vojtás, editors, *Proceedings of the 3rd International Workshop on Uncertainty Reasoning for the Semantic Web (URSW 2007)*. CEUR Workshop Proceedings, 2007.

[18] Fernando Bobillo, Miguel Delgado, and Juan Gómez-Romero. Crisp representation and reasoning for fuzzy ontologies. *International Journal of Uncertainty, Fuzziness and Knowledge-Based Systems*, 17(4):501–530, 2009.

[19] Fernando Bobillo, Miguel Delgado, and Juan Gómez-Romero. DeLorean: A reasoner for fuzzy OWL 2. *Expert Systems with Applications*, 39:258–272, 2012.

[20] Fernando Bobillo, Miguel Delgado, Juan Gómez-Romero, and Umberto Straccia. Fuzzy description logics under Gödel semantics. *International Journal of Approximate Reasoning*, 50(3):494–514, 2009.

[21] Fernando Bobillo, Miguel Delgado, Juan Gómez-Romero, and Umberto Straccia. Joining Gödel and Zadeh fuzzy logics in fuzzy description logics. *International Journal of Uncertainty, Fuzziness and Knowledge-Based Systems*, 20(4):475–508, 2012.

[22] Fernando Bobillo and Umberto Straccia. A fuzzy description logic with product t-norm. In *Proceedings of the 16th IEEE International Conference on Fuzzy Systems (FUZZ-IEEE 2007)*, pages 652–657. IEEE Computer Society, 2007.

[23] Fernando Bobillo and Umberto Straccia. fuzzyDL: An expressive fuzzy description logic reasoner. In *Proceedings of the 17th IEEE International Conference on Fuzzy Systems (FUZZ-IEEE 2008)*, pages 923–930. IEEE Xplore. Digital Library, 2008.

[24] Fernando Bobillo and Umberto Straccia. Towards a crisp representation of fuzzy description logics under Łukasiewicz semantics. In Aijun An, Stan Matwin, Zbigniew W. Raś, and Dominik Ślezak, editors, *Proceedings of the 17th International Symposium on Methodologies for Intelligent Systems (ISMIS 2008)*, volume 4994 of *Lecture Notes in Computer Science*, pages 309–318. Springer, 2008.

[25] Fernando Bobillo and Umberto Straccia. Fuzzy description logics with general t-norms and datatypes. *Fuzzy Sets and Systems*, 160(23):3382–3402, 2009.

[26] Fernando Bobillo and Umberto Straccia. Fuzzy ontology representation using OWL 2. *International Journal of Approximate Reasoning*, 52:1073–1094, 2011.

[27] Fernando Bobillo and Umberto Straccia. Reasoning with the finitely many-valued Łukasiewicz fuzzy description logic $\mathcal{SROIQ}$. *Information Sciences*, 181:758–778, 2011.

[28] Fernando Bobillo and Umberto Straccia. Aggregation operators for fuzzy ontologies. *Applied Soft Computing*, 13:3816–3830, 2013.

[29] Fernando Bobillo and Umberto Straccia. Finite fuzzy description logics and crisp representations. In Fernando Bobillo, Paulo Cesar G. da Costa, Claudia d'Amato, Nicola Fanizzi, Kathryn Laskey, Ken Laskey, Thomas Lukasiewicz, Matthias Nickles, and Michael Pool, editors, *Uncertainty Reasoning for the Semantic Web II*, volume 7123 of *Lecture Notes in Computer Science*, pages 102–121. Springer, 2013.

[30] Alex Borgida. On the relative expressiveness of description logics and predicate logics. *Artificial Intelligence*, 82:353–367, 1996.

[31] Stefa Borgwardt and Rafael Peñaloza. Description logics over lattices with multi-valued ontologies. In Toby Walsh, editor, *Proceedings of the 22nd International Joint Conference on Artificial Intelligence (IJCAI 2011)*, pages 768–773. AAAI Press, 2011.

[32] Stefan Borgwardt. *Fuzzy Description Logics with General Concept Inclusions*. PhD thesis, Technische Universität Dresden, 2014.

[33] Stefan Borgwardt, Marco Cerami, and Rafael Peñaloza. The complexity of subsumption in fuzzy $\mathcal{EL}$. In Qiang Yang and Michael Wooldridge, editors, *Proceedings of the 24th International Joint Conference on Artificial Intelligence (IJCAI 2015)*, pages 2812–2818. AAAI Press, 2015.

[34] Stefan Borgwardt, Felix Distel, and Rafael Peñaloza. Gödel negation makes unwitnessed consistency crisp. In Yevgeny Kazarov, Domenico Lembo, and Frank Wolter, editors, *Proceedings of the 25th International Workshop on Description Logics (DL 2012)*, volume 846, pages 103–113. CEUR Workshop Proceedings, 2012.

[35] Stefan Borgwardt, Felix Distel, and Rafael Peñaloza. Decidable Gödel description logics without the finitely-valued model property. In Chitta Baral, Giuseppe De Giacomo, and Thomas Eiter, editors, *Proceedings of the 14th Conference on Principles of Knowledge Representation and Reasoning (KR 2014)*, pages 228–237. AAAI Press, 2014.

[36] Stefan Borgwardt, Felix Distel, and Rafael Peñaloza. The limits of decidability in fuzzy description logics with general concept inclusions. *Artificial Intelligence*, 218:23–55, 2015.

[37] Stefan Borgwardt and Rafael Peñaloza. Finite lattices do not make reasoning in $\mathcal{ALCI}$ harder. In Fernando Bobillo, Rommel Carvalho, Paulo Cesar G. da Costa, Claudia d'Amato, Nicola Fanizzi, Kathryn B. Laskey, Kenneth J. Laskey, Thomas Lukasiewicz, Trevor Martin, Matthias Nickles, and Michael Pool, editors, *Proceedings of the 7th International Workshop on Uncertainty Reasoning for the Semantic Web (URSW 2011)*, volume 778, pages 51–62. CEUR Workshop Proceedings, 2011.

[38] Stefan Borgwardt and Rafael Peñaloza. Fuzzy ontologies over lattices with t-norms. In Riccardo Rosati, Sebastian Rudolph, and Michael Zakharyaschev, editors, *Proceedings of the 24th International Workshop on Description Logics (DL 2011)*, pages 70–80. CEUR Workshop Proceedings, 2011.

[39] Stefan Borgwardt and Rafael Peñaloza. Non-Gödel negation makes unwitnessed consistency undecidable. In Yevgeny Kazarov, Domenico Lembo, and Frank Wolter, editors, *Proceedings of the 25th International Workshop on Description Logics (DL 2012)*, volume 846, pages 411–421. CEUR Workshop Proceedings, 2012.

[40] Stefan Borgwardt and Rafael Peñaloza. A tableau algorithm for fuzzy description logics over residuated De Morgan lattices. In Markus Kroetzsch and Umberto Straccia, editors, *Proceedings of the 6th International Conference on Web Reasoning and Rule Systems (RR 2012)*, volume 7497 of *Lecture Notes in Computer Science*, pages 9–24. Springer, 2012.

[41] Stefan Borgwardt and Rafael Peñaloza. Undecidability of fuzzy description logics. In Gerhard Brewka, Thomas Eiter, and Sheila A. McIlraith, editors, *Proceedings of the 13th Conference on Principles of Knowledge Representation and Reasoning (KR 2012)*, pages 232–242. AAAI Press, 2012.

[42] Stefan Borgwardt and Rafael Peñaloza. The complexity of lattice-based fuzzy description logics. *Journal on Data Semantics*, 2(1):1–19, 2013.

[43] Stefan Borgwardt and Rafael Peñaloza. Positive subsumption in fuzzy $\mathcal{EL}$ with general t-norms. In Francesca Rossi, editor, *Proceedings of the 23rd International Joint Conference on Artificial Intelligence (IJCAI 2013)*, pages 789–795. AAAI Press, 2013.

[44] Stefan Borgwardt and Rafael Peñaloza. Consistency reasoning in lattice-based fuzzy description logics. *International Journal of Approximate Reasoning*, 55(9):1917–1938, 2014.

[45] Félix Bou, Marco Cerami, and Francesc Esteva. Concept satisfiability in finite-valued fuzzy description logics is PSPACE-complete. In *Logic, Algebra and Truth Degrees 2012*, pages 44–48, 2012.

[46] Ronald J. Brachman. On the epistemological status of semantic networks. In Nicholas V. Findler, editor, *Associative Networks*, pages 3–50. Academic Press, 1979.

[47] Ronald J. Brachman, Alexander Borgida, Deborah L. McGuinness, and Lori A. Resnick. CLASSIC: A structural data model for objects. In James Clifford, Bruce Lindsay, and David Maier, editors, *Proceedings of the 1989 ACM SIGMOD International Conference on Management of Data*, pages 59–67. Association for Computing Machinery Publication, 1989.

[48] Ronald J. Brachman, Richard E. Fikes, and Hector J. Levesque. KRYPTON: Integrating terminology and assertion. In *Proceedings of the 3rd Conference of the American Association for Artificial Intelligence (AAAI 1983)*, pages 31–35. AAAI Press, 1983.

[49] Ronald J. Brachman and Hector J. Levesque. The tractability of subsumption in frame based description languages. In *Proceedings of the 4th Conference of the American Association for Artificial Intelligence (AAAI 1984)*, pages 34–37. AAAI Press, 1984.

[50] Ronald J. Brachman and Hector J. Levesque, editors. *Readings in Knowledge Representation*. Morgan Kaufmann, 1985.

[51] Ronald J. Brachman and James G. Schmolze. An overview of the KL-ONE knowledge representation system. *Cognitive Science*, 9(2):171–216, 1985.

[52] Manuela Busaniche and Franco Montagna. Hájek's logic BL and BL-algebras. In Petr Cintula, Petr Hájek, and Carles Noguera, editors, *Handbook of Mathematical Fuzzy Logic - Volume 1*, volume 37 of *Studies in Logic, Mathematical Logic and Foundations*, pages 355–447. College Publications, London, 2011.

[53] Marco Cerami. Fuzzy Description Logics under a Mathematical Fuzzy Logic Point of View. To appear. `http://www.iiia.csic.es/files/pdfs/monografia_cerami.pdf`.

[54] Marco Cerami. *Fuzzy Description Logics from a Mathematical Fuzzy Logic point of view*. PhD thesis, Universitat de Barcelona, 2012. `http://www.iiia.csic.es/files/pdfs/tesi-cerami.pdf`.

[55] Marco Cerami, Francesc Esteva, and Félix Bou. Decidability of a description logic over infinite-valued product logic. In Fangzhen Lin, Ulrike Sattler, and Miroslaw Truszczynski, editors, *Principles of Knowledge Representation and Reasoning: Proceedings of KR 2010*, pages 203–213. AAAI Press, 2010.

[56] Marco Cerami, Francesc Esteva, and Àngel García-Cerdaña. On finitely valued fuzzy description logics: The Łukasiewicz case. In Salvatore Greco, Bernadette Bouchon-Meunier, Giulianella Coletti, Mario Fedrizzi, Benedetto Matarazzo, and Ronad R. Yager, editors, *Proceedings of the 14th International Conference on Information Processing and Management of Uncertainty in Knowledge-Based Systems (IPMU 2012)*, pages 235–244. Springer, 2012.

[57] Marco Cerami, Àngel García-Cerdaña, and Francesc Esteva. On the relations between fuzzy description logics and multi-valued modal logics. Forthcoming.

[58] Marco Cerami, Àngel García-Cerdaña, and Francesc Esteva. From classical description logic to $n$-graded fuzzy description logics. In *Proceedings of the 19th IEEE International Conference on Fuzzy Systems (FUZZ-IEEE 2010)*, pages 1506–1513. IEEE Xplore. Digital Library, 2010.

[59] Marco Cerami, Àngel García-Cerdaña, and Francesc Esteva. On finitely-valued fuzzy description logics. *International Journal of Approximate Reasoning*, 55(9):1890–1916, 2014.

[60] Marco Cerami and Umberto Straccia. On the (un)decidability of fuzzy description logics under Łukasiewicz t-norms. *Information Sciences*, 227:1–21, 2013.

[61] Marco Cerami and Umberto Straccia. Complexity sources in fuzzy description logic. In Meghyn Bienvenu, Magdalena Ortiz, Riccardo Rosati, and Mantas Simkus, editors, *Proceedings of the 27th International Workshop on Description Logics (DL 2014)*, volume 1193, pages 421–433. CEUR Workshop Proceedings, 2014.

[62] Roberto Cignoli and Antoni Torrens. Glivenko like theorems in natural expansions of BCK-logic. *Mathematical Logic Quarterly*, 50(2):111–125, 2003.

[63] Cristina Civili, Marco Console, Giuseppe De Giacomo, Domenico Lembo, Maurizio Lenzerini, Lorenzo Lepore, Riccardo Mancini, Antonella Poggi, Riccardo Rosati, Marco Ruzzi, Valerio Santarelli, and Domenico Fabio Savo. Mastro studio: Managing ontology-based data access applications. *Proceedings of VLDB Endowment*, 6(12):1314–1317, 2013.

[64] Bernardo Cuenca-Grau, Ian Horrocks, Boris Motik, Bijan Parsia, Peter F. Patel-Schneider, and Ulrike Sattler. OWL 2: The next step for OWL. *Journal of Web Semantics*, 6(4):309–322, 2008.

[65] Giuseppe De Giacomo and Maurizio Lenzerini. Boosting the correspondence between description logics and propositional dynamic logics. In *Proceedings of the 12th National Conference on Artificial Intelligence (AAAI 1994)*, pages 205–212. AAAI Press/MIT Press, 1994.

[66] Miguel Delgado, M. Dolores Ruiz, Daniel Sánchez, and M. Amparo Vila. Fuzzy quantification: a state of the art. *Fuzzy Sets and Systems*, 242:1–30, 2014.

[67] Premkumar Devanbu, Ronald J. Brachman, Peter J. Selfridge, and Bruce W. Ballard. LASSIE: A knowledge-based software information system. *Communications of the Association for Computing Machinery*, 34(5):36–49, 1991.

[68] Francesco M. Donini, Maurizio Lenzerini, Daniele Nardi, and Werner Nutt. The complexity of concept languages. In James Allen, Ronald J. Brachman, Erik Sandewall, Hector J. Levesque, Ray Reiter, and Richard Fikes, editors, *Proceedings of the 2nd International Conference on Principles of Knowledge Representation and Reasoning (KR 1991)*, pages 151–162. Morgan Kaufmann, 1991.

[69] Francesc Esteva, Lluís Godo, Petr Hájek, and Mirko Navara. Residuated fuzzy logics with an involutive negation. *Archive for Mathematical Logic*, 39(2):103–124, 2000.

[70] Francesc Esteva, Lluís Godo, and Enrico Marchioni. Fuzzy logics with enriched language. In Petr Cintula, Petr Hájek, and Carles Noguera, editors, *Handbook of Mathematical Fuzzy Logic - Volume 2*, volume 38 of *Studies in Logic, Mathematical Logic and Foundations*, pages 627–711. College Publications, London, 2011.

[71] Francesc Esteva, Lluís Godo, and Carles Noguera. Expanding the propositional logic of a t-norm with truth-constants: Completeness results for rational semantics. *Soft Computing*, 14(3):273–284, 2010.

[72] Enrico Franconi. CRACK. In Enrico Franconi, Giuseppe De Giacomo, Robert M. MacGregor, Werner Nutt, and Christopher A. Welty, editors, *Proceedings of the International Workshop on Description Logics (DL 1998)*, pages 58–59. CEUR Workshop Proceedings, 1998.

[73] Àngel García-Cerdaña, Eva Armengol, and Francesc Esteva. Fuzzy description logics and t-norm based fuzzy logics. *International Journal of Approximate Reasoning*, 51(6):632–655, 2010.

[74] Franco Giunchiglia and Roberto Sebastiani. A SAT-based decision procedure for ALC. In Luigia Calucci Aiello, Jon Doyle, and Stuart C. Shapiro, editors, *Proceedings of the 6th International Conference on the Principles of Knowledge Representation and Reasoning (KR 1996)*, pages 304–314. Morgan Kaufmann, 1996.

[75] Volker Haarslev, Kay Hidde, Ralf Möller, and Michael Wessel. The racerpro knowledge representation and reasoning system. *Semantic web*, 3(3):267–277, 2012.

[76] Volker Haarslev and Ralf Möller. RACER system description. In Rajeev Gore, Alexander Leitsch, and Tobias Nipkow, editors, *Proceedings of the 15th International Joint Conference on Automated Reasoning (IJCAR 2001)*, volume 2083 of *Lecture Notes in Artificial Intelligence*, pages 701–705. Springer, 2001.

[77] Reiner Hähnle. Advanced many-valued logics. In *Handbook of Philosophical Logic*, pages 297–395. Kluwer, 2 edition, 2001.

[78] Petr Hájek. *Metamathematics of Fuzzy Logic*, volume 4 of *Trends in Logic*. Kluwer, Dordrecht, 1998.

[79] Petr Hájek. On very true. *Fuzzy Sets and Systems*, 124(3):329–333, 2001.

[80] Petr Hájek. Making fuzzy description logic more general. *Fuzzy Sets and Systems*, 154(1):1–15, 2005.

[81] Petr Hájek. Computational complexity of t-norm based propositional fuzzy logics with rational truth constants. *Fuzzy Sets and Systems*, 157(5):677–682, 2006.

[82] Petr Hájek. What does mathematical fuzzy logic offer to description logic? In Elie Sanchez, editor, *Fuzzy Logic and the Semantic Web*, volume 1 of *Capturing Intelligence*, pages 91–100. Elsevier, Amsterdam, 2006.

[83] Petr Hájek, Franco Montagna, and Carles Noguera. Arithmetical complexity of first-order fuzzy logics. In Petr Cintula, Petr Hájek, and Carles Noguera, editors, *Handbook of Mathematical Fuzzy Logic - Volume 2*, volume 38 of *Studies in Logic, Mathematical Logic and Foundations*, pages 853–908. College Publications, London, 2011.

[84] Patrick J. Hayes. The logic of frames. In Dieter Metzing, editor, *Frame Concepts and Text Understanding*, pages 46–61. W. de Gruyter, Berlin, 1979. Reprinted in [50] pages 287–295.

[85] Bernhard Hollunder, Werner Nutt, and Manfred Schmidt-Schauss. Subsumption algorithms for concept description languages. In Luigia Calucci Aiello, editor, *Proceedings of the 9th European Conference on Artificial Intelligence (ECAI 1990)*, pages 348–353. Pitman, 1990.

[86] John E. Hopkroft and Richard M. Karp. An $n^5/2$ algorithm for maximum matchings in bipartite graphs. *SIAM Journal on Computing*, 2:225–231, 1973.

[87] Ian Horrocks. Using an expressive description logic: FaCT or fiction? In *Proceedings of the 8th International Conference on the Principles of Knowledge Representation and Reasoning (KR 1998)*, pages 636–647. Morgan Kaufmann, 1998.

[88] Ian Horrocks, Oliver Kutz, and Ulrike Sattler. The even more irresistible $\mathcal{SROIQ}$. In Patrick Doherty, John Mylopoulos, and Christopher Welty, editors, *Proceedings of the 10th International Conference of Knowledge Representation and Reasoning (KR 2006)*, pages 452–457. AAAI Press, 2006.

[89] Thomas S. Kaczmarek, Raymond Bates, and Gabriel Robins. Recent developments in NIKL. In *Proceedings of the 5th Conference of the American Association for Artificial Intelligence (AAAI 1986)*, pages 978–985. AAAI Press, 1986.

[90] Yevgeny Kazakov, Markus Krötzsch, and František Simančík. The incredible ELK - from polynomial procedures to efficient reasoning with $\mathcal{EL}$ ontologies. *Journal of Automated Reasoning*, 53(1):1–61, 2014.

[91] Erich Peter Klement, Radko Mesiar, and Endre Pap. *Triangular Norms*, volume 8 of *Trends in Logic*. Kluwer, Dordrecht, 2000.

[92] Thomas Lukasiewicz and Umberto Straccia. Description logic programs under probabilistic uncertainty and fuzzy vagueness. *International Journal of Approximate Reasoning*, 50(6):837–853, 2009.

[93] Robert MacGregor and R. Bates. The Loom knowledge representation language. Technical Report ISI/RS-87-188, University of Southern California, Information Science Institute, Marina del Rey, 1987.

[94] Gaspar Mayor and Joan Torrens. On a class of operators for expert systems. *International Journal of Intelligent Systems*, 8:771–778, 1993.

[95] Gaspar Mayor and Joan Torrens. Triangular norms in discrete settings. In Erich Peter Klement and Radko Messiar, editors, *Logical, Algebraic, Analytic, and Probabilistic Aspects of Triangular Norms*, pages 189–230. Elsevier, Amsterdam, 2005.

[96] Eric Mays, Robert Dionne, and Robert Weida. K-REP system overview. *SIGART Bulletin*, 2(3):93–97, 1991.

[97] George Metcalfe, Nicola Olivetti, and Dov M. Gabbay. *Proof Theory for Fuzzy Logics*, volume 36 of *Applied Logic Series*. Springer, 2008.

[98] Marvin Minsky. A framework for representing knowledge. In John Haugeland, editor, *Mind Design*. MIT Press, 1981.

[99] Paul S. Mostert and Allen L. Shields. On the structure of semigroups on a compact manifold with boundary. *The Annals of Mathematics, Second Series*, 65:117–143, 1957.

[100] Bernhard Nebel and Christof Peltason. Terminological reasoning and information management. In D. Karagianis, editor, *Artificial Intelligence and Information Systems: Integration Aspects*, volume 474 of *Lecture Notes in Computer Science*, pages 181–212. Springer, 1991.

[101] Bernhard Nebel and Kai von Luck. Hybrid reasoning in BACK. In Zbigniew W. Raś and Lorenza Saitta, editors, *Proceedings of the 3rd International Symposium on Methodologies for Intelligent Systems (ISMIS 1988)*, pages 260–269. North-Holland, Amsterdam, 1988.

[102] OWL 2 Web Ontology Language Document Overview. `http://www.w3.org/TR/2009/REC-owl2-overview-20091027/`. W3C, 2009.

[103] OWL 2 Web Ontology Language Profiles. `http://www.w3.org/TR/2009/REC-owl2-profiles-20091027/`. W3C, 2009.

[104] Peter F. Patel-Schneider. Small and be beautiful in knowledge representation. Technical Report 660, Fairchild, Palo Alto, CA, 1984.

[105] Peter F. Patel-Schneider. Undecidability of subsumption in NIKL. *Artificial Intelligence*, 39:263–272, 1989.

[106] Emil L. Post. A variant of a recursively unsolvable problem. *Bulletin of the American Mathematical Society*, 52:264–268, 1946.

[107] M. Ros Quillian. Word concepts: A theory and simulation of some basic capabilities. *Behavioral Science*, 12:410–430, 1967.

[108] Steven W. Reyner. An analysis of a good algorithm for the subtree problem. *SIAM Journal on Computing*, 6(4):730–732, 1977.

[109] Daniel Sánchez and Andrea G.B. Tettamanzi. Generalizing quantification in fuzzy description logics. In B. Reusch, editor, *Computational Intelligence, Theory and Applications. Proceedings 8th Fuzzy Days in Dortmund*, Advances in Soft Computing Series, pages 397–411. Springer, 2004.

[110] Daniel Sánchez and Andrea G.B. Tettamanzi. Fuzzy quantification in fuzzy description logics. In Elie Sanchez, editor, *Fuzzy Logic and the Semantic Web*, Capturing Intelligence, chapter 8, pages 135–159. Elsevier, 2006.

[111] Ulrike Sattler, Diego Calvanese, and Ralf Molitor. Relationship with other formalisms. In Franz Baader, Diego Calvanese, Deborah L. McGuinness, Daniele Nardi, and Peter F. Patel-Schneider, editors, *The Description Logic Handbook: Theory, Implementation and Applications*, pages 137–177. Cambridge University Press, 2003.

[112] Klaus Schild. A correspondence theory for terminological logics: Preliminary report. In John Mylopoulos and Raymond Reiter, editors, *Proceedings of the 12th International Joint Conference on Artificial Intelligence (IJCAI 1991)*, pages 466–471. Morgan Kaufmann, 1991.

[113] Manfred Schmidt-Schauß. Subsumption in KL-ONE is undecidable. In Ronald J. Brachman, Hector J. Levesque, and Raymond Reiter, editors, *Proceedings of the 1st International Conference on Principles of Knowledge Representation and Reasoning (KR 1989)*, pages 421–431. Morgan Kaufmann, 1989.

[114] Manfred Schmidt-Schauß and Gert Smolka. Attributive concept descriptions with complements. *Artificial Intelligence*, 48(1):1–26, 1991.

[115] Fabrizio Sebastiani and Umberto Straccia. A computationally tractable terminological logic. In Thore Danielsen, editor, *Proceedings of the 3rd Scandinavian Conference on Artificial Intelligence (SCAI 1991)*, pages 307–315. IOS, 1991.

[116] Rob Shearer, Boris Motik, and Ian Horrocks. HermiT: A highly-efficient owl reasoner. In Catherine Dolbear, Alan Ruttenberg, and Ulrile Sattler, editors, *Proceedings of the 5th International Workshop on OWL: Experiences and Directions (OWLED 2008)*, page 91. CEUR Workshop Proceedings, 2008.

[117] Evren Sirin, Bijan Parsia, Bernardo Cuenca Grau, Aditya Kalyanpur, and Yarden Katz. Pellet: A practical OWL-DL reasoner. *Web Semantics*, 5(2):51–53, 2007.

[118] Giorgios Stoilos, Nikos Simou, Giorgos Stamou, and Stefanos Kollias. Uncertainty and the semantic web. *IEEE Intelligent Systems*, 21(5):84–87, 2006.

[119] Giorgos Stoilos and Giorgos Stamou. A framework for reasoning with expressive continuous fuzzy description logics. In Bernardo Cuenca Grau, Ian Horrocks, Boris Motik, and Urike Sattler, editors, *Proceedings of the 22nd International Workshop on Description Logics (DL 2009)*. CEUR Workshop Proceedings, 2009.

[120] Giorgos Stoilos and Giorgos Stamou. Reasoning with fuzzy extensions of OWL and OWL 2. *Knowledge and Information Systems*, 40(1):205–242, 2014.

[121] Giorgos Stoilos, Giorgos Stamou, Vassilis Tzouvaras, Jeff Z. Pan, and Ian Horrocks. Fuzzy OWL: Uncertainty and the semantic web. In Bernardo Cuenca Grau, Ian Horrocks, Bijan Parsia, and Peter F. Patel-Schneider, editors, *Proceedings of the 2nd International Workshop of OWL (OWLED 2005)*. CEUR Workshop Proceedings, 2005.

[122] Giorgos Stoilos, Giorgos B. Stamou, Jeff Z. Pan, Vassilis Tzouvaras, and Ian Horrocks. The fuzzy description logic f-$\mathcal{SHIN}$. In Paulo Cesar G. da Costa, Kathryn B. Laskey, Kenneth J. Laskey, and Michael Pool, editors, *Proceedings of the 1st International Workshop on Uncertainty Reasoning For the Semantic Web (URSW 2005)*. CE, 2005.

[123] Giorgos Stoilos, Giorgos B. Stamou, Jeff Z. Pan, Vassilis Tzouvaras, and Ian Horrocks. Reasoning with very expressive fuzzy description logics. *Journal of Artificial Intelligence Research*, 30:273–320, 2007.

[124] Umberto Straccia. A fuzzy description logic. In Jack Mostow and Chuck Rich, editors, *Proceedings of the 15th National Conference on Artificial Intelligence (AAAI 1998)*, pages 594–599. AAAI Press, 1998.

[125] Umberto Straccia. Reasoning within fuzzy description logics. *Journal of Artificial Intelligence Research*, 14:137–166, 2001.

[126] Umberto Straccia. Transforming fuzzy description logic into classical description logics. In Jose J. Alferes and Joao Leite, editors, *Proceedings of the 9th European Conference on Logics in Artificial Intelligence (JELIA 2004)*, volume 3229 of *Lecture Notes in Computer Science*, pages 385–399. Springer, 2004.

[127] Umberto Straccia. Uncertainty in description logics: A lattice-based approach. In *Proceedings of the 10th International Conference on Information Processing and Managment of Uncertainty in Knowledge-Based Systems (IPMU 2004)*, pages 251–258. Editrice Università La Sapienza, 2004.

[128] Umberto Straccia. Description logics with fuzzy concrete domains. In Fahiem Bacchus and Tommi Jaakkola, editors, *Proceedings of the 21st Conference on Uncertainty in Artificial Intelligence (UAI 2005)*, pages 559–567. AUAI Press, 2005.

[129] Umberto Straccia. Fuzzy ALC with fuzzy concrete domains. In Ian Horrocks, Ulrike Sattler, and Frank Wolter, editors, *Proceedings of the International Workshop on Description Logics (DL 2005)*, pages 96–103. CEUR Workshop Proceedings, 2005.

[130] Umberto Straccia. Towards a fuzzy description logic for the semantic web (preliminary report). In Asunción Gómez-Pérez and Jérôme Euzenat, editors, *Proceedings of the 2nd European Semantic Web Conference (ESWC 2005)*, volume 3532 of *Lecture Notes in Computer Science*, pages 167–181. Springer, 2005.

[131] Umberto Straccia. A fuzzy description logic for the semantic web. In Elie Sanchez, editor, *Fuzzy Logic and the Semantic Web*, Capturing Intelligence, chapter 4, pages 73–90. Elsevier, 2006.

[132] Umberto Straccia. *Foundations of Fuzzy Logic and Semantic Web Languages*. CRC Studies in Informatics Series. Chapman and Hall/CRC, Boca Raton, FL, 2013.

[133] David Toman and Grant E. Weddell. On reasoning about structural equality in XML: A description logic approach. *Theoretical Computer Science*, 336(1):181–203, 2005.

[134] Vicenç Torra and Yasuo Narukawa. *Modeling decisions - information fusion and aggregation operators*. Springer, 2007.

[135] David S. Touretzky. *The Mathematics of Inheritance Systems*. Pitman, London, GB, 1986.

[136] Christopher B. Tresp and Ralf Molitor. A description logic for vague knowledge. In Henri Prade, editor, *Proceedings of the 13th European Conference on Artificial Intelligence (ECAI 1998)*, pages 361–365. Wiley, 1998.

[137] Dmitry Tsarkov and Ian Horrocks. Fact++ description logic reasoner: System description. In Ulrich Furbach and Natarajan Shankar, editors, *Proceedings of the 3rd International Joint Conference on Automated Reasoning (IJCAR 2006)*, pages 292–297, Berlin, Heidelberg, 2006. Springer.

[138] Marc Vilain. KL-TWO, a hybrid knowledge representation system. Technical Report 5694, BBN Laboratories, March 1984.

[139] William A. Woods. What's in a link: Foundations for semantic networks. In Daniel G. Bobrow and Allan Collins, editors, *Representation and Understanding: Studies in Cognitive Science*, pages 35–82. Academic Press, 1975. Reprinted in [50] pages 35–82.

[140] William A. Woods and James G. Schmolze. The KL-ONE family. *Computers & Mathematics with Applications*, 23(2–5):133–177, 1992.

[141] John Yen. Generalizing term subsumption languages to fuzzy logics. In John Mylopoulos and Raymond Reiter, editors, *Proceedings of the 12th International Joint Conference on Artificial Intelligence (IJCAI 1991)*, pages 472–477. Morgan Kaufmann, 1991.

[142] Lotfi A. Zadeh. Test-score semantics for natural languages. In Jan Horecký, editor, *Proceedings of the 9th Conference on Computational Linguistics (COLING 1982)*, pages 425–430. Academia, 1982.

FERNANDO BOBILLO
Department of Computer Science and Systems Engineering
University of Zaragoza
C. Maria de Luna 1
50018 Zaragoza, Spain
Email: fbobillo@unizar.es

MARCO CERAMI
Department of Computer Science
Palacký University Olomouc
17. listopadu 2
771 46 Olomouc, Czech Republic
Email: marco.cerami@upol.cz

FRANCESC ESTEVA
Artificial Intelligence Research Institute (IIIA)
Spanish National Research Council (CSIC)
Campus de la Universitat Autònoma de Barcelona s/n
08193 Bellaterra, Catalonia, Spain
Email: esteva@iiia.csic.es

ÀNGEL GARCÍA-CERDAÑA
Artificial Intelligence Research Institute (IIIA)
Spanish National Research Council (CSIC)
Campus de la Universitat Autònoma de Barcelona s/n
08193 Bellaterra, Catalonia, Spain
and
Information and Communication Technologies Department
Universitat Pompeu Fabra
Tànger 122-140
08018 Barcelona, Catalonia, Spain
Email: angel@iiia.csic.es

RAFAEL PEÑALOZA
KRDB Research Centre
Free University of Bozen-Bolzano
Piazza Domenicani 3
I-39100 Bozen-Bolzano BZ, Italy
Email: rafael.penaloza@unibz.it

UMBERTO STRACCIA
Istituto di Scienze e Tecnologie dell'Informazione (ISTI)
Consiglio Nazionalle delle Ricerche (CNR)
Via G. Moruzzi 1
56124 Pisa, Italy
Email: straccia@isti.cnr.it

# Chapter XVII: States of MV-algebras

TOMMASO FLAMINIO AND TOMÁŠ KROUPA

## 1 Introduction

This handbook series is devoted to many-valued logics and the associated mathematical structures. States of MV-algebras constitute the object of study of a rapidly growing discipline, which involves and takes inspiration from several areas of mathematics: Łukasiewicz logic, ordered algebraic structures, artificial intelligence, game theory and others.

States are functions from an MV-algebra into the real unit interval satisfying a normalization condition and a variant of the classical law of finite additivity of probability measures, making clear the intimate relation with finitely additive probability measures. In our exposition we underline several other interpretations of states. Namely the integral representation of states by regular Borel probability measures and the bookmaking theorem for infinite-valued events. We also discuss the relation of states to real homomorphisms of unital lattice ordered Abelian groups, finitely presented MV-algebras, piecewise linear geometry, and conditional probability in non-classical setting.

The chapter is organized as follows. In Section 2 we shall recall the basic notions and results used in the rest of this chapter. Namely we summarize the main mathematical tools involving MV-algebras, Łukasiewicz logic, probability measures and compact convex sets. In Section 3 we will introduce states of MV-algebras and discuss their elementary properties (Section 3.1). We will include basic examples of states (Section 3.2) and characterize completely the states of finitely presented MV-algebras (Section 3.3). In Section 3.4 the connection of states to lattice ordered groups is made clear.

One of the main results of this chapter, the integral representation theorem, is proved in Section 4. The set of all states (the state space) forms a Bauer simplex (Section 4.1). Section 4.2 is about the existence of invariant faithful states.

De Finetti's coherence criterion for many-valued events and states is discussed in Section 5: in Section 5.1 we concentrate on coherent books on free MV-algebras and in Section 5.2 we explore the computational complexity for the problem of deciding if a book on formulas of Łukasiewicz logic is coherent.

Section 6 introduces an algebraic setting for states of MV-algebras, the variety of SMV-algebras. In the ensuing subsections we study the algebraic properties of this variety and we present a way to characterize de Finetti's coherence criterion on many-valued events in a purely algebraic way. Conditional states and an extended coherence criterion for conditional many-valued events are the content of Section 7.

No survey chapter can deliver a fully accurate portrait of a given subject. Section 8 contains the brief discussion of omitted results and topics, accompanied with the detailed list of further reading about states and numerous references to the current literature.

## 2 Basic notions

We summarize basic notions and results which are used in this chapter. The scope of involved mathematical apparatus is somewhat extensive, ranging from algebraic semantics of Łukasiewicz logic (Section 2.1-2.2) over measure theory (Section 2.3) to infinite-dimensional convex sets (Section 2.4). Therefore we confine to discussing only the most essential concepts of those disciplines. The reader is invited to consult the cited references for details, if needed.

### 2.1 MV-algebras

Our notation and definitions are according to the Chapter VI of this Handbook. The standard reference about MV-algebras is [15].

DEFINITION 2.1.1. *An* MV-algebra *is a structure* $\boldsymbol{A} = \langle A, \oplus, \neg, 0\rangle$, *where* $\oplus$ *is a binary operation,* $\neg$ *is a unary operation and* $0$ *is a constant, such that the following conditions are satisfied for every* $a, b \in A$*:*

*(i)* $\langle A, \oplus, 0\rangle$ *is an Abelian monoid*

*(ii)* $\neg(\neg a) = a$

*(iii)* $\neg 0 \oplus a = \neg 0$

*(iv)* $\neg(\neg a \oplus b) \oplus b = \neg(\neg b \oplus a) \oplus a.$

The class of MV-algebras forms a variety that we shall denote by $\mathbb{MV}$. We introduce the new constant 1 and three additional operations $\odot$, $\ominus$ and $\to$ as follows:

$$1 = \neg 0 \qquad a \odot b = \neg(\neg a \oplus \neg b) \qquad a \ominus b = a \odot \neg b \qquad a \to b = \neg a \oplus b.$$

The *Chang distance* is the binary operation

$$d(a, b) = (a \ominus b) \oplus (b \ominus a). \tag{1}$$

In the rest of this chapter we shall always assume that any MV-algebra has at least two elements and thus $0 \neq 1$.

For every MV-algebra $\boldsymbol{A}$, the binary relation $\leq$ on $A$ given by

$$a \leq b \quad \text{whenever} \quad a \to b = 1$$

is a partial order. As a matter of fact, $\leq$ is a lattice order induced by the join $\vee$ and the meet $\wedge$ defined by

$$a \vee b = \neg(\neg a \oplus b) \oplus b \quad \text{and} \quad a \wedge b = \neg(\neg a \vee \neg b),$$

respectively. The lattice reduct of $\boldsymbol{A}$ then becomes a distributive lattice with the top element 1 and the bottom element 0. If the order $\leq$ of $\boldsymbol{A}$ is total, then $\boldsymbol{A}$ is said to be an MV-*chain*.

EXAMPLE 2.1.2. The basic example of an MV-algebra is the *standard* MV-*algebra* $[0,1]_Ł$, which is just the real unit interval $[0,1]$ equipped with the operations

$$a \oplus b = \min(1, a+b) \qquad a \odot b = \max(0, a+b-1) \qquad \neg a = 1-a. \tag{2}$$

The partial order of the standard MV-algebra coincides with the usual order of real numbers from $[0,1]$. It is worth mentioning that the standard MV-algebra $[0,1]_Ł$ is generic for the variety $\mathbb{MV}$, that is, $\mathbb{MV} = \mathbf{V}([0,1]_Ł)$.

EXAMPLE 2.1.3. Every Boolean algebra is an MV-algebra with respect to the operations $\oplus = \vee$, $\odot = \wedge$, and the complement $\neg$.

EXAMPLE 2.1.4. For every natural number $d$, the set $\mathbf{Ł}_d = \{0, 1/d, \ldots, (d-1)/d, 1\}$ endowed with the restriction of the operations of $[0,1]_Ł$ is a finite MV-chain.

MV-algebras generalize Boolean algebras in the following sense: an MV-algebra $\boldsymbol{A}$ is a Boolean algebra if and only if $a \oplus a = a$ for every $a \in A$. Hence MV-algebras are particular non-idempotent generalizations of Boolean algebras. For any MV-algebra $\boldsymbol{A}$, we denote

$$B(A) = \{a \in A \mid a \oplus a = a\}$$

and call $B(A)$ the *Boolean center* (or the *Boolean skeleton*) of $A$. It follows that the structure $\langle B(A), \vee, \wedge, \neg, 0, 1\rangle$ is a Boolean algebra.

Let $\boldsymbol{A} = \langle A, \oplus_A, \neg_A, 0_A\rangle$ and $\boldsymbol{B} = \langle B, \oplus_B, \neg_B, 0_B\rangle$ be MV-algebras. A *homomorphism* from $\boldsymbol{A}$ to $\boldsymbol{B}$ is a mapping $h\colon A \to B$ such that, for every $a_1, a_2 \in A$,

(i) $h(a_1 \oplus_A a_2) = h(a_1) \oplus_B h(a_2)$

(ii) $h(\neg_A a_1) = \neg_B h(a_1)$

(iii) $h(0_A) = 0_B$.

Let us define

$$\mathcal{H}(\boldsymbol{A}, \boldsymbol{B}) = \{h \mid h \text{ is a homomorphism from } \boldsymbol{A} \text{ to } \boldsymbol{B}\}.$$

In case that $\boldsymbol{B} = [0,1]_Ł$, we write simply $\mathcal{H}(\boldsymbol{A})$ in place of $\mathcal{H}(\boldsymbol{A}, [0,1]_Ł)$. An *isomorphism* $h \in \mathcal{H}(\boldsymbol{A}, \boldsymbol{B})$ is a bijective homomorphism.

Let $X$ be a nonempty set. The set $[0,1]^X$ of all functions $X \to [0,1]$ becomes an MV-algebra if the operations $\oplus$, $\neg$, and the element 0 as in (2) are defined pointwise. The corresponding lattice operations $\vee$ and $\wedge$ are then the pointwise maximum and the pointwise minimum of two functions $X \to [0,1]$, respectively.

DEFINITION 2.1.5. *Let $X$ be a nonempty set. A* clan *over $X$ is an* MV-*algebra $\boldsymbol{A}_X = \langle A_X, \oplus, \neg, 0\rangle$, where $A_X \subseteq [0,1]^X$ is a nonempty set of functions $X \to [0,1]$, endowed with the pointwise defined operations of the standard* MV-*algebra.*

We say that a clan $\boldsymbol{A}_X$ is *separating* whenever the following condition is satisfied: if $x, y \in X$ with $x \neq y$, then there exists $a \in A_X$ such that $a(x) \neq a(y)$.

It turns out that the clans of $[0,1]$-valued continuous functions over some compact Hausdorff space are the prototypes of an important class of MV-algebras. We will introduce the necessary algebraic machinery in order to formulate the corresponding representation theorem. An *ideal* in an MV-algebra $\boldsymbol{A}$ is a subset $I \subseteq A$ such that

(i) $0 \in I$.

(ii) If $a, b \in I$, then $a \oplus b \in I$.

(iii) If $b \in I$ and $b \geq a \in A$, then $a \in I$.

An ideal $I$ is *proper* if $I \neq A$. We say that a proper ideal $M$ is *maximal* if $M$ is not strictly included in any proper ideal of $\boldsymbol{A}$. If $\boldsymbol{A}$ is an MV-algebra and $S$ is a subset of $A$, then the ideal generated by $S$ coincides with

$$(S] = \{a \in A \mid a \leq i_1 \oplus \cdots \oplus i_n \text{ for some } n \in \mathsf{N} \text{ and } i_1, \ldots, i_n \in S\}.$$

An ideal $I$ is said to be *finitely generated* if there exists a finite subset $S$ of $A$ such that $I = (S]$. A *filter* in an MV-algebra $\boldsymbol{A}$ is a subset $F \subseteq A$ closed with respect to $\odot$ and such that $1 \in F$, and if $b \in F$ and $b \leq a \in A$, then $a \in F$. Ideals and filters are dual objects in the setting of MV-algebras. Indeed, there is a one-to-one correspondence between ideals and filters: if $I$ is an ideal of an MV-algebra $\boldsymbol{A}$, the set

$$F_I = \{a \in A \mid \neg a \in I\}$$

is a filter in $\boldsymbol{A}$. Conversely, for every filter $F$,

$$I_F = \{b \in A \mid \neg b \in F\}$$

is an ideal in $\boldsymbol{A}$. The notions of proper and maximal filter, respectively, are defined as usual.

Let $\operatorname{Max}(\boldsymbol{A})$ be the nonempty set of all maximal ideals of $\boldsymbol{A}$, which we call the *maximal ideal space of* $\boldsymbol{A}$. Given any ideal $I$ of $\boldsymbol{A}$, put

$$O_I = \{M \in \operatorname{Max}(\boldsymbol{A}) \mid I \nsubseteq M\}.$$

THEOREM 2.1.6 ([36, Theorem 3.6.10]). *For every* MV-*algebra* $\boldsymbol{A}$*, its maximal ideal space* $\operatorname{Max}(\boldsymbol{A})$ *is a compact Hausdorff space whose family of open sets coincides with* $\{O_I \mid I \text{ is an ideal of } \boldsymbol{A}\}$.

Let us consider the set

$$\operatorname{Rad}(\boldsymbol{A}) = \bigcap\{M \mid M \in \operatorname{Max}(\boldsymbol{A})\}$$

called the *radical of* $\boldsymbol{A}$. We say that $\boldsymbol{A}$ is *semisimple* if $\operatorname{Rad}(\boldsymbol{A}) = \{0\}$. It can be shown [36, Lemma 4.2.3] that semisimplicity is equivalent to non-existence of infinitesimal elements in $\boldsymbol{A}$, that is, for every $a \in A$ and every $n \in \mathsf{N}$, the condition $\bigoplus_{i=1}^{n} a \leq \neg a$ implies $a = 0$.

The representation of semisimple MV-algebras is one of the crucial tools employed in the study of states.

THEOREM 2.1.7 ([36, Theorem 5.4.7]). *Let $\boldsymbol{A}$ be a semisimple* MV*-algebra. Then $\boldsymbol{A}$ is isomorphic to a separating clan of* $[0,1]$*-valued continuous functions defined on the compact Hausdorff space* $\mathrm{Max}(\boldsymbol{A})$.

## 2.2 Łukasiewicz logic

In this section we provide a survey of Łukasiewicz infinite-valued propositional logic and its associated Lindenbaum algebra. For further details see the Chapter VI and XI of this Handbook, or consult the book [15].

We use the algebraic semantics based on MV-algebras, which enables us to use the same notation for the logical connectives and the corresponding MV-algebraic operations in the following paragraphs. Formulas $\phi, \psi, \ldots$ are constructed from propositional variables $A_1, A_2, \ldots$ by applying the standard rules known in Boolean logic. The connectives are negation $\neg$, disjunction $\oplus$, and conjunction $\odot$. The implication $\phi \to \psi$ can be defined as $\neg\phi \oplus \psi$. The set of all propositional formulas is denoted by $\mathsf{Form}$.

The standard semantics for connectives of Łukasiewicz logic is defined by the corresponding operations of the standard MV-algebra $[0,1]_{Ł}$. A *valuation* is a mapping $V\colon \mathsf{Form} \to [0,1]$ such that, for each $\phi, \psi \in \mathsf{Form}$,

$$V(\neg\phi) = 1 - V(\phi)$$
$$V(\phi \oplus \psi) = V(\phi) \oplus V(\psi)$$
$$V(\phi \odot \psi) = V(\phi) \odot V(\psi).$$

For every two formulas $\phi, \psi \in \mathsf{Form}$, we define the relation $\equiv$ by the following stipulation: we say that $\phi \equiv \psi$ iff $V(\phi) = V(\psi)$ for every valuation $V$. It turns out that $\equiv$ is an equivalence relation on $\mathsf{Form}$. For every $\phi \in \mathsf{Form}$ we denote $[\phi]$ the equivalence class of $\phi$ with respect to $\equiv$.

The Lindenbaum algebra of Łukasiewicz logic is the MV-algebra $\boldsymbol{F}$ of equivalence classes $\{[\phi] \mid \phi \in \mathsf{Form}\}$ endowed with the canonical operations $\neg$, $\oplus$, and $\odot$:

$$\neg[\phi] = [\neg\phi]$$
$$[\phi] \oplus [\psi] = [\phi \oplus \psi]$$
$$[\phi] \odot [\psi] = [\phi \odot \psi].$$

By Chang's completeness theorem [36, Theorem 5.3.7], we may identify $\boldsymbol{F}$ with the free MV-algebra over countably many generators, which is a sub-MV-algebra of $[0,1]^{[0,1]^{\mathsf{N}}}$.

Assume that $\mathsf{Form}_n$ is the set of all formulas $\phi \in \mathsf{Form}$ containing only the propositional variables from the list $A_1, \ldots, A_n$. Let $\boldsymbol{F}_n$ be the corresponding Lindenbaum algebra, which coincides with the free $n$-generated MV-algebra. By McNaughton's theorem [1, Theorem 2.1.20], we know that $\boldsymbol{F}_n$ is isomorphic to the clan of functions $f\colon [0,1]^n \to [0,1]$ such that each $f$ is

(i) continuous,

(ii) piecewise linear, and

(iii) each linear piece has integer coefficients only.

We call each function $[0,1]^n \to [0,1]$ satisfying (i)–(iii) a *McNaughton function.*

REMARK 2.2.1. *In this context we use the term "linear" as a synonym for "affine". There are more non-equivalent definitions of a piecewise linear function appearing in literature. For the purposes of this chapter, we adopt the weaker but the usual definition used in* MV*-algebraic context; see [1]. A function* $f\colon [0,1]^n \to [0,1]$ *is* piecewise linear *if there exist linear functions* $f_1, \dots, f_m$ *on* $[0,1]^n$ *such that for each* $\mathbf{x} \in [0,1]^n$ *there is some* $j \in \{1, \dots, m\}$ *with* $f_j(\mathbf{x}) = f(\mathbf{x})$. *Other authors may use a stronger definition that implies continuity; cf. [7, Definition 7.10].*

An MV-algebra $\boldsymbol{A}$ is said to be *finitely presented* if it is isomorphic to the quotient $\boldsymbol{F}_n/I$ for some natural number $n$ and some finitely generated ideal $I$ of $\boldsymbol{F}_n$.

Łukasiewicz logic is an algebraizable logic in the sense of Blok and Pigozzi. As a consequence, a finitely presented MV-algebra is the Lindenbaum algebra of a finitely axiomatizable theory[1] in Łukasiewicz logic and, moreover, it has a completely geometric characterization—see [48, Theorem 6.3]. Indeed, let $P \subseteq [0,1]^n$ be a nonempty rational polyhedron[2] and $\boldsymbol{M}(P)$ be the MV-algebra of all the restrictions to $P$ of $n$-variable McNaughton functions in $\boldsymbol{F}_n$.

PROPOSITION 2.2.2. *Let* $P \subseteq [0,1]^n$. *Then the following are equivalent:*

*(i)* $P$ *is a rational polyhedron.*

*(ii)* $P = f^{-1}(1)$ *for some McNaughton function* $f \in \boldsymbol{F}_n$.

The following characterizations of a finitely presented MV-algebra are used freely in this chapter.

THEOREM 2.2.3. *Let* $\boldsymbol{A}$ *be an* MV*-algebra. Then the following are equivalent:*

*(i)* $\boldsymbol{A}$ *is a finitely presented* MV*-algebra.*

*(ii)* $\boldsymbol{A}$ *is isomorphic to the Lindenbaum algebra of a theory with a single axiom.*

*(iii)* $\boldsymbol{A}$ *is isomorphic to* $\boldsymbol{M}(P)$, *where* $P$ *is a nonempty rational polyhedron in* $[0,1]^n$, *for some* $n \in \mathsf{N}$.

## 2.3 Probability measures

We will need the basic notions of measure theory as appearing in [10] and [54], for example. The definitions given below apply to the case of Boolean algebras of sets, which is sufficient for most of our purposes.

Let $X$ be a nonempty set. An *algebra of sets* $\mathfrak{A}$ is any Boolean algebra of subsets of $X$. A *finitely additive probability* is a function $\mu\colon \mathfrak{A} \to [0,1]$ such that:

(i) If $A, B \in \mathfrak{A}$, where $A \cap B = \emptyset$, then $\mu(A \cup B) = \mu(A) + \mu(B)$.

(ii) $\mu(\emptyset) = 0$ and $\mu(X) = 1$.

[1] In Łukasiewicz logic, we can replace any finite number of axioms with a single one. It has the effect that a finitely generated ideal $I$ corresponds, through the algebraizability of Łukasiewicz logic, to a finitely axiomatizable theory $T_I$, which, in turn, corresponds to a single formula $\vartheta$. Thus, algebraically, finitely generated and principal ideals in MV-algebras coincide.

[2] By a *rational polyhedron* we mean a finite point-set union of simplices with rational vertices in $\mathsf{R}^n$; see Section 2.4.

REMARK 2.3.1. *The terminology oscillates: for example, Rao and Rao [54] call the above function $\mu$ a* probability charge. *Moreover, we will see that $\mu$ is at the same time a special example of state of an* MV*-algebra; cf. Definition 3.0.1.*

The condition (i), which is called *finite additivity*, must be strengthened in most applications. To this end, we require $\mathfrak{A}$ to be closed with respect to countable point-set unions. A *$\sigma$-algebra* is an algebra of sets $\mathfrak{A}$ such that the condition of *$\sigma$-completeness* holds true:

$$\text{if } A_1, A_2, \ldots \in \mathfrak{A}, \text{ then } \bigcup_{n=1}^{\infty} A_n \in \mathfrak{A}.$$

Let $\mathfrak{A}$ be a $\sigma$-algebra. A finitely additive probability $\mu\colon \mathfrak{A} \to [0,1]$ is a *probability measure* if it is *$\sigma$-additive*, that is:

$$\text{if } A_1, A_2, \ldots \in \mathfrak{A} \text{ with } A_i \cap A_j = \emptyset \text{ for } i \neq j, \text{ then } \mu\left(\bigcup_{n=1}^{\infty} A_n\right) = \sum_{n=1}^{\infty} \mu(A_n).$$

In case that the universe $X$ is a topological space, the most natural choice of the $\sigma$-algebra over $X$ is the *Borel $\sigma$-algebra* $\mathfrak{B}(X)$. Namely, the family $\mathfrak{B}(X)$ is the smallest $\sigma$-algebra over $X$ containing every open set in $X$. Let $\mu$ be a probability measure defined on $\mathfrak{B}(X)$. Then there is no ambiguity in referring to $\mu$ as a *Borel probability measure* on $X$. Assume that $X$ is a compact Hausdorff space. The *support* of Borel probability measure $\mu$ on $X$ is the closed set $\bigcap\{A \subseteq X \mid \mu(A) = 1, A \text{ closed}\}$. Let $x \in X$. Then the Borel probability measure defined by

$$\delta_x(A) = \begin{cases} 1 & x \in A \\ 0 & x \notin A \end{cases}$$

is called the *Dirac measure* at $x$. The support of $\delta_x$ is the singleton $\{x\}$. We say that a Borel probability measure $\mu$ on $X$ is *regular* whenever for every Borel set $A \subseteq X$,

$$\mu(A) = \sup\{\mu(B) \mid B \subseteq A, B \text{ compact}\}.$$

In general, not every Borel probability measure on a compact Hausdorff space $X$ is regular. On the other hand, if every open set in the compact Hausdorff space $X$ is an *$F_\sigma$ set*, that is, a countable union of closed sets, then every Borel probability measure is necessarily regular. In particular, this applies to a compact Hausdorff space $X$ that is metrizable.

## 2.4 Compact convex sets

The detailed exposition on compact convex sets in a space with weak topology is the subject of [5] or [27]. Our goal is to introduce the notion of infinite-dimensional simplex, the so-called Choquet simplex, that is a faithful generalization of an $n$-dimensional simplex.

Let $E$ be a real linear space. A *convex set* in $E$ is any subset $K$ of $E$ that is closed under *convex combinations:* if $x_1, \ldots, x_n \in K$ and $\alpha_i \geq 0$ with $\sum_{i=1}^{n} \alpha_i = 1$, then $\alpha_1 x_1 + \cdots + \alpha_n x_n \in K$. Given any set $X \subseteq E$, the *convex hull of $X$* is the set $\mathrm{co}(X)$ of all convex combinations of elements in $X$.

By an *affine combination* we mean a linear combination $\alpha_1 x_1 + \cdots + \alpha_n x_n$ with $\alpha_1 + \cdots + \alpha_n = 1$, where $\alpha_i \in \mathsf{R}$. A subset $A$ of $E$ is said to be *affinely independent* if there does not exist an element $a \in A$ that can be expressed as an affine combination of elements from $A \setminus \{a\}$. An *affine subspace* of $E$ is any subset of $E$ that is closed under affine combinations. An *$n$-dimensional simplex* is the convex hull of $n+1$ affinely independent points in $E$.

Let $K_1$ and $K_2$ be convex sets in linear spaces $E_1$ and $E_2$, respectively. A mapping $f : K_1 \to K_2$ is said to be *affine* if $f(\alpha_1 x_1 + \cdots + \alpha_n x_n) = \alpha_1 f(x_1) + \cdots + \alpha_n f(x_n)$, for every convex combination $\alpha_1 x_1 + \cdots + \alpha_n x_n \in K_1$. If $f$ is a bijection, then the inverse mapping $f^{-1}$ is also affine and we call $f$ an *affine isomorphism* of $K_1$ and $K_2$.

A *convex cone* in $E$ is a subset $C$ of $E$ such that $0 \in C$ and $\alpha_1 x_1 + \alpha_2 x_2 \in C$ for any non-negative $\alpha_1, \alpha_2 \in \mathsf{R}$ and $x_1, x_2 \in C$. A *strict convex cone* is any convex cone $C$ that satisfies this condition: if both $x \in C$ and $-x \in C$, then $x = 0$. A *base* for a convex cone $C$ is any convex subset $K$ of $C$ such that every non-zero element $y \in C$ may be uniquely expressed as $y = \alpha x$ for some $\alpha \geq 0$ and some $x \in K$. The bases of cones can be easily visualized geometrically: the following characterization is true even for infinite-dimensional spaces.

PROPOSITION 2.4.1. *Let $K$ be a non-empty convex subset of a linear space $E$ and $C = \{\alpha x \mid \alpha \geq 0,\ x \in K\}$. Then $C$ is a convex cone in $E$ and the following conditions are equivalent:*

*(i) $K$ is a base for $C$.*

*(ii) $K$ is contained in an affine subspace $A$ of $E$ such that $0 \notin A$.*

Strict convex cones determine a partial order on linear spaces: if $C$ is a strict convex cone in a linear space $E$, then the relation $\leq_C$ defined by $x \leq_C y$ iff $y - x \in C$ for any $x, y \in E$ makes $E$ into a partially ordered linear space and, moreover, we have that $C = \{x \in E \mid 0 \leq_C x\}$. A *lattice cone* is any strict convex cone $C$ in $E$ such that the set $C$ endowed with a partial order $\leq_C$ is a lattice. A *simplex* in a linear space $E$ is any convex subset $S$ of $E$ that is affinely isomorphic to a base for a lattice cone in some linear space.

The deepest results about convex sets in infinite-dimensional spaces are attained when we assume that the linear space $E$ is a locally convex Hausdorff space. We say that a subset $S$ of $E$ is a *Choquet simplex* if it is a simplex that is a compact set. Clearly, since $E = \mathsf{R}^k$ is locally convex, every $n$-dimensional simplex is a Choquet simplex.

An *extreme point* of a convex set $K$ is a point $e \in K$ such that the set $K \setminus \{e\}$ remains convex. The set

$$\partial(K) = \{e \in K \mid e \text{ is an extreme point of } K\}$$

is called an *extreme boundary of $K$*. For any set $X \subseteq E$, let $\overline{\mathrm{co}}(X)$ be the topological closure in $E$ of the convex hull $\mathrm{co}(X)$. The next theorem is the well-known characterization [29] of every compact convex set by its extreme boundary.

THEOREM 2.4.2 (Krein-Milman). *If $K$ is a compact convex subset of a locally convex Hausdorff space $E$, then $K = \overline{\mathrm{co}}\, \partial(K)$.*

Let $K_1$ and $K_2$ be two compact convex subsets of $E$. An *affine homeomorphism* $h$ of $K_1$ and $K_2$ is an affine isomorphism $h\colon K_1 \to K_2$ that is simultaneously a homeomorphism.

Choquet simplices whose extreme boundaries satisfy additional topological conditions are of particular interest.

DEFINITION 2.4.3. *Let $E$ be a locally convex Hausdorff space. A Choquet simplex $S \subseteq E$ is called a* Bauer simplex *if its extreme boundary $\partial(S)$ is a closed subset of $E$.*

Every $n$-dimensional simplex $S$ is a Bauer simplex since its extreme boundary $\partial S$ has only finitely many extreme points. More generally, let $X$ be a compact Hausdorff space and let $\mathcal{M}(X)$ denote the convex set of all regular Borel probability measures over $X$. We endow the set $\mathcal{M}(X)$ with the so-called *weak* topology*: a net $(\mu_\gamma)$ in $\mathcal{M}(X)$ weak* converges to $\mu \in \mathcal{M}(X)$ if and only if

$$\int_X f \, \mathrm{d}\mu_\gamma \to \int_X f \, \mathrm{d}\mu \quad \text{for every continuous function } f\colon X \to \mathsf{R}. \tag{3}$$

The properties of $\mathcal{M}(X)$ are collected in the next theorem for a future reference; see [5, Corollary II.4.2] for further details.

THEOREM 2.4.4. *Let $X$ be a compact Hausdorff space.*

*(i) The set of all Borel probability measures $\mathcal{M}(X)$ endowed with the weak* topology is a Bauer simplex.*

*(ii) The extreme boundary of $\mathcal{M}(X)$ is $\partial\mathcal{M}(X) = \{\delta_x \mid x \in X\}$.*

*(iii) The mapping $x \mapsto \delta_x$ is a homeomorphism of $X$ and $\partial\mathcal{M}(X)$.*

As we recalled in the end of Section 2.3, if $X$ is a compact set in a finite-dimensional Euclidean space, then every Borel probability measure in $\mathcal{M}(X)$ is regular. Moreover, the compact space $\mathcal{M}(X)$ is metrizable in its weak* topology.

# 3 States

DEFINITION 3.0.1. *Let **A** be an* MV*-algebra. A mapping $s\colon A \to [0,1]$ is a* state *of **A** whenever $s(1) = 1$ and for every $a, b \in A$ the following condition is satisfied:*

$$\textit{if } a \odot b = 0, \textit{ then } s(a \oplus b) = s(a) + s(b). \tag{4}$$

The condition (4) means *additivity* with respect to Łukasiewicz sum $\oplus$. Indeed, the requirement $a \odot b = 0$ is analogous to disjointness of a pair of elements in a Boolean algebra. Thus states can be thought of as generalizations of finitely additive probabilities: every finitely additive probability on a Boolean algebra is a state as a special case of the above definition. In particular, every Borel probability measure is a state as well.

REMARK 3.0.2. *For historical reasons going back to the relation between states of $\ell$-groups and* MV*-algebraic states, which in turns can be traced back to quantum mechanics, we always refer to states* of *an* MV*-algebra and not to states* on *it.*

### 3.1 Basic properties

We will summarize the basic properties of states with respect to the operations and the lattice order of an MV-algebra $\boldsymbol{A}$.

PROPOSITION 3.1.1. *Let $s$ be a state of an* MV*-algebra $\boldsymbol{A}$. For every $a, b \in A$:*

*(i)* $s(0) = 0$.

*(ii)* $s(\neg a) = 1 - s(a)$.

*(iii) If $a \leq b$, then $s(b \ominus a) = s(b) - s(a)$.* *(subtractivity)*

*(iv)* $s(a \oplus b) + s(a \odot b) = s(a) + s(b)$. *(strong modularity)*

*(v)* $s(a \vee b) + s(a \wedge b) = s(a) + s(b)$. *(weak modularity)*

*(vi) If $a \leq b$, then $s(a) \leq s(b)$.* *(monotonicity)*

*(vii)* $s(a \vee b) \leq s(a \oplus b) \leq s(a) + s(b)$. *(subadditivity)*

*(viii) If $a \in \mathrm{Rad}(\boldsymbol{A})$, then $s(a) = 0$.*

*Proof.* (i) We have $s(1) = s(1 \oplus 0) = s(1) + s(0)$, which yields $s(0) = 0$. The identity (ii) follows from $s(a) + s(\neg a) = s(a \oplus \neg a) = s(1) = 1$ since in every MV-algebra, $a \odot \neg a = 0$ holds true. In order to prove (iii), observe that $a \leq b$ implies $(b \ominus a) \oplus a = b$. Since $(b \ominus a) \odot a = b \odot \neg a \odot a = 0$, we get $s(b) = s((b \ominus a) \oplus a) = s(b \ominus a) + s(a)$. Property (iv): as a consequence of [36, Proposition 2.1.2(h)], the identity $(a \oplus b) \odot \neg b = (\neg a \oplus \neg b) \odot a$ holds true. This, together with subtractivity, yields

$$
\begin{aligned}
s(a \oplus b) - s(b) &= s((a \oplus b) \ominus b) = s((a \oplus b) \odot \neg b) = s((\neg a \oplus \neg b) \odot a) \\
&= s(\neg(a \odot b) \odot a) = s(a \ominus (a \odot b)) = s(a) - s(a \odot b).
\end{aligned}
$$

Weak modularity (v) is a consequence of (iv) and [36, Proposition 2.1.2(d)–(e)], which says that $a \oplus b = (a \vee b) \oplus (a \wedge b)$ and $a \odot b = (a \vee b) \odot (a \wedge b)$. Thus

$$
\begin{aligned}
s(a) + s(b) &= s(a \oplus b) + s(a \odot b) \\
&= s((a \vee b) \oplus (a \wedge b)) + s((a \vee b) \odot (a \wedge b)) \\
&= s(a \vee b) + s(a \wedge b).
\end{aligned}
$$

Monotonicity (vi) is a direct consequence of subtractivity (iii). Subadditivity (vii) results from (vi) by considering that $a \vee b \leq a \oplus b$ together with modularity (iv) and nonnegativity of state. (viii) Let $a \in \mathit{Rad}(\boldsymbol{A})$. Reasoning by contradiction, assume that $s(a) > 0$. Then [36, Lemma 3.5.2(a)] gives $a \odot a = 0$. Hence $s(a \oplus a) = 2s(a)$. Proceeding by induction on the number of summands $a$, we can analogously derive the identity

$$
s\left(\bigoplus_{i=1}^{n} a\right) = ns(a), \quad \text{for every } n \in \mathsf{N}.
$$

But this means that there exists some $n_0 \in \mathsf{N}$ for which we obtain

$$s\left(\bigoplus_{i=1}^{n_0} a\right) = n_0 s(a) > 1,$$

a contradiction. □

COROLLARY 3.1.2. *Let $\boldsymbol{A}$ be an* MV*-algebra. The following are equivalent for a function $s\colon A \to [0,1]$ with $s(1) = 1$:*

*(i) $s$ is a state and*

*(ii) $s(a \oplus b) + s(a \odot b) = s(a) + s(b)$, for every $a, b \in A$.*

*Proof.* An easy consequence of Proposition 3.1.1(iv). □

REMARK 3.1.3. *Every state satisfies the two modularity laws* (iv)–(v) *whose combination gives $s(a \oplus b) - s(a \vee b) = s(a \wedge b) - s(a \odot b)$. In particular, the weak modularity* (v) *expresses the fact that any state is a* monotone valuation *in the sense of lattice theory [11, Chapter V]. The converse clearly fails: not every monotone valuation on the lattice reduct of an* MV*-algebra $\boldsymbol{A}$ is a state of $\boldsymbol{A}$. Indeed, consider $\boldsymbol{A} = [0,1]_{Ł}$ and a lattice homomorphism $h\colon [0,1] \to [0,1]$ such that $h(a) = 1$ if $a = 1$, and $h(a) = 0$ otherwise. Then $h$ is a monotone valuation but not a state of $[0,1]_{Ł}$.*

The next assertion is an immediate consequence of the properties of the Boolean center of any MV-algebra and the definition of state.

PROPOSITION 3.1.4. *For every state $s$ of an* MV*-algebra $\boldsymbol{A}$, the restriction of $s$ to the Boolean center $B(A)$ is a finitely additive probability.*

The restriction of $s$ to $B(A)$ may carry little information about the state $s$. Indeed, when $\boldsymbol{A} = \boldsymbol{F}_n$ is the MV-algebra of $n$-variable McNaughton functions, its Boolean center contains only two elements: the functions 0 and 1.

There is no stateless MV-algebra.

PROPOSITION 3.1.5. *Every* MV*-algebra $\boldsymbol{A}$ carries at least one state $s$.*

*Proof.* By [36, Proposition 3.4.5], the collection $\mathrm{Max}(\boldsymbol{A})$ of all maximal ideals of $\boldsymbol{A}$ is nonempty. Let $M \in \mathrm{Max}(\boldsymbol{A})$. Then the quotient MV-algebra $\boldsymbol{A}/M$ is simple [36, Proposition 4.2.10] and thus isomorphic to a subalgebra of the standard MV-algebra $[0,1]_{Ł}$ by [36, Proposition 5.4.1]. Hence we can compose the natural epimorphism $e\colon \boldsymbol{A} \to \boldsymbol{A}/M$ with the embedding $\iota\colon \boldsymbol{A}/M \to [0,1]$ and put $s = \iota \circ e$. Since both $e$ and $\iota$ are homomorphisms, their composition $s$ is a homomorphism of $\boldsymbol{A}$ into $[0,1]_{Ł}$ and thus a state. □

REMARK 3.1.6. *Notice that the proof of Proposition 3.1.5 uses the fact that homomorphisms into the standard* MV*-algebra are states. This simple observation will play a key role in the characterization of the state space; cf. Example 3.2.2 and Section 4.1.*

The set of all states on an MV-algebra, which can be endowed with a topology, is called the *state space*. We will establish its basic geometric properties and show that semisimple MV-algebras have the largest possible state spaces among all MV-algebras.

PROPOSITION 3.1.7. *Let $\mathcal{S}(\boldsymbol{A})$ be the family of all states of an* MV*-algebra $\boldsymbol{A}$. Then:*

*(i) $\mathcal{S}(\boldsymbol{A})$ is a nonempty compact convex subset of $[0,1]^A$.*

*(ii) The state spaces $\mathcal{S}(\boldsymbol{A})$ and $\mathcal{S}(\boldsymbol{A}/\operatorname{Rad}(\boldsymbol{A}))$ are affinely homeomorphic.*

*Proof.* (i) That $\mathcal{S}(\boldsymbol{A}) \neq \emptyset$ is a consequence of Proposition 3.1.5. By Tychonoff's theorem (see [6, Theorem 2.61], for example), the product space $[0,1]^A$ is a compact subspace of the locally convex space $\mathsf{R}^A$. It can routinely be verified that $\mathcal{S}(\boldsymbol{A}) \subseteq [0,1]^A$ is closed and thus compact. To check that a convex combination of two states is a state is straightforward as well.

(ii) The assertion is trivial if $\boldsymbol{A}$ is semisimple. Assume that $\boldsymbol{A}$ is not semisimple so that $\operatorname{Rad}(\boldsymbol{A}) \neq \{0\}$. Let $s \in \mathcal{S}(\boldsymbol{A})$. First, we will show that $s(a) = s(b)$, whenever $a, b \in A$ are such that $a/\operatorname{Rad}(\boldsymbol{A}) = b/\operatorname{Rad}(\boldsymbol{A})$. The last condition implies $d(a,b) \in \operatorname{Rad}(\boldsymbol{A})$, where $d$ is the Chang distance given by (1), and Proposition 3.1.1(viii) yields $s(d(a,b)) = 0$. Strong modularity together with monotonicity give

$$
\begin{aligned}
s(a \ominus b) + s(b \ominus a) &= s((a \ominus b) \odot (b \ominus a)) + s((a \ominus b) \oplus (b \ominus a)) \\
&= s((a \ominus b) \odot (b \ominus a)) + s(d(a,b)) = 0.
\end{aligned}
$$

Hence necessarily $s(a \ominus b) = s(b \ominus a) = 0$. Using strong modularity again, we can now write

$$
s(\neg a) + s(b) = s(\neg a \oplus b) = 1 = s(a \oplus \neg b) = s(a) + s(\neg b),
$$

from which results $s(a) = s(b)$. It is thus correct to define $s' \colon \boldsymbol{A}/\operatorname{Rad}(\boldsymbol{A}) \to [0,1]$ by $s'(a/\operatorname{Rad}(\boldsymbol{A})) = s(a)$, for every $a \in A$. Then $s' \in \mathcal{S}(\boldsymbol{A}/\operatorname{Rad}(\boldsymbol{A}))$. It is easily seen that the mapping $s \mapsto s'$ is an affine isomorphism. In order to show that the mapping $s \mapsto s'$ is a homeomorphism, we need only check that it is continuous. Let $(s_\gamma)$ be a net (generalized sequence) of elements in $\mathcal{S}(\boldsymbol{A})$ such that $s_\gamma \to s$ in $\mathcal{S}(\boldsymbol{A})$. This means that

$$
\lim_\gamma s_\gamma(a) = s(a), \quad \text{for every } a \in A.
$$

Then, for every $a \in A$, we have $s'_\gamma(a/\operatorname{Rad}(\boldsymbol{A})) = s_\gamma(a) \to s(a) = s'(a/\operatorname{Rad}(\boldsymbol{A}))$ in $[0,1]$. Hence $s \mapsto s'$ is continuous. □

We say that a state $s$ of $\boldsymbol{A}$ is *faithful* if $s$ is strictly positive, that is, we have $s(a) > 0$ whenever $a \in A$ is nonzero.

PROPOSITION 3.1.8. *Let $\boldsymbol{A}$ be an* MV*-algebra. Then:*

*(i) A state $s$ of $\boldsymbol{A}$ is faithful iff $s(a) < s(b)$, for every $a, b \in A$ with $a < b$.*

*(ii) If $\boldsymbol{A}$ carries a faithful state, then $\boldsymbol{A}$ is semisimple.*

*Proof.* (i) If $s$ is faithful, then the condition $a < b$ implies $b \ominus a > 0$ and subtractivity gives $s(b) - s(a) = s(b \ominus a) > 0$. Conversely, if $a \neq 0$, then $s(a) = s(a \ominus 0) = s(a) - s(0) > 0$.

(ii) Suppose that $\boldsymbol{A}$ is not semisimple. Then there exists a nonzero $a \in \operatorname{Rad}(\boldsymbol{A})$. However, Proposition 3.1.1(viii) says that $s(a) = 0$ for every state $s$ of $\boldsymbol{A}$. □

### 3.2 Examples of states

We will discuss basic examples of states of various MV-algebras. The next sections then reveal the general pattern common to all of those examples. We have already proved (Proposition 3.1.5) that an arbitrary MV-algebra carries at least one state; in fact, the examples suggest that MV-algebras are abundant in states.

EXAMPLE 3.2.1 (Finitely-additive probability). We know that any Boolean algebra $\boldsymbol{B}$ is an MV-algebra in which the MV-operations $\oplus$ and $\odot$ coincide with the lattice operations $\vee$ and $\wedge$, respectively. Thus every state $s$ of $\boldsymbol{B}$ is a finitely additive probability since the condition (4) reads as follows:

$$\text{if } a \wedge b = 0, \text{ then } s(a \vee b) = s(a) + s(b).$$

EXAMPLE 3.2.2 (Homomorphism). Every homomorphism $h$ of an MV-algebra $\boldsymbol{A}$ into the standard MV-algebra $[0,1]_Ł$ is a state of $\boldsymbol{A}$. In particular, whenever $\boldsymbol{A}$ is a subalgebra of the MV-algebra $[0,1]^X$ of all functions $X \to [0,1]$. For any $x \in X$ the evaluation mapping $s_x \colon A \to [0,1]$ given by

$$s_x(f) = f(x), \quad f \in A, \tag{5}$$

is a state of $\boldsymbol{A}$.

Both examples above are special cases of the following construction. Let $\boldsymbol{A}$ be a semisimple MV-algebra with the maximal ideal space $\mathrm{Max}(\boldsymbol{A})$ and assume that $\boldsymbol{A}^*$ is the separating clan of continuous functions $\mathrm{Max}(\boldsymbol{A}) \to [0,1]$ that is isomorphic to $\boldsymbol{A}$ (Theorem 2.1.7). For every $a \in A$, let $a^* \in A^*$ be the function corresponding to $a$ via the isomorphism. Consider any regular Borel probability measure $\mu$ on the compact Hausdorff space $\mathrm{Max}(\boldsymbol{A})$. Put

$$s_\mu(a) = \int_{\mathrm{Max}(\boldsymbol{A})} a^* \,\mathrm{d}\mu, \quad \text{for every } a \in A. \tag{6}$$

It can be routinely checked that $s_\mu$ is a state of $\boldsymbol{A}$. Observe that (5) is a special case of (6) upon putting $\mu = \delta_M$, where $\delta_M$ is the Dirac measure supported by a maximal ideal $M \in \mathrm{Max}(\boldsymbol{A})$.

EXAMPLE 3.2.3 (Lebesgue state). As an important special case of (6), consider $\boldsymbol{A}$ to be a finitely presented MV-algebra. Then there exists $n \in \mathsf{N}$ and a nonempty rational polyhedron $P \subseteq [0,1]^n$, $\dim(P) = n$, such that $\boldsymbol{A}$ is isomorphic to the MV-algebra $\boldsymbol{M}(P)$ of restrictions of McNaughton functions $f \in \boldsymbol{F}_n$ to $P$ (Theorem 2.2.3). Let $\lambda$ be the $n$-dimensional Lebesgue measure on $[0,1]^n$. Since $\lambda(P) > 0$ it makes sense to define a state of $\boldsymbol{M}(P)$ by putting

$$s_\lambda(f) = \frac{\int_P f \,\mathrm{d}\lambda}{\lambda(P)}, \quad f \in \boldsymbol{M}(P). \tag{7}$$

The MV-algebra $\boldsymbol{M}(P)$ is isomorphic to the Lindenbaum algebra of theory $\{\vartheta\}$ for some satisfiable formula $\vartheta \in \mathsf{Form}_n$, so we arrive at the following interpretation of $s_\lambda$. The number (7) is the average truth value (with respect to all the models of $\{\vartheta\}$) of any formula $\varphi$ such that $[\varphi]$ coincides with the restriction of $f$ to $P$. Indeed, $s_\lambda(f)$ is the expected value of $f$ with respect to the uniform distribution $\lambda$ over the possible worlds in $P$.[3]

EXAMPLE 3.2.4 (States of Lindenbaum algebra). Another example interesting from the logical viewpoint is the Lindenbaum algebra $\boldsymbol{F}$ of Łukasiewicz logic over countably-many propositional variables. For every $f \in \boldsymbol{F}$ there exist $n \in \mathsf{N}$, $g \in \boldsymbol{F}_n$ and a nonempty set $I \subseteq \{1, \ldots, n\}$ such that $f(\langle x_i \rangle_{i \in \mathsf{N}}) = g(\langle x_i \rangle_{i \in I})$, for every sequence $\langle x_i \rangle_{i \in \mathsf{N}} \in [0,1]^{\mathsf{N}}$. Thus every element of $\boldsymbol{F}$ is a function of $n$ variables only and coincides with some McNaughton function.

This enables us to simplify the computation of states of $\boldsymbol{F}$ defined by (6) since, for any Borel probability measure $\mu$ on the compact Hausdorff space (in the product topology) $[0,1]^{\mathsf{N}}$, we have

$$s_\mu(f) = \int_{[0,1]^{\mathsf{N}}} f \, \mathrm{d}\mu = \int_{[0,1]^n} g \, \mathrm{d}\mu',$$

where $\mu'$ is the Borel measure on $[0,1]^n$ given by $\mu'(A) = \mu\left(\pi^{-1}(A)\right)$, $A$ is a Borel set in $[0,1]^n$ and $\pi \colon [0,1]^{\mathsf{N}} \to [0,1]^n$ is the projection function. In conclusion, when it comes to computing the average truth value (6), there is no loss of generality in replacing the Lindenbaum algebra $\boldsymbol{F}$ of Łukasiewicz logic with the MV-algebra of McNaughton functions $\boldsymbol{F}_n$, for some $n \in \mathsf{N}$.

EXAMPLE 3.2.5. Let ${}^*[0,1]_{Ł}$ be a non-trivial ultraproduct of the standard MV-algebra $[0,1]_{Ł}$ and choose a positive infinitesimal $c \in {}^*[0,1]$. The *Chang* MV-*algebra* $\boldsymbol{C}$ is (up to an isomorphism) the MV-subalgebra of ${}^*[0,1]_{Ł}$ generated by the set $\{0, c\}$ (see [36, Example 2.4.5] for a detailed analysis of Chang MV-algebra). It is known that $\boldsymbol{C}$ has the universe $\mathrm{Rad}(\boldsymbol{C}) \cup \mathrm{Rad}(\boldsymbol{C})^*$, where $\mathrm{Rad}(\boldsymbol{C}) = \{0, c, c \oplus c, c \oplus c \oplus c, \ldots\}$ is the radical of $\boldsymbol{C}$ and $\mathrm{Rad}(\boldsymbol{C})^* = \{1, 1-c, 1-(c \oplus c), 1-(c \oplus c \oplus c), \ldots\}$ is the co-radical of $\boldsymbol{C}$.

It is worth to notice that, by Proposition 3.1.1 (viii) and (ii), Chang algebra has only one trivial state. Namely $\boldsymbol{C}$ has a unique state $s$ such that $s(x) = 0$ for every $x \in \mathrm{Rad}(\boldsymbol{C})$ and $s(x) = 1$ for every $x \in \mathrm{Rad}(\boldsymbol{C})^*$.

In general, having only one trivial state is a property that characterizes a wide class of MV-algebras, the so-called *perfect* MV-*algebras*, which are MV-algebras $\boldsymbol{A}$ having $\mathrm{Rad}(\boldsymbol{A}) \cup \mathrm{Rad}(\boldsymbol{A})^*$ as the universe (see [36, Section 4.3]).

## 3.3 States of finitely presented MV-algebras

In this section we completely characterize the states of any finitely presented MV-algebra. Although the result proved herein is a special case of integral representation theorem developed below (Section 4), we consider its separate treatment a worthwhile digression on the way to understanding the structure and the properties of states of any MV-algebra.

[3] With a small abuse of notation, we will identify the possible worlds of an MV-algebra $\boldsymbol{A}$ with the elements of $\mathcal{H}(\boldsymbol{A})$. This identification will be made explicit in Section 5.

First, we will recall the basic results concerning Schauder hats and bases; see [1] or [48]. Let $T \subseteq [0,1]^n$ be an $n$-dimensional simplex with rational vertices. Let $\mathbf{x} = \langle a_1/d, \ldots, a_n/d \rangle$ be a vertex of $T$, for uniquely determined relatively prime integers $a_1, \ldots, a_n, d$ with $d \geq 1$. Call $\langle a_1, \ldots, a_n, d \rangle$ the *homogeneous coordinates* of $\mathbf{x}$, and let $\mathrm{den}(\mathbf{x}) = d$ be the *denominator* of $\mathbf{x}$. We say that simplex $T$ is *unimodular* if the determinant of the integer square matrix having the homogeneous coordinates of all the vertices of $T$ in its rows is equal to $\pm 1$. An $r$-dimensional simplex with $r \leq n$ is unimodular if it is a face of some unimodular $n$-dimensional simplex.

Let $p_1, \ldots, p_l$ be the linear pieces of an $n$-variable McNaughton function $f$. For every permutation $\pi$ of $\{1, \ldots, l\}$, put:

$$P_\pi = \{\mathbf{x} \in [0,1]^n \mid p_{\pi(1)}(\mathbf{x}) \leq p_{\pi(2)}(\mathbf{x}) \leq \cdots \leq p_{\pi(l)}(\mathbf{x})\}, \tag{8}$$

$$C = \{P_\pi \mid P_\pi \text{ is } n\text{-dimensional}\}. \tag{9}$$

Clearly $C$ is a finite set of $n$-dimensional polytopes with rational vertices, that is, every $P_\pi \in C$ is the convex hull of a finite set of rational points in $[0,1]^n$. It is well-known that $C$ can be refined to a *unimodular triangulation* of $[0,1]^n$ that linearizes $f$, which is a finite set $\Sigma$ of $n$-dimensional unimodular simplices over the rational vertices such that:

(i) the union of all simplices in $\Sigma$ is equal to $[0,1]^n$,

(ii) any two simplices in $\Sigma$ intersect in a common face and

(iii) for each simplex $T \in \Sigma$, there exists $j = 1, \ldots, l$ such that the restriction of $f$ to $T$ coincides with $p_j$ (we also say that $f$ is *linear* over $\Sigma$).

Let $\mathrm{V}_\Sigma = \{\mathbf{x}_1, \ldots, \mathbf{x}_m\}$ be the set of vertices of all simplices in $\Sigma$. The *Schauder hat* at $\mathbf{x}_i \in \mathrm{V}_\Sigma$ is the McNaughton function $h_{\mathbf{x}_i} = h_i$ linearized by $\Sigma$ such that $h_i(\mathbf{x}_i) = 1/\mathrm{den}(\mathbf{x}_i)$ and $h_i(\mathbf{x}_j) = 0$ for every vertex $\mathbf{x}_j$ distinct from $\mathbf{x}_i$ in $\Sigma$. The set $H_\Sigma = \{h_1, \ldots, h_m\}$ is called a *Schauder basis for* $\boldsymbol{F}_n$. The *normalized* Schauder hat at $\mathbf{x}_i$ is the McNaughton function

$$\hat{h}_{\mathbf{x}_i} = \hat{h}_i = \mathrm{den}(\mathbf{x}_i) \cdot h_i.$$

We denote by $\hat{H}_\Sigma = \{\hat{h}_1, \ldots, \hat{h}_m\}$ the subset of $\boldsymbol{F}_n$ consisting of all the normalized Schauder hats. The set $\hat{H}_\Sigma$ is also called a *normalized Schauder basis.*

Note that every McNaughton function that is linear over $\Sigma$ is a linear combination of the family of Schauder hats corresponding to $\Sigma$, where each hat has a uniquely determined integer coefficient between 0 and $\mathrm{den}(\mathbf{x}_i)$. Thus

$$f = \sum_{\mathbf{x}_i \in \mathrm{V}_\Sigma} a_{\mathbf{x}_i} \cdot h_i, \tag{10}$$

for uniquely determined integers $0 \leq a_{\mathbf{x}_i} \leq \mathrm{den}(\mathbf{x}_i)$.

This argument can be easily generalized to the case of finitely many McNaughton functions. In particular, if $f_1, \ldots, f_k$ are McNaughton functions on the $n$-cube $[0,1]^n$, we can find a unimodular triangulation $\Sigma$ of $[0,1]^n$ with vertices $\mathrm{V}_\Sigma = \{\mathbf{x}_1, \ldots, \mathbf{x}_m\}$ such that each $f_i$ is linear over each simplex of $\Sigma$. In what follows we will need the following result.

LEMMA 3.3.1. *Let $f_1, \ldots, f_k$, $\Sigma$, $\mathrm{V}_\Sigma$ and $\hat{H}_\Sigma$ be as above. Then:*

*(i) For distinct $\hat{h}_i, \hat{h}_j \in \hat{H}_\Sigma$, $\hat{h}_i \odot \hat{h}_j = 0$.*

*(ii) $\bigoplus_{t=1}^{m} \hat{h}_t = 1$.*

*(iii) $f_i = \bigoplus_{t=1}^{m} f_i(\mathbf{x}_t) \cdot \hat{h}_t$, for each $i = 1, \ldots, k$.*

For the proof of Theorem 3.3.4 we prepare two more lemmas.

LEMMA 3.3.2. *Let $P \subseteq [0,1]^n$ be a nonempty rational polyhedron. If $\mu$ and $\nu$ are Borel probability measures on $P$ such that $\mu \neq \nu$, then $s_\mu \neq s_\nu$, where $s_\mu$ and $s_\nu$ are given by (6).*

*Proof.* By way of contradiction, suppose $s_\mu = s_\nu$. The Borel subsets of $P$ are generated by the collection of all closed (in the subspace Euclidean topology of $P$) rational polyhedra. Indeed, every closed subset in $P$ can be written as a countable intersection of such polyhedra. Since the set of all rational polyhedra is closed under finite intersections, [10, Theorem 3.3] yields the existence of a rational polyhedron $R \subseteq P$ with $\mu(R) \neq \nu(R)$.

There is $f \in M(P)$ such that $R = f^{-1}(1)$ by Proposition 2.2.2. Let $\chi_R$ denote the characteristic function of $R$. Then we obtain

$$\chi_R = \bigwedge_{m \in \mathsf{N}} \bigodot_{i=1}^{m} f. \tag{11}$$

For every $m \in \mathsf{N}$, the function $\bigodot_{i=1}^{m} f$ belongs to $M(P)$. The contradiction now follows from (11) together with the Lebesgue's dominated convergence theorem [6, Theorem 11.21]:

$$\begin{aligned}
\mu(R) = \int_P \chi_R \, \mathrm{d}\mu &= \bigwedge_{m \in \mathsf{N}} \int_P \bigodot_{i=1}^{m} f \, \mathrm{d}\mu = \bigwedge_{m \in \mathsf{N}} s_\mu \left( \bigodot_{i=1}^{m} f \right) \\
&= \bigwedge_{m \in \mathsf{N}} s_\nu \left( \bigodot_{i=1}^{m} f \right) = \bigwedge_{m \in \mathsf{N}} \int_P \bigodot_{i=1}^{m} f \, \mathrm{d}\nu = \int_P \chi_R \, \mathrm{d}\nu = \nu(R).
\end{aligned}$$

□

LEMMA 3.3.3. *Let $P \subseteq [0,1]^n$ be a rational polyhedron, $\Sigma$ be any unimodular triangulation of $P$ and $\hat{H}_\Sigma$ be a normalized Schauder basis. If $a\colon \hat{H}_\Sigma \to [0,1]$ is a function such that $\sum_{\mathbf{x} \in \mathrm{V}_\Sigma} a(\hat{h}_\mathbf{x}) = 1$, then there exists a Borel probability measure $\delta$ with finite support on $P$ satisfying $a(\hat{h}_\mathbf{x}) = s_\delta(\hat{h}_\mathbf{x})$, for each $\mathbf{x} \in \mathrm{V}_\Sigma$.*

*Proof.* Let $\delta_\mathbf{x}$ be the Dirac measure concentrated at a vertex $\mathbf{x} \in \mathrm{V}_\Sigma$. Put

$$\delta = \sum_{\mathbf{x} \in \mathrm{V}_\Sigma} a(\hat{h}_\mathbf{x}) \delta_\mathbf{x}.$$

Then, for each vertex $\mathbf{x} \in \mathrm{V}_\Sigma$, we get

$$\begin{aligned} s_\delta(\hat{h}_\mathbf{x}) = \int_P \hat{h}_\mathbf{x} \,\mathrm{d}\delta &= \sum_{\mathbf{x}' \in \mathrm{V}_\Sigma} \int_P a(\hat{h}_{\mathbf{x}'})\hat{h}_\mathbf{x} \,\mathrm{d}\delta_{\mathbf{x}'} \\ &= \sum_{\mathbf{x}' \in \mathrm{V}_\Sigma} a(\hat{h}_{\mathbf{x}'})\hat{h}_\mathbf{x}(\mathbf{x}') = a(\hat{h}_\mathbf{x})\hat{h}_\mathbf{x}(\mathbf{x}) = a(\hat{h}_\mathbf{x}). \end{aligned}$$

□

THEOREM 3.3.4. *Let $s$ be a state of the finitely presented* MV*-algebra $\boldsymbol{M}(P)$, where $P$ is a nonempty rational polyhedron. Then there exists a unique Borel probability measure $\mu$ on $P$ such that $s = s_\mu$.*

*Proof.* Let $s$ be a state of $\boldsymbol{M}(P)$. For any normalized Schauder basis $\hat{H}_\Sigma$, let us define the set of Borel probability measures

$$\mathcal{M}_\Sigma = \{\mu \mid s(\hat{h}_\mathbf{x}) = s_\mu(\hat{h}_\mathbf{x}), \text{ for each } \mathbf{x} \in \mathrm{V}_\Sigma\}.$$

Note that $\mathcal{M}_\Sigma \neq \emptyset$ by Lemma 3.3.3. It follows directly from the definition of weak* topology (see Section 2.4) on the set of all Borel probability measures $\mathcal{M}(P)$ over $P$ that $\mathcal{M}_\Sigma$ is weak* closed in $\mathcal{M}(P)$. Indeed, consider any net $(\mu_\gamma)$ in $\mathcal{M}_\Sigma$ weak* converging to some $\mu \in \mathcal{M}(P)$. For each $\hat{h}_\mathbf{x} \in \hat{H}_\Sigma$, this means by (3) that

$$s(\hat{h}_\mathbf{x}) = s_{\mu_\gamma}(\hat{h}_\mathbf{x}) \to s_\mu(\hat{h}_\mathbf{x}),$$

and $\mathcal{M}_\Sigma$ is thus weak* closed.

Let $\mathfrak{T}$ denote the family of all unimodular triangulations of polyhedron $P$. We are going to show that

$$\bigcap_{\Sigma \in \mathfrak{T}} \mathcal{M}_\Sigma \neq \emptyset. \tag{12}$$

Since $\mathcal{M}(P)$ is weak* compact by Theorem 2.4.4, $\mathcal{M}_\Sigma$ is weak* compact too, so it suffices to prove $\bigcap_{\Sigma' \in \mathfrak{T}'} \mathcal{M}_{\Sigma'} \neq \emptyset$ for every finite subset $\mathfrak{T}' \subseteq \mathfrak{T}$. It follows from [1, Lemma 2.1.7] that finitely-many unimodular triangulations—ergo Schauder bases—can always be jointly refined. Specifically, this means that there exists a Schauder basis $H_\Sigma$ such that: for every $\Sigma' \in \mathfrak{T}'$ with $\mathfrak{T}'$ finite and for each normalized Schauder hat $\hat{h}_\mathbf{x} \in \hat{H}_{\Sigma'}$, there exists a uniquely determined nonnegative integer vector $(\beta_\mathbf{y})_{\mathbf{y} \in \mathrm{V}_\Sigma}$ such that $\hat{h}_\mathbf{x} = \sum_{\mathbf{y} \in \mathrm{V}_\Sigma} \beta_\mathbf{y} h_\mathbf{y}$. Let $\delta = \sum_{\mathbf{y} \in \mathrm{V}_\Sigma} s(\hat{h}_\mathbf{y})\delta_\mathbf{y}$. Clearly, $\delta \in \mathcal{M}(P)$. Linearity of Lebesgue integral gives

$$s_\delta(\hat{h}_\mathbf{x}) = \int_P \hat{h}_\mathbf{x} \,\mathrm{d}\delta = \sum_{\mathbf{y}' \in \mathrm{V}_\Sigma} s(\hat{h}_{\mathbf{y}'}) \int_P \sum_{\mathbf{y} \in \mathrm{V}_\Sigma} \frac{\beta_\mathbf{y}}{\mathrm{den}(\mathbf{y})} \hat{h}_\mathbf{y} \,\mathrm{d}\delta_{\mathbf{y}'}. \tag{13}$$

For every $\mathbf{y} \in \mathrm{V}_\Sigma$,

$$\frac{s(\hat{h}_\mathbf{y})}{\mathrm{den}(\mathbf{y})} = \frac{s(\mathrm{den}(\mathbf{y}) h_\mathbf{y})}{\mathrm{den}(\mathbf{y})} = s(h_\mathbf{y}).$$

Then additivity of states enables us to express the right-hand side of (13) as

$$\sum_{\mathbf{y}'\in V_\Sigma} s(\hat{h}_{\mathbf{y}'}) \sum_{\mathbf{y}\in V_\Sigma} \frac{\beta_{\mathbf{y}}}{\mathrm{den}(\mathbf{y})}\hat{h}_{\mathbf{y}}(\mathbf{y}') = \sum_{\mathbf{y}'\in V_\Sigma} s(\hat{h}_{\mathbf{y}'})\frac{\beta_{\mathbf{y}'}}{\mathrm{den}(\mathbf{y}')} = \sum_{\mathbf{y}'\in V_\Sigma} s\left(\frac{\beta_{\mathbf{y}'}}{\mathrm{den}(\mathbf{y}')}\hat{h}_{\mathbf{y}'}\right)$$

$$= s\left(\sum_{\mathbf{y}\in V_\Sigma} \beta_{\mathbf{y}} h_{\mathbf{y}}\right) = s(\hat{h}_{\mathbf{x}}).$$

Thus we have shown that $\delta \in \bigcap_{\Sigma'\in\mathfrak{T}'} \mathcal{M}_{\Sigma'}$, which implies (12), as $\mathfrak{T}'$ was an arbitrary finite set of unimodular triangulations of $P$.

We will prove that $s_\mu = s$ for every $\mu \in \bigcap_{\Sigma\in\mathfrak{T}} \mathcal{M}_\Sigma$. Indeed, given a function $f \in \boldsymbol{M}(P)$, find $\Pi \in \mathfrak{T}$ and a Schauder basis $H_\Pi$ such that $f = \sum_{\mathbf{x}\in V_\Pi} \alpha_{\mathbf{x}} h_{\mathbf{x}}$, for uniquely determined nonnegative integers $\alpha_{\mathbf{x}}$ [1, Lemma 2.1.19]. It results that

$$s(f) = s\left(\sum_{\mathbf{x}\in V_\Pi} \alpha_{\mathbf{x}} h_{\mathbf{x}}\right) = \sum_{\mathbf{x}\in V_\Pi} \alpha_{\mathbf{x}} s(h_{\mathbf{x}}) = \sum_{\mathbf{x}\in V_\Pi} \alpha_{\mathbf{x}} s_\mu(h_{\mathbf{x}})$$

$$= s_\mu\left(\sum_{\mathbf{x}\in V_\Pi} \alpha_{\mathbf{x}} h_{\mathbf{x}}\right) = \sum_{\mathbf{x}\in V_\Pi} \alpha_{\mathbf{x}} s(h_{\mathbf{x}}) = s_\mu(f).$$

Finally, the set $\bigcap_{\Sigma\in\mathfrak{T}} \mathcal{M}_\Sigma$ contains a single element by Lemma 3.3.2. □

REMARK 3.3.5. *The unique Borel probability measure $\mu$ on the rational polyhedron $P$ from Theorem 3.3.4 is regular since $P \subseteq [0,1]^n$.*

The previous proof highlights the central role of Schauder bases as the basic building blocks of any element in the finitely presented algebra. As a matter of fact, the resulting measure $\mu$ such that $s = s_\mu$ is the "finest" probability measure among all the probabilities agreeing with $s$ over all Schauder bases or, equivalently, over the collection of all unimodular triangulations of the rational polyhedron $P$. Theorem 3.3.4 is substantially generalized into an integral representation for states of any MV-algebra in Section 4.

### 3.4 States of $\ell$-groups

The states of MV-algebras are closely related to normalized positive real homomorphisms of lattice ordered Abelian groups (further abbreviated as $\ell$-groups); see [27]. We will use the construction of unital $\ell$-group associated with an MV-algebra—the details can be found in [36, Section 5]. Let $\langle G, 1\rangle$ be an Abelian $\ell$-group with strong unit 1 and neutral element 0 (a unital $\ell$-group, for short). Then the order interval $\Gamma(G,1) = \{a \in G \mid 0 \leq a \leq 1\}$ becomes an MV-algebra with the induced operations $a \oplus b = (a+b) \wedge 1$ and $\neg a = 1 - a$. The group operation $+$ of $\langle G, 1\rangle$ and the MV-algebraic operations of $\Gamma(G,1)$ are related as follows:

$$a + b = (a \oplus b) + (a \odot b), \quad \text{for every } a, b \in \Gamma(G,1). \tag{14}$$

Conversely, given some MV-algebra $\boldsymbol{A}$, Mundici [43] constructed the unital Abelian $\ell$-group $\langle G_{\boldsymbol{A}}, 1\rangle$ such that $\boldsymbol{A}$ is isomorphic to $\Gamma(G_{\boldsymbol{A}}, 1)$ and showed that $\Gamma$ provides the categorical equivalence between the category of MV-algebras and that of unital $\ell$-groups.

By $G^+$ we denote the partially ordered monoid of all positive elements ($a \geq 0$) in a unital $\ell$-group $\langle G, 1\rangle$. A *state of a unital $\ell$-group* $\langle G, 1\rangle$ is a group homomorphism $s\colon G \to \mathsf{R}$ such that $s(a) \geq 0$, for every $a \in G^+$, and $s(1) = 1$. By $\mathcal{S}(G, 1)$ we denote the set of all states of $\langle G, 1\rangle$. It turns out that measuring the elements of an MV-algebra $\boldsymbol{A}$ with a state $s$ is essentially the same as specifying a state of the corresponding unital $\ell$-group $\langle G_{\boldsymbol{A}}, 1\rangle$.

PROPOSITION 3.4.1. *Let $\langle G, 1\rangle$ be a unital $\ell$-group and $\boldsymbol{A} = \Gamma(G, 1)$ be the associated* MV*-algebra. Then:*

*(i) For every state $s$ of $\langle G, 1\rangle$, the restriction $\bar{s}$ of $s$ to $\boldsymbol{A}$ is a state of $\boldsymbol{A}$.*

*(ii) The mapping $s \mapsto \bar{s}$ is an affine isomorphism of $\mathcal{S}(G, 1)$ onto $\mathcal{S}(\boldsymbol{A})$.*

*Proof.* (i) Clearly, $\bar{s}(1) = s(1) = 1$. Let $a, b \in A$ be such that $a \odot b = 0$. By (14) this means that $a \oplus b = a + b$, where $+$ is the addition in $(G, 1)$. Therefore

$$\bar{s}(a \oplus b) = \bar{s}(a + b) = s(a + b) = s(a) + s(b) = \bar{s}(a) + \bar{s}(b).$$

(ii) We need to invert the mapping $s \mapsto \bar{s}$ sending a state $s$ of $\langle G, 1\rangle$ to its restriction $\bar{s}$ on $\boldsymbol{A}$. To this end, assume that $r$ is a state of $\boldsymbol{A}$. First, we extend state $r$ to a monoid homomorphism $\tilde{r}\colon G^+ \to [0, \infty)$ in a unique way as follows. We can identify $G^+$ with the monoid of good sequences [36, Proposition 5.1.14]. Specifically, for every $a \in G^+$ there exists a unique tuple $\langle g_1, \ldots, g_n\rangle \in A^n$ (up to appending a finite sequence of 0s) such that

(i) $a = g_1 + \cdots + g_n$,

(ii) $g_i \oplus g_{i+1} = g_i$,

(iii) $g_i \odot g_{i+1} = g_{i+1}$.

Define $\tilde{r}(a) = \tilde{r}(\langle g_1, \ldots, g_n\rangle) = r(g_1) + \cdots + r(g_n)$. Clearly $\tilde{r}(0) = r(0) = 0$. We need only show that $\tilde{r}(a + b) = \tilde{r}(a) + \tilde{r}(b)$ for every $a \in G^+$ and every $b \in A$. Put

$$\begin{aligned}
g'_1 &= g_1 \oplus b,\\
g'_2 &= g_2 \oplus g_1 \odot b,\\
g'_3 &= g_3 \oplus g_2 \odot b,\\
&\;\vdots\\
g'_n &= g_n \oplus g_{n-1} \odot b,\\
g'_{n+1} &= g_n \odot b.
\end{aligned}$$

It can be shown that the good sequence $\langle g'_1, \ldots, g'_{n+1}\rangle$ represents $a + b \in G^+$. Then

$$\begin{aligned}\tilde{r}(a+b) &= \tilde{r}(\langle g'_1, \ldots, g'_{n+1}\rangle) = r(g'_1) + r(g'_2) + \cdots + r(g'_{n+1})\\ &= r(g_1 \oplus b) + r(g_2 \oplus g_1 \odot b) + \cdots + r(g_n \odot b).\end{aligned}$$

The application of strong modularity (Proposition 3.1.1(iv)) to each summand makes it possible to write the last sum as

$$\begin{aligned}& r(g_1) + r(b) - r(g_1 \odot b)\\ &+ r(g_2) + r(g_1 \odot b) - r(\underbrace{g_2 \odot g_1}_{g_2} \odot b)\\ &+ r(g_3) + r(g_2 \odot b) - r(\underbrace{g_3 \odot g_2}_{g_3} \odot b)\\ &+ \cdots\\ &+ r(g_n) + r(g_{n-1} \odot b) - r(\underbrace{g_n \odot g_{n-1}}_{g_n} \odot b)\\ &+ r(g_n \odot b)\\ &= r(g_1) + \cdots + r(g_n) + r(b)\\ &= \tilde{r}(a) + \tilde{r}(b).\end{aligned}$$

This shows that $\tilde{r}$ is a monoid homomorphism. Since $G = G^+ - G^+$, we can put $\hat{r}(a-b) = \tilde{r}(a) - \tilde{r}(b)$ for every $a, b \in G$ and routinely show that the definition is correct. The state $\hat{r}$ is the sought unique extension of $r$. Consequently, there exists a one-to-one correspondence between $\mathcal{S}(G, 1)$ and $\mathcal{S}(\boldsymbol{A})$ given by the restriction map $s \mapsto \bar{s}$. This map is easily seen to be affine. □

Thus every state of an MV-algebra $\boldsymbol{A}$ can be lifted to the unique state of the enveloping Abelian $\ell$-group $G_{\boldsymbol{A}}$. This fact has, among others, the following interesting consequence derived from the known results about states of $\ell$-groups.

COROLLARY 3.4.2. *Let $\boldsymbol{A}$ be an* MV*-algebra and $\boldsymbol{B}$ be a sub-*MV*-algebra. For every state $s$ of $\boldsymbol{B}$ there exists a state $s'$ of $\boldsymbol{A}$ such that $s'(b) = s(b)$, $b \in B$.*

*Proof.* The $\ell$-group $G_{\boldsymbol{B}}$ is a subgroup of the $\ell$-group $G_{\boldsymbol{A}}$. The state $s$ of $\boldsymbol{B}$ lifts to a unique state (also denoted by $s$) of $G_{\boldsymbol{B}}$. It suffices to show that there exists a state $s'$ of $G_{\boldsymbol{A}}$ such that $s'(b) = s(b)$, $b \in G_{\boldsymbol{B}}$. The last claim is, however, the content of [27, Corollary 4.3]. □

# 4 Integral representation

In this section we are going to prove one of the main results of this chapter: the one-to-one correspondence of states to regular Borel probability measures over the maximal ideal space of an MV-algebra. This correspondence is realized via Lebesgue integral and turns out to have very strong geometrical and topological properties. Specifically, we will show that the integral states of the form $s_\mu$ introduced in (6) are the most general examples of states.

From now on, we always assume that $\mathcal{M}(\mathrm{Max}(\boldsymbol{A}))$, the Bauer simplex of all regular Borel probability measures over $\mathrm{Max}(\boldsymbol{A})$, is equipped with the weak* topology (see Section 2.4).

THEOREM 4.0.1. *Let $\boldsymbol{A}$ be an* MV*-algebra and $\mathcal{M}(\mathrm{Max}(\boldsymbol{A}))$ be the set of all regular Borel probability measures on* $\mathrm{Max}(\boldsymbol{A})$. *Then there is an affine homeomorphism*

$$\Phi\colon \mathcal{S}(\boldsymbol{A}) \to \mathcal{M}(\mathrm{Max}(\boldsymbol{A}))$$

*such that, for every $a \in A$,*

$$s(a) = \int_{\mathrm{Max}(\boldsymbol{A})} a^*(M)\,\mathrm{d}\mu_s(M), \qquad \textit{where } \mu_s = \Phi(s). \tag{15}$$

*Proof.* In the light of Proposition 3.1.7(ii), we may assume that $\boldsymbol{A}$ is semisimple, without loss of generality. In particular, $\boldsymbol{A}$ is isomorphic to a separating MV-algebra $\boldsymbol{A}^*$ of continuous functions $\mathrm{Max}(\boldsymbol{A}) \to [0,1]$ (see Theorem 2.1.7). Let $s \in \mathcal{S}(\boldsymbol{A}^*)$. We are going to extend $s$ uniquely to a bounded linear functional over the Banach space $C(\mathrm{Max}(\boldsymbol{A}))$ of all continuous functions $\mathrm{Max}(\boldsymbol{A}) \to \mathsf{R}$ endowed with the supremum norm $\|\cdot\|$.

By Proposition 3.4.1, the state $s$ of $\boldsymbol{A}^*$ uniquely corresponds to a state $s'$ of unital $\ell$-group $G_{\boldsymbol{A}^*}$ such that $\Gamma(G_{\boldsymbol{A}^*},1) = \boldsymbol{A}^*$. The unital $\ell$-group $G_{\boldsymbol{A}^*}$ embeds in its divisible hull, i.e., the rational sub-vector lattice $H_{\boldsymbol{A}^*} = \{qa \mid a \in G_{\boldsymbol{A}^*}, q \in \mathsf{Q}\}$ (see, e.g. [9, Sections 1.6.8–1.6.9]). Putting $s''(qa) = qs'(a)$ for every $q \in \mathsf{Q}$ and $a \in G_{\boldsymbol{A}^*}$, it is easy to check that $s''\colon H_{\boldsymbol{A}^*} \to \mathsf{R}$ is a positive linear functional uniquely extending state $s'$. The functional $s''$ is bounded: if $b \in H_{\boldsymbol{A}^*}$ is such that $\|b\| \leq 1$, then $|s''(b)| \leq s''(1) = 1$. Hence we obtain a continuous linear functional $s''$ on the vector lattice $H_{\boldsymbol{A}^*}$.

The lattice version of Stone–Weierstrass Theorem [6, Theorem 9.12] says that $H_{\boldsymbol{A}^*}$ is a norm-dense subspace of the Banach space $C(\mathrm{Max}(\boldsymbol{A}))$. For every $b \in C(\mathrm{Max}(\boldsymbol{A}))$, there exists a sequence $\langle b_n\rangle \in H_{\boldsymbol{A}^*}^{\mathsf{N}}$ such that $\|b - b_n\| \to 0$ whenever $n \to \infty$. Therefore, we can uniquely extend $s''$ onto $C(\mathrm{Max}(\boldsymbol{A}))$ by letting $\hat{s}(b) = \lim_{n\to\infty} s(b_n)$. Since for $b \geq 0$ we can find the converging sequence with elements $b_n \geq 0$, the unique extension $\hat{s}$ is a positive linear functional on $C(\mathrm{Max}(\boldsymbol{A}))$.

In order to complete the proof, it suffices to apply the Riesz representation theorem [6, Theorem 14.14] to $\hat{s}$. This yields a unique regular Borel probability measure $\mu_s$ such that

$$\hat{s}(b) = \int_{\mathrm{Max}(\boldsymbol{A})} b\,\mathrm{d}\mu_s, \qquad \text{for every } b \in C(\mathrm{Max}(\boldsymbol{A})),$$

so that

$$s(a^*) = \hat{s}(a^*) = \int_{\mathrm{Max}(\boldsymbol{A})} a^*\,\mathrm{d}\mu_s. \tag{16}$$

Consider the mapping $\Phi\colon \mathcal{S}(\boldsymbol{A}) \to \mathcal{M}(\mathrm{Max}(\boldsymbol{A}))$ sending each $s$ to a unique $\mu_s = \Phi(s)$ such that (16) holds. It is easy to see that $\Phi$ is an affine mapping onto $\mathcal{M}(\mathrm{Max}(\boldsymbol{A}))$. Let $\mu,\nu \in \mathcal{M}(\mathrm{Max}(\boldsymbol{A}))$ and $\mu = \nu$. Then

$$s_\mu(a^*) = \int_{\mathrm{Max}(\boldsymbol{A})} a^*\,\mathrm{d}\mu = \int_{\mathrm{Max}(\boldsymbol{A})} a^*\,\mathrm{d}\nu = s_\nu(a^*),$$

for every $a \in A$. Thus $s_\mu = s_\nu$ and the mapping $\Phi$ is one-to-one. To finish the proof, we only need to check that the affine isomorphism $\Phi$ is also a homeomorphism. Since $\Phi$ is a bijection between compact Hausdorff spaces, it suffices to check that $\Phi^{-1}$ is continuous. Let $(\mu_\gamma)$ be a net in $\mathcal{M}(\mathrm{Max}(\boldsymbol{A}))$ weak* converging to some $\mu \in \mathcal{M}(\mathrm{Max}(\boldsymbol{A}))$. This means that

$$\int_{\mathrm{Max}(\boldsymbol{A})} f \, \mathrm{d}\mu_\gamma \to \int_{\mathrm{Max}(\boldsymbol{A})} f \, \mathrm{d}\mu \quad \text{for every } f \in C(\mathrm{Max}(\boldsymbol{A})). \tag{17}$$

However, setting $f = a^*$ for $a \in A$, formula (17) reads as

$$s_{\mu_\gamma}(a) \to s_\mu(a) \quad \text{for every } a \in A.$$

In other words, the net $(\Phi^{-1}(\mu_\gamma))$ converges to $\Phi^{-1}(\mu)$ in the state space $\mathcal{S}(\boldsymbol{A})$. This concludes the proof. □

COROLLARY 4.0.2. *Let $\boldsymbol{A}_X$ be a separating clan of continuous functions over a compact Hausdorff space $X$. Then for every state $s$ of $\boldsymbol{A}_X$ there exists a unique regular Borel probability measure $\mu_s$ on $X$ such that $s(a) = \int_X a \, \mathrm{d}\mu_s$, for every $a \in A_X$.*

REMARK 4.0.3. *The special case of Theorem 4.0.1 appeared in [32], where the formula* (15) *was proved with the assumption that $\boldsymbol{A}$ is semisimple and without the explicit proof of the uniqueness of the representing Borel probability measure $\mu_s$. The main idea of the proof of Theorem 4.0.1 presented above belongs to Panti [52]. Independently, the uniqueness of representing probability measure $\mu_s$ for any state $s$ of the semisimple* MV*-algebra $\boldsymbol{A}$ was established in [31], where also the proof that $\Phi$ is an affine homeomorphism appears. The result presented in [31] in fact shows that Theorem 4.0.1 is equivalent to characterizing the state space of $\boldsymbol{A}$ as a Bauer simplex.*

REMARK 4.0.4. *If $X$ is a metrizable space, then every Borel probability measure on $X$ is regular and we may thus drop the word "regular" in the statement of the above theorems. A case in point is the maximal ideal space* $\mathrm{Max}(\boldsymbol{A})$ *of every countable* MV*-algebra $\boldsymbol{A}$ since the corresponding* $\mathrm{Max}(\boldsymbol{A})$ *is second countable and thus metrizable.*

A special case of Theorem 4.0.1 for a Boolean algebra $\boldsymbol{B}$ shows somewhat unexpected result: every finitely additive probability corresponds to a unique regular Borel probability measure on the Stone space of $\boldsymbol{B}$.

COROLLARY 4.0.5. *Let $\boldsymbol{B}$ be a Boolean algebra. For every finitely additive probability $\nu$ of $\boldsymbol{B}$ there exists a unique regular Borel probability measure $\mu_\nu$ on the Stone space* $\mathrm{Max}(\boldsymbol{B})$ *of $\boldsymbol{B}$ such that*

$$\nu(a) = \mu_\nu(a^*), \quad \textit{for every } a \in \boldsymbol{B},$$

*where $a^*$ is the clopen subset of* $\mathrm{Max}(\boldsymbol{B})$ *corresponding to $a$.*

A word of caution is in order here: the statement of Corollary 4.0.5 does not express the wrong claim "every finitely additive probability is a probability measure". As in the proof of Theorem 4.0.1, the real content of Corollary 4.0.5 is the extension of the dual representation of finitely additive probability to a continuous linear functional on the Banach space of continuous functions over the Stone space.

REMARK 4.0.6. *To the best of the authors' knowledge, there is only one reference where the result of Corollary 4.0.5 is explicitly formulated. Namely, Nešetřil and Ossona de Mendez [50] use the correspondence of finitely additive probabilities and regular Borel probability measures in their model-theoretic study of graph limits.*

## 4.1 Characterization of state space

Using integral representation, we will slightly refine the description of state space provided by Proposition 3.1.7. Recall that we always consider the state space $\mathcal{S}(\boldsymbol{A})$ with the relative topology of the product space $[0,1]^A$.

THEOREM 4.1.1. *Let $\boldsymbol{A}$ be an* MV*-algebra. Then:*

*(i) The state space $\mathcal{S}(\boldsymbol{A})$ is a Bauer simplex.*

*(ii) The extreme boundary $\partial\mathcal{S}(\boldsymbol{A})$ is compact and*

$$\partial\mathcal{S}(\boldsymbol{A}) = \{h \in \mathcal{S}(\boldsymbol{A}) \mid h \text{ is a homomorphism } A \to [0,1]\}.$$

*(iii) $\mathcal{S}(\boldsymbol{A}) = \overline{\mathrm{co}}(\mathcal{H}(\boldsymbol{A}))$.*

*Proof.* Without loss of generality, we may assume that the algebra $\boldsymbol{A}$ is semisimple (Proposition 3.1.7(ii)). By Proposition 3.1.7(i), $\mathcal{S}(\boldsymbol{A})$ is a compact convex set in $[0,1]^A$. Hence, it suffices to show that $\mathcal{S}(\boldsymbol{A})$ is a Choquet simplex with $\partial\mathcal{S}(\boldsymbol{A})$ compact. By Theorem 4.0.1, the mapping $\Phi\colon \mathcal{S}(\boldsymbol{A}) \to \mathcal{M}(\mathrm{Max}(\boldsymbol{A}))$ is an affine homeomorphism. This implies that $\Phi^{-1}$ is an affine homeomorphism also. Theorem 2.4.4 yields

$$\partial\mathcal{M}(\mathrm{Max}(\boldsymbol{A})) = \{\delta_M \mid M \in \mathrm{Max}(\boldsymbol{A})\}$$

so that every element $s \in \Phi^{-1}(\partial\mathcal{M}(\mathrm{Max}(\boldsymbol{A})))$ is of the form

$$s_M\colon a \in A \mapsto a^*(M) \in [0,1].$$

Thus $s_M$ is a homomorphism. Conversely, every homomorphism of $\boldsymbol{A}$ into $[0,1]$ arises in this way. It follows that the extreme boundary $\partial\mathcal{S}(\boldsymbol{A})$ must be compact since $\Phi$ is a homeomorphism. Since $\Phi$ is an affine isomorphism and $\mathcal{M}(\mathrm{Max}(\boldsymbol{A}))$ is a simplex, the image $\Phi^{-1}(\mathcal{M}(\mathrm{Max}(\boldsymbol{A}))) = \mathcal{S}(\boldsymbol{A})$ is also a simplex. This establishes (i) and (ii). The part (iii) directly follows from (i)–(ii) and Krein-Milman theorem (Theorem 2.4.2). □

Theorem 4.1.1 says that these are the only examples of a state $s \in \mathcal{S}(\boldsymbol{A})$:

(i) A homomorphism $h \in \partial\mathcal{S}(\boldsymbol{A})$

(ii) A convex combination $g = \sum_{i=1}^{n} \alpha_i h_i$ for some $h_1, \dots, h_n \in \partial\mathcal{S}(\boldsymbol{A})$ and non-negative reals $\alpha_1, \dots, \alpha_n$ satisfying $\sum_{i=1}^{n} \alpha_i = 1$

(iii) the limit in $[0,1]^A$ of a generalized sequence $(g_\gamma)$, where every $g_\gamma$ is as in (ii).

Recall that an MV-algebra $\boldsymbol{A}$ is said to be *simple* whenever the only maximal ideal of $\boldsymbol{A}$ is $\{0\}$.

COROLLARY 4.1.2. *Let $\boldsymbol{A}$ be an* MV*-algebra. Then:*

*(i) There are $n$ maximal ideals in $\boldsymbol{A}$ if and only if $\mathcal{S}(\boldsymbol{A})$ is affinely homeomorphic to the $(n-1)$-dimensional standard simplex.*

*(ii) In particular, if $\boldsymbol{A}$ is simple, then the only state of $\boldsymbol{A}$ is the unique embedding of $\boldsymbol{A}$ into the real unit interval $[0,1]$.*

In conclusion, simpliciality and the topology of state space enables us to claim that states of an MV-algebra and regular Borel probability measures over the compact Hausdorff maximal ideal space can be identified up to an affine homeomorphism. The affine homeomorphism is determined by the integral formula (15), which is frequently used in applications.

### 4.2 Existence of invariant states and faithful states

The integral representation theorem for states enables us to study MV-algebraic dynamics in analogy with the probabilistic dynamics. A *measure-theoretic dynamical system* is a quadruple $\langle X, \mathfrak{A}, \mu, T\rangle$, where $X$ is a nonempty set, $\mathfrak{A}$ is a $\sigma$-algebra of subsets of $X$, a mapping $T\colon X \to X$ is $\mathfrak{A}$-$\mathfrak{A}$ measurable ($T^{-1}(A) \in \mathfrak{A}$, for every $A \in \mathfrak{A}$), and $\mu\colon \mathfrak{A} \to [0,1]$ is a probability measure invariant with respect to $T$, i.e,

$$\mu(T^{-1}(A)) = \mu(A), \quad \text{for every } A \in \mathfrak{A}.$$

Analogously, let $e$ be an endomorphism of an MV-algebra $\boldsymbol{A}$ and $s \in \mathcal{S}(\boldsymbol{A})$. Put

$$s^e(a) = s(e(a)), \quad a \in A. \tag{18}$$

Then $s^e$ is a state of $\boldsymbol{A}$. We call $s$ an *$e$-invariant state* whenever $s^e = s$. Every endomorphism $e$ of $\boldsymbol{A}$ induces a continuous transformation $T_e\colon \operatorname{Max}(\boldsymbol{A}) \to \operatorname{Max}(\boldsymbol{A})$:

$$T_e(M) = \{a \in A \mid e(a) \in M\}, \quad \text{for every } M \in \operatorname{Max}(\boldsymbol{A}).$$

In particular, the function $T_e$ is $\mathfrak{B}(\operatorname{Max}(\boldsymbol{A}))$-$\mathfrak{B}(\operatorname{Max}(\boldsymbol{A}))$ measurable.

PROPOSITION 4.2.1. *Let $e$ be an endomorphism of an* MV*-algebra $\boldsymbol{A}$ and $s \in \mathcal{S}(\boldsymbol{A})$. Then $s$ is an $e$-invariant state if and only if $\langle \operatorname{Max}(\boldsymbol{A}), \mathfrak{B}(\operatorname{Max}(\boldsymbol{A})), \mu_s, T_e\rangle$ is a measure-theoretic dynamical system.*

*Proof.* Clearly, $e$ is an endomorphism of the isomorphic image $\boldsymbol{A}^*$ of $\boldsymbol{A}$ and we have $e(a^*) = a^* \circ T_e$. Then Theorem 4.0.1 and the change of variables in the Lebesgue integral (see [6, Theorem 13.46]) yield

$$s^e(a) = s(e(a)) = \int\limits_{\operatorname{Max}(\boldsymbol{A})} e(a^*)\,\mathrm{d}\mu_s = \int\limits_{\operatorname{Max}(\boldsymbol{A})} a^* \circ T_e\,\mathrm{d}\mu_s = \int\limits_{\operatorname{Max}(\boldsymbol{A})} a^*\,\mathrm{d}\left(\mu_s \circ T_e^{-1}\right).$$

Hence the equality

$$\int\limits_{\operatorname{Max}(\boldsymbol{A})} a^*\,\mathrm{d}\left(\mu_s \circ T_e^{-1}\right) = \int\limits_{\operatorname{Max}(\boldsymbol{A})} a^*\,\mathrm{d}\mu_s = s(a)$$

holds if and only if $\mu_s$ is invariant with respect to $T_e$. □

PROPOSITION 4.2.2. *Let $e$ be an endomorphism of an* MV*-algebra $\boldsymbol{A}$. Then there exists an $e$-invariant state $s$ of $\boldsymbol{A}$.*

*Proof.* Consider the mapping $\bar{e}\colon \mathcal{S}(\boldsymbol{A}) \to \mathcal{S}(\boldsymbol{A})$ defined by $\bar{e}(s) = s^e$, where $s^e$ is as in (18). Then $\bar{e}$ is continuous. Indeed, for every $a \in A$ and every net $(s_\gamma)$ in $\mathcal{S}(\boldsymbol{A})$ such that $s_\gamma \to s \in \mathcal{S}(\boldsymbol{A})$, we have

$$\bar{e}(s_\gamma)(a) = s_\gamma(e(a)) \to s(e(a)) = \bar{e}(s)(a).$$

Because $\bar{e}$ is a continuous mapping of the compact convex set $\mathcal{S}(\boldsymbol{A})$ into itself, the Brouwer–Schauder–Tychonoff fixed point theorem [6, Theorem 17.56] says that there must exist a fixed point of the mapping $\bar{e}$, a state $s \in \mathcal{S}(\boldsymbol{A})$ such that $\bar{e}(s) = s$. □

The existence of a state $s$ invariant with respect to a single endomorphism $e$ is the consequence of purely geometrical-topological properties of the state space. The situation becomes more interesting if we require invariance of $s$ with respect to all the automorphisms of the MV-algebra: we call a state $s$ of $\boldsymbol{A}$ *invariant* if it is $\alpha$-invariant for every automorphism $\alpha$ of $\boldsymbol{A}$. The existence of invariant states possibly satisfying additional properties can directly be proved for the free $n$-generated MV-algebra.

THEOREM 4.2.3. *The free $n$-generated* MV*-algebra $\boldsymbol{F}_n$ has an invariant faithful state with rational values.*

*Proof.* The natural candidate for a state from the statement is the Lebesgue state $s_\lambda$ (Example 3.2.3) given by the Riemann integral

$$s_\lambda(f) = \int_{[0,1]^n} f(\mathbf{x})\, \mathrm{d}\mathbf{x}, \quad f \in F_n.$$

Let $f \in F_n$. By [1, Lemma 2.1.4], there exists a polyhedral complex $\Sigma$ and $n$-dimensional convex polytopes $P_1, \dots, P_m \in \Sigma$ with rational vertices such that $f$ is linear over each $P_i$ and $\bigcup_{i=1}^m P_i = [0,1]^n$. For every $i = 1, \dots, m$, put

$$v_i(f) = \int_{P_i} f(\mathbf{x})\, \mathrm{d}\mathbf{x}$$

and observe that $s_\lambda(f) = \sum_{i=1}^m v_i(f)$. Since each polytope $P_i$ has rational vertices and the linear function $f$ over $P_i$ has Z coefficients, the value $v_i(f)$ is rational and thus $s_\lambda(f) \in [0,1] \cap \mathrm{Q}$.

State $s_\lambda$ is faithful. Indeed, if $f \neq 0$, then there exists a polytope $P_i \in \Sigma$ of dimension $n$ such that $f$ is nonzero in the interior of $P_i$. Thus $s_\lambda(f) \geq v_i(f) > 0$.

Let $\alpha$ be an automorphism of $\boldsymbol{F}_n$. Then $\alpha$ is determined by its action on the free generators of $\boldsymbol{F}_n$, the $i$-th coordinate projection functions $\pi_i\colon [0,1]^n \to [0,1]$. Denote $q_i = \alpha(\pi_i)$, for each $i = 1, \dots, n$, and put $T_\alpha(\mathbf{x}) = (q_1(\mathbf{x}), \dots, q_n(\mathbf{x}))$, $\mathbf{x} \in [0,1]^n$. The two mappings $\alpha$ and $T_\alpha$ are related by the formula

$$\alpha(f) = f \circ T_\alpha, \quad f \in F_n.$$

It can be shown that $T_\alpha$ is a Z-homeomorphism of $[0,1]^n$, that is, $T_\alpha$ is a homeomorphism of $[0,1]^n$ such that the scalar components of $T_\alpha$ and $T_\alpha^{-1}$ are in $F_n$. This implies existence of a polyhedral complex $\Theta$ such that $Q_1,\dots,Q_r \in \Theta$ are convex polytopes with rational vertices, $\bigcup_{i=1}^r Q_i = [0,1]^n$, and both $T_\alpha$ and $T_\alpha^{-1}$ are linear over each $Q_i$. Since the components of $T_\alpha$ and $T_\alpha^{-1}$ restricted to $Q_i$ are linear functions with Z coefficients, the corresponding Jacobian matrix satisfies $|J_{T_\alpha}(\mathbf{x})| = 1$ for every $\mathbf{x}$ in the interior of $Q_i$. Thus, for every $f \in F_n$,

$$s_\lambda(\alpha(f)) = \int_{[0,1]^n} \alpha(f)(\mathbf{x})\,\mathrm{d}\mathbf{x} = \int_{[0,1]^n} f(T_\alpha(\mathbf{x}))\,\mathrm{d}\mathbf{x} = \sum_{i=1}^r \int_{Q_i} f(T_\alpha(\mathbf{x}))\,\mathrm{d}\mathbf{x}$$
$$= \sum_{i=1}^r \int_{\text{int } Q_i} f(T_\alpha(\mathbf{x})) \cdot |J_{T_\alpha}(\mathbf{x})|\,\mathrm{d}\mathbf{x}.$$

Using the change of variable [6, Theorem 13.49], the last expression is equal to

$$\sum_{i=1}^r \int_{\text{int } Q_i} f(\mathbf{x})\,\mathrm{d}\mathbf{x} = \int_{[0,1]^n} f(\mathbf{x})\,\mathrm{d}\mathbf{x} = s_\lambda(f). \qquad \square$$

The theorem underlines the importance of Lebesgue state for Łukasiewicz logic: the average truth value of a formula is invariant under all substitutions $A_i \mapsto \alpha(A_1,\dots,A_n)$, $i = 1,\dots,n$ such that the equivalence classes $[\alpha(A_1,\dots,A_n)]$ generate $\boldsymbol{F}_n$. Moreover, the statement of Theorem 4.2.3 can easily be extended onto the class of all finitely presented MV-algebras.

## 5 De Finetti's coherence criterion for many-valued events

De Finetti's foundation of subjective probability theory is based on the notion of *coherent betting odds*. Two players, a *bookmaker* and a *gambler*, wager money on the occurrence of some *events of interest* $e_1,\dots,e_k$. At the very first stage of the game, the bookmaker publishes a book $\beta$ assigning a betting odd $\beta_i \in [0,1]$ to each event $e_i$ and the gambler, once the book has been published, places *stakes* $\sigma_1,\dots,\sigma_k \in$ R, one for each event $e_i$, and pays to the bookmaker the amount of $\sum_{i=1}^k \sigma_i \cdot \beta_i$ euros. Notice that each stake $\sigma_i$ is positive for the gambler whenever the bet is placed, while it is negative for a bet accepted. In other words, the effect of gambler's decision to pay a negative amount $\sigma_i$ on $e_i$, is that she will receive, already in this first stage of the game, $\sigma_i \cdot \beta_i$ euros from the bookmaker. We call this assumption *reversibility.*

At this stage of the game, the bookmaker and the gambler are obviously uncertain about the truth value of the events involved in the game. However, once a future possible world $w$ is reached, every $e_i$ is known to be either true or false in this possible world. For every event $e_i$, the bookmaker pays back to the gambler $\sigma_i$ euros if $e_i$ turns out to be true in $w$, or nothing if $e_i$ is false in $w$.

Formally, if we denote by $w(e_i)$ the truth-value of $e_i$ in the world $w$, the total balance for the bookmaker at the end of the game is calculated by $\sum_{i=1}^k \sigma_i(\beta_i - w(e_i))$ and hence, in the world $w$, the bookmaker gains money if his balance is positive, or he looses money if it is negative. Notice that the balance is calculated with respect to a specific possible world $w$, but if the bookmaker arranges his book $\beta$ in a such a way that he is going to loose money in every possible world, then we say he incurred a *sure loss* and the book $\beta$ is called *Dutch*, or *incoherent*. Conversely, the book $\beta$ is called *coherent* if it does not ensure bookmaker to incur a sure loss, i.e., for every choice of stakes $\sigma_1, \ldots, \sigma_k$, there exists a world $w$ in which the bookmaker's total balance is not negative.

A suitable formalization of classical de Finetti's betting game consists in interpreting events, books and possible worlds by the following stipulations:

- *events* are elements of an arbitrary Boolean algebra $\boldsymbol{B}$,
- a *book* on a finite subset $\{e_1, \ldots, e_k\} \subseteq B$ is a map $\beta\colon e_i \mapsto \beta_i \in [0,1]$, and
- a *possible world* is a Boolean homomorphism of $\boldsymbol{B}$ into the two element Boolean chain $\mathbf{2}$, that is, any element of $\mathcal{H}(\boldsymbol{B}, \mathbf{2})$.

Within this framework, de Finetti's coherence criterion reads as follows:

**Classical Coherence Criterion.** *Let $\boldsymbol{B}$ be a Boolean algebra and let $\{e_1, \ldots, e_k\}$ be a finite subset of $B$. A book $\beta\colon e_i \mapsto \beta_i$ is said to be* coherent *iff for each choice of $\sigma_1, \ldots, \sigma_k \in \mathsf{R}$, there exists $w \in \mathcal{H}(\boldsymbol{B}, \mathbf{2})$ such that*

$$\sum_{i=1}^{k} \sigma_i(\beta_i - w(e_i)) \geq 0.$$

The celebrated de Finetti's theorem can be stated as follows.

THEOREM 5.0.1 (de Finetti). *For every Boolean algebra $\boldsymbol{B}$, for every finite subset $\{e_1, \ldots, e_k\}$ and for every book $\beta$, the following are equivalent:*

*(i) $\beta$ is coherent.*

*(ii) There exists a finitely additive probability measure $\mu\colon \boldsymbol{B} \to [0,1]$ such that $\mu(e_i) = \beta_i$, for every $i = 1, \ldots, k$.*

It is not difficult to generalize the Classical Coherence Criterion to events being elements of an MV-algebra $\boldsymbol{A}$ and possible worlds being MV-homomorphisms of $\boldsymbol{A}$ into the standard MV-algebra $[0,1]_{\text{Ł}}$. Within the MV-algebraic setting, the game played by the bettor and the bookmaker must take into account that events are evaluated by every possible world $w$ taking values in $[0,1]$. Therefore, in a possible world $w$, the amount of money that the gambler will receive back from the bookmaker is calculated by *weighting* each stake $\sigma_i$ with the truth-value $w(e_i) \in [0,1]$ of $e_i$. This leads to the many-valued version of the Classical Coherence Criterion.

**Many-valued Coherence Criterion.** *Let $\boldsymbol{A}$ be an* MV*-algebra and $A' = \{e_1, \ldots, e_k\}$ be a finite subset of $A$. We say that a book $\beta\colon e_i \mapsto \beta_i$ is* coherent *iff for each choice of $\sigma_1, \ldots, \sigma_k \in \mathsf{R}$, there exists $w \in \mathcal{H}(\boldsymbol{A})$ such that*

$$\sum_{i=1}^{k} \sigma_i(\beta_i - w(e_i)) \geq 0. \tag{19}$$

The aim of the first part of this section is to generalize de Finetti's theorem to many-valued events. First, we need some preliminary results.

LEMMA 5.0.2. *Let* $\boldsymbol{A}$ *be an* MV*-algebra and* $A' = \{e_1, \ldots, e_k\}$ *be a finite subset of* $A$*. Then, for every* $\beta \in \mathrm{co}(\mathcal{H}(\boldsymbol{A}))$*, its restriction to* $A'$ *is a coherent book.*

*Proof.* Let $A' = \{e_1, \ldots, e_k\}$ be a finite subset of $A$. Let $\lambda_1, \ldots, \lambda_r \in \mathsf{R}^+$ be such that $\sum_{j=1}^{r} \lambda_j = 1$ and let $w_1, \ldots, w_r \in \mathcal{H}(\boldsymbol{A})$ be such that, for all $i = 1, \ldots, k$,

$$\beta_i = \beta(e_i) = \sum_{j=1}^{r} \lambda_j w_j(e_i). \tag{20}$$

Assume, by way of contradiction, that for some $\sigma_1, \ldots, \sigma_k \in \mathsf{R}$, it holds that, for all $w \in \mathcal{H}(\boldsymbol{A})$, $\sum_{i=1}^{k} \sigma_i(\beta_i - w(e_i)) < 0$. Then, in particular, for all $j = 1, \ldots, r$, one has

$$\sum_{i=1}^{k} \sigma_i(\beta_i - w_j(e_i)) = \sum_{i=1}^{k} \sigma_i \beta_i - \sum_{i=1}^{k} \sigma_i w_j(e_i) < 0.$$

Therefore, since $\sum_{j=1}^{r} \lambda_j = 1$,

$$\sum_{i=1}^{k} \sigma_i \beta_i - \sum_{j=1}^{r} \lambda_j \left( \sum_{i=1}^{k} \sigma_i w_j(e_i) \right) < 0,$$

that is, from (20),

$$\sum_{i=1}^{k} \sigma_i \left( \sum_{j=1}^{r} \lambda_j w_j(e_i) \right) - \sum_{j=1}^{r} \lambda_j \left( \sum_{i=1}^{k} \sigma_i w_j(e_i) \right) < 0,$$

a contradiction. □

LEMMA 5.0.3. *Let* $\boldsymbol{A}$ *be an* MV*-algebra, let* $A' = \{e_1, \ldots, e_k\}$ *be a finite subset of* $A$ *and let* $\beta$ *be a book on* $A'$*. If* $\beta$ *is coherent, then there exists* $\gamma \in \mathrm{co}(\mathcal{H}(\boldsymbol{A}))$ *such that* $\beta(e_i) = \gamma(e_i)$ *for all* $e_i \in A'$*.*

*Proof.* Let

$$\mathcal{H} \restriction_{A'} = \{\mathbf{x} \in [0,1]^k \mid \mathbf{x} = (h(e_1), \ldots, h(e_k)) \text{ for } h \in \mathcal{H}(\boldsymbol{A})\}. \tag{21}$$

Since $\mathcal{H}(\boldsymbol{A})$ is closed in $[0,1]^A$, then both $\mathcal{H} \restriction_{A'}$ and $\mathrm{co}(\mathcal{H} \restriction_{A'})$ are compact subsets of $[0,1]^k$. Let $\mathbf{y} = \langle \beta(e_1), \ldots, \beta(e_k) \rangle \in [0,1]^k$ and assume, by way of contradiction, that $\mathbf{y} \notin \mathrm{co}(\mathcal{H} \restriction_{A'})$. Then, since $\mathrm{co}(\mathcal{H} \restriction_{A'})$ is convex, by the Separating Hyperplane Theorem [22, Lemma 3.5] there exist $\mathbf{p} \in \mathsf{R}^k$ and $r \in \mathsf{R}$ such that the affine hyperplane $H = \{\mathbf{a} \mid \mathbf{p} \circ \mathbf{a} = r\}$ strongly separates $\mathbf{y}$ and $\mathrm{co}(\mathcal{H} \restriction_{A'})$, meaning that the scalar product $\mathbf{p} \circ \mathbf{y} < r$ and, for all $\mathbf{x} \in \mathcal{H} \restriction_{A'}$, we have $\mathbf{p} \circ \mathbf{x} > r$. In particular, for every $\mathbf{x} \in \mathcal{H} \restriction_{A'}$, it follows that $\mathbf{p} \circ (\mathbf{y} - \mathbf{x}) < 0$ and hence, letting $\mathbf{p} = \langle \sigma_1, \ldots, \sigma_k \rangle$,

$$\sum_{i=1}^{k} \sigma_i(\beta(e_i) - h(e_i)) < 0$$

for every $h \in \mathcal{H}(\boldsymbol{A})$, contradicting the coherence of $\beta$. □

THEOREM 5.0.4. *Let $\boldsymbol{A}$ be an* MV*-algebra, $A' = \{e_1, \ldots, e_k\}$ be a finite subset of $A$ and let $\beta$ be a book on $A'$. Then the following are equivalent:*

*(i) $\beta$ is coherent.*

*(ii) There exists a state $s \in \mathcal{S}(\boldsymbol{A})$ such that $s$ coincides with $\beta$ over $A'$.*

*(iii) $\beta$ can be extended to a convex combination of at most $k+1$ elements of $\mathcal{H}(A)$.*

*Proof.* The fact that the last two claims are equivalent is a consequence of Carathéodory theorem [22, Theorem 2.3]. To prove the equivalence of the first two we shall use the fact that, for every MV-algebra $\boldsymbol{A}$, $\mathcal{S}(\boldsymbol{A}) = \overline{\text{co}}(\mathcal{H}(\boldsymbol{A}))$ (Theorem 4.1.1(iii)).

(ii) $\Rightarrow$ (i): Let $s \in \overline{\text{co}}(\mathcal{H}(\boldsymbol{A}))$ and assume, by way of contradiction, that $\beta$ is not $A'$-coherent for some finite subset $A' = \{e_1, \ldots, e_k\}$ of $A$. So there are $\sigma_1, \ldots, \sigma_k \in \mathsf{R}$ such that, for every $w \in \mathcal{H}(\boldsymbol{A})$, $\sum_{i=1}^{k} \sigma_i(\beta_i - w(e_i)) < 0$, which gives

$$\sum_{i=1}^{k} \sigma_i \beta_i < \sum_{i=1}^{k} \sigma_i w(e_i).$$

From Theorem 4.1.1(ii), $\mathcal{H}(\boldsymbol{A})$ is closed and therefore, by the continuity of addition and multiplication,

$$\begin{aligned} \min_{w \in \mathcal{H}(\boldsymbol{A})} \sum_{i=1}^{k} \sigma_i w(e_i) &= \sum_{i=1}^{k} \min_{w \in \mathcal{H}(\boldsymbol{A})} \sigma_i w(e_i) \\ &= \sum_{i=1}^{k} \sigma_i \min_{w \in \mathcal{H}(\boldsymbol{A})} w(e_i) \\ &= \sum_{i=1}^{k} \sigma_i \overline{w}(e_i) \end{aligned}$$

for some $\overline{w} \in \mathcal{H}(\boldsymbol{A})$. Further, let

$$z = \sum_{i=1}^{k} \sigma_i \overline{w}(e_i) - \sum_{i=1}^{k} \sigma_i \beta_i > 0.$$

Since by hypothesis $s$ is a state and so it belongs to $\overline{\text{co}}(\mathcal{H}(\boldsymbol{A}))$, it holds

$$\forall \varepsilon > 0\ \exists \beta' \in \text{co}(\mathcal{H}(\boldsymbol{A})) \text{ such that, } \forall i = 1, \ldots, k,\ |\beta_i - \beta'(e_i)| < \varepsilon.$$

Therefore, in particular, for all sufficiently small $\varepsilon$, there is $\beta' \in \text{co}(\mathcal{H}(\boldsymbol{A}))$, such that, for all $w \in \mathcal{H}(\boldsymbol{A})$,

$$\sum_{i=1}^{k} \sigma_i \beta'(e_i) < \frac{z}{2} + \sum_{i=1}^{k} \sigma_i \beta_i < \sum_{i=1}^{k} \sigma_i \overline{w}(e_i) \leq \sum_{i=1}^{k} \sigma_i w(e_i)$$

and thus $\sum_{i=1}^{k} \sigma_i(\beta'(e_i) - w(e_i)) < 0$. Therefore $\beta'$ is not coherent, contradicting Lemma 5.0.2.

(i) $\Rightarrow$ (ii): Let $\beta$ be coherent. Then, for every finite subset $A' \subseteq A$, Lemma 5.0.3 ensures that the set

$$\mathcal{D}(A') = \{\gamma \in \overline{\mathrm{co}}(\mathcal{H}(\boldsymbol{A})) \mid \beta = \gamma \text{ on } A'\}$$

is a nonempty closed subset of $\overline{\mathrm{co}}(\mathcal{H}(\boldsymbol{A}))$. Notice that, for every choice of finite subsets $A'_1, \ldots, A'_m \subseteq A$,

$$\mathcal{D}(A'_1) \cap \cdots \cap \mathcal{D}(A'_m) = \mathcal{D}(A'_1 \cup \cdots \cup A'_m)$$

and hence the family $\{\mathcal{D}(A') \mid A' \text{ is a finite subset of } A\}$ has the finite intersection property. Therefore there exists $\gamma \in \overline{\mathrm{co}}(\mathcal{H}(\boldsymbol{A}))$ such that

$$\gamma \in \bigcap \{\mathcal{D}(A') \mid A' \text{ is a finite subset of } A\}.$$

Thus $\gamma = \beta$ on $A$ and $\beta \in \overline{\mathrm{co}}(\mathcal{H}(\boldsymbol{A}))$ and so the claim follows by setting $\gamma = s$. □

## 5.1 Betting on formulas of Łukasiewicz logic

It is common to think about an *event* as a formula in the language of a logical calculus. For this reason, a suitable setting for many-valued events is the Lindenbaum algebra of Łukasiewicz propositional logic. By Form ($\mathsf{Form}_n$) we denote the class of all formulas in Łukasiewicz logic (the class of all formulas containing at most the first $n$ propositional variables), respectively. In Section 2.2 we recalled that the Lindenbaum algebra of Łukasiewicz propositional calculus with $n$ variables is the free $n$-generated MV-algebra $\boldsymbol{F}_n$ which, in turn, is isomorphic to the MV-algebra of McNaughton functions on $[0,1]^n$ with pointwise operations of $\oplus$ and $\neg$.

In this section we are going to provide an alternative proof of Theorem 5.0.4 considering coherent assignments on finite subsets of the free $n$-generated MV-algebra $\boldsymbol{F}_n$. Moreover, we will show that the problem of establishing the coherence for a rational-valued book on formulas of Łukasiewicz logic is **NP**-complete.

We now prepare some terminology and notation in view of the main theorem of this section. We assume a reasonably compact binary encoding of $\phi \in \mathsf{Form}$, such that the number size($\phi$) of bits in the encoding of $\phi$ is bounded above by a polynomial $e_1 \colon \mathsf{N} \to \mathsf{N}$ of the number $c(\phi)$ of symbols $\odot, \to$ occurring in $\phi$, that is,

$$\mathrm{size}(\phi) \le e_1(c(\phi_i)).$$

Analogously, we assume that the length in bits of the encoding of a finite set of formulas $\{\phi_1, \ldots, \phi_k\} \subseteq \mathsf{Form}$, in symbols size($\{\phi_1, \ldots, \phi_k\}$), satisfies

$$\mathrm{size}(\{\phi_1, \ldots, \phi_k\}) \le e_2(\mathrm{size}(\phi_1) + \cdots + \mathrm{size}(\phi_k))$$

for some polynomial $e_2 \colon \mathsf{N} \to \mathsf{N}$. Also, letting $\beta \colon \{\phi_1, \ldots, \phi_k\} \to [0,1]$ be a rational book such that $\beta(\phi_i) = n_i/d_i$ with $n_i$ and $d_i$ relatively prime integers for all $i$ in $\{1, \ldots, k\}$, we assume a binary encoding of $\beta$ such that the number of bits in the encoding of $\beta$, in symbols, size($\beta$), satisfies

$$\mathrm{size}(\beta) \le e_3\left(\mathrm{size}(\{\phi_1, \ldots, \phi_k\}) + k \cdot \log_2 \max\{d_1, \ldots, d_k\}\right)$$

for some polynomial $e_3 \colon \mathsf{N} \to \mathsf{N}$.

PROPOSITION 5.1.1. *Let $s$ be a state of $\boldsymbol{F}_n$ and let $H_\Sigma$ be a Schauder basis for $\boldsymbol{F}_n$. If $h_i \in H_\Sigma$ is the Schauder hat at $\mathbf{x}_i$, then $s(\mathrm{den}(\mathbf{x}_i) \cdot h_i) = \mathrm{den}(\mathbf{x}_i) \cdot s(h_i)$.*

*Proof.* This is a direct consequence of additivity of $s$ and the definition of a (normalized) Schauder hat. □

PROPOSITION 5.1.2. *Let $\phi_1, \ldots, \phi_k \in \mathsf{Form}_n$ for some $k \geq 1$. Then there exist a unary polynomial $q\colon \mathsf{N} \to \mathsf{N}$ and a unimodular triangulation $\Sigma$ of $[0,1]^n$ linearizing $[\phi_1], \ldots, [\phi_k]$, such that each rational vertex $\mathbf{x}$ of $\Sigma$ satisfies*

$$\log_2 \mathrm{den}(\mathbf{x}) \leq q(\mathrm{size}(\{\phi_1, \ldots, \phi_k\})).$$

*Proof.* For all $i \in \{1, \ldots, k\}$, let $f_i$ be the $n$-ary McNaughton function $[\phi_i]$. Let $p_1, \ldots, p_l$ be the list of all the linear pieces of the functions $f_1, \ldots, f_k$, together with the projection functions $x_1, \ldots, x_n$ and the constants $0, 1$, and define $P_\pi$ as in (8) and $C$ as in (9) based on these pieces. Let $\Sigma$ be a unimodular triangulation produced from $C$ without adding new vertices, as explained in Section 3.3. We show that $\Sigma$ satisfies the statement.

First, since $C$ includes all the linear domains of all the functions $f_1, \ldots, f_k$ and $\Sigma$ is a subdivision of $C$, it follows that $\Sigma$ linearizes all the functions $f_1, \ldots, f_k$. Second, by the definition of McNaughton function each piece $p_i$ has the form

$$p_i(x_1, \ldots, x_n) = c_{i,1}x_1 + \cdots + c_{i,n}x_n + d_i$$

with $c_{i,1}, \ldots, c_{i,n}, d_n \in \mathsf{Z}$. Thus, by inspection of (8), each vertex $\mathbf{x} \in \mathrm{V}_\Sigma$ is the rational solution of a system of $n$ linear equations in $n$ unknowns, each equation having one of the following forms

$$\begin{aligned} p_h(x_1, \ldots, x_n) &= p_i(x_1, \ldots, x_n) \\ p_h(x_1, \ldots, x_n) &= 0 \\ p_h(x_1, \ldots, x_n) &= 1 \end{aligned}$$

for $h, i \in \{1, \ldots, l\}$, or $x_i = 0$, $x_i = 1$ for $i \in \{1, \ldots, n\}$.

Suppose that $p_i$ is a linear piece of $f_j$. A routine induction on $\phi_j$ shows that

$$|c_{i,1}|, \ldots, |c_{i,n}| \leq \mathrm{size}(\phi_j).$$

Hence, the largest coefficient (in absolute value) of any linear piece amongst $p_1, \ldots, p_l$ is bounded above by

$$\max\{\mathrm{size}(\phi_j) \mid j \in \{1, \ldots, k\}\} \leq \mathrm{size}(\{\phi_1, \ldots, \phi_k\}),$$

so that the $m$-th equation in the linear system having $\mathbf{x}$ as solution has the form

$$a_{m,1}x_1 + \cdots + a_{m,n}x_n = b_m$$

with

$$|a_{m,1}|, \ldots, |a_{m,n}| \leq 2 \cdot \mathrm{size}(\{\phi_1, \ldots, \phi_k\}). \tag{22}$$

Since

$$\mathbf{x} = \begin{pmatrix} a_{1,1} & \dots & a_{1,n} \\ \vdots & \ddots & \vdots \\ a_{n,1} & \dots & a_{n,n} \end{pmatrix}^{-1} \begin{pmatrix} b_1 \\ \vdots \\ b_n \end{pmatrix} = A^{-1}\boldsymbol{B},$$

it follows that $\mathrm{den}(\mathbf{x}) \leq |\det(A)|$ by elementary linear algebra [56]. In light of (22), the application of Hadamard's inequality now yields the desired bound:

$$\begin{aligned} |\det(A)| &\leq \prod_{i\in\{1,\dots,m\}} (a_{i,1}^2 + \cdots + a_{i,n}^2)^{1/2} \\ &\leq \prod_{i\in\{1,\dots,n\}} |a_{i,1}| + \cdots + |a_{i,n}| \\ &\leq \prod_{i\in\{1,\dots,n\}} 2n \cdot \mathrm{size}(\{\phi_1,\dots,\phi_k\}) \\ &\leq 2^{2n \log_2 n \cdot \mathrm{size}(\{\phi_1,\dots,\phi_k\})} \\ &\leq 2^{q(\mathrm{size}(\{\phi_1,\dots,\phi_k\}))}. \end{aligned}$$

It is enough to put

$$q(m) = m^2$$

and notice that $n \leq \mathrm{size}(\{\phi_1,\dots,\phi_k\})$, since the size of a set of formulas over $n$ distinct variables is greater than or equal to $n$. □

THEOREM 5.1.3. *Let* $\phi_1,\dots,\phi_k \in \mathsf{Form}_n$ *and* $\beta\colon [\phi_i] \mapsto \beta_i \in [0,1]\cap \mathsf{Q}$ *be a book. The following are equivalent:*

*(i)* $\beta$ *is coherent.*

*(ii) There exist a unary polynomial* $p\colon \mathsf{N} \to \mathsf{N}$ *and* $l \leq k+1$ *homomorphisms* $q_1,\dots,q_l$ *in* $\mathcal{H}(\boldsymbol{F}_n)$ *satisfying the following. For all* $i \in \{1,\dots,l\}$, $q_i$ *ranges in the finite* MV*-chain* $\mathbf{Ł}_{d_i}$*, where*

$$\log_2 d_i \leq p(\mathrm{size}(\beta)),$$

*and* $\langle \beta(\phi_i)\rangle_{i\in\{1,\dots,k\}}$ *is a convex combination of*

$$\langle q_1(\phi_i)\rangle_{i\in\{1,\dots,k\}}, \dots, \langle q_l(\phi_i)\rangle_{i\in\{1,\dots,k\}}.$$

*Proof.* (i) ⇒ (ii) Let $s$ be a state of $\boldsymbol{F}_n$ satisfying Theorem 5.0.4 and let $\Sigma$ be a unimodular triangulation satisfying Proposition 5.1.2. Let $\mathbf{x}_1,\dots,\mathbf{x}_m$ be the rational vertices of $\Sigma$ and put $d_1 = \mathrm{den}(\mathbf{x}_1),\dots,d_n = \mathrm{den}(\mathbf{x}_m)$. Let $\hat{h}_i$ be the normalized Schauder hat at vertex $\mathbf{x}_i$, $i \in \{1,\dots,m\}$, and define $\lambda_1,\dots,\lambda_m \in \mathsf{R}^+$ by putting, for every $i \in \{1,\dots,m\}$,

$$\lambda_i = s(\hat{h}_i).$$

Then

$$\begin{aligned}
\lambda_1 + \cdots + \lambda_m &= s(\hat{h}_1) + \cdots + s(\hat{h}_m) \\
&= s(\hat{h}_1 \oplus \cdots \oplus \hat{h}_m) && \text{by Lemma 3.3.1(i) and additivity of } s \\
&= s(1) && \text{by Lemma 3.3.1(ii)} \\
&= 1.
\end{aligned}$$

Let $h = \hat{h}_i/d_i$ be the Schauder hat at vertex $\mathbf{x}_i$, $i \in \{1, \ldots, m\}$. For all $i \in \{1, \ldots, k\}$, $f_i$ is a McNaughton function linearized by $\Sigma$, and for $j \in \{1, \ldots, m\}$, $h_j$ is the Schauder hat at vertex $\mathbf{x}_j$ of $\Sigma$. Thus by (10) there is a unique choice of integers $0 \leq a_{i,j} \leq \text{den}(\mathbf{x}_j) \leq 1$ such that

$$f_i = \sum_{j=1}^{n} a_{i,j} \cdot h_j.$$

For all $i \in \{1, \ldots, m\}$, let $t_i$ be the homomorphism from $\boldsymbol{F}_n$ to $[0,1]_Ł$ defined by putting, for every $f \in \boldsymbol{F}_n$,

$$t_i(f) = f(\mathbf{x}_i).$$

Note that $t_i$ ranges in $\{0, 1/d_i, \ldots, (d_i - 1)/d_i, 1\}$, and by Proposition 5.1.2,

$$\log_2 d_i = \log_2 \text{den}(\mathbf{x}_i) \leq q(\text{size}(\{\phi_1, \ldots, \phi_k\})) \leq p(\text{size}(\beta)),$$

letting

$$p(n) = q(n) = n^2,$$

as $\text{size}(\{\phi_1, \ldots, \phi_k\}) \leq \text{size}(\beta)$. For every $j \in \{1, \ldots, k\}$:

$$\begin{aligned}
\sum_{i=1}^{m} \lambda_i \cdot t_i(\phi_j) &= \sum_{i=1}^{m} \lambda_i \cdot f_j(\mathbf{x}_i) \\
&= \sum_{i=1}^{m} s(\hat{h}_i) \cdot a_{j,i}/d_i \\
&= \sum_{i=1}^{m} s(d_i \cdot h_i) \cdot a_{j,i}/d_i \\
&= \sum_{i=1}^{m} s(a_{j,i} \cdot h_i) && \text{by Proposition 5.1.1} \\
&= \sum_{i=1}^{m} s(a_{j,i}/d_i \cdot \hat{h}_i) \\
&= s\left(\bigoplus_{i=1}^{m} a_{j,i}/d_i \cdot \hat{h}_i\right) && \text{by Lemma 3.3.1(1) and additivity of } s \\
&= s(f_j).
\end{aligned}$$

As $\lambda_1 + \cdots + \lambda_n = 1$, the point $\langle\beta(\phi_i)\rangle_{i\in\{1,\ldots,k\}}$ is a convex combination of points $\langle t_1(\phi_i)\rangle_{i\in\{1,\ldots,k\}}, \langle t_2(\phi_i)\rangle_{i\in\{1,\ldots,k\}}, \ldots, \langle t_m(\phi_i)\rangle_{i\in\{1,\ldots,k\}}$. By Carathéodory theorem (see, e.g. [22, Theorem 2.3]), there exists a choice of $l \leq k+1$ homomorphisms $q_1, \ldots, q_l$ amongst $t_1, \ldots, t_m$ such that $(\beta(\phi_i))_{i\in\{1,\ldots,k\}}$ is a convex combination of

$$\langle q_1(\phi_i)\rangle_{i\in\{1,\ldots,k\}}, \ldots, \langle q_l(\phi_i)\rangle_{i\in\{1,\ldots,k\}},$$

and we are done.

(ii) $\Rightarrow$ (i) By Theorem 4.1.1, for every $\lambda_1, \ldots, \lambda_m \in \mathsf{R}^+$ satisfying $\sum_{i=1}^m \lambda_i = 1$, the map $s$ from $\boldsymbol{F}_n$ to $[0,1]$ defined by putting

$$s([\varphi]) = \sum_{i=1}^{m} \lambda_i \cdot q_i([\varphi])$$

for every $\varphi \in \mathsf{Form}_n$, is a state. This means that $\beta$ is coherent by Theorem 5.0.4. □

## 5.2 Complexity

Let $\langle\beta\rangle$ denote the binary encoding of some rational Łukasiewicz assessment $\beta$. The problem of deciding coherence of rational Łukasiewicz assessments is defined as:

$$\mathsf{LUK\text{-}COH} = \{\langle\beta\rangle \mid \beta \text{ is a coherent book on Łukasiewicz formulas}\}.$$

THEOREM 5.2.1. $\mathsf{LUK\text{-}COH}$ *is* $\mathbf{NP}$*-complete.*

In the next two paragraphs we prove that $\mathsf{LUK\text{-}COH}$ is in $\mathbf{NP}$ (Lemma 5.2.2) and is $\mathbf{NP}$-hard (Lemma 5.2.3), thus proving Theorem 5.2.1.

**Upper Bound.** It is known that the feasibility problem of linear systems is decidable in polynomial time in the size of the binary encoding of the linear system [56]. Therefore, Theorem 5.1.3 directly furnishes a nondeterministic polynomial time algorithm for the coherence problem as follows.

LEMMA 5.2.2. $\mathsf{LUK\text{-}COH}$ *is in* $\mathbf{NP}$.

*Proof.* Let $\beta\colon \{\phi_1, \ldots, \phi_k\} \to [0,1] \cap \mathsf{Q}$ be a rational-valued book on Łukasiewicz formulas $\phi_1, \ldots, \phi_k$ over variables $A_1, \ldots, A_m$. Following Lemma 5.1.3, the algorithm guesses a natural number $l \leq k+1$ and, for all $i \in \{1, \ldots, l\}$, the algorithm guesses the denominator $d_i$, the restriction of homomorphism $q_i$ to variables $A_1, \ldots, A_m$, and eventually checks the feasibility of the following linear system:

$$\begin{aligned} x_1 + \cdots + x_{l-1} + x_l &= 1 \\ q_1(\phi_1)x_1 + \cdots + q_{l-1}(\phi_1)x_{l-1} + q_l(\phi_1)x_l &= \beta(\phi_1) \\ &\vdots \\ q_1(\phi_k)x_1 + \cdots + q_{l-1}(\phi_k)x_{l-1} + q_l(\phi_k)x_l &= \beta(\phi_k). \end{aligned}$$

By Lemma 5.1.3, for all $i \in \{1, \ldots, l\}$, the denominator $d_i$ has a polynomial-space encoding. It follows that the restriction of $q_i$ to $A_1, \ldots, A_m$, as well as the coefficients $q_1(\phi_1), \ldots, q_l(\phi_k)$, are in $\{0, 1/d_i, \ldots, (d_i-1)/d_i, 1\}$. So the size of the system is polynomial in size($\beta$), and the algorithm terminates in time polynomial in size($\beta$). Notice that the linear system is feasible if and only if $\beta$ is a convex combination of $q_1, \ldots, q_l$ if and only if $\beta$ is coherent. □

**Lower Bound.** Let $\langle\phi\rangle$ denote the binary encoding of the formula $\phi \in \mathsf{Form}$. In [44] it is proved that the problem

$$\mathsf{LUK\text{-}SAT} = \{\langle\phi\rangle \mid \phi \text{ is satisfiable in Łukasiewicz logic}\}$$

is **NP**-complete.

LEMMA 5.2.3. LUK-COH *is* **NP**-*hard.*

*Proof.* We provide a logarithmic-space reduction from the **NP**-hard problem LUK-SAT to LUK-COH.

Let $\phi \in \mathsf{Form}_n$. Let $\beta$ be the book sending formulas $A_1 \oplus \neg A_1, \ldots, A_m \oplus \neg A_m$, and $\phi$ to 1, that is,

$$\beta(A_1 \oplus \neg A_1) = \ldots = \beta(A_m \oplus \neg A_m) = \beta(\phi) = 1.$$

The construction of the assessment $\beta$ is feasible in space logarithmic in size($\phi$). We show that $\beta$ is coherent if and only if $\phi$ is satisfiable in Łukasiewicz logic.

($\Rightarrow$) Suppose that $\beta$ is coherent. Let $b_i = -1$ for all $i \in [m+1]$, and let $q$ be a homomorphism from $\boldsymbol{F}_n$ to $[0,1]_{Ł}$ such that (19) holds, that is,

$$\beta(\phi) - q(\phi) \leq \sum_{i=1}^{m}(q(A_i \oplus \neg A_i) - \beta(A_i \oplus \neg A_i)).$$

As $q(A_i \oplus \neg A_i) = 1 = \beta(A_i \oplus \neg A_i)$ for every $i \in \{1, \ldots, m\}$, the right-hand side vanishes so that

$$1 = \beta(\phi) \leq q(\phi) \leq 1.$$

($\Leftarrow$) Let $q \in \mathcal{H}(\boldsymbol{F}_n)$ be such that $q(\phi) = 1$. Then $q$ is a state of $\boldsymbol{F}_n$ satisfying

$$q([\phi]) = 1 = \beta(\phi)$$

and for every $i \in \{1, \ldots, m\}$,

$$q([A_i \oplus \neg A_i]) = 1 = \beta(A_i \oplus \neg A_i).$$

Hence $\beta$ is coherent by Example 3.2.2 and Theorem 5.0.4. □

# 6 MV-algebras with internal states

The content presented in the preceding sections shows that states of MV-algebras are tightly connected to Borel probability measures on the maximal spectral spaces. Moreover, de Finetti's theorem about coherent betting on Boolean events can be generalized to states and many-valued events. In this section we are going to enrich the existing perspectives with a purely algebraic approach to states. Namely we are going to introduce a class of algebras, called SMV-*algebras*, which provide a universal-algebraic framework for states. SMV-algebras are defined by expanding the signature of MV-algebras with a fresh unary symbol $\sigma$, which is equationally described in order to preserve the basic properties of a state. The mapping $\sigma$ is called an *internal state* of an MV-algebra $\boldsymbol{A}$ and, as we will see, it can be successfully applied to cope with de Finetti's coherence criterion in a purely algebraic setting, among other applications.

DEFINITION 6.0.1. *An* MV-algebra with internal state *(*SMV-algebra, *for short) is an algebra* $\langle \boldsymbol{A}, \sigma \rangle = \langle A, \oplus, \neg, \sigma, 0 \rangle$, *where* $\langle A, \oplus, \neg, 0 \rangle$ *is an* MV-*algebra and* $\sigma$ *is a unary operator on* $\boldsymbol{A}$ *satisfying the following conditions for every* $x, y \in A$*:*

($\sigma$1) $\sigma(0) = 0$

($\sigma$2) $\sigma(\neg x) = \neg(\sigma(x))$

($\sigma$3) $\sigma(x \oplus y) = \sigma(x) \oplus \sigma(y \ominus (x \odot y))$

($\sigma$4) $\sigma(\sigma(x) \oplus \sigma(y)) = \sigma(x) \oplus \sigma(y)$.

*An* SMV-*algebra* $\langle \boldsymbol{A}, \sigma \rangle$ *is said to be* faithful *if it satisfies the following quasi-equation:* $\sigma(x) = 0$ *implies* $x = 0$.

Clearly the class of SMV-algebras constitutes a variety, which is denoted by $\mathbb{SMV}$.

EXAMPLE 6.0.2. (a) We start with a trivial example. Let $\boldsymbol{A}$ be any MV-algebra and $\sigma$ be the identity on $\boldsymbol{A}$. Then $\langle \boldsymbol{A}, \sigma \rangle$ is an SMV-algebra.

(b) Let $\sigma$ be an idempotent endomorphism of an MV-algebra $\boldsymbol{A}$. For example, we may take $\boldsymbol{A}$ to be a non-trivial ultrapower of the standard MV-algebra $[0,1]_Ł$ and $\sigma$ to be the standard part function. Then $\langle \boldsymbol{A}, \sigma \rangle$ is an SMV-algebra.

(c) This is a sufficiently general example for our purposes. Let $\boldsymbol{A}$ be the MV-algebra of all continuous and piecewise linear functions with real coefficients from $[0,1]^n$ into $[0,1]$. Then $\boldsymbol{A}$ is an MV-algebra endowed with the pointwise application of the operations $\oplus$ and $\neg$. For every $f \in A$ let $\sigma(f)$ be the function from $[0,1]^n$ to $[0,1]$ which is constantly equal to

$$\int_{[0,1]^n} f(x)\, \mathrm{d}x.$$

It follows that $\langle \boldsymbol{A}, \sigma \rangle$ is an SMV-algebra. It will become clear from the results of the next section that $\langle \boldsymbol{A}, \sigma \rangle$ is simple and thus subdirectly irreducible, but it is not totally ordered. This algebra is faithful, i.e., it satisfies the quasi-equation $\sigma(x) = 0$ implies $x = 0$.

LEMMA 6.0.3. *In any* SMV*-algebra* $\langle \boldsymbol{A}, \sigma \rangle$ *the following properties hold:*

*(i)* $\sigma(1) = 1$.

*(ii) If* $x \leq y$*, then* $\sigma(x) \leq \sigma(y)$.

*(iii)* $\sigma(x \oplus y) \leq \sigma(x) \oplus \sigma(y)$*; and if* $x \odot y = 0$*, then* $\sigma(x \oplus y) = \sigma(x) \oplus \sigma(y)$.

*(iv)* $\sigma(x \ominus y) \geq \sigma(x) \ominus \sigma(y)$*; and if* $y \leq x$*, then* $\sigma(x \ominus y) = \sigma(x) \ominus \sigma(y)$.

*(v)* $d(\sigma(x), \sigma(y)) \leq \sigma(d(x, y))$*, where d is the Chang distance.*

*(vi)* $\sigma(x) \odot \sigma(y) \leq \sigma(x \odot y)$*. Thus if* $x \odot y = 0$*, then* $\sigma(x) \odot \sigma(y) = 0$.

*(vii)* $\sigma(\sigma(x)) = \sigma(x)$.

*(viii) The image* $\sigma(A)$ *of* $A$ *under* $\sigma$ *is the domain of an* MV*-subalgebra* $\sigma(\boldsymbol{A})$ *of* $\boldsymbol{A}$.

*Proof.* (i) A direct consequence of ($\sigma 1$) and ($\sigma 2$).

(ii) If $x \leq y$, then $y = x \oplus (y \ominus x)$, and hence $\sigma(y) = \sigma(x \oplus (y \ominus x))$. Since $x \odot (y \ominus x) = 0$, by ($\sigma 3$) we get $\sigma(y) = \sigma(x \oplus (y \ominus x)) = \sigma(x) \oplus \sigma(y \ominus x) \geq \sigma(x)$.

(iii) By (ii), $\sigma(y) \geq \sigma(y \ominus (x \odot y))$, so

$$\sigma(x \oplus y) = \sigma(x) \oplus \sigma(y \ominus (x \odot y)) \leq \sigma(x) \oplus \sigma(y).$$

If $(x \odot y) = 0$, then $\sigma(x \oplus y) = \sigma(x) \oplus \sigma(y \ominus (x \odot y)) = \sigma(x) \oplus \sigma(y)$.

(iv) Using ($\sigma 2$), (iii) and the order-reversing property of $\neg$, we obtain:

$$\begin{aligned}\sigma(x \ominus y) &= \sigma(\neg(\neg x \oplus y)) = \neg(\sigma(\neg x \oplus y)) \geq \neg(\neg\sigma(x) \oplus \sigma(y)) \\ &= \neg(\neg(\sigma(x)) \oplus \sigma(y)) = \sigma(x) \ominus \sigma(y).\end{aligned}$$

Moreover, if $y \leq x$, then $\neg x \odot y = 0$. Hence again by (c),

$$\sigma(x \ominus y) = \neg(\sigma(\neg x \oplus y)) = \neg(\sigma(\neg x) \oplus \sigma(y)) = \sigma(x) \ominus \sigma(y).$$

(v) Since $(x \ominus y) \odot (y \ominus x) = 0$, by (iv) and (iii) we get $\sigma(d(x, y)) = \sigma(x \ominus y) \oplus \sigma(y \ominus x) \geq (\sigma(x) \ominus \sigma(y)) \oplus (\sigma(y) \ominus \sigma(x)) = d(\sigma(x), \sigma(y))$.

(vi) We have $x \odot y = x \ominus \neg y$ and thus (iv) and ($\sigma 2$) yield

$$\sigma(x \odot y) = \sigma(x \ominus \neg y) \geq \sigma(x) \ominus \sigma(\neg y) = \sigma(x) \ominus \neg(\sigma(y)) = \sigma(x) \odot \sigma(y).$$

If $x \odot y = 0$, then $0 = \sigma(x \odot y) \geq \sigma(x) \odot \sigma(y)$. Therefore, $\sigma(x \odot y) = 0$.

(vii) By (i), $\sigma(0) = 0$, and using ($\sigma 4$) we get

$$\sigma(\sigma(x)) = \sigma(\sigma(x) \oplus \sigma(0)) = \sigma(x) \oplus \sigma(0) = \sigma(x).$$

(viii) By (vii), the range of $\sigma$ consists of all the fixed points of $\sigma$. Therefore, it is sufficient to prove that the set of the fixed points is closed under $\oplus$ and under $\neg$. Closure under $\oplus$ follows from ($\sigma 4$). Concerning closure under $\neg$, using ($\sigma 2$) and (vii), we get $\sigma(\neg(\sigma(x))) = \neg(\sigma(\sigma(x)) = \neg(\sigma(x))$. □

### 6.1 Subdirectly irreducible SMV-algebras

A *$\sigma$-filter* of an SMV-algebra $\langle \boldsymbol{A}, \sigma\rangle$ is a filter in the MV-algebra $\boldsymbol{A}$ closed under $\sigma$. Given a congruence $\theta$ of an SMV-algebra $\langle \boldsymbol{A}, \sigma\rangle$, we define

$$F_\theta = \{x \in A \mid \langle x, 1\rangle \in \theta\}.$$

Conversely, given a $\sigma$-filter $F$ of $\langle \boldsymbol{A}, \sigma\rangle$, we define

$$\theta_F = \{\langle x, y\rangle \mid \neg(d(x,y)) \in F\},$$

where $d$ is the Chang distance.

THEOREM 6.1.1. *The maps $F \mapsto \theta_F$ and $\theta \mapsto F_\theta$ are mutually inverse isomorphisms between the lattice of congruences of an* SMV*-algebra $\langle \boldsymbol{A}, \sigma\rangle$, and the lattice of $\sigma$-filters of $\langle \boldsymbol{A}, \sigma\rangle$.*

*Proof.* The above defined maps are mutually inverse isomorphisms between the lattice of MV-congruences and the lattice of MV-filters. Therefore it suffices to prove that $F$ is a $\sigma$-filter iff $\theta_F$ is an SMV-congruence. Since $\sigma(1) = 1$, the congruences classes of 1 are $\sigma$-filters. Conversely, let $F$ be a $\sigma$-filter of an SMV-algebra $\langle \boldsymbol{A}, \sigma\rangle$. If $\langle x, y\rangle \in \theta_F$, then $\neg(d(x,y)) \in F$, so that $\sigma(\neg d(x,y)) \in F$, where $F$ is a $\sigma$-filter. From Lemma 6.0.3 (v) we obtain $d(\sigma(x), \sigma(y)) \leq \sigma(d(x,y))$, and thus by the order-reversing property of $\neg$ it follows that $\neg(d(\sigma(x), \sigma(y))) \geq \neg(\sigma(d(x,y))) \in F$. Hence $\neg(d(\sigma(x), \sigma(y))) \in F$, and $\langle \sigma(x), \sigma(y)\rangle \in \theta_F$. In conclusion, $\theta_F$ is a congruence of $\langle \boldsymbol{A}, \sigma\rangle$. □

For every positive integer $n$ and $a \in A$ we define

$$a^n = \underbrace{a \odot \cdots \odot a}_{n}.$$

LEMMA 6.1.2. *Let $\langle \boldsymbol{A}, \sigma\rangle$ be an* SMV*-algebra. Then the $\sigma$-filter $F_{\sigma(x)}$ generated by a single element $\sigma(x) \in \sigma(A)$ is $F_{\sigma(x)} = \{y \in A \mid \exists n \in \mathsf{N}(y \geq \sigma(x)^n)\}$.*

*Proof.* Let $H = \{y \in A \mid \exists n \in \mathsf{N}(y \geq \sigma(x)^n)\}$. By the definition of $\sigma$-filter, every element of $H$ also belongs to $F_{\sigma(x)}$. Thus $H \subseteq F_{\sigma(x)}$. For the converse inclusion, it is sufficient to prove that $H$ is a $\sigma$-filter and $\sigma(x) \in H$. Let us show that $H$ is closed under $\sigma$. If $y \in H$, then there is $n \in \mathsf{N}$ such that $y \geq \sigma(x)^n$. By Lemma 6.0.3 (ii), (vi) and (vii), we get

$$\sigma(y) \geq \sigma(\sigma(x)^n) \geq (\sigma(\sigma(x)))^n \geq \sigma(x)^n.$$

Thus $\sigma(y) \in H$. That $\sigma(x) \in H$ is trivial. □

Now we are ready to prove the main result of this section.

THEOREM 6.1.3. *(i) If $\langle \boldsymbol{A}, \sigma\rangle$ is a subdirectly irreducible* SMV*-algebra, then $\sigma(\boldsymbol{A})$ is linearly ordered.*

*(ii) If $\langle \boldsymbol{A}, \sigma\rangle$ is faithful, then $\langle \boldsymbol{A}, \sigma\rangle$ is a subdirectly irreducible* SMV*-algebra iff $\sigma(\boldsymbol{A})$ is a subdirectly irreducible* MV*-algebra.*

*Proof.* (i) Let $H$ be the smallest non-trivial $\sigma$-filter of $\langle \boldsymbol{A}, \sigma\rangle$ and let $x \in H \setminus \{1\}$. Suppose by contradiction that $\sigma(\boldsymbol{A})$ is not linearly ordered, and let $\sigma(a), \sigma(b) \in \sigma(A)$ be such that $\sigma(a) \not\leq \sigma(b)$ and $\sigma(b) \not\leq \sigma(a)$. Then the filters $F_{\sigma(a)\to\sigma(b)}$ and $F_{\sigma(b)\to\sigma(a)}$ generated by $\sigma(a) \to \sigma(b)$ and $\sigma(b) \to \sigma(a)$, respectively, are non-trivial. Hence they both contain $H$. In particular, $x \in F_{\sigma(a)\to\sigma(b)}$ and $x \in F_{\sigma(b)\to\sigma(a)}$. Since we know that $\sigma(a) \to \sigma(b) \in \sigma(A)$ and $\sigma(b) \to \sigma(a) \in \sigma(A)$, by Lemma 6.1.2 there is $n \in \mathsf{N}$ such that $x \geq (\sigma(a) \to \sigma(b))^n$ and $x \geq (\sigma(b) \to \sigma(a))^n$. Therefore,

$$x \geq (\sigma(a) \to \sigma(b))^n \vee (\sigma(b) \to \sigma(a))^n = 1.$$

Hence $x = 1$, which is a contradiction.

(ii) If $\langle \boldsymbol{A}, \sigma\rangle$ is faithful, then by definition $\sigma(x) = 0$ implies $x = 0$ and $\sigma(x) = 1$ implies $x = 1$. It follows that the intersection of a non-trivial $\sigma$-filter $H$ of $\langle \boldsymbol{A}, \sigma\rangle$ with $\sigma(A)$ is a non-trivial $\sigma$-filter of $\sigma(\boldsymbol{A})$. Moreover, every filter of $\sigma(\boldsymbol{A})$ is closed under $\sigma$. Then every MV-filter of $\sigma(\boldsymbol{A})$ is indeed a $\sigma$-filter. Hence, if $H$ is a minimal $\sigma$-filter of $\langle \boldsymbol{A}, \sigma\rangle$, then $H \cap \sigma(A)$ is a minimal non-trivial $\sigma$-filter of $\sigma(\boldsymbol{A})$. In fact, if $F$ is another non-trivial filter of $\sigma(\boldsymbol{A})$, then the $\sigma$-filter $F'$ of $\langle \boldsymbol{A}, \sigma\rangle$ generated by $F$ contains $H$, and

$$F = F' \cap \sigma(A) \supseteq H \cap \sigma(A).$$

Hence $H \cap \sigma(A)$ is minimal. Therefore, if $\langle \boldsymbol{A}, \sigma\rangle$ is subdirectly irreducible, so is $\sigma(\boldsymbol{A})$.

Conversely, if $H$ is the minimal non-trivial filter of $\sigma(\boldsymbol{A})$, then the $\sigma$-filter $F$ of $\langle \boldsymbol{A}, \sigma\rangle$ generated by $H$ is the minimal non-trivial $\sigma$-filter of $\langle \boldsymbol{A}, \sigma\rangle$. Indeed, if $G$ is another non-trivial $\sigma$-filter of $\langle \boldsymbol{A}, \sigma\rangle$, then $G \cap \sigma(A) \supseteq F \cap \sigma(A) = H$. Then $G$ contains the $\sigma$-filter generated by $H$, that is, $F \subseteq G$ and $F$ is minimal. Thus $\langle \boldsymbol{A}, \sigma\rangle$ is subdirectly irreducible. □

REMARK 6.1.4. *In addition to the results contained in the previous theorem, the subdirectly irreducible* SMV*-algebras have been completely characterized by Dvurečenskij, Kowalski and Montagna. Their proof requires techniques from universal algebra which are out of the scope of this chapter. We invite the interested reader to consult the last section of this document, where we include an extensive list of references.*

REMARK 6.1.5. *The variety* $\mathbb{SMV}$ *is not generated by its linearly ordered algebras. Indeed, the equation* $\sigma(x \vee y) = \sigma(x) \vee \sigma(y)$ *is valid in any linearly ordered* SMV*-algebra, but it does not hold in general. It is enough to consider Example 6.0.2 (c) with* $f(x) = x$ *and* $g(x) = 1 - x$*. Then* $\sigma(f \vee g) = \frac{3}{4} > \sigma(f) \vee \sigma(g) = \frac{1}{2}$.

## 6.2 States of MV-algebras and internal states

In this section we relate the notion of an SMV-algebra and that of state of an MV-algebra. We will show that, starting from an SMV-algebra $\langle \boldsymbol{A}, \sigma\rangle$, one can define a state $s$ of the MV-algebra $\boldsymbol{A}$. Conversely, starting from a state $s$ of an MV-algebra $\boldsymbol{A}$, we shall recover an MV-algebra $\boldsymbol{T}$ containing $\boldsymbol{A}$ as an MV-subalgebra together with an internal state $\sigma$ of $\boldsymbol{T}$.

Let us start with an SMV-algebra $\langle \boldsymbol{A}, \sigma\rangle$. By Lemma 6.0.3 (viii), $\langle \sigma(\boldsymbol{A}), \oplus, \neg, 0\rangle$ is an MV-subalgebra of $\boldsymbol{A}$, where $\oplus$ and $\neg$ denote respectively the restrictions of MV-algebraic operations of $\boldsymbol{A}$ to $\sigma(\boldsymbol{A})$. If $M$ is a maximal filter of $\sigma(\boldsymbol{A})$, then the quotient MV-algebra $\sigma(\boldsymbol{A})/M$ is simple and thus there has to be a unique embedding of

$\sigma(\boldsymbol{A})/M$ into the standard MV-algebra $[0,1]_Ł$; see the proof of Proposition 3.1.5. Let $i\colon \sigma(\boldsymbol{A})/M \hookrightarrow [0,1]_Ł$ be such an embedding and let $\eta_M\colon \sigma(\boldsymbol{A}) \to \sigma(\boldsymbol{A})/M$ be the canonical MV-homomorphism induced by the maximal filter $M$. Finally, let us call $s$ the map obtained by the composition

$$i \circ \eta_M \circ \sigma\colon \boldsymbol{A} \to [0,1]_Ł. \tag{23}$$

Then $s$ is a state of $\boldsymbol{A}$ as the following theorem shows.

THEOREM 6.2.1. *Let $\langle \boldsymbol{A}, \sigma\rangle$ be any* SMV*-algebra and let $s\colon \boldsymbol{A} \to [0,1]_Ł$ be defined by (23). Then $s$ is a state of $\boldsymbol{A}$.*

*Proof.* Since $\sigma$, $i$ and $\eta_M$ preserve 1, it is clear that $s(1) = 1$. To show that $s$ is additive, let $x, y \in A$ be such that $x \odot y = 0$. By Lemma 6.0.3 (iii) one has $\sigma(x \oplus y) = \sigma(x) \oplus \sigma(y)$. Moreover, by Lemma 6.0.3 (f), $\sigma(x) \odot \sigma(y) = 0$, thus $s(x) \odot s(y) = 0$. Hence $s(x \oplus y) = s(x) \oplus s(y) = s(x) + s(y) - (s(x) \odot s(y)) = s(x) + s(y)$. □

Conversely, we shall obtain an SMV-algebra from an MV-algebra equipped with a state. To this purpose, we need to introduce an MV-*algebraic tensor product* (or simply a *tensor product*) between MV-algebras. Let $\boldsymbol{A}$, $\boldsymbol{B}$ and $\boldsymbol{C}$ be MV-algebras. A *bimorphism* from the direct product $\boldsymbol{A} \times \boldsymbol{B}$ of $\boldsymbol{A}$ and $\boldsymbol{B}$ into $\boldsymbol{C}$, is a map $\Upsilon$ satisfying the following list of properties:

(i) $\Upsilon(1,1) = 1$ and for all $a \in A$ and $b \in B$, $\Upsilon(a,0) = \Upsilon(0,b) = 0$.

(ii) For all $a, a_1, a_2 \in A$ and $b, b_1, b_2 \in B$, $\Upsilon(a_1 \wedge a_2, b) = \Upsilon(a_1, b) \wedge \Upsilon(a_2, b)$, $\Upsilon(a_1 \vee a_2, b) = \Upsilon(a_1, b) \vee \Upsilon(a_2, b)$ and $\Upsilon(a, b_1 \wedge b_2) = \Upsilon(a, b_1) \wedge \Upsilon(a, b_2)$, $\Upsilon(a, b_1 \vee b_2) = \Upsilon(a, b_1) \vee \Upsilon(a, b_2)$.

(iii) For all $a, a_1, a_2 \in A$ and $b, b_1, b_2 \in B$, if $a_1 \odot a_2 = 0$, then $\Upsilon(a_1 \odot a_2, b) = \Upsilon(a_1, b) \odot \Upsilon(a_2, b)$ and $\Upsilon(a_1 \oplus a_2, b) = \Upsilon(a_1, b) \oplus \Upsilon(a_2, b)$. If $b_1 \odot b_2 = 0$, then $\Upsilon(a, b_1 \odot b_2) = \Upsilon(a, b_1) \odot \Upsilon(a, b_2)$ and $\Upsilon(a, b_1 \oplus b_2) = \Upsilon(a, b_1) \oplus \Upsilon(a, b_2)$.

Then the *tensor product* $\boldsymbol{A} \otimes \boldsymbol{B}$ of two MV-algebras $\boldsymbol{A}$ and $\boldsymbol{B}$ is an MV-algebra (unique up to an isomorphism) such that there is a universal bimorphism $\Upsilon$ from $A \times B$ into $\boldsymbol{A} \otimes \boldsymbol{B}$. Universality means that for any bimorphism $\Upsilon'\colon \boldsymbol{A} \times \boldsymbol{B} \to \boldsymbol{C}$, where $\boldsymbol{C}$ is an MV-algebra, there exists a unique homomorphism $\lambda\colon \boldsymbol{A} \otimes \boldsymbol{B} \to \boldsymbol{C}$ such that $\Upsilon' = \lambda \circ \Upsilon$.

In algebraic terms, $\boldsymbol{A} \otimes \boldsymbol{B}$ is constructed in the following way: let $\boldsymbol{F}(A \times B)$ be the free MV-algebra over the free generating set $A \times B$. Let $I_T$ be the ideal of $\boldsymbol{F}(A \times B)$ generated by the following elements for every $a, a_1, a_2 \in A$ and $b, b_1, b_2 \in B$:

(i) $d(\langle 1,1\rangle, 1)$

(ii) $d(\langle a,0\rangle, 0)$

(iii) $d(\langle 0,b\rangle, 0)$

(iv) $d(\langle a_1 \wedge a_2, b\rangle, \langle a_1, b\rangle \wedge \langle a_2, b\rangle)$

(v) $d(\langle a_1 \vee a_2, b\rangle, \langle a_1, b\rangle \vee \langle a_2, b\rangle)$

(vi) $d(\langle a, b_1 \wedge b_2\rangle, \langle a, b_1\rangle \wedge \langle a, b_2\rangle))$

(vii) $d(\langle a, b_1 \vee b_2\rangle, \langle a, b_1\rangle \vee \langle a, b_2\rangle)$,

(viii) $d(\langle a_1 \odot a_2, b\rangle, 0)$ whenever $a_1 \odot a_2 = 0$

(ix) $d(\langle a_1 \oplus a_2, b\rangle, \langle a_1, b\rangle \oplus \langle a_2, b\rangle)$ whenever $a_1 \odot a_2 = 0$

(x) $d(\langle a, b_1 \odot b_2\rangle, 0)$ whenever $b_1 \odot b_2 = 0$

(xi) $d(\langle a, b_1 \oplus b_2\rangle, \langle a, b_1\rangle \oplus \langle a, b_2\rangle)$ whenever $b_1 \odot b_2 = 0$.

Then we define $\boldsymbol{A} \otimes \boldsymbol{B}$ to be the MV-algebra $\boldsymbol{F}(A \times B)/I_T$.

In the following we consider the tensor products of the form $\boldsymbol{T} = [0,1]_{\text{Ł}} \otimes \boldsymbol{A}$, where $\boldsymbol{A}$ is an arbitrary MV-algebra. For every $\alpha \in [0,1]$ and $a \in A$, we shall denote $\Upsilon(\alpha, a)$ by $\alpha \otimes a$.

PROPOSITION 6.2.2. *Let $\boldsymbol{T} = [0,1]_{\text{Ł}} \otimes \boldsymbol{A}$. Then the following conditions hold for any $\alpha, \alpha_1, \alpha_2 \in [0,1]$ and any $a, a_1, a_2 \in A$:*

*(i)* $(\alpha_1 \oplus \alpha_2) \otimes 1 = (\alpha_1 \otimes 1) \oplus (\alpha_2 \otimes 1)$, *and* $1 \otimes (a_1 \oplus a_2) = (1 \otimes a_1) \oplus (1 \otimes a_2)$.

*(ii)* $\neg(\alpha \otimes 1) = (1 - \alpha) \otimes 1$, *and* $\neg(1 \otimes a) = 1 \otimes \neg a$.

*(iii) The maps* $\Phi\colon \alpha \mapsto (\alpha \otimes 1)$, *and* $\Psi\colon a \mapsto (1 \otimes a)$ *are respectively embeddings of* $[0,1]_{\text{Ł}}$ *and* $\boldsymbol{A}$ *into* $\boldsymbol{T}$.

*(iv) If* $\alpha_1 \odot \alpha_2 = 0$, *then* $(\alpha_1 + \alpha_2) \otimes a = (\alpha_1 \otimes a) \oplus (\alpha_2 \otimes a)$, *and if* $a_1 \odot a_2 = 0$, *then* $\alpha \otimes (a_1 \oplus a_2) = (\alpha \otimes a_1) \oplus (\alpha \otimes a_2)$.

*(v)* $\alpha \otimes (a_1 \ominus a_2) = (\alpha \otimes a_1) \ominus (\alpha \otimes a_2)$, *and* $(\alpha_1 \ominus \alpha_2) \otimes a = (\alpha_1 \otimes a) \ominus (\alpha_2 \otimes a)$.

*(vi)* $1 \otimes 1$ *is the top element of* $\boldsymbol{T}$, *while for every* $a \in A$ *and every* $\alpha \in [0,1]$, $0 \otimes a$ *and* $\alpha \otimes 0$ *coincide with the bottom element of* $\boldsymbol{T}$.

*Proof.* All the properties but (iii) are straightforward consequences of the tensor product construction explained above. For the proof of (iii) see Proposition 2.1 in [35]. □

Due to Proposition 6.2.2 (iii), for any $\alpha \in [0,1]$, we may denote $\alpha \otimes 1$ by $\alpha$.

THEOREM 6.2.3. *Let $\boldsymbol{A}$ be an* MV*-algebra and $s$ be a state of $\boldsymbol{A}$. Then there exists a state $\hat{s}\colon [0,1]_{\text{Ł}} \otimes A \to [0,1]$ making $\langle \boldsymbol{T}, \hat{s}\rangle$ an* SMV*-algebra. Moreover, if $\Phi$ and $\Psi$ are the embeddings of $[0,1]_{\text{Ł}}$ and $\boldsymbol{A}$ into $\boldsymbol{T}$, respectively, then $\Phi(s(a)) = \hat{s}(\Psi(a))$ for each $a \in A$.*

*Proof.* Let us define the state $s_1$ of $\Psi(\boldsymbol{A})$ by the stipulation $s_1(\Psi(a)) = s(a)$ for every $a \in A$. Then, since $\Psi(\boldsymbol{A})$ is an MV-subalgebra of $\boldsymbol{T}$, by Corollary 3.4.2 the state $s_1$ can be extended to a state $s_2\colon [0,1]_{\text{Ł}} \otimes A \to [0,1]$. Finally, the map $\hat{s}\colon [0,1]_{\text{Ł}} \otimes A \to [0,1]_{\text{Ł}} \otimes A$ defined by

$$\hat{s}(t) = s_2(t) \otimes 1$$

makes $(\boldsymbol{T}, \hat{s})$ into an SMV-algebra. Moreover, for every $a \in A$,

$$\hat{s}(\Psi(a)) = s_2(\Psi(a)) \otimes 1 = \Phi(s(a))$$

and the claim is settled. □

### 6.3 Dealing with coherent books in SMV-algebraic theory

Let $\phi_1, \ldots, \phi_k$ be formulas of Łukasiewicz logic and let $\beta : [\phi_i] \mapsto \beta_i$ be a rational-valued book, that is, let us assume that the $\beta_i$'s are rational numbers, say $\beta_i = \frac{n_i}{m_i}$. Moreover, let $x_1, \ldots, x_k$ be fresh variables, and consider for each $i = 1, \ldots, k$, the equations

$$\varepsilon_i \colon (m_i - 1)x_i = \neg x_i \qquad\qquad \delta_i \colon \sigma(\phi) = n_i x_i.$$

Then we can prove the following:

THEOREM 6.3.1. *Let $\beta \colon [\phi_i] \mapsto \frac{n_i}{m_i}$ be a rational book on the Łukasiewicz formulas $\phi_1, \ldots, \phi_k$. Then the following are equivalent:*

*(i) $\beta$ is coherent.*

*(ii) The equations $\varepsilon_i$ and $\delta_i$ (for $i = 1, \ldots, k$) are satisfied in some non-trivial* SMV-*algebra.*

*Proof.* As in Example 3.2.4, let $\boldsymbol{F}$ be the Lindenbaum algebra of Łukasiewicz logic over countably-many variables. By Theorem 5.0.4 it is sufficient to prove that (ii) is equivalent to the existence of a state $s$ on $\boldsymbol{F}$ such that, for all $i = 1, \ldots, k$, $s([\phi_i]) = \frac{n_i}{m_i}$.

(i) $\Rightarrow$ (ii). Let $s$ be a state on $\boldsymbol{F}$ extending $\beta$. Recalling the tensor product construction (see Theorem 6.2.3), let $\boldsymbol{T} = [0,1]_Ł \otimes \boldsymbol{F}$ and let $\sigma \colon \boldsymbol{T} \to \boldsymbol{T}$ be defined as in the proof of Theorem 6.2.3. Since $1 \otimes [\phi_i] \in [0,1]_Ł \otimes \boldsymbol{F}$ and $\sigma(1 \otimes [\phi_i]) = s([\phi_i])$, for each $i = 1, \ldots, n$, it is clear that $\sigma$ extends $s$ (up to an isomorphism).

Let $V$ be a valuation on $\langle \boldsymbol{T}, \sigma \rangle$ such that $V(x_i) = \frac{1}{m_i}$ for each $i = 1, \ldots, n$ (notice that $\frac{1}{m_i} = \frac{1}{m_i} \otimes [1] \in [0,1]_Ł \otimes \boldsymbol{F}$, whence $V$ is a valuation on $\langle \boldsymbol{T}, \sigma \rangle$). Then $V$ satisfies the equations $\varepsilon_i$ because

$$(m_i - 1)V(x_i) = \frac{m_i - 1}{m_i} = 1 - \frac{1}{m_i} = V(\neg x_i).$$

Moreover, $V$ satisfies the equations $\delta_i$:

$$\sigma(1 \otimes [\phi_i]) = 1 \cdot s([\phi_i]) = s([\phi_i]) = \frac{n_i}{m_i} = n_i V(x_i).$$

Thus the equations $\varepsilon_i$ and $\delta_i$ are satisfied in a non-trivial SMV-algebra as required.

(ii) $\Rightarrow$ (i). Let $\langle \boldsymbol{A}, \sigma \rangle$ be an SMV-algebra and $V$ be a valuation on $\langle \boldsymbol{A}, \sigma \rangle$ satisfying the equations $\varepsilon_i$ and $\delta_i$ for each $i = 1, \ldots, k$. Without loss of generality, we may assume that $\boldsymbol{A}$ is finitely (or even countably) generated, so that there is an epimorphism $h_V \colon \boldsymbol{F} \to \boldsymbol{A}$ such that $h_V([x]) = V([x])$ for every propositional variable $x$. Then

$$(m_i - 1)h_V([x]) = \neg(h_V([x_i])) \quad \text{and} \quad \sigma(h_V([\phi_i])) = n_i h_V([x_i]).$$

As in the proof of Theorem 6.2.1, let $M$ be a maximal MV-filter of $\sigma(\boldsymbol{A})$ and define, for each $[\psi] \in \boldsymbol{F}$,

$$s([\psi]) = \sigma(h_V([\psi]))/M.$$

Since quotients preserve identities, one has

$$s([\phi_i]) = (n_i h_V([x_i]))/M \quad \text{and} \quad (m_i - 1)(h_V([x_i])/M) = \neg(h_V([x_i])/M).$$

Hence the MV-homomorphism $\eta_M : \sigma(\boldsymbol{A})/M \to [0,1]_{\text{Ł}}$ maps $h_V([x_i])/M$ to $\frac{1}{m_i}$ and $s([\phi_i])$ to $\frac{n_i}{m_i}$, respectively.

It remains to be proved that $s$ is a state of $\boldsymbol{F}$. First of all it is clear that $s([1]) = 1$. As for additivity, let $[\psi_1], [\psi_2] \in \boldsymbol{F}$ such that $[\psi_1] \odot [\psi_2] = 0$. Then

$$\begin{aligned} s([\psi_1] \oplus [\psi_2]) &= (\sigma(h_V([\psi_1]) \oplus h_V([\psi_2])))/M \\ &= (\sigma(h_V([\psi_1])))/M \oplus (\sigma(h_V([\psi_2])))/M \\ &= s([\psi_1]) \oplus s([\psi_2]) \\ &= s([\psi_1]) + s([\psi_2]), \end{aligned}$$

where the last equality follows from the following fact: if $[\psi_1] \odot [\psi_2] = 0$, then $h_V([\psi_1]) \odot h_V([\psi_2]) = 0$ in $A$, and so

$$s([\psi_1]) \odot s([\psi_2]) = (\sigma(h_V([\psi_1]))) \odot \sigma(h_V([\psi_2]))/M = 0/M = 0$$

and $s([\psi_1]) \oplus s([\psi_2]) = s([\psi_1]) + s([\psi_2])$. This implies that $s$ is a state of $\boldsymbol{F}$ extending the assessment $\beta$. Therefore $\beta$ is coherent. □

## 7 Conditional probability and Dutch Book argument

One of the main motivations for dealing with conditional probability in the classical Boolean setting is to quantify the uncertainty degree of an "event given an event". In this scenario, given two elements $a, b$ of a Boolean algebra $\boldsymbol{B}$, the conditional probabilistic value of the conditional event "$a$ given $b$"—denoted $a|b$—is computed with the help of a given unconditional finitely additive probability measure $\mu : B \to [0,1]$ as follows:

$$\mu(a|b) = \frac{\mu(a \wedge b)}{\mu(b)}. \tag{24}$$

Clearly, the value $\mu(a|b)$ is defined only when $\mu(b) \neq 0$. This means that we can define a conditional probability $\mu(\cdot\,|\,\cdot)$ as a partial map on the product $B \times B$ via (24). Every conditional probability satisfies the following fundamental property, which also motivates the definition of conditional states: whenever $b \in B$ is such that $\mu(b) \neq 0$, then the function $\mu(\cdot\,|b) : B \to [0,1]$ is a finitely additive probability measure.

In order to define conditional states, it is worth noticing that the direct generalization of (24) by substituting the operation $\wedge$ with the MV-operation $\odot$ fails to satisfy the additivity property of states. Indeed, if $a_1, a_2, b$ are elements of an MV-algebra $\boldsymbol{A}$ such that $a_1 \odot a_2 = 0$ and $s$ is a state of $\boldsymbol{A}$ with $s(b) > 0$, then

$$\frac{s((a_1 \oplus a_2) \odot b)}{s(b)} \neq \frac{s(a_1 \odot b) + s(a_2 \odot b)}{s(b)}$$

since $\odot$ does not distribute over $\oplus$.

A possible solution is to introduce a new MV-algebraic operation whose standard behavior has the features of the usual product between real numbers. The resulting algebraic structures are called $\mathrm{PMV}^{+}$-algebras.

DEFINITION 7.0.1. *A* $\mathrm{PMV}^{+}$*-algebra is a pair* $\langle \boldsymbol{A}, \cdot \rangle$ *such that* $\boldsymbol{A}$ *is an* MV*-algebra and* $\cdot$ *is a binary operation on* $A$ *satisfying the following properties for all* $x, y, z \in A$*:*

*(i)* $\langle A, \cdot, 1 \rangle$ *is a commutative monoid.*

*(ii)* $x \cdot (y \ominus z) = (x \cdot y) \ominus (x \cdot z)$.

*(iii) If* $x \cdot x = 0$*, then* $x = 0$.

The class of $\mathrm{PMV}^{+}$-algebras forms a quasivariety, which can be generated by the standard algebra $[0,1]_{\mathrm{PMV}^{+}} = \langle [0,1]_{Ł}, \cdot \rangle$, where $\cdot$ is the ordinary product of real numbers in $[0,1]$. Let $\boldsymbol{A}$ and $\boldsymbol{B}$ be $\mathrm{PMV}^{+}$-algebras. A map $h \colon A \to B$ is a $\mathrm{PMV}^{+}$-*homomorphism* if $h$ is an MV-homomorphism and $h(a \cdot b) = h(a) \cdot h(b)$. For every $\mathrm{PMV}^{+}$-algebra $\boldsymbol{A}$, we denote by $\mathcal{H}^{+}(\boldsymbol{A})$ the set of homomorphisms of $\boldsymbol{A}$ in $[0,1]_{\mathrm{PMV}^{+}}$. A *filter* of a $\mathrm{PMV}^{+}$-algebra $\boldsymbol{A}$ is a subset $F$ of $A$ such that $F$ is a filter of the MV-reduct $\boldsymbol{A}^{-}$ of $\boldsymbol{A}$ and for every $a, b \in F$, the condition $a \cdot b \in F$ holds. By a *state* of a $\mathrm{PMV}^{+}$-algebra $\boldsymbol{A}$ we mean a state of its MV-reduct.

REMARK 7.0.2. *If* $\boldsymbol{A}$ *is a* $\mathrm{PMV}^{+}$*-algebra, it is known that* $\boldsymbol{A}$ *and its* MV*-reduct* $\boldsymbol{A}^{-}$ *have the same congruences. In particular,* $\mathrm{Max}(\boldsymbol{A}) = \mathrm{Max}(\boldsymbol{A}^{-})$*. Let* $s$ *be a state of a* $\mathrm{PMV}^{+}$*-algebra* $\boldsymbol{A}$*. Then, since* $s$ *is a state of* $\boldsymbol{A}^{-}$*, by Theorem 4.0.1 there exists a unique probability measure* $\mu \in \mathcal{M}(\mathrm{Max}(\boldsymbol{A}^{-}))$ *such that, for every* $a \in A$*,*

$$s(a) = \int_{\mathrm{Max}(\boldsymbol{A}^{-})} a^{*}(M) \,\mathrm{d}\mu(M) = \int_{\mathrm{Max}(\boldsymbol{A})} a^{*}(M) \,\mathrm{d}\mu(M).$$

*Therefore, states of* $\mathrm{PMV}^{+}$*-algebras corresponds to integrals with respect to regular Borel probability measures.*

Analogously to the Many-valued Coherence Criterion for MV-algebras, we introduce the following notions. For every $\mathrm{PMV}^{+}$-algebra $\boldsymbol{A}$ and a finite subset $A' = \{e_1, \ldots, e_k\}$ of $A$, a *book* on $A'$ is defined to be a map from $A'$ into $[0,1]$. An assessment $\beta$ is called *coherent* if the bookmaker does not lose money in every possible world $w \in \mathcal{H}^{+}(\boldsymbol{A})$.

LEMMA 7.0.3. *Let* $e_1, \ldots, e_k$ *be elements of a* $\mathrm{PMV}^{+}$*-algebra* $\boldsymbol{A}$ *and let* $\beta \colon e_i \mapsto \beta_i$ *be a book. Then the following are equivalent:*

*(i)* $\beta$ *is coherent.*

*(ii)* $\beta$ *extends to a state of* $\boldsymbol{A}$*.*

*Proof.* Adopting the same notation as in Remark 7.0.2, let $\boldsymbol{A}^{-}$ be the MV-reduct of $\boldsymbol{A}$. Then the book $\beta$ (regarded as a partial map on $A^{-}$) is coherent if and only if there exists a state of $\boldsymbol{A}^{-}$ that extends it (cf. Theorem 5.0.4) if and only if, by the above Remark 7.0.2, there is a state of $\boldsymbol{A}$ that extends it. □

Similarly to the Boolean setting, any state $s$ of a $\mathrm{PMV}^+$-algebra $\boldsymbol{A}$ defines a conditional state, which can also be regarded as a partial map on $A \times A$ by the stipulation: for every $a, b \in A$,

$$s(a|b) = \frac{s(a \cdot b)}{s(b)}, \quad \text{whenever } s(b) > 0. \tag{25}$$

We may leave $s(a|b)$ undefined otherwise.

PROPOSITION 7.0.4. *Let $\boldsymbol{A}$ be any $\mathrm{PMV}^+$-algebra, $s$ be a state of $\boldsymbol{A}$ and let $b \in A$ be such that $s(b) \neq 0$. Then the map $s(\cdot\,|b) : A \to [0,1]$ is a state of $\boldsymbol{A}$.*

*Proof.* We have

$$s(1|b) = \frac{s(1 \cdot b)}{s(b)} = 1.$$

Let $a_1, a_2 \in A$ be such that $a_1 \odot a_2 = 0$. Then $(a_1 \cdot b) \odot (a_2 \cdot b) = 0$ and $(a_1 \oplus a_2) \cdot b = (a_1 \cdot b) \oplus (a_2 \cdot b)$. Therefore

$$s(a_1 \oplus a_2|b) = \frac{s((a_1 \oplus a_2) \cdot b)}{s(b)} = \frac{s((a_1 \cdot b) \oplus (a_2 \cdot b))}{s(b)} = \frac{s(a_1 \cdot b) + s(a_2 \cdot b)}{s(b)}$$
$$= s(a_1|b) + s(a_2|b).$$

This means that $s(\cdot\,|b)$ is a state of the MV-reduct of $\boldsymbol{A}$ and our claim is settled. □

### 7.1 Bookmaking on many-valued conditional events

The classical coherence criterion discussed in Section 5 was extended by de Finetti [17] to a class $\{e_1|h_1, \ldots, e_k|h_k\}$ of conditional events by introducing an additional rule on which a bookmaker and a gambler must agree: any bet on a conditional event $e_i|h_i$ is ruled out in a possible world $w$, when $w$ falsifies $h_i$, that is, $w(h_i) = 0$.

When moving from classical to many-valued events, it is reasonable to assume that the truth value $w(h_i)$ is neither 0 nor 1. Consider for instance the following example introduced by Franco Montagna: suppose that we are betting on the conditional event "The Barcelona soccer team will win the next match, provided that Messi plays". For convenience, let us denote by $\phi$ the event "the Barcelona soccer team will win" and by $\psi$ the antecedent of the previous statement: "Messi will play", so that the above conditional event can be written as $\phi|\psi$. Assume that, during the soccer match (and hence in the possible world $w$), Messi plays the whole match except for the last 30 seconds. It would not make sense to completely invalidate the bet; instead it would be meaningful to think that the bet on that many-valued conditional event is true to the degree $w(\psi)$. Thus, if $w(\psi) = 1$, then the bet is completely valid. If $w(\psi) = 0$, then the bet is called off. In all the intermediate cases $0 < w(\psi) < 1$ the bet is partially valid with degree $w(\psi)$. Obviously, in order to cope with the partial validity of bets, we shall require our book to be *complete*, meaning that if the bookmaker chooses many-valued conditional events $e_1|h_1, \ldots, e_k|h_k$ to assign a betting odd, he will also assign a betting odd to the antecedent $h_1, \ldots, h_k$ of each conditional event.

REMARK 7.1.1. *In this section many-valued events will be identified with elements of any* PMV$^+$*-algebra. Therefore, when we will speak about a many-valued* conditional *event, we always refer to an ordered pair* $\langle e_i, h_i \rangle$ *(denoted by* $e_i|h_i$*) of elements of a* PMV$^+$*-algebra.*

Formally, let $C = \{e_1|h_1, \ldots, e_k|h_k\}$ be a set of many-valued conditional events and $U = \{u_1, \ldots, u_l\}$ (with $l \geq k$) be a set of many-valued unconditional events such that for all $i = 1, \ldots, k$, there is $j = 1, \ldots, l$ such that $h_i = u_j$. Further, let there be a complete book $\beta$ such that $\beta(e_i|h_i) = \beta_i$ and $\beta(u_j) = \gamma_j$; if the gambler bets $\sigma_i$ on $e_i|h_i$ and $\lambda_j$ on $u_j$, respectively, then the bookmaker's balance with respect to the possible world $w$ is computed as

$$\sum_{i=1}^{k} \sigma_i w(h_i)(\beta_i - w(e_i)) + \sum_{i=j}^{l} \lambda_j(\gamma_j - w(u_j)).$$

A many-valued coherence criterion can be formulated in the following way.

**Many-valued Conditional Coherence Criterion.** Let $\boldsymbol{A}$ be a PMV$^+$-algebra and let $C$ and $U$ be defined as above. A complete book such that $\beta(e_i|h_i) = \beta_i$ and $\beta(u_j) = \gamma_j$ is said to be *conditionally coherent* if and only if for every choice of $\sigma_1, \ldots, \sigma_k, \lambda_1, \ldots, \lambda_l \in \mathsf{R}$, there exists $w \in \mathcal{H}^+(\boldsymbol{A})$ that does not cause a sure loss, that is,

$$\sum_{i=1}^{k} \sigma_i w(h_i)(\beta_i - w(e_i)) + \sum_{j=1}^{l} \lambda_j(\gamma_j - w(u_j)) \geq 0. \tag{26}$$

In the rest of this section, we will always assume that $\beta_i$ and $\gamma_j$ are rational numbers and, moreover, we will use the following notation without danger of confusion:

(i) For all $i = 1, \ldots, k$, $\beta_i$ denotes the value that a complete book $\beta$ assigns to the conditional events $e_i|h_i$ (for $i = 1, \ldots, k$), while, for every $j = 1, \ldots, l$, $\gamma_j$ denotes the value $\beta(u_j)$.

(ii) When we will refer to a class $C$ of many-valued conditional events and a class of many-valued unconditional events $U$, we will always understand that for each $e_i|h_i \in C$ there is $u_j \in U$ such that $h_i = u_j$. Therefore, we shall speak about a conditional book $\beta$ on $C \cup U$ without loss of generality and in particular, unless otherwise specified, we will always assume that $C = \{e_1|h_1, \ldots, e_k|h_k\}$ and $U = \{u_1, \ldots, u_l\}$.

We are going to characterize complete coherent books in terms of conditional states. Clearly, since a conditional state is not defined for any conditional event $e|h$, where $s(h) = 0$, we have to ensure that all the antecedents $h_i$'s were assigned positive betting odds in a complete book $\beta$. In this case, i.e., when $\gamma_i > 0$ for all $h_i$'s, we will say that the complete book $\beta$ is *positive*. In what follows we will show that this is not so restrictive. Nevertheless, for the sake of clarity, let us start by considering the case of positive complete books.

LEMMA 7.1.2. *Let* $\boldsymbol{A}$ *be a* PMV$^+$ *algebra and let* $\beta$ *be a positive complete book on* $C \cup U$*. Then* $\beta$ *avoids sure loss iff* $\beta' \colon h_i \mapsto \gamma_i, e_i \cdot h_i \mapsto \beta_i\gamma_i$ $(i \leq k)$ *avoids sure loss.*

*Proof.* ($\Leftarrow$) We argue contrapositively. Suppose that betting $\lambda_1, ..., \lambda_k$, $\sigma_1, \ldots, \sigma_k$ on $h_1, \ldots, h_k$, $e_1|h_1, \ldots, e_k|h_k$ causes a sure loss. Then

$$\sum_{i=1}^{k} \lambda_i(\beta_i - w(h_i)) + \sum_{i=1}^{k} \sigma_i w(h_i)(\gamma_i - w(e_i)) < 0$$

for every valuation $w$. Adding and subtracting $\sigma_i\gamma_i\beta_i$ yield

$$\sum_{i=1}^{k} \lambda_i(\beta_i - w(h_i)) + \sum_{i=1}^{k} \sigma_i(\gamma_i\beta_i - w(e_i \cdot h_i)) + \sum_{i=1}^{k} \sigma_i\gamma_i(w(h_i - \beta_i) < 0,$$

so that

$$\sum_{i=1}^{k} (\lambda_i - \sigma_i\gamma_i)(\beta_i - w(h_i)) + \sum_{i=1}^{k} \sigma_i(\gamma_i\beta_i - w(e_i \cdot h_i)) < 0.$$

Therefore by betting $\lambda_i - \sigma_i\gamma_i$ on $h_i$ and $\sigma_i$ on $e_ih_i$ we cause a sure loss and $\beta'$ is not coherent.

($\Rightarrow$) Conversely, if $\beta'$ is not coherent, then there are $\delta_i, \mu_i$ $(i = 1, \ldots, k)$ such that

$$\sum_{i=1}^{k} \delta_i \cdot (\beta_i - w(h_i)) + \sum_{i=1}^{k} \mu_i(\gamma_i\beta_i - w(e_i \cdot h_i)) < 0$$

for every valuation $w$. Adding and subtracting $\mu_i\gamma_i(\beta_i - w(h_i))$ give

$$\sum_{i=1}^{k} (\delta_i + \mu_i\gamma_i)(\beta_i - w(h_i)) + \sum_{i=1}^{k} \mu_i(\gamma_i\beta_i - w(e_i) \cdot w(h_i)) - \sum_{i=1}^{k} \mu_i\gamma_i(\beta_i - w(h_i)) < 0,$$

which implies

$$\sum_{i=1}^{k} (\delta_i + \mu_i\gamma_i)(\beta_i - w(h_i)) + \sum_{i=1}^{k} \mu_i \cdot w(h_i)(\gamma_i - w(e_i)) < 0.$$

It follows that betting $\delta_i + \mu_i\gamma_i$ on $h_i$ and $\mu_i$ on $e_i|h_i$ causes a sure loss and hence $\beta$ is not conditionally coherent. □

THEOREM 7.1.3. *Let* $\boldsymbol{A}$ *be a* PMV$^+$*-algebra and let* $\beta$ *a positive complete book on* $C \cup U$. *Then the following are equivalent:*

*(i)* $\beta$ *is conditionally coherent.*

*(ii) There is a state* $s$ *of* $\boldsymbol{A}$ *such that, for all* $i = 1, \ldots, k$, $\beta_i s(h_i) = s(e_i \cdot h_i)$ *and* $\gamma_i = s(h_i)$, *i.e., for all* $i = 1, \ldots, k$, $\beta_i = s(e_i|h_i)$ *and* $\gamma_i = s(h_i)$.

*Proof.* It follows from Lemma 7.1.2 that $\beta$ is conditionally coherent if and only if the book $\beta'$ on $\{e_i \cdot h_i, h_i | i = 1, \ldots, k\} \subset A$, which assigns $\beta'(e_i \cdot h_i) = \beta_i\gamma_i$ and $\beta'(h_i) = \gamma_i$, is coherent as well. By Lemma 7.0.3, there is a state $s$ of $\boldsymbol{A}$ such that, for all $i = 1, \ldots, k$, $s(e_i \cdot h_i) = \beta_i\gamma_i$ and $s(h_i) = \gamma_i$. Thus $s(h_i)\beta_i = s(e_i \cdot h_i)$ and the claim is settled. □

Let us now analyze the case of a not necessarily positive complete book $\beta$ on $C \cup U$. We will assume that for some $h_i$'s, $\beta(h_i) = 0$. Without loss of generality, let $h_1, \ldots, h_t$ (for $t < k$) be such that $\beta$ assigns a strictly positive value to them, while $\beta(h_{t+1}) = \cdots = \beta(h_k) = 0$. In what follows, we shall denote by $\beta^-$ the complete book obtained by removing from $\beta$ all the occurrences of $e_i|h_i$ for which $\beta(h_i) = 0$. In other words, $\beta^-$ will denote the complete book on $C' \cup U = \{e_1|h_1, \ldots, e_t|h_t, u_1, \ldots, u_l\}$ obtained from $\beta$ by restriction.

LEMMA 7.1.4. *Let $\boldsymbol{A}$ be a* PMV$^+$*-algebra and $\beta$ be a complete book on $C \cup U$. Then:*

*(i) $\beta$ is conditionally coherent if and only if so is $\beta^-$ on $C' \cup U$.*

*(ii) There is a s state of $\boldsymbol{A}$ such that for all $i = 1, \ldots, k$, $s(e_i \cdot h_i) = \beta(e_i|h_i)s(h_i)$ if and only if there is a state $s'$ of $\boldsymbol{A}$ such that for all $i = 1, \ldots, t$, $s'(e_i \cdot h_i) = \beta^-(e_i|h_i)s'(h_i)$.*

*Proof.* (i) It is easy to see that if $\beta$ is conditionally coherent, then so is $\beta^-$. Conversely, let us assume $\beta^-$ to be conditionally coherent. Then, since $\beta^-$ is positive, by Lemma 7.1.2 and Lemma 7.0.3 there are $k + l + 1 \geq t + l + 1$ homomorphisms $w_s$ of $\boldsymbol{A}$ into $[0, 1]_{\mathrm{PMV}^+}$ and positive real numbers $\alpha_1, \ldots, \alpha_{k+l+1}$ such that $\sum_{s=1}^{k+l+1} \alpha_s = 1$ and the following holds:

(c1) For all $u_j$ such that for some $i = 1, \ldots, k$, $u_j = h_i$ and $i \leq t$, we have

$$\beta^-(h_i) = \sum_{s=1}^{k+l+1} \alpha_s w_s(h_i).$$

(c2) If $i > t$, then $\beta^-(h_i) = \sum_{s=1}^{k+l+1} \alpha_s w_s(h_i) = 0$.

(c3) For all $i = 1, \ldots, t$, $\beta^-(e_i|h_i) = \dfrac{\sum_{s=1}^{k+l+1} \alpha_s w_s(e_i) w_s(h_i)}{\sum_{s=1}^{k+l+1} \alpha_s w_s(h_i)}$.

Let $\sigma_1, \ldots, \sigma_k, \lambda_1, \ldots, \lambda_l$ be any system of bets on $\beta$. Then there exists $y$ with $0 \leq y \leq k + l + 1$ and such that the homomorphism $w_y$ of $\boldsymbol{A}$ into $[0, 1]_{\mathrm{PMV}^+}$ satisfies

$$\sum_{i=1}^{t} \sigma_i w_y(h_i)(\beta(e_i|h_i) - w_y(e_i)) + \sum_{j=1}^{l} \lambda_j(\beta(u_j) - w_y(u_j)) \geq 0.$$

By way of contradiction, assume that for all $s = 1, \ldots, k + l + 1$,

$$\sum_{i=1}^{t} \sigma_i w_s(h_i)(\beta(e_i|h_i) - w_s(e_i)) + \sum_{j=1}^{l} \lambda_j(\beta(u_j) - w_s(u_j)) < 0.$$

Then, letting

$$w'(a) = \sum_{s=1}^{k+l+1} \alpha_s w_s(a)$$

we obtain

$$\sum_{i=1}^{t} \sigma_i w'(h_i)\beta(e_i|h_i) - \sum_{i=1}^{t} \sigma_i w'(h_i) w'(e_i) + \sum_{j=1}^{l} \lambda_j(\beta(u_j) - w'(u_j)) < 0.$$

On the other hand, from the above (c1), (c3) and the definition of $w'$, we have

$$\sum_{i=1}^{t} \sigma_i w'(h_i)\beta(e_i|h_i) = \sum_{i=1}^{t} \sigma_i w'(h_i) w'(e_i) \text{ and } \sum_{j=1}^{l} \lambda_j(\beta(u_j) - w'(u_j)) = 0,$$

a contradiction.

Moreover, from (c2) is follows that for all $i \geq t$, $w_y(h_i) = \sum_{s=1}^{k+l+1} \alpha_s w_s(h_i) = 0$ and hence

$$\sum_{i=1}^{k} \sigma_i w_y(h_i)(\beta(e_i|h_i) - w_y(e_i)) + \sum_{j=1}^{l} \lambda_j(\beta(u_j) - w_y(u_j)) \geq 0$$

and $\beta$ is conditionally coherent.

(ii) Since $\beta$ extends $\beta^-$, each state extending $\beta$ extends $\beta^-$ as well. Conversely, assume that $s$ is a state extending $\beta^-$. Then, for every $i > t$, we have $s(h_i) = 0$ and thus $s(e_i \cdot h_i) = 0$. Therefore $s$ satisfies $s(e_i \cdot h_i) = \beta(e_i|h_i)s(h_i)$ for all $i = 1, \dots, k$ and the claim is proved. □

The expected characterization theorem follows from Lemma 7.0.3 and 7.1.4.

THEOREM 7.1.5. *Let* $\boldsymbol{A}$ *be a* PMV$^+$*-algebra and* $\beta$ *be a complete book on* $C \cup U$*. Then the following are equivalent:*

*(i)* $\beta$ *is conditionally coherent.*

*(ii) There is a state of* $\boldsymbol{A}$ *such that for all* $i = 1, \dots, k$,

$$s(e_i \cdot h_i) = \beta(e_i|h_i)s(h_i).$$

*(iii) There are homomorphisms* $w_1, \dots, w_{k+l+1}$ *and positive reals* $\alpha_1, \dots, \alpha_{k+l+1}$ *such that* $\sum_{s=1}^{k+l+1} \alpha_s = 1$*, and*

*(a) for all* $i \leq t$, $\gamma_i = \sum_{s=1}^{k+l+1} \alpha_s w_s(h_i)$,

*(b) for all* $i = 1, \dots, k$, $\beta(e_i|h_i) \sum_{s=1}^{k+l+1} \alpha_s w_s(h_i) = \sum_{s=1}^{k+l+1} \alpha_s w_s(e_i) w_s(h_i)$.

# 8 Historical remarks and further reading

The states of MV-algebras were defined and studied by Mundici in [45] as averaging processes for truth values in Łukasiewicz logic. Since then the topic attracted a number of researchers in many-valued logics. In this chapter we made an effort to include state-of-the-art results and, in the same time, to present self-contained proofs together with some useful techniques for dealing with states. This approach unavoidably led to omitting some important developments; otherwise the scope of mathematical prerequisites would become too broad, ranging from piecewise linear topology to geometric measure theory. Thus many highly interesting parts of the theory are not discussed in this chapter, such as the rational measure of rational polyhedra, Rényi invariant conditional in Łukasiewicz logic, and the properties of Lebesgue state. The interested reader is referred to Mundici's recent book [48] for an in-depth treatment of those topics.

Three chapters of *Handbook of measure theory* published in 2002 are related to states of MV-algebras. Barbieri and Weber [8] studied MV-algebraic measures, which are bounded additive real functions on MV-algebras. The set of all such functions forms a Dedekind complete vector lattice such that the state space is the base of a lattice cone made of bounded additive and positive functions. *Probability on* MV*-algebras* is the chapter [55] by Riečan and Mundici in which an MV-algebraic counterpart of Boolean probability is thoroughly explored. The notion of a probability MV-algebra, which is a $\sigma$-complete MV-algebra equipped with a $\sigma$-order continuous state, is the framework for developing point-free versions of the central limit theorem, individual ergodic theorem, and Kolmogorov's construction of an infinite-dimensional sequence space. Butnariu and Klement [14] provide a survey on $\sigma$-continuous measures over the families of functions called $T$-tribes, where $T$ is a t-norm. Since a $T$-tribe with Łukasiewicz t-norm $T$ is a $\sigma$-complete MV-algebra, Butnariu and Klement deal with $\sigma$-states in particular. The main focus of their work is on integral representations of $T$-norm-based measures, which is of chief importance in theory of cooperative games with fuzzy coalitions [13]. A relatively recent treatment of tribes and their measures can be found in [49] by Navara.

The states of finitely presented algebras have attracted special attention. The results concerning invariant and faithful states presented in Section 4.2 are only scratching the surface of a rapidly developing subject—the dynamics of Z-homeomorphisms of the unit hypercube. The main results in this area include, but are not limited to: Panti's purely algebraic characterization of Lebesgue state and his study of Bernoulli automorphisms of the free finitely generated MV-algebra [51, 52]; the Haar theorem for lattice-ordered Abelian groups with order-unit, which can be directly applied to MV-algebras [47]; Marra's characterization of Lebesgue state [37]. The interpretation of state as a probability operator on formulas (or many-valued events) is discussed also by Marra [38].

Conditional probability over MV-algebras has been studied in several directions. The definition of conditioning involving the notion of algebraic product (25) was first used in [30]. PMV$^{+}$-algebras discussed in Section 7 were introduced in [39] and further studied in [40]. The approach to de Finetti theorem based on conditional events developed in Section 7 is based on Montagna's paper [41]. Further concepts of conditional probability were developed by Mundici [48, Chapter 15] and Montagna et al. [42].

The proof of Theorem 3.3.4—the integral representation for finitely presented MV-algebras—was published in [33]. The core of its proof is the refinement technique, which is used to recover a unique representing probability measure over all the Schauder bases. This idea goes back to a construction appearing in Pták's paper about extension of states on quantum logics [53].

The algebraic structures we discussed in Section 6, namely SMV-algebras, were introduced in [26] and they have been intensively studied since then. A particular attention has also been devoted to the case in which the internal state of an MV-algebra $\boldsymbol{A}$ is an MV-endomorphism of $\boldsymbol{A}$. The latter structures, called *state-morphism* MV*-algebras*, were introduced by Di Nola and Dvurečenskij in [18]. As we have already mentioned in Remark 6.1.4, subdirectly irreducible SMV-algebras and subdirectly irreducible state-morphisms MV-algebras were fully characterized in [21, Theorem 3.4] by Dvurečenskij, Kowalski, and Montagna.

The chapter was about states of MV-algebras, which are the algebras associated with Łukasiewicz infinite-valued logic. The systematic development of other many-valued logics was made possible by the pioneering work of Hájek [28]. The efforts to study states in other logics than Łukasiewicz are complicated by non-existence of the natural notion of addition and the discontinuity of logical operations, among other things. Convincing results were achieved mainly for Gödel and nilpotent minimum logics by Aguzzoli, Gerla, and Marra; see [2, 3]. The integral representation of states in Gödel logic was proved by the same authors in [4]. De Finetti style-theorem for the integral states was exhibited for the whole class of many-valued logics with continuous connectives in [34] by Kühr and Mundici. States on pseudo MV-algebras, non-commutative generalizations of MV-algebras, were introduced by Dvurečenskij [19]. The same author studied integral representation for a large class of algebras (including BL-algebras and effect algebras) [20]. Ciungu devotes several chapters of her book [16] to states of non-commutative structures, providing extensive bibliography.

The decision problem of coherence for rational books in infinite-valued Łukasiewicz logic was shown to be decidable by Mundici [46]. The NP-completeness result was achieved by Bova and Flaminio in [12].

Probability theory and states belong to the colorful mosaic consisting of calculi for uncertainty modeling and reasoning such as Dempster-Shafer theory or possibility theory, which are based on non-additive functions on Boolean algebras. Some classes of uncertainty measures have been generalized to MV-algebras. Since they are not the topic of this exposition, we confine ourselves to pointing the interested reader to the references [23–25] for the survey of current results.

## Acknowledgements

We are grateful to Vincenzo Marra and Franco Montagna for reading the chapter and providing us with many valuable comments. The work of Tommaso Flaminio was supported by the Italian project FIRB 2010 (RBFR10DGUA_002). The work of Tomáš Kroupa was supported by grant GA ČR P402/12/1309. The authors acknowledge partial financial support of the project MaToMUVI—Mathematical Tools for the Management of Uncertain and Vague Information (FP7-PEOPLE-2009-IRSES).

# BIBLIOGRAPHY


[1] Stefano Aguzzoli, Simone Bova, and Brunella Gerla. Free algebras and functional representation for fuzzy logics. In Petr Cintula, Petr Hájek, and Carles Noguera, editors, *Handbook of Mathematical Fuzzy Logic - Volume 2*, volume 38 of *Studies in Logic, Mathematical Logic and Foundations*, pages 713–791. College Publications, London, 2011.

[2] Stefano Aguzzoli and Brunella Gerla. Probability measures in the logic of Nilpotent Minimum. *Studia Logica*, 94(2):151–176, 2010.

[3] Stefano Aguzzoli, Brunella Gerla, and Vincenzo Marra. De Finetti's no-Dutch-book criterion for Gödel logic. *Studia Logica*, 90(1):25–41, 2008.

[4] Stefano Aguzzoli, Brunella Gerla, and Vincenzo Marra. Defuzzifying formulas in Gödel logic through finitely additive measures. In *Fuzzy Systems, 2008. FUZZ-IEEE 2008. IEEE International Conference on Computational Intelligence*, pages 1886–1893, June 2008.

[5] Erik M. Alfsen. *Compact Convex Sets and Boundary Integrals*, volume 57 of *Ergebnisse der Mathematik und ihrer Grenzgebiete*. Springer, New York, 1971.

[6] Charalambos D. Aliprantis and Owen Burkinshaw. *Infinite Dimensional Analysis: a Hitchhiker's Guide*. Springer, Berlin, third edition, 2006.

[7] Charalambos D. Aliprantis and Rabee Tourky. *Cones and Duality*, volume 84 of *Graduate Studies in Mathematics*. American Mathematical Society, Providence, RI, 2007.

[8] Giuseppina Barbieri and Hans Weber. Measures on clans and on MV-algebras. In *Handbook of Measure Theory, Vol. I, II*, pages 911–945. North-Holland, Amsterdam, 2002.

[9] Alain Bigard, Klaus Keimel, and Samuel Wolfenstein. *Groupes at Anneaux Reticulés*, volume 608 of *Lectures Notes in Mathematics*. Springer, Columbia, 1977.

[10] Patrick Billingsley. *Probability and Measure*. Wiley Series in Probability and Mathematical Statistics. John Wiley & Sons, New York, third edition, 1995.

[11] Garrett Birkhoff. *Lattice Theory*. American Mathematical Society, Providence, RI, 1948.

[12] Simone Bova and Tommaso Flaminio. The coherence of Łukasiewicz assessments is NP-complete. *International Journal of Approximate Reasoning*, 51(3):294–304, 2010.

[13] Dan Butnariu and Erich Peter Klement. *Triangular Norm Based Measures and Games with Fuzzy Coalitions*. Kluwer, Dordrecht, 1993.

[14] Dan Butnariu and Erich Peter Klement. Triangular norm-based measures. In *Handbook of Measure Theory, Vol. I, II*, pages 947–1010. North-Holland, Amsterdam, 2002.

[15] Roberto Cignoli, Itala M.L. D'Ottaviano, and Daniele Mundici. *Algebraic Foundations of Many-Valued Reasoning*, volume 7 of *Trends in Logic*. Kluwer, Dordrecht, 1999.

[16] Lavinia Corina Ciungu. *Non-Commutative Multiple-Valued Logic Algebras*. Springer Monographs in Mathematics. Springer, 2014.

[17] Bruno de Finetti. *Wahrscheinlichkeitstheorie: Einführende Synthese mit Kritischem Anhang*. Oldenbourg, 1981.

[18] Antonio Di Nola and Anatolij Dvurečenskij. State-morphism MV-algebras. *Annals of Pure and Applied Logic*, 161:161–173, 2009.

[19] Anatolij Dvurečenskij. States on pseudo MV-algebras. *Studia Logica*, 68:301–327, 2001.

[20] Anatolij Dvurečenskij. States on quantum structures versus integrals. *International Journal of Theoretical Physics*, 50(12):3761–3777, 2011.

[21] Anatolij Dvurečenskij, Tomasz Kowalski, and Franco Montagna. State morphism MV-algebras. *International Journal of Approximate Reasoning*, 52(8):1215–1228, 2011.

[22] Günter Ewald. *Combinatorial Convexity and Algebraic Geometry*, volume 168 of *Graduated Texts in Mathematics*. Springer, New York, 1996.

[23] Tommaso Flaminio, Lluís Godo, and Tomáš Kroupa. Belief functions on MV-algebras of fuzzy sets: An overview. In V. Torra, Y. Narukawa, and M. Sugeno, editors, *Non-Additive Measures: Theory and Applications*, volume 310 of *Studies in Fuzziness and Soft Computing*, pages 173–200. Springer, 2014.

[24] Tommaso Flaminio, Lluís Godo, and Enrico Marchioni. On the logical formalization of possibilistic counterparts of states over $n$-valued Łukasiewicz events. *Journal of Logic and Computation*, 21(3): 429–446, 2011.

[25] Tommaso Flaminio, Lluís Godo, and Enrico Marchioni. Reasoning about uncertainty of fuzzy events: An overview. In Petr Cintula, Christian G. Fermüller, Lluís Godo, and Petr Hájek, editors, *Understanding Vagueness: Logical, Philosophical and Linguistic Perspectives*, volume 36 of *Studies in Logic*, pages 367–400. College Publications, London, 2011.

[26] Tommaso Flaminio and Franco Montagna. MV-algebras with internal states and probabilistic fuzzy logics. *International Journal of Approximate Reasoning*, 50(1):138–152, 2009.

[27] Ken R. Goodearl. *Partially Ordered Abelian Groups With Interpolation*, volume 20 of *Mathematical Surveys and Monographs*. American Mathematical Society, Providence, RI, 1986.

[28] Petr Hájek. *Metamathematics of Fuzzy Logic*, volume 4 of *Trends in Logic*. Kluwer, Dordrecht, 1998.

[29] Mark Krein and David Milman. On extreme points of regular convex sets. *Studia Mathematica*, 9: 133–138, 1940.

[30] Tomáš Kroupa. Conditional probability on MV-algebras. *Fuzzy Sets and Systems*, 149(2):369–381, 2005.

[31] Tomáš Kroupa. Every state on semisimple MV-algebras is integral. *Fuzzy Sets and Systems*, 157(20): 2771–2782, 2006.

[32] Tomáš Kroupa. Representation and extension of states on MV-algebras. *Archive for Mathematical Logic*, 45:381–392, 2006.

[33] Tomáš Kroupa. Note on construction of probabilities on many-valued events via Schauder bases and inverse limits. In *Multiple-Valued Logic (ISMVL), 2010 40th IEEE International Symposium on Multiple-Valued Logic*, pages 185–188, 2010.

[34] Jan Kühr and Daniele Mundici. De Finetti Theorem and Borel states in $[0, 1]$-valued logic. *International Journal of Approximate Reasoning*, 46(3):605–616, 2007.

[35] Serafina Lapenta and Ioana Leuştean. Scalar extensions for algebraic structures of Łukasiewicz logic. Journal of Pure and Applied Algebra, 220(4):1538–1553, 2016.

[36] Ioana Leuştean and Antonio Di Nola. Łukasiewicz logic and MV-algebras. In Petr Cintula, Petr Hájek, and Carles Noguera, editors, *Handbook of Mathematical Fuzzy Logic - Volume 2*, volume 38 of *Studies in Logic, Mathematical Logic and Foundations*, pages 469–583. College Publications, London, 2011.

[37] Vincenzo Marra. The Lebesgue state of a unital Abelian lattice-ordered group, II. *Journal of Group Theory*, 12(6):911–922, 2009.

[38] Vincenzo Marra. Is there a probability theory of many-valued events? In *Probability, Uncertainty and Rationality*, volume 10 of *CRM Series*, pages 141–166. Ed. Norm., Pisa, 2010.

[39] Franco Montagna. An algebraic approach to propositional fuzzy logic. *Journal of Logic, Language and Information*, 9(1):91–124, 2000.

[40] Franco Montagna. Subreducts of MV-algebras with product and product residuation. *Algebra Universalis*, 53(1):109–137, 2005.

[41] Franco Montagna. A notion of coherence for books on conditional events in many-valued logic. *Journal of Logic and Computation*, 21(5):829–850, 2011.

[42] Franco Montagna, Martina Fedel, and Giuseppe Scianna. Non-standard probability, coherence and conditional probability on many-valued events. *International Journal of Approximate Reasoning*, 54(5):573–589, 2013.

[43] Daniele Mundici. Interpretations of AF $C^{\star}$-algebras in Łukasiewicz sentential calculus. *Journal of Functional Analysis*, 65(1):15–63, 1986.

[44] Daniele Mundici. Satisfiability in many-valued sentential logic is NP-complete. *Theoretical Computer Science*, 52(1–2):145–153, 1987.

[45] Daniele Mundici. Averaging the truth-value in Łukasiewicz logic. *Studia Logica*, 55(1):113–127, 1995.

[46] Daniele Mundici. Bookmaking over infinite-valued semantics. *International Journal of Approximate Reasoning*, 43:223–240, 2006.

[47] Daniele Mundici. The Haar theorem for lattice-ordered Abelian groups with order unit. *Discrete and Continuous Dynamical Systems*, 21(2):537–549, 2008.

[48] Daniele Mundici. *Advanced Łukasiewicz Calculus and MV-Algebras*, volume 35 of *Trends in Logic*. Springer, New York, 2011.

[49] Mirko Navara. Triangular norms and measures of fuzzy sets. In Erich Peter Klement and Radko Mesiar, editors, *Logical, Algebraic, Analytic, and Probabilistic Aspects of Triangular Norms*, pages 345–390. Elsevier, 2005.

[50] Jaroslav Nešetřil and Patrice Ossona De Mendez. A model theory approach to structural limits. *Commentationes Mathematicae Universitatis Carolinae*, 53(4):581–603, 2012.

[51] Giovanni Panti. Bernoulli automorphisms of finitely generated free MV-algebras. *Journal of Pure and Applied Algebra*, 208(3):941–950, 2007.

[52] Giovanni Panti. Invariant measures in free MV-algebras. *Communications in Algebra*, 36(8):2849–2861, 2009.

[53] Pavel Pták. Extension of states on logics. *Polish Academy of Sciences. Bulletin. Mathematics*, 33(9-10): 493–497, 1985.
[54] Bhaskara K. P. S. Rao and Bhaskara M. Rao. *Theory of Charges*, volume 109 of *Pure and Applied Mathematics*. Academic Press, Inc. [Harcourt Brace Jovanovich, Publishers], New York, London, 1983.
[55] Beloslav Riečan and Daniele Mundici. Probability on MV-algebras. In Endre Pap, editor, *Handbook of Measure Theory*, volume II, pages 869–909. North-Holland, Amsterdam, 2002.
[56] Alexander Schrijver. *Theory of Linear and Integral Programming*. Wiley-Interscience Series in Discrete Mathematics and Optimization. John Wiley & Sons, Chichester, 1998.

TOMMASO FLAMINIO
Dipartimento di Scienze Teoriche e Applicate
Universitá dell'Insubria
Via Mazzini 5
21100 Varese, Italy
Email: tommaso.flaminio@uninsubria.it

TOMÁŠ KROUPA
Institute of Information Theory and Automation
Czech Academy of Sciences
Pod Vodárenskou věží 4
182 08 Prague, Czech Republic
Email: kroupa@utia.cas.cz

# Chapter XVIII: Fuzzy Logics in Theories of Vagueness

NICHOLAS J.J. SMITH

## 1 Introduction

Vagueness is centrally a property of *predicates*—for example 'bald', 'tall', 'heavy' and 'interesting'. Predicates that are not vague are said to be *precise*—for example 'under 1.4m in height', 'weighs at least 500 grams' and 'north of the equator'.[1] However we cannot properly delineate vagueness just by presenting examples—for the examples generally exhibit various other properties as well, in addition to vagueness. So to delineate our topic properly, we shall do two things: first (Section 1.1) give three positive identifying characteristics of vague predicates and second (Section 1.2) distinguish some other phenomena which often accompany vagueness in natural language but are not our central concern here.

The chapter then proceeds as follows. In Section 2 we discuss what a theory of vagueness should do and introduce the main non-fuzzy theories of vagueness. In Section 3 we introduce the central topic of this entry: fuzzy theories of vagueness. In Section 4 we consider some major arguments in favour of such theories and in Section 5 we present—and reply to—some major objections to fuzzy theories of vagueness.

### 1.1 Characteristics of vague predicates

Vague predicates are generally picked out as those possessing three characteristics: *blurry boundaries, borderline cases and sorites susceptibility.*

**Blurry boundaries:** The *extension* of a predicate is the set of things to which the predicate applies. The boundaries of the extensions of precise predicates are sharp, whereas the boundaries of the extensions of vague predicates are blurry. If you place a pin on a map, you can draw a sharp circle around the points that are within one kilometre of the point hit by the pin; you cannot draw a sharp circle around the points that are *near* it. If you take a large crowd of persons—say, an audience at a concert—you can sharply separate the seat numbers of persons whose height is at least six feet from the seat numbers of all other persons; not so for the seat numbers of the tall persons. In both

[1] Discussions of vagueness are not always limited to predicates: sometimes expressions other than predicates are held to be vague [106, §3.4.6] and [120, p. 124]—and sometimes objects, properties and other constituents of the world itself (as opposed to expressions in language) are held to be vague [95], [106, p. 158] and [4]. Furthermore, some treatments of vagueness in these other areas appeal to fuzzy logics [104]. Nevertheless, the bulk of discussions of vagueness is focussed on predicates—and predicates will be our focus here.

cases ('at least six feet in height' and 'tall') some persons are definitely in the extension and some are definitely not in it—but in the former case the line between them is sharp while in the latter case it is blurry.

**Borderline cases:** In general, predicates have clear positive cases—things of which we confidently assert that they fall under the predicate—and clear negative cases—things of which we confidently deny that they fall under the predicate. Vague predicates, unlike precise ones, also have (persistent) borderline cases: things of which we will neither confidently assert nor confidently deny that they fall under the predicate. Consider again the predicates 'at least six feet in height' and 'tall'. Simply on the basis of looking at them, you might not be able to separate a group of people into those who are at least six feet in height and those who are under six feet in height—but once you know everyone's height, you can do so (without remainder). The same cannot be said for 'tall': even once you know everyone's height, there will still be borderline cases.

**Sorites susceptibility:** A *sorites series* for a predicate $F$ is a series of objects with the following characteristics:

1. $F$ definitely applies to the first object in the series.
2. $F$ definitely does not apply to the last object in the series.
3. Each object in the series[2] is extremely similar to the object after it in all respects relevant to the application of $F$.

For example, a series of men ranging in height from seven feet to four feet in increments of a thousandth of an inch is a sorites series for 'tall' and for 'at least six feet in height'; a series of points one millimetre apart on a straight line from a point $p$ to a point $q$ one thousand kilometres away is a sorites series for 'far from $q$' and for 'more than one kilometre from $q$'; and so on. Given a sorites series, we can generate an associated *sorites argument*:[3]

1. $x_1$ is $F$
2. For every $x$, if $x$ is $F$ then $x'$ is $F$.

   (Or: There is no $x$ such that $x$ is $F$ and $x'$ isn't $F$. Or all of the following: If $x_1$ is $F$ then $x_2$ is $F, \ldots,$ If $x_{n-1}$ is $F$ then $x_n$ is $F$. Or all of the following: It is not the case both that $x_1$ is $F$ and $x_2$ isn't $F, \ldots,$ It is not the case both that $x_{n-1}$ is $F$ and $x_n$ isn't $F$.)
3. $\therefore x_n$ is $F$.

Now, the difference between vague and precise predicates is as follows. Where $F$ is a precise predicate, the sorites argument will be obviously mistaken: there is indeed some $x$ such that $x$ is $F$ and $x'$ is not $F$.[4] Where $F$ is a vague predicate, on the other hand, the sorites argument will seem genuinely paradoxical: the premisses will all seem to be true, the reasoning will seem to be correct, and yet the conclusion will seem to be false. Thus vague predicates, unlike precise ones, generate sorites paradoxes.

[2] Except the last one, which has no object after it.

[3] $x_1, \ldots, x_n$ denote the objects in the series from first ($x_1$) through to last ($x_n$). $x$ ranges over all objects in the sorites series except the last object and $x'$ denotes the object immediately after $x$ in the series.

[4] We don't need to know which object it is: the point is just that there is such an object.

## 1.2 Other phenomena

In natural language, vagueness tends to come packaged with other phenomena. So as to bring vagueness into clear relief, we distinguish four of these other phenomena now: *uncertainty, context sensitivity, ambiguity and generality.*

**Uncertainty:** Given an object $x$ and a predicate $P$, a speaker may be uncertain whether or not $x$ is in the extension of $P$. This is an epistemic phenomenon: the speaker lacks information; she does not know whether or not $x$ is $P$. In this case, $x$ is a kind of borderline case—and a predicate $P$ may even have persistent borderline cases of this kind. For example, suppose we define the predicate 'bearfast' to apply to all and only objects that are moving faster than any polar bear moved on 11th January 1904. We know that some objects are bearfast (e.g. the jet plane flying overhead) and that some are not (e.g. the parked car across the street) but there are many things of which we will neither confidently assert nor confidently deny that they are bearfast—and it seems that we will never be able to gain the information required to classify these cases one way or the other. Such uncertainty also brings with it a kind of blurry boundary. In itself, of course, the boundary of the extension of 'bearfast' is perfectly sharp: but it is shrouded by an epistemic blur. Nevertheless, uncertainty is distinct from vagueness: 'bearfast' is not a vague predicate.[5] In particular, it does not generate sorites paradoxes. We can set up a sorites series for 'bearfast', beginning with an object moving at great speed and progressing by tiny increments to a stationary object—but the associated sorites argument will be obviously mistaken: there is indeed some object $x$ in the series such that $x$ is bearfast and $x'$ is not. Of course we do not—cannot—know which object it is: but this will not make us think that the second premise of the sorites argument is *true*.[6]

**Context sensitivity:** A predicate can have different extensions in different contexts. For example, in a discussion of professional jockeys, a certain individual might count as 'tall', while in a discussion of professional basketball players, the same individual might not count as 'tall'. Many standard examples of vague predicates—'tall', 'loud', 'heavy', 'small' etc.—are also context sensitive. Nevertheless, vagueness and context sensitivity are distinct phenomena—for vagueness arises even on a single occasion of use. Consider a vague predicate—say 'tall'—as uttered on some occasion, in some particular context. Let the predicate $P$ have as its extension the extension that 'tall' had on that single occasion of use. Then $P$ has blurry boundaries and borderline cases—and it generates a sorites paradox.[7]

[5] It is uncontroversial that the existence of an object and a speaker, such that the speaker is uncertain whether the object is in the extension of $P$, does not render $P$ vague. The aim of the present discussion of 'bearfast' is to establish a stronger claim: the existence of a whole *class* of objects such that *no* speaker *can* know whether these objects are in the extension of $P$ does not in and of itself render $P$ vague. The aim is not to rule out in advance a view that we shall encounter in Section 2.2: epistemicism about vagueness, according to which vagueness essentially involves a certain kind of ignorance or uncertainty.

[6] Recall that on some formulations of the sorites argument there is a single second premise that makes a general claim, while on other formulations this single premise is replaced by a multitude of premisses each of which makes a particular claim. For simplicity, I shall generally write of 'the second premise'. Relative to formulations involving a multitude of premises after the first one, talk of 'the second premise' being true (false) should be interpreted as talk of all (some) of them being true (false); talk of 'the second premise' being accepted should be interpreted as talk of all of them being accepted; and so on.

[7] The example involving 'tall' given in the text above illustrates a simple and familiar kind of context sensitivity: relativity to a comparison class. It is uncontroversial that vagueness and this kind of context

**Ambiguity:** An ambiguous expression is one that is susceptible of multiple interpretations that give rise to different truth conditions. A classic example is 'bank', which can mean (among other things) the edge of a river or a financial institution—and the truth conditions of 'John went to the bank' depend on which interpretation is given. Some vague expressions are also ambiguous—for example, 'heavy' can mean 'of great weight' or 'very important or serious'—but we can see that vagueness is distinct from ambiguity by noting that vagueness can still be present after disambiguation. For example, even once we specify that we are using it in the sense of 'of great weight', 'heavy' is still vague: it has blurry boundaries and borderline cases and it generates a sorites paradox.

**Generality:** Colloquially, someone might describe a person as giving a vague response if, when asked his age, he replies 'I was born last century'. This reply—in contrast to 'I was born in 1926'—leaves open many possibilities. However 'born last century' is not vague in our sense:[8] for example, it does not generate a sorites paradox (it is obviously not true that for any persons born, say, one minute apart, if one was born last century then so was the other).

## 2 Non-fuzzy theories of vagueness

We shall gain a better understanding of theories of vagueness that employ fuzzy logics if we are familiar with the main alternatives. In this section, therefore, we look at prominent theories of vagueness that do not employ fuzzy logic.[9] First, in Section 2.1, we make some brief general remarks about what a theory of vagueness (fuzzy or non-fuzzy) is supposed to do.

### 2.1 Theory of vagueness

One fundamental question that a theory of vagueness should answer is: What is the meaning (semantic value) of a vague predicate? Another is: How should we reason in the presence of vagueness? Thus a theory of vagueness needs to cover two bases: semantics and logic. If (as is standard) we adopt a truth conditional approach to semantics and hold that correct reasoning must be truth preserving then the two goals are closely connected.

A more specific requirement on any theory of vagueness is that it solve the sorites paradox. There are two aspects to such a solution. One task is to locate the error in the sorites argument: the premise that isn't true or the step of reasoning that is incorrect. This part of the solution should fall out of what a theory of vagueness says in general about semantics and logic. There is also a further task, that goes beyond pure logic and semantics. A satisfying solution to the sorites should explain why it is a paradox rather than a simple mistake. Thus, as well as locating the error in the argument, a

sensitivity are distinct phenomena (because 'tall' is still vague, even when we fix the comparison class—e.g. to NBA players in 2014). In Section 2.5 we shall encounter contextualism about vagueness: a view according to which vagueness essentially involves some other form of context sensitivity. Whether or not the same kind of argument also defeats contextualism about vagueness (i.e. vagueness cannot be any kind of context sensitivity because vagueness remains even once the entire context—not just a comparison class—is fixed) is a topic of debate; see e.g. [3].

[8] At least not if we suppose, for the sake of argument, that each person is born at a particular instant.

[9] As our main topic here is fuzzy theories of vagueness, we give only a brief presentation of these other views; furthermore, the choice of aspects to focus on in this presentation is partly guided by considerations of what needs to be on the table in order best to understand fuzzy theories when we get to them later. For a much more detailed presentation of these views see [106, ch. 2].

theory of vagueness must provide an explanation of why competent speakers find the argument compelling but not convincing: why they do not spot the error immediately and yet—even in the absence of a clear idea of what the error is—are not inclined to accept the conclusion.

## 2.2 Epistemicism

On this approach, the semantics and logic of vagueness are both entirely classical. Vague predicates—like precise ones—have crisp sets as their extensions. Classical reasoning is correct in the presence of vagueness just as it is in the precise realm of mathematics. This is not to say that there is no distinction at all between vague predicates and precise ones: while they are the same from the semantic and logical points of view, there is an epistemological difference between them. Although the extensions of all predicates are crisp sets, with vague predicates we cannot *know* where the borders lie. Thus, according to the epistemicist, the blurriness of the boundaries of vague predicates is of an epistemic sort: in themselves the boundaries are perfectly sharp but they are hidden behind a veil of ignorance. For all objects $x$, $Px$ is true or false; the borderline cases are the objects $x$ for which we cannot *know* whether $Px$ is true or false.

The first part of the epistemicist solution to the sorites paradox—saying what is wrong with the argument—is straightforward. According to the epistemicist, there is a sharp cut-off in the sorites series between the last object that is $P$ and the first object that isn't—so the second premise is false. The second part of the solution—saying why we are taken in—is trickier. According to the epistemicist, we cannot know where the cut-off is—and so we mistakenly think that there is no such cut-off. This is why we are inclined to accept the second premise even though it is in fact false.

Note that the second part of the solution involves a departure from the usual modus operandi in formal semantics, in which the semantic theory one develops is taken to be implicit in the ordinary usage of competent speakers [101]. In the case of the epistemic theory of vagueness, the explanation of ordinary competent speakers' reactions to the sorites argument turns on their being fundamentally mistaken about the semantics of the predicates they are using. For suppose that a speaker did realise that a predicate $P$—say, 'bearfast'—had sharp but unknowable boundaries. Then she would not think for a moment that the second premise of a sorites argument for $P$ was actually *true*—even though she would indeed be unable to mark the cut-off point in the sorites series between the $P$'s and the non-$P$'s.

A major problem for epistemicists is what I have elsewhere dubbed the *location problem* [106]. According to the epistemicist, in a sorites series for $P$, there is a last object $x$ that is $P$—and the very next object $x'$ in the series is not $P$. More generally, the epistemicist thinks that the boundaries of the extensions of vague predicates are (in themselves) perfectly sharp (even if they are hidden beneath a blur of ignorance). The problem is to explain how these boundaries get to be exactly where they are. Why is $x$ the last object in the series that is $P$—why not some other object that is similar to $x$? What is it that makes our vague term 'tall' (say) have *this* crisp set $S$ as its extension rather than some other crisp set $S'$ (which might differ from $S$ in only very small ways—for example $S$ includes Bill and excludes Ben, who is a nanometre shorter than Bill, while $S'$ includes them both)?

It is generally accepted that language is a human artefact. The sounds we make mean what they do because of the kinds of situations in which we, and earlier speakers, have made those sounds (e.g. had we always used the word 'dog' where we in fact used 'cat' and vice versa, then 'dog' would have meant what 'cat' in fact means and vice versa). Hence there should be some sort of connection between *meaning* and *use*. More precisely, consider the following kinds of facts:

- All the facts as to what speakers actually say and write, including the circumstances in which these things are said and written, and any causal relations obtaining between speakers and their environments.
- All the facts as to what speakers are disposed to say and write in all kinds of possible circumstances.
- All the facts concerning the eligibility as referents of objects and sets.

There is widespread (not universal) agreement in the literature that if these facts are insufficient to determine (unique) meanings for some utterances, then those utterances have no (unique) meanings. That is: semantic facts are never primitive or brute—they are always determined by the meaning-determining facts; and the meaning-determining facts are the ones just itemised. This generates a constraint on any theory of vagueness: if the theory says that vague predicates have meanings of such-and-such a kind (e.g. crisp sets), then we must be able to satisfy ourselves that the meaning-determining facts itemized above could indeed determine such meanings for actual vague predicates. To the extent that the meaning-determining facts do *not* appear sufficient to determine meanings for vague predicates of the kind posited by some theory of vagueness, that theory is undermined. This is precisely where the epistemicist theory faces a problem.

To see this, let's start with the following proposal concerning the connection between meaning and use (for the case of predicates):

(MU) The claim $Pa$ is true if and only if most competent speakers would confidently assent if presented with $a$ in normal conditions and asked whether it was $P$, and is false if and only if most competent speakers would confidently dissent if presented with $a$ in normal conditions and asked whether it was $P$.

There are obvious counterexamples to (MU), for example natural kind terms, and technical terms subject to the division of linguistic labour [87]. However the principle is prima facie plausible for common, everyday, non-specialist vocabulary.

The epistemicist must reject (MU): for any borderline case $x$ of baldness (say), '$x$ is bald' is either true, or false (on the epistemicist view)—yet most competent speakers would neither assent to nor dissent from '$x$ is bald' (we hedge over such cases). So within the borderline cases there is a failure of match-up between meaning and use. We can picture the situation—say for the predicate 'is tall'—as in Figure 1. Epistemicism entails a mismatch between use (on the left) and meaning (on the right).

Of course (MU) is just one proposal connecting meaning and use and denying it does not necessarily mean denying the general principle that use determines meaning. (MU) tells us that in Figure 1 the left hand side and the right hand side must *match*. The

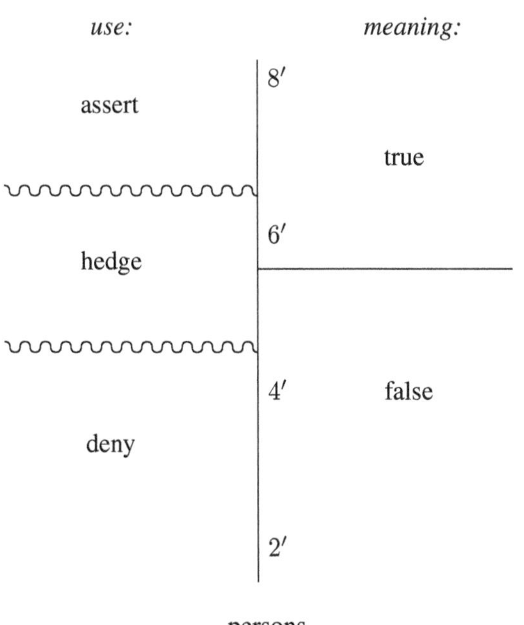


Figure 1. Use and meaning

claim that use determines meaning—that meaning *supervenes* on use, that there can be no difference in meaning without a difference in use—is the claim that the left hand side *determines* the right hand side: that if the right hand side of the picture were different, the left hand side would be different too. It is *consistent* to claim that the left hand side and the right hand side do not match *and* that the right hand side could not be different without the left hand side being different. This is the sort of line taken by Williamson [122, §4] and [123, §7.5]. The idea is that use does determine meaning—just not in any straightforward way. We have no idea of the mechanisms intervening between use and meaning: we do not know how (if at all) the right hand side of our picture would alter if we altered this or that bit of usage on the left hand side. Yet Williamson can still maintain that the right hand side could not be different unless the left hand side was too.

This is a consistent position but it does not provide a real response to the location problem—the problem of *how* precise extensions for vague predicates are determined: it leaves the connection between use and meaning mysterious. Hence it is by nature unsatisfying: we are told merely *that* meaning is determined by use when we want to know *how* this could possibly be the case (assuming—as the epistemicist thinks—that meaning is cleanly bipartite, and given that usage is fuzzily tripartite). In the absence of any explanation of how usage determines sharp boundaries for vague predicates, the most reasonable conclusion seems to be that the meaning-determining facts do *not* suffice to pick out meanings for vague predicates of the kind the epistemicist says they have.

## 2.3 Gaps and third values

In contrast to epistemicism, there are views according to which vague predicates have alethic borderline cases: objects $x$ such that $Px$ is neither true nor false (as opposed to being one or the other, we know not which). One way of spelling out this idea is to say that the extension of a predicate is a partial set: a *partial* function from objects to the classical truth values 1 and 0. Objects sent to neither value are the borderline cases. Where $x$ is a borderline case of $P$, $Px$ will then lack a truth value. Another way is to posit a third truth value in addition to the two classical truth values and say that the extension of a predicate is a function from objects to the set of three truth values. Objects sent to the new third value are the borderline cases. Where $x$ is a borderline case of $P$, $Px$ will then have the third truth value. For ease of exposition, I shall use the term 'tripartite' to cover both partial two-valued and three-valued approaches.

Tripartite approaches can accept (MU) and hence respond to the location problem in a straightforward way. The things sent to 1 by the extension of $P$ are the things that competent speakers would confidently classify as $P$; the things sent to 0 by the extension of $P$ are the things that competent speakers would confidently classify as non-$P$; and the rest are the cases over which speakers would hedge. This is progress, admittedly—but of course a residual problem remains: usage is *fuzzily* tripartite (there is no precise boundary between assertion and hedging, nor between denial and hedging) while meaning (on tripartite views) is *sharply* tripartite.

Given tripartite extensions for predicates, the question now arises how to handle compound statements—conjunctions, negations and so on—where some components have a value other than 1 or 0.[10] One option is the recursive route, where we extend the two-valued truth tables of classical logic to cover the cases where one or both components lack a classical value. There are many options here; some prominent ones are shown in Figures 2(a), 2(b) and 2(c) (in which the $*$ can be interpreted as a third truth value—on three-valued views—or no value—on partial two-valued views).[11]

Another option is to proceed via classical sharpenings.[12] Begin with a three-valued valuation $V_3$ (i.e. a mapping from propositional constants to the set of three truth values). Now instead of extending this to a three-valued model (i.e. an assignment of truth values to all well formed formulas) using truth tables, say that a classical valuation $V_2$ (i.e. a mapping from propositional constants to the set of two classical truth values) *extends* $V_3$ iff whenever $V_3$ assigns 1 (respectively, 0) to a proposition, $V_2$ does too.[13] Consider all the classical valuations that extend $V_3$. Each one of them determines (via the classical truth tables) a classical model $\mathfrak{M}_2$; call these models *extensions* of $V_3$. The rule for extending $V_3$ to a three-valued model $\mathfrak{M}_3$ (which we shall call an s-valuational model) is now as follows (where $[\alpha]$ is the truth value of $\alpha$):

- $[\alpha] = 1$ (respectively, 0) on $\mathfrak{M}_3$ iff $[\alpha] = 1$ (resp., 0) on all classical extensions.
- $[\alpha] = *$ on $\mathfrak{M}_3$ otherwise (i.e. iff $[\alpha] = 1$ on some classical extension and $[\alpha] = 0$ on some classical extension).

[10] For simplicity I focus on connectives here; similar remarks apply to quantifiers.

[11] For further details on these logics see [108].

[12] For simplicity, I shall present the main idea using three truth values in the context of propositional logic, but the approach extends readily to first order logics and can also be formulated using two values and gaps.

[13] So $V_2$ differs from $V_3$ only where $V_3$ assigns $*$ to some basic proposition $\alpha$.

| $\alpha$ | $\beta$ | $\neg\alpha$ | $\alpha\wedge\beta$ | $\alpha\vee\beta$ | $\alpha\to\beta$ | $\alpha\leftrightarrow\beta$ |
|---|---|---|---|---|---|---|
| 1 | 1 | 0 | 1 | 1 | 1 | 1 |
| 1 | * | | * | * | * | * |
| 1 | 0 | | 0 | 1 | 0 | 0 |
| * | 1 | * | * | * | * | * |
| * | * | | * | * | * | * |
| * | 0 | | * | * | * | * |
| 0 | 1 | 1 | 0 | 1 | 1 | 0 |
| 0 | * | | * | * | * | * |
| 0 | 0 | | 0 | 0 | 1 | 1 |

(a) Bočvar (aka Kleene weak) truth tables

| $\alpha$ | $\beta$ | $\neg\alpha$ | $\alpha\wedge\beta$ | $\alpha\vee\beta$ | $\alpha\to\beta$ | $\alpha\leftrightarrow\beta$ |
|---|---|---|---|---|---|---|
| 1 | 1 | 0 | 1 | 1 | 1 | 1 |
| 1 | * | | * | 1 | * | * |
| 1 | 0 | | 0 | 1 | 0 | 0 |
| * | 1 | * | * | 1 | 1 | * |
| * | * | | * | * | * | * |
| * | 0 | | 0 | * | * | * |
| 0 | 1 | 1 | 0 | 1 | 1 | 0 |
| 0 | * | | 0 | * | 1 | * |
| 0 | 0 | | 0 | 0 | 1 | 1 |

(b) Kleene (strong) truth tables

| $\alpha$ | $\beta$ | $\neg\alpha$ | $\alpha\wedge\beta$ | $\alpha\vee\beta$ | $\alpha\to\beta$ | $\alpha\leftrightarrow\beta$ |
|---|---|---|---|---|---|---|
| 1 | 1 | 0 | 1 | 1 | 1 | 1 |
| 1 | * | | * | 1 | * | * |
| 1 | 0 | | 0 | 1 | 0 | 0 |
| * | 1 | * | * | 1 | 1 | * |
| * | * | | * | * | 1 | 1 |
| * | 0 | | 0 | * | * | * |
| 0 | 1 | 1 | 0 | 1 | 1 | 0 |
| 0 | * | | 0 | * | 1 | * |
| 0 | 0 | | 0 | 0 | 1 | 1 |

(c) Łukasiewicz truth tables

Figure 2. Prominent tripartite truth tables

We can get a logic from this framework by picking a set of designated values: a wff is valid iff there is no s-valuational model on which it has a non-designated value and an argument is valid iff there is no s-valuational model on which the premisses all have designated values and the conclusion has a non-designated value. If we set 1 as the only designated value we get *supervaluationist* logic; if we set 1 and $*$ as designated we get *subvaluationist* logic.[14] Note that a formula gets a designated value on a model $\mathfrak{M}_3$ (based on a valuation $V_3$) in the supervaluationist framework if it is true on every classical model that extends $V_3$; such a formula is sometimes called 'supertrue'. In the subvaluationist framework, a formula gets a designated value on a model $\mathfrak{M}_3$ (based on a valuation $V_3$) if it is true on at least one classical model that extends $V_3$.

In the context of vagueness, it is standard to add a further refinement to this general framework. The base valuation—which is three-valued or partial—is taken to correspond to ordinary usage of vague language: clear cases/noncases of a predicate $P$ are mapped to 1/0 by the extension of the predicate in the base model and the borderline cases are neither mapped to 1 nor to 0. A classical extension of the base model then represents a way of *sharpening* or *precisifying* all vague predicates: classifying their borderline cases as positive or negative cases. (Think for example of the way 'adult' might be sharpened in a certain legal context so that it applies to all and only the persons over 18.) Instead of considering *all* such extensions, we consider only the *admissible* ones: the ones that represent acceptable ways of sharpening all vague predicates at once.

[14] This unified presentation of supervaluationism and subvaluationism in terms of s-valuational models is inspired by [93].

So, for example, suppose that Bill and Ben are borderline tall, and Bill is slightly taller than Ben. Considering just 'Bill' and the predicate 'tall' it might be acceptable to put him in or out of the extension; likewise for the predicate 'short'. The same might hold for Ben. Yet it is not acceptable to put Bill in *both* the extensions of 'tall' and 'short' and it is not acceptable to put 'Bill' in the extension of 'short' and 'Ben' in the extension of 'tall'. In other words, borderline cases cannot always be classified independently when we are precisifying a predicate—and different predicates cannot always be precisified independently of one another.

The first part of the supervaluationist solution to the sorites paradox—saying what is wrong with the argument—is that the second premise is false no matter how we precisify and hence comes out as false (in the tripartite base model). The second part of the solution—saying why we are taken in—is as follows. Where $x$ or $x'$ is a borderline case of $P$, each statement of the form '$x$ is $P$ and $x'$ is not $P$' is true on one admissible extension and false on the others—and hence comes out neither true nor false (in the tripartite base model). So we cannot truly say of any object in the series that it is the last $P$. From here—the story goes—we (mistakenly) conclude that the second premise of the sorites argument is true.

Note that the second part of the solution involves the same sort of departure from the usual modus operandi in formal semantics that we saw in the case of epistemicism: the explanation of ordinary competent speakers' reactions to the sorites argument turns on their being fundamentally mistaken about the semantics of the predicates they are using [101]. If a speaker thought that her language works in the way the supervaluationist says it does then she would have no tendency to move from the non-truth of '$x$ is the last $P$' to the truth of 'there is no last $P$'. Consider an analogous case. We are rolling a die. We know that we can only say 'it is certainly the case that $\Phi$' if $\Phi$ will be true no matter how the die falls. So we cannot say 'it is certainly the case that we will roll 1'—or 2, 3, 4, 5 or 6. Yet we have no tendency to infer from this that 'we will roll one of the numbers 1 through 6' is false. On the contrary, it is certainly true: for it will hold however the die lands. According to the supervaluationist, however, this is precisely the kind of mistake ordinary speakers make in relation to the sorites paradox.

Moving from sorites susceptibility to borderline cases and blurry boundaries: unlike epistemicism, supervaluationism holds that vague predicates have genuinely alethic borderline cases—but it does not capture the idea that vague predicates draw blurry boundaries. On the contrary, the divisions between the clear cases and the borderline cases and between the clear noncases and the borderline cases are perfectly sharp. Supervaluationists have explored various approaches to this problem. One that we should mention here—because it bears a superficial similarity to fuzzy approaches—is the degree-theoretic form of supervaluationism. The machinery of a tripartite base model together with its admissible classical extensions is augmented with a normalised measure over the classical extensions. Formulas are then assigned not one of three truth values in the base model—but one of the infinitely many values in $[0, 1]$, according to the rule that the value of $\alpha$ in the base model is the measure of the set of extensions in which $\alpha$ is true. We can think of what is going on here as follows [110]. The values assigned to formulas in the base model are *probabilities* of full truth under complete precisification of the language. The original supervaluationist focusses on formulas that have 100% probability of truth and the subvaluationist focusses on formulas that have non-zero probability of truth; the new degree-theoretic form of supervaluationism, on the other hand, is interested in all intermediate probabilities as well. Earlier we saw that, according to the epistemicist, the blurriness of the boundaries of vague predicates is of an epistemic sort: in themselves the boundaries are perfectly sharp but they are hidden behind a veil of ignorance. In degree supervaluationism, the blurriness of the boundaries of vague predicates is probabilistic: in themselves (as things stand in the base model, which represents language in its actual vague state) the boundaries are sharply tripartite (the extension of a predicate in the base model is a three-valued function or a partial two-valued function)—but the probability that a given object would end up in the extension *were* the language to be fully precisified can in general be anywhere between 0 and 100%.

## 2.4 Plurivaluationism

In order to understand plurivaluationism—and how it differs from both epistemicism and supervaluationism—we need to distinguish pure model theory and model-theoretic semantics (MTS). MTS requires an additional notion that does not figure in pure model theory: its role is to distinguish one (or some) of the infinity of models of a given formal language countenanced in pure model theory as the one(s) relevant to questions of the (actual) meaning and truth (simpliciter) of utterances in some discourse. Such questions are of central interest in natural language semantics—but pure model theory cannot (fully) answer them: for it tells us only that a formula is true on this model and false on that one (etc.). If we want to know whether a given statement is true (simpliciter), then we need to single out a particular model (or a class of models): truth simpliciter will then be truth relative to this model. So we get a system of MTS by combining a system of pure model theory with some notion that plays the role of distinguishing some model(s) as the ones relevant to questions of (actual) meaning and truth (simpliciter) of utterances in some discourse. The simplest choice of model theory is classical model theory. The simplest choice of auxiliary notion is the idea that for each discourse, there is a *unique* relevant model: the 'intended model'. Combining these two choices yields the 'classical semantic picture' [106, §1.2]. It is the version of MTS that underlies epistemicism.

Supervaluationism differs from epistemicism by retaining the idea of a unique intended model of vague discourse but rejecting the idea that it should be a classical model: instead it should be an s-valuational model. Plurivaluationism on the other hand countenances only classical models—but instead of holding that a vague discourse is associated with a single intended model it holds that it is associated with multiple equally acceptable models. These models are precisely the classical models that figure as admissible extensions of the (uniquely intended) tripartite base model in the supervaluationist picture. This can lead to plurivaluationism and supervaluationism being confused or conflated—but they are really quite different. On the supervaluationist picture, there is just one model that is relevant to questions of the (actual) meaning and truth (simpliciter) of utterances in a given vague discourse, and it is nonclassical. On the plurivaluationist picture, there are many models that are all equally relevant to questions of the (actual) meaning and truth (simpliciter) of utterances in a given vague discourse, and each of them is classical.

One natural motivation for plurivaluationism would be a fondness for classical models coupled with a conviction that epistemicism cannot adequately answer the location problem: the epistemicist cannot explain how usage singles out one model as the unique intended one (the model $\mathfrak{M}$ such that the unique sharp extension of 'tall' that the epistemicist countenances is its extension on $\mathfrak{M}$)—and so we should not believe that there is a single intended model. Rather, there are many acceptable models: all the ones that are not ruled out as incorrect by the meaning-determining facts.

As the plurivaluationist lacks the idea of a unique intended model of vague discourse, he cannot follow the route of identifying truth simpliciter for statements in that discourse with truth on the intended model. When it comes to truth, he can only say, of each acceptable model, whether a statement is true or false on that model. As there is no further semantic machinery beyond the individual classical models, there can be no further overarching semantic fact, such as that the statement is 'true simpliciter'. However, the plurivaluationist can say something about assertability that is analogous to what the supervaluationist says about truth. For the supervaluationist, truth simpliciter is truth on the unique intended tripartite base model—and (in the s-valuationist semantics) this means truth on all admissible extensions of the base model. The plurivaluationist can say that when a statement is true on all acceptable models, we can simply assert it; when a statement is false on all acceptable models, we can simply deny it; and when a statement is true relative to some acceptable models and false relative to others, we can neither simply assert it nor simply deny it.

Here's a useful way of thinking about the plurivaluationist view. When I utter 'Bob is tall' (say), I say many things at once: one claim for each acceptable model. Thus we have semantic indeterminacy—or equally, semantic plurality. However, if all the claims I make are true (or false) then we can talk as if I make only one claim, which is true (or false). Figuratively, think of a shotgun fired (once) at a target: many pellets are expelled, not just one bullet; but if all the pellets go through the bullseye, then we can harmlessly *talk as if* there was just one bullet, which went through.

This approach will lead to a solution to the sorites that is analogous to the supervaluationist solution: the second premise is false on every acceptable model (recall that each such model is classical) but speakers fail to see this because where $x$ or $x'$ is a borderline case of $P$, each statement of the form '$x$ is $P$ and $x'$ is not $P$' is true on one acceptable model and false on the others—and hence cannot be asserted. Again, this involves a departure from the usual modus operandi in formal semantics: the explanation of ordinary competent speakers' reactions to the sorites turns on their not realising that the semantics of their discourse is plurivaluationist [101]. Consider an analogous case. We are canvassing opinions about football and (suppose for the sake of argument) there is a convention that one can say 'the man in the street believes $\Phi$' only when *everyone* canvassed believes $\Phi$. Now suppose that each person canvassed has a favourite team but there is no single team that is everyone's favourite. So we cannot say 'the man on the street's favourite team is $X$' (or $Y$ or $Z$ etc., through all the teams). Yet we have no tendency to infer from this that we cannot say 'the man on the street has a favourite team'. On the contrary, this is clearly something that we can assert (in the imagined circumstances) and it may well convey important information (e.g. it excludes the possibilities that some people just don't care about football or have several equally favoured teams).

Turning to borderline cases and blurry boundaries: like epistemicism and unlike supervaluationism, plurivaluationism gives a non-alethic account of borderline cases. On each acceptable model, '$x$ is $P$' is true or false. There is no further machinery beyond the classical models—and so there is no level at which '$x$ is $P$' lacks one of the two classical truth values. It is just that when it has different truth values on different acceptable models, we can neither simply assert nor simply deny that $x$ is $P$. As for blurry boundaries: like supervaluationism, plurivaluationism makes no room for them; the divisions between the clear cases and the borderline cases and between the clear noncases and the borderline cases are perfectly sharp.

## 2.5 Contextualism

Combining a recursive tripartite approach with some additional machinery yields contextualist theories of vagueness, which have played a significant role in the recent literature. Let's suppose that at any particular time, a vague discourse has a unique intended model, which (in the most prominent versions of contextualism) is tripartite and recursive. In supervaluationism, the idea of a complete precisification of the language plays a central role. In contextualism, a more local and less idealised form of precisification plays a central role: the sort of precisification where we partially precisify a predicate, by classifying one of its borderline cases as a positive or negative case. For example, we might partially precisify 'tall' by deeming that Bob—a borderline case—is to count as tall. We can understand what is going on here as a change in the intended model: the intended model $\mathfrak{M}$ of the discourse at some time $t$ is one in which Bob is sent neither to 1 nor to 0 by the extension of 'tall'; the intended model $\mathfrak{M}'$ of the discourse at some slightly later time $t'$—where this extended discourse now includes the act of stipulating that Bob is tall—is one in which Bob is sent to 1 by the extension of 'tall'. According to contextualists, this will typically not be the only difference between $\mathfrak{M}$ and $\mathfrak{M}'$: other persons who are similar in height to Bob and who were also borderline cases of 'tall' in $\mathfrak{M}$ will be positive cases in $\mathfrak{M}'$. Exactly what prompts the change from $\mathfrak{M}$ to $\mathfrak{M}'$ and exactly how $\mathfrak{M}'$ differs from $\mathfrak{M}$ are matters over which different contextualists differ—but one prominent way to change the intended model is to stipulate that a (hitherto) borderline case is $P$ or that it is not $P$. Note that even after such an act of precisification the intended model will still be tripartite: it will generally not be a classical model of the sort that supervaluationists take as extensions of their tripartite base models. Note also that because truth simpliciter is truth on the intended model, 'Bob is tall' will be neither true nor false as uttered at $t$ and true as uttered at $t'$: at $t'$, this sentence is still neither true nor false on model $\mathfrak{M}$ (the model itself does not change) but model $\mathfrak{M}$ is no longer the intended model of the discourse; $\mathfrak{M}'$ is now the intended model—the one that is relevant to questions of the (actual) meaning and truth (simpliciter) of utterances in the discourse.

Contextualists think that this combination of a tripartite approach within each model with a dynamic story about how acts of stipulation change which model is intended yields a satisfying theory of vagueness. In particular, the solution to the sorites will be along the following lines. The second premise fails to be true on the intended model (at any stage of the discourse). We nevertheless find it plausible because first, it is not false, and second, if we suppose that some object $x$ in the sorites series is $P$, we thereby affect which model is intended in such a way that 'it is not the case that ($x$ is $P$ and $x'$ is not $P$)'

and ‘if $x$ is $P$ then $x'$ is $P$’ become true (even if they were not so beforehand). Again, however, note that this solution to the sorites turns on speakers not believing that the contextualist story is the correct account of vague language [101]. A speaker who believed that classifying a borderline case $x$ as $P$ or as not $P$ changes the intended model in such a way as to render these classifications—and analogous statements about objects similar to $x$—true would still have no reason to think that the second premise of the sorites argument was true (relative to any model that might be the intended one at any point in time).

Contextualism gives an alethic account of borderline cases. As for blurry boundaries, the contextualist will say that the kind of blurriness involved is a matter of rapid shifting. At any point in time, the extension of a vague predicate is a sharply tripartite three-valued or partial two-valued set—but *which* such set is the correct extension tends to shift as a conversation proceeds. In particular, attempts to locate the boundary between the $P$'s and the borderline cases—by saying that $x$ is $P$ and $x'$ is not—cause the boundary to move elsewhere: the act of deeming $x$ to be $P$ triggers a shift to a model in which $x$ *and* things sufficiently similar to it—which includes $x'$—are in the extension of $P$.

# 3 Fuzzy theories of vagueness

We noted earlier that a system of MTS has two components: the models themselves (of some particular sort) and some notion like that of the ‘intended model’ that plays the role of distinguishing one (or some) of the infinity of models as the one(s) relevant to questions of the (actual) meaning and truth (simpliciter) of utterances in some discourse. We'll touch on the second component in Section 3.3 and then come back to it in more detail later. For now, let's look at the models. In fuzzy theories of vagueness, the kind of models involved will be fuzzy models. Here—as in any kind of model—we can distinguish two components: a part that deals with the nonlogical vocabulary and a part that treats the logical vocabulary. We shall discuss these in turn—under the headings ‘degrees of truth’ (Section 3.1) and ‘truth rules’ (Section 3.2) respectively. For each component, there are many choices available: so there are many kinds of fuzzy model.

## 3.1 Degrees of truth

It will be helpful to start by considering classical models—say of a standard first order language, which has the usual logical vocabulary (connectives, quantifiers, variables) and the following nonlogical vocabulary:

- Individual constants: $a, b, c, \ldots$
- $n$-place predicates (for each $n > 0$): $P^n, Q^n, R^n, \ldots$

A classical model has some machinery dealing with the nonlogical vocabulary. Each primitive nonlogical symbol is assigned a value: for an individual constant $a$ its value $[a]$ is an object in the domain of the model; for an $n$-place predicate $P$ its value $[P]$ is a function from the set of $n$-tuples of members of the domain to the set $\{0, 1\}$ of classical truth values (such a function can be thought of as a *set* of $n$-tuples: the $n$-tuples sent to 1 are in the set and the $n$-tuples sent to 0 are not in the set). This machinery suffices to determine a truth value for each closed wff that involves no logical vocabulary, via the following principle:

$$[P^n a_1 \dots a_n] = [P^n]([a_1], \dots, [a_n]).$$

That is, the truth value of $P^n a_1 \dots a_n$ is the value (0 or 1) to which the value of the predicate sends the $n$-tuple containing the values of the names (in the order in which they appear after the predicate).

There is more to classical models: there is the part that determines the truth value on a model of a closed formula that contains logical vocabulary. But we shall return to this below. Already, at the stage of treating the nonlogical vocabulary, fuzzy models will differ from classical models. Specifically, the guiding idea of fuzzy models is that representing the extension of a predicate as a crisp set—a function from objects to $\{0, 1\}$ (where objects sent to 1 are in the set and objects sent to 0 are out of the set)—is inadequate for vague predicates: because vague predicates have borderline cases and blurry boundaries. The thought is to replace the set $\{0, 1\}$ of classical truth values with a set $D$ of *degrees of truth* so that the extension of a vague predicate can be represented as a function from objects to $D$ in a way that does justice to its vagueness: to its blurriness and to its possession of borderline cases. In particular, this will mean that $D$ contains an element corresponding to the classical value 1—representing complete truth—and an element corresponding to the classical value 0—representing complete falsity—*and* many other elements in between. The gradual fading off of possession of the property $P$ as we move along the sorites series is then modelled by the mapping of successive elements in the series to values in $D$ that get gradually further from 1 and closer to 0. The elements of $D$ (like the classical values 0 and 1) thus play two roles: they serve as degrees of property possession and degrees of truth. The value to which the extension of $P$ sends the object $x$ represents the degree to which $x$ possesses the property picked out by $P$. This will then also be the degree of truth of the sentence $Pa$, when $x$ is the referent of $a$. So the degree of truth of $Pa$ is the degree to which the referent of $a$ possesses the property picked out by $P$. This is exactly the kind of relationship we saw in the classical case: the truth value of $Pa$ is the membership value of the referent of $a$ in the extension of $P$. The only difference is that now these values do not have to be full-on (1) or full-off (0): $D$ contains a host of values in between 1 and 0; this models the idea that objects can possess properties to intermediate degrees and (correspondingly) sentences can be true to intermediate degrees.

So far we have spoken in a general way about the set $D$ of degrees of truth. When it comes time to be more specific about the exact content and structure of $D$, there are many options. Facts about the structure of $D$ will show up in relationships amongst sentences. For example, if $D$ is equipped only with an ordering and not with anything like a metric structure or a notion of distance between values, then we will be able to say things like '$\alpha$ is truer than $\beta$' but we will not be able to say things like '$\alpha$ is just a little bit truer than $\beta$, while $\beta$ is a lot more true than $\gamma$' or '$\alpha$ and $\beta$ are very close in truth value'. For another example, if $D$ is equipped with a linear ordering then for any two sentences $\alpha$ and $\beta$, either one will be truer than the other or they will have exactly the same degree of truth—while if $D$ is only partially ordered, then it might be that $\alpha$ and $\beta$ have distinct degrees of truth *without* either of them being truer than the other. So, views about the appropriate structure of the set $D$ of degrees of truth will be influenced by intuitions about such matters. Different options have been explored in the literature. Here it is useful to distinguish between a more concrete kind of approach and a more abstract kind of approach.

On the concrete sort of approach we pick some particular antecedently known structure as $D$. The most common option here is to let $D$ be $[0, 1]$, comprising all the real numbers between 0 and 1 inclusive (together with all the usual structure: e.g. arithmetical operations such as multiplication, notions of distance between elements, and so on). On this approach, each primitive nonlogical symbol is assigned a value as follows: for an individual constant $a$ its value $[a]$ is an object in the domain of the model (as in the classical view); for an $n$-place predicate $P$ its value $[P]$ is a function from the set of $n$-tuples of members of the domain to the set $[0, 1]$ of degrees of truth (as in the classical view—except that the set $\{0, 1\}$ of classical truth values has been replaced by the set $[0, 1]$ of degrees of truth). As before, such a function can be thought of as a *set* of $n$-tuples: but this time it is not a crisp set (where every object is in—sent to 1—or out—sent to 0) but a *fuzzy set*: the number $x \in [0, 1]$ to which an $n$-tuple is sent represents the *degree* of membership of that $n$-tuple in the set. As before, this machinery suffices to determine a degree of truth for each closed wff that involves no logical vocabulary, via the following principle:

$$[P^n a_1 \dots a_n] = [P^n]([a_1], \dots, [a_n]).$$

That is, the degree of truth of $P^n a_1 \dots a_n$ is the value to which the value of the predicate sends the $n$-tuple containing the values of the names (in the order in which they appear after the predicate). The statement of the principle is exactly the same as in the classical case—but this time, the truth value of $P^n a_1 \dots a_n$ may be any real number between 0 and 1 inclusive.

On the abstract sort of approach we do not take a particular set of objects (together with a known structure thereon) as $D$—we just impose some structural properties and suppose that $\boldsymbol{D}$ is some structure satisfying these properties. For example, we might specify that $\boldsymbol{D} = \langle D, \vee, \wedge, \&, \rightarrow, 0, 1\rangle$ is to be a *residuated lattice*,[15] that is:

1. $\langle D, \vee, \wedge, 0, 1\rangle$ is a lattice with least element 0 and greatest element 1.

2. $\langle D, \&, 1\rangle$ is a commutative monoid (i.e. $\&$ is associative and commutative and for all $x \in D$, $x \,\&\, 1 = x$).

3. $\rightarrow$ is the residuum of $\&$ (i.e. for all $x, y, z \in D$, $x \,\&\, y \leq z$ iff $x \leq y \rightarrow z$).

Or we might specify that $\boldsymbol{D}$ is to be an MTL-*algebra* (a residuated lattice satisfying also the condition of *prelinearity*: for all $x, y \in D$, $(x \rightarrow y) \vee (y \rightarrow x) = 1$), or a BL-*algebra* (an MTL-algebra satisfying also the condition of *divisibility*: for all $x, y \in D$, $x \,\&\, (x \rightarrow y) = x \wedge y$) or a BL-*chain* (a BL-algebra where the underlying lattice is linearly ordered). There are also other options besides these ones that are studied in the literature (e.g. MV-algebras, product algebras, Gödel algebras...; see e.g. [10] and elsewhere in this

[15] Properly, a *bounded integral commutative* residuated lattice or $\mathrm{FL}_{\mathrm{ew}}$-algebra; see [34]. This simplification of terminology is common in the fuzzy logic literature; see [10, p. 22].

Handbook). Note that on this abstract kind of approach we don't care what particular objects are in $D$: we just care about the way that they are structured. Nevertheless, it would be odd to call the elements of $D$ *degrees of truth* if there were (say) only two of them: 0 and 1. Yet the Boolean algebra of classical truth values can be taken as an example of many of these abstract structures. Hence when considering vagueness one tends to have in mind particular examples of these structures—for example, $[0, 1]$.

### 3.2 Truth rules

Let's return to classical models for our first order language. We now have enough machinery to determine a truth value for each closed wff that involves no logical vocabulary—but in order to determine a truth value for each closed wff that involves logical vocabulary we need further machinery: one piece per logical operator (connective or quantifier). For each $n$-place connective $\triangledown$ we specify a corresponding operation $\blacktriangledown$ (of arity $n$) on the classical truth values and then stipulate that

$$[\triangledown(\alpha_1, \ldots, \alpha_n)] = \blacktriangledown([\alpha_1], \ldots, [\alpha_n]).$$

For example, corresponding to the negation connective $\neg$ we have the unary operation $x - 1$ on $\{0, 1\}$ and corresponding to the conjunction and disjunction connectives $\wedge$ and $\vee$ we have the binary operations $\min\{x, y\}$ and $\max\{x, y\}$ on $\{0, 1\}$; thus $[\neg\alpha] = 1 - [\alpha]$, $[\alpha \wedge \beta] = \min\{[\alpha], [\beta]\}$ and $[\alpha \vee \beta] = \max\{[\alpha], [\beta]\}$. For the universal (existential) quantifier, the truth value of the closed formula $\forall x\alpha$ ($\exists x\alpha$) is determined by considering all the truth values that one gets by taking the free variable $x$ in $\alpha$ to denote some object in the domain—one truth value per object in the domain—and then applying an infinitary analogue of the conjunction (disjunction) operation to these truth values.

In the fuzzy case, one option is to proceed in an analogous way: to specify an operation on the set $D$ of degrees of truth corresponding to each logical operator. How specifically one does this will depend on the structure imposed on $D$ at the first stage of specifying fuzzy models (i.e. the stage of specifying the degrees of truth, examined in Section 3.1). But the general idea is the same in all cases: one uses the structure in the set of truth degrees to define logical operators.

In the case where one has taken $[0, 1]$ as the set of truth degrees—together with (as we mentioned) all its usual structure (e.g. arithmetical operations such as multiplication, notions of distance between elements, and so on)—there is plenty to work with. For universal and existential quantification, it is standard to define these using the infimum and supremum operations respectively (which are indeed infinitary analogues of one kind of conjunction and disjunction: the min conjunction and max disjunction, which we shall encounter shortly). For the connectives of propositional logic, on the other hand, there are multiple live options; let me just mention a few which have played a prominent role in theories of vagueness. First consider Zadeh logic (Figure 3). Negation, conjunction and disjunction are defined precisely as above in classical logic—although this time the operations on truth values take all reals in $[0, 1]$ as inputs and outputs, not just 1 and 0. The conditional is defined in terms of negation and conjunction—or equivalently negation and disjunction—in precisely the way that is familiar from classical logic. Likewise, the biconditional is defined in terms of conditional and conjunction in the familiar classical way.

$$\begin{array}{rcl}[\neg\alpha] & = & 1-[\alpha] \\ [\alpha\wedge\beta] & = & \min\{[\alpha],[\beta]\} \\ [\alpha\vee\beta] & = & \max\{[\alpha],[\beta]\} \\ [\alpha\rightarrow\beta] & = & [\neg\alpha\vee\beta] \quad = \quad [\neg(\alpha\wedge\neg\beta)] \\ [\alpha\leftrightarrow\beta] & = & [(\alpha\rightarrow\beta)\wedge(\beta\rightarrow\alpha)]\end{array}$$

Figure 3. Zadeh logic

$$\begin{array}{rcl}[\neg\alpha] & = & 1-[\alpha] \\ [\alpha\wedge\beta] & = & \min\{[\alpha],[\beta]\} \\ [\alpha\vee\beta] & = & \max\{[\alpha],[\beta]\} \\ [\alpha\rightarrow\beta] & = & \begin{cases}1 & \text{if } [\alpha]\leq[\beta] \\ 1-[\alpha]+[\beta] & \text{if } [\alpha]>[\beta]\end{cases} \\ [\alpha\leftrightarrow\beta] & = & [(\alpha\rightarrow\beta)\wedge(\beta\rightarrow\alpha)]\end{array}$$

Figure 4. Philosophers' Fuzzy Logic

Second, consider what we may call Philosophers' Fuzzy Logic or PFL (Figure 4).[16] Negation, conjunction and disjunction are the same as in Zadeh logic. The definition of the biconditional looks the same as in Zadeh logic but note that the result is different, because the conditional featuring in the definition is different. As for the conditional, the idea is this: if the antecedent isn't truer than the consequent, then the conditional is true to degree 1; but if the antecedent is truer than the consequent, then whatever the difference between them, the conditional falls precisely *that* far short of complete truth.

Third, consider t-norm fuzzy logics. A t-norm is a binary function $\wedge$ on $[0,1]$ satisfying the conditions:

$$\begin{array}{rcl}x\wedge y & = & y\wedge x \\ (x\wedge y)\wedge z & = & x\wedge(y\wedge z) \\ x_1\leq x_2 & \Rightarrow & x_1\wedge y\leq x_2\wedge y \\ y_1\leq y_2 & \Rightarrow & x\wedge y_1\leq x\wedge y_2 \\ 1\wedge x & = & x \\ 0\wedge x & = & 0.\end{array}$$

A t-norm logic is specified by picking a t-norm and taking it to be the conjunction operation, and then defining the other operations (conditional, negation and so on) in certain specific ways. Notably, the conditional is taken to be the residuum of the t-norm (the residuum exists iff the t-norm is left-continuous) and the negation the precomplement of the conditional:

$$x\rightarrow y=\max\{z : x\wedge z\leq y\} \qquad\qquad \neg x = x\rightarrow 0.$$

[16] The reason for the name is that many philosophers write as if 'fuzzy logic' just *is* PFL.

| | Łukasiewicz logic | Gödel logic | Product logic |
|---|---|---|---|
| $x \wedge y$ | $= \max(0, x+y-1)$ | $= \min(x, y)$ | $= x \cdot y$ |
| $x \to y$ | $= \begin{cases} 1 & \text{if } x \leq y \\ 1-x+y & \text{if } x > y \end{cases}$ | $= \begin{cases} 1 & \text{if } x \leq y \\ y & \text{if } x > y \end{cases}$ | $= \begin{cases} 1 & \text{if } x \leq y \\ y/x & \text{if } x > y \end{cases}$ |
| $\neg x$ | $= 1 - x$ | $= \begin{cases} 1 & \text{if } x = 0 \\ 0 & \text{otherwise} \end{cases}$ | $= \begin{cases} 1 & \text{if } x = 0 \\ 0 & \text{otherwise} \end{cases}$ |

Figure 5. Three prominent t-norm logics

Figure 5 shows the conjunctions, conditionals and negations in three prominent t-norm logics.

It is common in these logics to define a second, 'weak' (or 'lattice') conjunction (with the t-norm conjunction then termed 'strong'). In all these logics, the weak conjunction is the same as the min operation used to define conjunction in Zadeh logic.[17]

Instead of taking an antecedently known structure such as $[0, 1]$ as degrees of truth, one might have specified a structure of degrees of truth in an abstract way—for example, as $\langle D, \vee, \wedge, \&, \to, 0, 1\rangle$ with certain constraints imposed on $\vee, \wedge, \&$ and so on. In this case the obvious way to proceed will be to associate a connective in the logical language with an operator of the same arity in the algebra of truth degrees: either one directly given in the definition of the algebra, or one defined in terms of those. So, in the examples discussed above, one can again get *two* conjunctions: one defined as $\wedge$ and one as $\&$.

I said near the beginning of this section that when it comes to specifying truth values for closed wffs that involve logical vocabulary, *one* approach is to proceed in an analogous way to the classical approach: to specify an operation on the set of degrees of truth corresponding to each logical operator. We have just explored some particular options within this approach. Another kind of approach is also possible: one where we assign truth values in a non-degree-functional way. For example, given a fuzzy valuation $V$—an assignment of a referent to each name and a fuzzy set of $n$-tuples to each $n$-place predicate—we can define the notion of a classical extension of this valuation in an obvious way: it is a classical model $\mathfrak{M}$ that assigns the same referents to names as $V$, and assigns extensions to predicates in such a way that whenever the value of $P$ on $V$ sends an $n$-tuple to 1 (or to 0), its value on $\mathfrak{M}$ also sends that $n$-tuple to 1 (or, respectively, to 0). We can then assign closed formulas degrees of truth in—for example—the supervaluationist way, the subvaluationist way or the degree supervaluationist way.

[17] So in Gödel logic, there is no difference between the strong and weak conjunction. Note that the min conjunction of Figure 4 is definable using the operations for Łukasiewicz logic of Figure 5 as $\alpha \wedge (\alpha \to \beta)$ *and* the Łukasiewicz t-norm conjunction of Figure 5 is definable using the negation and conditional of Figure 4 (which are the same as the negation and conditional for Łukasiewicz logic of Figure 5) as $\neg(\alpha \to \neg\beta)$. So 'PFL' and 'Łukasiewicz logic' pick out different perspectives (rather than different logics): from the PFL perspective, the Łukasiewicz t-norm conjunction is ignored (it is not even defined, let alone put to any use: only min conjunction is considered); from the Łukasiewicz logic perspective, the Łukasiewicz t-norm conjunction is of central importance (although not necessarily to the *exclusion* of min conjunction, which may *also* be considered).

### 3.3 The intended model

I have distinguished two parts to fuzzy MTS: the part that tells us about fuzzy models and the part that tells us which models are relevant to questions of the (actual) meaning and truth (simpliciter) of utterances in a given vague discourse. Within the first part, I have again distinguished two aspects: the 'degrees of truth' part—discussed in Section 3.1—which provides the machinery for modelling degrees of property possession and for assigning degrees of truth to closed wffs involving no logical vocabulary; and the 'truth rules' part—discussed in Section 3.2—which provides the machinery for assigning degrees of truth to closed wffs involving logical vocabulary.

What about the second part of fuzzy MTS—the part that introduces some notion like that of the 'intended model' that plays the role of distinguishing one (or some) of the infinity of models as the one(s) relevant to questions of the (actual) meaning and truth (simpliciter) of utterances in some discourse? For the moment we shall work with the simplest possible assumption: each vague discourse is associated with a unique intended fuzzy model. Later we shall return to this issue and find reason to explore a plurivaluationist alternative.

## 4 Arguments for

In this section we look at some of the major considerations in favour of adopting a theory of vagueness based on fuzzy logic.[18]

### 4.1 The nature of vagueness

All the theories of vagueness examined in Section 2 revolve in one way or another around the idea of sharpening the boundaries of vague predicates: either the models involved are already classical (in the case of epistemicism and plurivaluationism) or else the theory gives a central place to the idea of removing borderline cases (either completely, in the case of supervaluationism, or partially, in the case of contextualism). Now of course *one* thing we can do with vague predicates is sharpen them. But we also employ them in their natural, unsharpened state—blurry boundaries and all. We do so very successfully all the time. One motivation for fuzzy theories is that they offer an account which does justice to vagueness *as it is*—without one eye (or both) on how it could be *removed* (or reduced). Of course at this stage this does not amount to an argument for fuzzy theories. It does, however, capture part of the motivation of many who are attracted to fuzzy theories—and furthermore something like this line of thought can be turned into a real argument to the conclusion that only fuzzy theories can do justice to the phenomenon of vagueness. We examine this argument now.

In Section 1.1 above we introduced vague predicates via three characteristics: blurry boundaries, borderline cases and sorites susceptibility. This can be compared to explaining what water is by saying it's a clear potable liquid that falls as rain and boils at 100°C. This will help someone who doesn't know what water is to identify samples of it—but it still leaves open the question of the underlying nature or essence of water: of what water fundamentally *is*, that explains why it has these characteristics. The same goes for the

[18] More detailed versions of the arguments in Section 4.1 and Section 4.2.1 can be found in [106].

case of vagueness. It would be desirable to know the underlying essence of vagueness: to understand its fundamental nature and explain why it has these surface characteristics (blurry boundaries, borderline cases and sorites susceptibility).

Let's briefly consider some proposals for a fundamental definition of vagueness that do not work—before getting to a proposal that does work. For a start, one might think that perhaps one of the three characteristics of vague predicates is not merely a surface feature but is in fact the underlying essence of vagueness. However this thought does not pan out. Having borderline cases cannot be fundamental to vagueness, because a predicate could have *sharply delineated* borderline cases—borderline cases without blurry boundaries—and then it would not be vague. Generating sorites paradoxes cannot itself be what vagueness fundamentally consists in: for surely there is some feature that vague predicates possess and precise ones lack that *explains why* the former generate sorites paradoxes—and it is then this feature (rather than sorites-susceptibility, which this feature explains) that is fundamental to vagueness. Nor will it do to say that having blurry boundaries is the essence of vagueness—in the absence of a further explanation of what we mean by 'blurry boundaries'. Do we mean the purely epistemic sort of blur that the epistemicist countenances, or the probabilistic kind of blur that the degree supervaluationist countenances—or something else?

A different proposal that has been mentioned in the literature is that vagueness is semantic indeterminacy of the sort involved in plurivaluationism. However, such indeterminacy cannot be fundamental to vague predicates, because we can easily imagine predicates that exhibit such indeterminacy but are not vague—for example 'gavagai' or 'mass'. If Quine [88, ch. 2] and Field [29] are right, then these predicates exhibit semantic indeterminacy—but they do not generate sorites paradoxes, nor do their extensions have blurred boundaries: hence they are not vague.

A proposal that has received considerable discussion in the literature is that vague predicates are *tolerant*. There are various ways of formulating tolerance but for our purposes the following will be most useful. Recall that in a sorites series for the predicate $P$, each object is extremely similar to the object after it in all respects relevant to the application of $P$ (for short, in '$P$-relevant respects'). A predicate $P$ is tolerant iff it satisfies the following principle (for any objects $a$ and $b$):

**Tolerance** If $a$ and $b$ are very similar in $P$-relevant respects then $Pa$ and $Pb$ are identical in respect of truth.

(Two sentences are identical in respect of truth if they have the same truth value, or both lack a truth value, or in general have exactly the same truth status—where the possible truth statuses of a sentence will depend on the system of MTS in question.) The problem with Tolerance as a definition of vagueness is that it leads to contradiction. Suppose that we have a sorites series for $P$—so '$x$ is $P$' is true when $x$ is the first object in the series and false when $x$ is the last object in the series, and adjacent pairs of objects in the series are very similar in $P$-relevant respects. Then, given Tolerance, for every object $x$ in the series, the claim '$x$ is $P$' is both true and false.

This brings us to a proposal which, I shall argue, does work: that vagueness is fundamentally Closeness. That is, a predicate $P$ is vague iff it satisfies the following principle (for any objects $a$ and $b$):

**Closeness** If $a$ and $b$ are very similar in $P$-relevant respects then $Pa$ and $Pb$ are very similar in respect of truth.

This yields what we wanted in a fundamental definition of vagueness: an explanation of why vague predicates have the three characteristics mentioned at the outset. We'll return to sorites susceptibility in Section 4.2.1; here let's consider the other two characteristics.

Blurry boundaries: Suppose that we have a range of (possible) objects, some of which are $F$ and some of which are not, and where we can get from any object to any other by passing between objects that are very similar in $F$-relevant respects. (For example, consider a sorites series for $F$.) If $F$ conforms to Closeness, then its extension cannot consist in a sharp line between the $F$'s and the non-$F$'s: for then we would have two objects $a$ and $b$ which are very similar in $F$-relevant respects, with $a$ on one side of the line and $b$ on the other, so that $Fa$ is true and $Fb$ false—and this would violate Closeness. Rather, $F$-ness must gradually fade away as one travels further from the definite $F$ objects. To take a concrete example, consider the term 'red', and suppose that it conforms to Closeness. This term does not cut a sharp band out of the rainbow: as one moves across the points of the rainbow, small steps in red-relevant respects—which in this case correspond to small steps in space—can never, given Closeness, make for big changes in the truth of the claim that the point one is considering is red. By small steps one can move from full-fledged red points to full-fledged non-red points: but there is no sharp boundary between them that can be crossed in one small step. Thus Closeness yields an explanation of the blurred boundaries phenomenon.

Borderline cases: If a predicate satisfies Closeness then it admits of borderline cases. Consider a predicate $F$ that conforms to Closeness, and a sorites series $x_1, \ldots, x_n$ for $F$. $Fx_1$ is true and $Fx_n$ is false; but given Closeness, it cannot be that there is an $i$ such that $Fx_i$ is true and $Fx_{i+1}$ is false. There must then be sentences $Fx_i$ which are neither true nor false—and the corresponding objects $x_i$ are borderline cases for $F$.

Thus, if we define vagueness in terms of Closeness, we can explain why vague predicates have blurry boundaries and admit of borderline cases—and also why they generate sorites paradoxes (although we shall not discuss that until Section 4.2.1). Other advantages of the Closeness definition are that it accommodates the intuitions that have been taken to support the idea that vague predicates satisfy Tolerance, without generating contradictions; and it accommodates, within the definition of vagueness itself, the intuitions that lead those who wish to define vagueness in terms of possession of borderline cases to posit an additional phenomenon—'higher-order vagueness'—over and above mere vagueness.[19] So there are good reasons to think that conforming to Closeness is the essence of vagueness. But if that is so, then it leads to an argument in favour of fuzzy theories of vagueness. In brief, the argument is that no other kind of theory can allow for the existence of vague predicates.

First consider the epistemicist view. It will be helpful to consider a concrete example. Suppose we have a strip of paper that is red at the left end and orange at the right end, and in between gradually changes colour from red to orange. Now consider the sentence 'Point $x$ is red', for each point $x$ on the strip. According to the epistemicist,

[19] For details see [105] and [106, ch. 3].

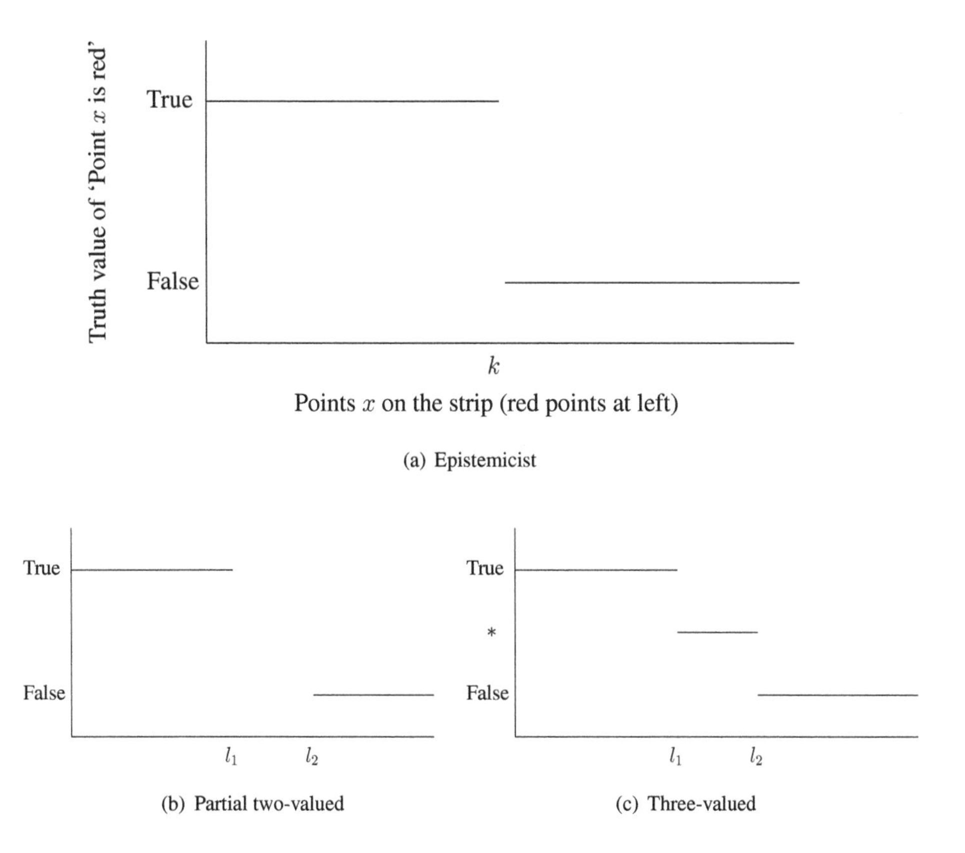


(a) Epistemicist

(b) Partial two-valued

(c) Three-valued

Figure 6. The strip according to epistemicist, partial two-valued and three-valued views

every one of these sentences is either true or false. Figure 6(a) represents the situation according to the epistemicist. This flouts Closeness: there are points $k$ (shown on the diagram) and $k + \Delta$ (a point to the right of, and arbitrarily close to, $k$) whose colours are *very* close together, while the sentence 'Point $k$ is red' is True and the sentence 'Point $k + \Delta$ is red' is False. This sort of flouting of Closeness is unavoidable for the epistemic theory. The epistemicist, then, cannot allow the existence of predicates $F$ that both conform to Closeness and have associated sorites series. Thus—given that vagueness is Closeness—he cannot allow for the existence of vague predicates (that have associated sorites series).

Going plurivaluationist does not solve the problem. Any acceptable model must make $F$ true of the things at the beginning of the sorites series for $F$ and false of the things at the end. But (on the plurivaluationist view) every acceptable model is classical. Thus, on *every* acceptable model, $F$ will violate Closeness (for the reasons just seen). But if Closeness is violated on every acceptable model, then—on the plurivaluationist picture—it is violated everywhere. There are only the acceptable models, and so there simply is nowhere else for Closeness to be accommodated.

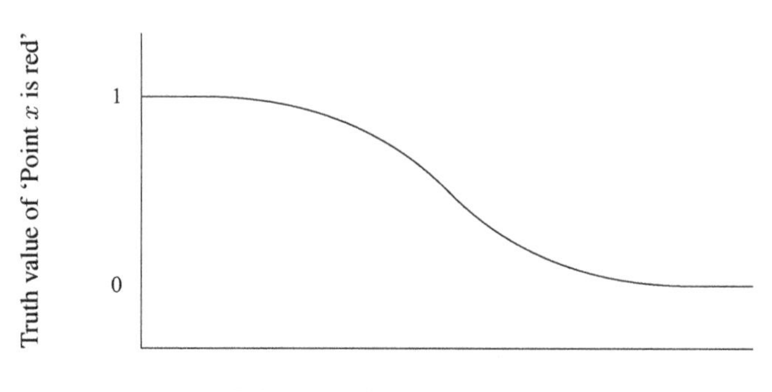


Figure 7. The strip according to the fuzzy view

Tripartite views fare no better. Consider again our strip of paper. Figure 6(b) represents the situation according to the truth gap view and Figure 6(c) represents the situation according to the third value view. According to these views, there are points $l_1$ and $l_1 + \Delta$ whose colours are very close together, while the sentence 'Point $l_1$ is red' is True and the sentence 'Point $l_1 + \Delta$ is red' has the third status (i.e. a third value, or a gap); and there are points $l_2$ and $l_2 + \Delta$ whose colours are very close together, while the sentence 'Point $l_2$ is red' has the third status and the sentence 'Point $l_2 + \Delta$ is red' is False. But if one sentence is True and another False, then they are very far apart in respect of truth—and so no third truth status can be *very close* to *both* of them. Thus, to the extent that a sentence is very close to True, it is *not* very close to False, and vice versa. Hence Closeness must be violated by at least one of the pairs of points $l_1/l_1 + \Delta$ or $l_2/l_2 + \Delta$—and given the natural assumption that the third status is symmetric with respect to truth and falsity, Closeness will be violated at both places.

This problem arises for all tripartite views: going supervaluationist (rather than recursive) makes no difference. Even if 'Point $l_1$ is red' is classically true/false on almost exactly the same admissible extensions as 'Point $l_1 + \Delta$ is red', at best this makes the two sentences very similar in the respects that determine truth. If one sentence is True and the other has the third status, then they are not very similar in respect of truth.[20]

If Closeness gives the essence of vagueness, then to allow that there are vague predicates (with associated sorites series) we need to be able to accommodate the idea of a predicate $F$ where $Fx$ is true when $x$ is the first object in the series and false when $x$ is the last object in the series—*and* $Fx$ and $Fx'$ are always very similar in respect of truth, when $x$ and $x'$ are adjacent objects in the series. This requires *degrees of truth*. For example, consider a view on which the set of truth values is $[0, 1]$. In this case we can easily accommodate Closeness; for example, we could handle the strip of paper as in Figure 7. Here there are no points $x$ and $x'$ that are very close together on the strip but where $Fx$ and $Fx'$ are not very similar in respect of truth.

[20] Or more precisely: the third status cannot be very similar in respect of truth *both* to True and to False.

Taking $[0, 1]$ as the set of truth degrees would, then, be one way to accommodate Closeness. It is not the only way—and furthermore, not *any* set of degrees of truth will suffice. For example, if the only structure on the degrees of truth was an ordering—if there was no meaningful way to say that two truth values were *very close* or that the *difference* between them was *very small* or something else along these lines—then (even if there were infinitely many of these degrees of truth) we would not be able to accommodate Closeness.

A natural response on the part of non-fuzzy theorists to the foregoing argument for degrees of truth from the Closeness definition of vagueness is to say that the definition is subtly wrong. Suppose that we defined vagueness not in terms of Closeness—which features the notion of similarity in respect of truth—but in terms of a variant principle in which truth is replaced by, say, warranted assertability. In that case we could not conclude that degrees of truth are required if vague predicates are to be allowed. We shall return to this line of thought in Section 4.2.1. In the meantime, to sum up the present section: the first big advantage of fuzzy theories of vagueness (at least fuzzy theories that take $[0, 1]$ as the set of degrees of truth—or some other structure that has the right properties to accommodate predicates that satisfy Closeness and have associated sorites series) is that they are motivated by the best account we have of what vagueness fundamentally consists in—viz., conformity to Closeness.

## 4.2 Solving the sorites

The second big advantage of fuzzy theories is that they alone are in a position to solve the sorites paradox without departing from the basic modus operandi of formal semantics, which assumes that speakers' linguistic behaviour flows from their semantic competence (modulated by performance factors) [101]. It is not assumed that speakers have a full understanding of the semantics of their language. However, it is assumed that speakers have some sort of implicit grasp of this semantics—and that this grasp manifests itself in their behaviour. That is why speakers' linguistic behaviour is evidence for formal semantic theories. It goes entirely against the grain of this kind of approach to posit a semantic theory $T$ for a language which is such that if speakers thought that $T$ was the correct semantics for their language, then they would use their language quite differently from the way they actually use it. This is, as we saw, the kind of position that non-fuzzy theories of vagueness find themselves in when it comes to the sorites paradox. Ordinary speakers respond to sorites paradoxes in the following sort of way: they find the first premise undeniable; they are strongly inclined to accept the second premise; they find no fault in the reasoning leading from the premisses to the conclusion; and yet they find the conclusion unacceptable. However, a speaker who thought that epistemicism (for example) gives the correct semantics of the predicate $P$ would have no inclination at all to think that the second premise was true (and—as we have seen—similar remarks apply to other non-fuzzy theories of vagueness). Fuzzy theories, on the other hand, can offer a semantic theory $T$ for vague language which does have the following desired feature: if speakers thought that theory $T$ gave the correct semantics for a vague predicate $P$, they would respond to a sorites argument for $P$ in just the way they *do* ordinarily respond to sorites paradoxes.

There are two different kinds of fuzzy approach to the sorites in the literature. One solves the paradox using only resources already available at the 'degrees of truth' stage of formulating a fuzzy theory; the other posits specific resources at the 'truth rules' stage and uses those to solve the paradox. Let's examine the two approaches in turn.[21]

### 4.2.1 Sorites via degrees of truth

We said in Section 4.1 that if we define vagueness as Closeness then we can explain why vague predicates generate sorites paradoxes. Let's now see why this is so. Distinguish two readings of the second premise in a sorites argument (i.e. two different claims that one might understand the second premise to be making): the Closeness reading and the Tolerance reading. On the Closeness reading, the second premise expresses (a particular consequence of) the claim that $F$ conforms to Closeness: $x$ and $x'$ are very similar in $F$-relevant respects, so if $x$ is $F$ then for all practical purposes we can just say that $x'$ is $F$ too. On the Tolerance reading, the second premise expresses (a particular consequence of) the claim that $F$ conforms to Tolerance: $x$ and $x'$ are very similar in $F$-relevant respects, so if $x$ is $F$ then—not just for all practical purposes but without qualification—$x'$ is $F$ too (the two claims $Fx$ and $Fx'$ are *exactly the same* in respect of truth). Now note the following:

1. On the Closeness reading, the conclusion does not follow from the premises.
2. If one thinks that $F$ conforms to Closeness (and not to Tolerance), one will nevertheless tend to go along with claims that $F$ is Tolerant in normal situations.
3. On the Tolerance reading, the conclusion does follow from the premises.

Claim 3 is obvious. The basic idea behind claim 1 is this: each successive statement $Fx$ must be *very similar* in respect of truth to the one before, but need not be *exactly the same* in respect of truth—and so by the end, the final statement $Fx_n$ may be simply false. In fact one can formalise the sorites argument (under a Closeness reading) and show that it is invalid.[22] This leaves Claim 2. Suppose we accept that a predicate $F$ satisfies Closeness, but not Tolerance. The very fact that we accept Closeness will mean that in many ordinary circumstances, we act as if we believe Tolerance. For Closeness tells us that a negligible or insignificant difference between $a$ and $b$ in $F$-relevant respects makes for at most a negligible or insignificant difference between $Fa$ and $Fb$ in respect of truth—and a negligible or insignificant difference is one that we are entitled to ignore for practical purposes. So for practical purposes, when there is a negligible or insignificant difference between $a$ and $b$ in $F$-relevant respects, we will simply ignore any difference between $Fa$ and $Fb$ in respect of truth, and so treat them as being identical in respect of truth. This practice is licensed by our acceptance of Closeness. However the practice has its limits: sometimes we find ourselves in a situation in which Tolerance cannot be accepted as a useful approximation of Closeness—and a sorites series is precisely such a context. Consider an analogy. We do not believe that dust particles are weightless: we believe that the weight of a dust particle is negligible. (This is the analogue for dust particles of Closeness without Tolerance.) But this very

[21] Historically, the 'truth rules' kind of approach came first.

[22] This is done in one way in [106, p. 271] and in several different ways in [42].

belief licenses us to act as if we believe that dust particles are weightless: given that the weight of a dust particle is negligible, we do well to ignore it! We do not demand that the delicatessen assistant remove all specks of dust from the scale arms before weighing our smallgoods, and we do not wash and dry our hair (to remove all dust particles) before weighing ourselves. So our belief that the weight of a dust particle is negligible licenses us to accept the claim that dust particles are weightless as a useful approximation of our real belief, in ordinary circumstances. (This is the analogue for dust particles of Tolerance.) Nevertheless, the claim that dust particles are weightless is revealed as a mere approximation to our real belief—something that we act as if we believe, in ordinary circumstances, but not something we actually believe—in certain situations, for example when we are arranging to empty the bag from the dust extraction system at our carpentry shop, which weighs 85kg when full (of nothing but dust particles). The claim that a dust particle weighs nothing is a useful approximation to our true belief, except when we come across many dust particles together, at which time we see clearly that the claim is just an *approximation* to what we really believe, which is that the weight of a dust particle is extremely small. Similarly, the claim that the predicate 'heap' (say) applies equally to two things that differ negligibly in heap-relevant respects is a useful approximation of the real belief of someone who accepts Closeness (but not Tolerance), except when she encounters many pairs of things that differ negligibly in heap-relevant respects put together—a sorites series—at which time it becomes clear that the claim is just an *approximation* of what she really believes, which is that the difference in truth value between '$x$ is a heap' and '$y$ is a heap' is at most very small, when $x$ and $y$ are very similar in heap-relevant respects.

Putting together the ingredients just laid out, we can now see why a predicate $F$ that conforms to Closeness generates a sorites paradox. The sorites argument is compelling because when the second premise is taken to express Tolerance, (a) the argument is valid and (b) the second premise is (initially) acceptable (to one who accepts Closeness but not Tolerance) because accepting Closeness licenses one to accept Tolerance as a useful approximation of what one believes, in ordinary situations. Nevertheless, the argument is ultimately flawed because it leads one to see that in situations involving sorites series one cannot happily use Tolerance as an approximation of Closeness, and must work with Closeness itself—and when the major premise is taken to express Closeness (and not Tolerance), the argument is invalid. Thus, someone who believes that $F$ conforms to Closeness (and not Tolerance) can be expected to react to a sorites argument in precisely the way ordinary speakers do react: he will initially find the premisses acceptable and will agree that the conclusion follows from them—but he will not ultimately go on to accept that the conclusion is actually true.

We have now explained why predicates that satisfy Closeness will generate sorites paradoxes. This adds a piece that was earlier left missing to the argument that Closeness is the correct fundamental definition of vagueness. In light of this, let's turn now to our current task, which is to show that fuzzy theories can solve the sorites paradox in a way that does not require ordinary speakers to lack an understanding of the semantics of their own vague predicates. We saw in Section 4.1 that accommodating vagueness (defined as Closeness) requires positing degrees of truth (with a certain sort of structure). Now any fuzzy theory that employs degrees of truth with the right kind of structure—for example $[0, 1]$—has the resources to distinguish predicates that satisfy Closeness (and not Tolerance) from predicates that satisfy Tolerance. But once we can meaningfully distinguish Closeness and Tolerance, a solution to the sorites paradox is ready to hand. A sorites argument (as presented in English) can be understood in two ways: the second premise might express Closeness or Tolerance. On the Tolerance reading the argument is unsound (vague predicates are not Tolerant: they merely conform to Closeness) and on the Closeness reading it is invalid. This is why the argument is mistaken. However the argument is also compelling (to one who thinks that the predicate involved in the sorites argument conforms to Closeness but not Tolerance)—for the reason seen above. Believing that $F$ conforms to Closeness, one is thereby inclined to accept the second premise on the Tolerance reading (as a useful approximation); and as the argument is valid on that reading, one is led to the conclusion. One does not then simply accept the conclusion, however: the argument is compelling, but it is not ultimately convincing. At that point one realises that one is in a special situation in which Tolerance is not a useful approximation of Closeness—and when one goes back and reads the second premise as expressing Closeness, the conclusion no longer follows.

Before turning to the second approach to solving the sorites within fuzzy theories, we need to tie up one more loose end. As indicated earlier, proponents of non-fuzzy theories of vagueness might try to block the argument for fuzzy theories presented in Section 4.1 by claiming that vagueness should be defined in terms not of Closeness but of some variant principle that differs from Closeness by substituting something such as the notion of *warranted assertability* in place of the notion of *truth*—for example:

**A-Closeness** If $a$ and $b$ are very similar in $P$-relevant respects then $Pa$ and $Pb$ are very similar in respect of warranted assertability.

Note that there is no argument from A-Closeness to degrees of truth: epistemicists, for example, should have no trouble accommodating A-Closeness. But we are now in a position to see why A-Closeness (unlike Closeness) does not provide an adequate definition of vagueness: there is no connection between A-Closeness and the sorites. Suppose that we have a predicate $F$ that conforms to A-Closeness and suppose that we have a sorites series for $F$. Will the corresponding sorites argument be compelling? We have no reason to think so. Given A-Closeness, and the fact that adjacent objects in the sorites series are very close in $F$-relevant respects, we know that for any object in the series, to whatever extent it is justifiable, reasonable or warranted to assert, believe or judge that it is $F$, it is to a very similar extent justifiable, reasonable or warranted to assert, believe or judge that the next object is $F$. But we have no reason at all to conclude from this that the second premise of the sorites argument is true. Whether or not the second premise is compelling depends crucially on *why* we think A-Closeness holds for $F$. If it holds because $F$ satisfies Closeness, then the second premise does become compelling, as we have seen. However, if $F$ satisfies A-Closeness but not Closeness—say because $F$ works in the way the epistemicist thinks vague predicates work (it draws sharp but unknowable boundaries)—then we have no reason at all to accept the second premise, and so the sorites argument will not be compelling in the slightest. This shows that A-Closeness—by itself—yields no account of sorites-susceptibility. If we define vague predicates in terms of A-Closeness, we are left with no understanding of why they generate sorites paradoxes. Hence A-Closeness—unlike Closeness—does not provide an adequate definition of vagueness.

### 4.2.2 Sorites via truth rules

In the previous section we saw that simply positing a structure of degrees of truth (such as $[0, 1]$) that allows for predicates that satisfy Closeness without satisfying Tolerance opens the way to a solution to the sorites paradox. Now we turn to a different strategy for solving the sorites within fuzzy theories: one that turns on truth rules—in particular, on positing particular truth conditions for the connectives used in formulating the sorites argument.

Consider a version of the paradox that concerns a series of piles of sand 1 through 10,000, where pile $i$ has $i$ grains of sand and each pile is of a very similar shape to its neighbour(s). Consider the following sorites argument:

Pile 10,000 is a heap.

If pile 10,000 is a heap then pile 9,999 is a heap.

If pile 9,999 is a heap then pile 9,998 is a heap.

$\vdots$

If pile 2 is a heap then pile 1 is a heap.

$\therefore$ Pile 1 is a heap.

Let's suppose that 'if... then...' here is read as the Łukasiewicz conditional and that we define validity as follows: on every model on which every premise is true to degree 1, the conclusion is true to degree 1. Then we get the following solution to the sorites. The problem with the argument is that, although it is valid, it is unsound (i.e. it is not the case that every premise is true to degree 1). The first premise is true to degree 1. As for the conditionals, at first both antecedent and consequent are true to degree 1, and so are the conditionals. As we move along the series, we get to a point at which the antecedents are ever so slightly more true than the consequents. In this region, the conditionals are true to a degree ever so slightly less than 1. This continues for a while until both antecedent and consequent are true to degree 0, and hence the conditionals are true to degree 1 again. So why is the argument compelling? Because all the premises are *very nearly* true to degree 1. In normal contexts, we are naturally inclined simply to accept something as true when it is very nearly true—this is a useful approximation. Of course, once we see where the argument leads, we may well reconsider.[23]

[23] See [31, pp. 243–4], [32, pp. 171–2], [123, pp. 123–4], [82, p. 365] and [81]. Note that the explanation of why the sorites is compelling is sometimes put in terms of ordinary speakers being fooled—mistaking near truth for full truth. But this just gives away the advantage of fuzzy theories, which is that they can explain speakers' reactions to the sorites paradox without resorting to the view that speakers are mistaken about the semantic facts. In the explanation given in the text above, we do not suppose that speakers mistakenly think

Note that the solution depends on a particular choice of truth conditions for the conditional and a particular definition of validity. Suppose we instead define validity as follows: on every model, the truth value of the conclusion is greater than or equal to the infimum of the truth values of the premisses. Then modus ponens (for the Łukasiewicz conditional) and the above sorites argument are invalid—and so we lose the explanation given above of why the argument is compelling.[24] Or suppose we read 'if... then...' in the argument as (say) the Zadeh or Gödel conditional: in that case some premisses would have degrees of truth of around 0.5, and so again we lose the explanation given above of why the argument is compelling.

So far we have considered just one formulation of the sorites argument: one involving multiple conditional premisses. But what about other formulations? If we can solve the sorites only when it is formulated in one particular way, then we have not really solved the underlying problem. Crispin Wright took this to be a problem for fuzzy views because he thought that they could not handle conjunctive formulations of the paradox [124, pp. 251–2]—for example:

Pile 10,000 is a heap.

It is not the case that (pile 10,000 is a heap and pile 9,999 is not a heap).

It is not the case that (pile 9,999 is a heap and pile 9,998 is not a heap).

⋮

It is not the case that (pile 2 is a heap and pile 1 is not a heap).

∴ Pile 1 is a heap.

However, Wright was assuming that fuzzy logic is PFL. In this logic, some of the premisses of the argument just given have degrees of truth of around 0.5, and so indeed we cannot run the kind of explanation given above of why the argument is compelling. However, the problem dissolves in Łukasiewicz logic: if we take the 'and' to be strong conjunction, then 'If pile 2 is a heap then pile 1 is a heap' is equivalent to 'It is not the case that (pile 2 is a heap and pile 1 is not a heap)' [81]. Thus the only moral here is the one we already saw: this kind of solution to the sorites paradox is not universally applicable but requires a careful choice of fuzzy logic.

## 5 Arguments against

In this section we look at—and respond to—the two major objections to fuzzy theories of vagueness.[25]

---

that the premisses are fully true: rather we exploit the fact that someone who takes a statement to be extremely close to fully true would naturally just go along with the statement, at least in normal contexts and until trouble was seen to arise.

[24] See [72, pp. 69–75], [123, pp. 123–4] and [85, p. 332].

[25] More detailed versions of the arguments in Section 5.1 can be found in [106] and [107]. More detailed versions of the arguments in Section 5.2 can be found in [106] and [102]. Further objections are considered and responded to in [106].

## 5.1 Artificial precision

The first objection to fuzzy theories is that they involve *artificial precision*:[26]

> [Fuzzy logic] imposes artificial precision... [T]hough one is not obliged to require that a predicate either definitely applies or definitely does not apply, one *is* obliged to require that a predicate definitely applies to such-and-such, rather than to such-and-such other, degree (e.g. that a man 5 ft 10 in tall belongs to *tall* to degree 0.6 rather than 0.5) — Haack [38, p. 443]

> One immediate objection which presents itself to [the fuzzy] line of approach is the extremely artificial nature of the attaching of precise numerical values to sentences like '73 is a large number' or 'Picasso's *Guernica* is beautiful'. In fact, it seems plausible to say that the nature of vague predicates precludes attaching precise numerical values just as much as it precludes attaching precise classical truth values.— Urquhart [119, p. 108]

> [T]he degree theorist's assignments impose precision in a form that is just as unacceptable as a classical true/false assignment. In so far as a degree theory avoids determinacy over whether $a$ is $F$, the objection here is that it does so by enforcing determinacy over the *degree* to which $a$ is $F$. All predications of "is red" will receive a unique, exact value, but it seems inappropriate to associate our vague predicate "red" with any particular exact function from objects to degrees of truth. For a start, what could determine which is the correct function, settling that my coat is red to degree 0.322 rather than 0.321? — Keefe [57, p. 571]

In a nutshell, the problem is that it is artificial/implausible/inappropriate to associate each vague *predicate* in natural language with a particular function that assigns one particular real number between 0 and 1 to each object (the object's degree of possession of the property picked out by that predicate); likewise, it is artificial/implausible/inappropriate to associate each vague *sentence* in natural language with a particular real number between 0 and 1 (the sentence's degree of truth).

The first thing to note is that the problem only arises for fuzzy theories with the following two features: they take $[0,1]$ as the set of truth degrees; and they posit a unique intended model of each vague discourse. For it is only theories of this kind that *do* associate each vague predicate with a particular function from objects to reals between 0 and 1 (i.e. the extension of the predicate on the unique intended model) and each sentence with a particular real between 0 and 1 (i.e. the sentence's degree of truth on the unique intended model). For ease of exposition, let's call such theories 'basic fuzzy theories'. (Note that there are many of them, not just one: they differ amongst themselves over the truth conditions of the logical operators.) We'll also refer to $[0,1]$ as the 'standard fuzzy truth values' or sftv's for short.

The next thing to do is determine the true nature and source of the problem. Haack offers no diagnosis; Urquhart maintains that the *nature of vague predicates* precludes

[26] For further statements of the problem see [19, pp. 521–2], [36, p. 332], [37, p. 54], [64, p. 462, p. 481], [72, p. 187], [74], [94, pp. 223–4], [98, p. 46], [118, p. 11], [123, pp. 127–8] and [58, pp. 113–4].

attaching precise numerical values; Keefe asks what could *determine* which is the correct function, settling that her coat is red to degree 0.322 rather than 0.321. Assuming that vagueness is correctly defined in terms of Closeness, we can see that Keefe is on the right track and Urquhart is not: the problem with the fuzzy view turns not on considerations having to do with the nature of vagueness, but rather on considerations having to do with the way in which the meanings of our terms are fixed. In order to see this, it will help to consider epistemicism again. As we saw in Section 4.1, epistemicism conflicts with the nature of vagueness: it cannot allow for the existence of predicates that satisfy Closeness (and have associated sorites series). As we saw in Section 2.2, epistemicism also runs into problems concerning the determination of meaning: it seems that the meaning-determining facts itemized in Section 2.2 do not suffice to pick out a particular height dividing the tall from the non-tall, and so on. The epistemicist therefore faces two distinct problems:

1. The *existence* of a sharp drop-off from true to false in a sorites series: this conflicts with the nature of vagueness.
2. The *particular location* of the drop-off: this conflicts with our best views about how meaning is determined.

I refer to these two problems as the *jolt problem* and the *location problem* respectively.

Now let's return to basic fuzzy theories. As we saw in Section 4.1, they do not face a version of the jolt problem: they can allow for the existence of predicates that satisfy Closeness (and have associated sorites series) and so they do not conflict with the nature of vagueness. However they *do* face a version of the location problem. It seems that the meaning-determining facts itemized in Section 2.2 do *not* suffice to pick out a particular function from objects to $[0,1]$ representing the extension of 'is tall' (and similarly for other vague predicates). So the fundamental problem underlying the artificial precision worry is that basic fuzzy theories conflict with our best available views about how the meanings of our expressions are determined.

In light of this diagnosis, let's consider some responses to the problem that have been proposed in the literature. We'll see several that don't work before getting to one that does. Recall that the problem arises for basic fuzzy theories, which have two key ingredients: they take sftv's as truth degrees; and they posit a unique intended model of each vague discourse. Responses to the problem can be categorised according to whether they abandon the first ingredient or the second.

The first response we shall consider—*fuzzy epistemicism*—actually tries to avoid the problem without abandoning either ingredient. This view is modelled after epistemicism, which attempts to retain the classical semantic picture (i.e. classical model theory plus a unique intended model) in the face of vagueness. The fuzzy epistemicist holds that each vague sentence does indeed have a unique sftv as its truth value—but we do not (cannot) know which value it is.[27] However, once we are clear that the artificial precision problem concerns the determination of meaning, this approach evidently fails

[27] Fuzzy epistemicism is mentioned by Copeland [19, p. 522] and developed in more detail by MacFarlane [71]. Machina [72, p. 187, n. 8] could also be interpreted as hinting at such a view when he writes of "difficulties about how to assign degrees of truth to propositions"; Keefe in [57, p. 571] and [58, p. 115] interprets him in this way and criticises his view on this basis.

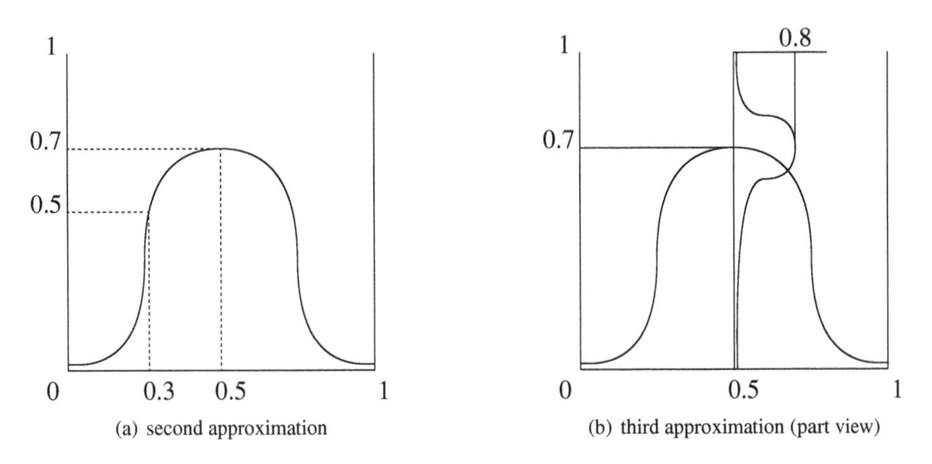


(a) second approximation

(b) third approximation (part view)

Figure 8. Bob's degree of tallness

to solve the problem. The problem concerns how there could *be* a unique function that is the extension of 'is tall' (say), given that our usage (etc.) does not suffice to pick out a unique such function. Saying that we do not or cannot *know* which function it is misses the point of the problem.

Let's turn then to a second proposal, due to [103]—the *blurry sets* view, which tries to solve the artificial precision problem by moving from $[0,1]$ to a different set of truth values: *degree functions*, which are functions from $[0,1]^*$ to $[0,1]$.[28] Suppose $f\colon [0,1]^* \to [0,1]$ is the truth value of 'Bob is tall' (B). The idea is that the value that $f$ assigns to the empty sequence—say, 0.5—is a *first approximation* to Bob's degree of tallness/the degree of truth of (B). The values assigned by $f$ to sequences of length 1 then play two roles. First, they rate possible first approximations. The higher the value assigned to $\langle x\rangle$, the better $x$ is as a first approximation to Bob's degree of tallness/the degree of truth of (B). If $f(\langle 0.3\rangle) = 0.5$, then we say that it is 0.5 true that Bob is tall to degree 0.3; if $f(\langle 0.5\rangle) = 0.7$, then we say that it is 0.7 true that Bob is tall to degree 0.5; and so on. Second, the assignments to sequences of length 1 jointly constitute a second level of approximation to Bob's degree of tallness/the degree of truth of (B). Together, these assignments determine a function $f_{\langle\rangle}\colon [0,1] \to [0,1]$. We regard this as encoding a density function over [0,1], and we require that its centre of mass is at $f(\langle\rangle)$ (Figure 8(a)).[29] The same thing happens again when we move to the values assigned to sequences of length 2: these values play two roles. First, they rate possible ratings of first approximations. The higher the value assigned to $\langle x, y\rangle$, the better $y$ is as a rating of $x$ as a first approximation to Bob's degree of tallness/the degree of truth of (B). If $f(\langle 0.5, 0.7\rangle) = 0.8$, then we say that it is 0.8 true that it is 0.7 true that Bob is tall to degree 0.5; if $f(\langle 0.4, 0.5\rangle) = 0.3$, then we say that it is 0.3 true that it is 0.5 true that Bob is tall to degree 0.4; and so on. Second, the assignments to sequences of length 2

[28] On notation: where $S$ is a set, $S^*$ is the set of all finite sequences of elements of $S$ (including the empty or null sequence).

[29] I am here giving a rough picture of the view: in order to make it precise, various definitions and assumptions are required—see [103] for the details.

jointly constitute a third level of approximation to Bob's degree of tallness/the degree of truth of (B). Together, the assignments made by $f$ to sequences $\langle a, x\rangle$ of length 2 whose first member is $a$ determine a function $f_{\langle a\rangle} : [0, 1] \rightarrow [0, 1]$. This can be seen as encoding a density function, and we require that its centre of mass is at $f(\langle a\rangle)$ (Figure 8(b)). And so the story goes, ad infinitum. Figuratively, we can picture a degree (of truth or property-possession) as a region of varying shades of grey spread between 0 and 1 on the real line. If you focus on any point in this region, you see that what appeared to be a point of a particular shade of grey is in fact just the centre of a further such grey region. The same thing happens if you focus on a point in this further region, and so on. The region is blurry all the way down: no matter how much you increase the magnification, it will not come into sharp focus. On this view, statements of the form 'The degree of truth of "Bob is tall" is 0.4' need not be simply true or false: they may themselves have intermediate degrees of truth. So rather than exactly one sentence of the form 'The degree of truth of "Bob is tall" is $x$' being true and the others false, many of them might be true to various degrees. Thus there is a sense in which sentences in natural language which predicate vague properties of objects are *not* each assigned just one particular sftv. Be that as it may, however, the artificial precision problem has not been solved. In just the way that they fail to determine a unique *classical* set (function from objects to classical truth values) or a unique *fuzzy* set (function from objects to fuzzy truth values) as the extension of 'is tall', the meaning-determining facts do not suffice to pick out a unique *blurry set* (function from objects to degree functions) as the extension of 'is tall'.

The third proposal that we shall consider—the *ordinal approach*—is that when we assign sftv's to sentences, the only thing about the assignments that is meaningful is the relative *ordering* of the values assigned. Views of this sort have been advocated by—amongst others—Goguen, Machina and Hyde:[30]

> We certainly do not want to claim there is some *absolute* [fuzzy] set representing 'short'.... It appears that many arguments about fuzzy sets do not depend on particular values of functions... This raises the problem of *measuring* fuzzy sets... Probably we should not expect particular numerical values of shortness to be meaningful (except 0 and 1), but rather their *ordering*... degree of membership may be measured by an *ordinal scale.* [36, pp. 331–2]
>
> the assignment of exact values usually doesn't matter much... what is of importance instead is the ordering relation between the values of various propositions. [72, p. 188]
>
> The foregoing account... requires only a totally-ordered dense set of values. The choice of a specific value from among the infinitely many possible... is arbitrary except in so far as it preserves ordering requirements... No significance attaches to the choice of value apart from these ordering requirements. [46, p. 207]

There are actually two quite different ways of spelling out this idea in detail.[31]

[30] See also [96, p. 29], [37, p. 59], [40, pp. 162–3] and [121].
[31] For a more detailed discussion see [109].

On the first way of looking at things, the truth degrees are not sftv's. The real interval $[0, 1]$ comprises some entities, together with some structure—an order structure, a metric structure—and some operations—addition, subtraction, and so on. Now suppose we retain the entities and the order structure, but discard the metric structure and any operations definable in terms of it or vice versa (e.g. subtraction). This gives us a new structure—and its elements are the truth values of the new sort of model theory now under consideration. Figuratively, one can think of the new structure as a rubbery unit interval, fixed at each end: its end-points have fixed positions, but between them, none of the other elements has a fixed position. They can be squeezed or stretched left or right at will—but they can never leapfrog one another: their *order* is fixed. Let us refer to the elements of this structure—the rubbery unit interval—as rtv's, and to models in which the truth degrees are rtv's as rubbery models (i.e. models that assign rubbery sets to predicates—where a rubbery set is a function from the domain to the rtv's—and rtv's to sentences). On the first way of spelling out the ordinal approach, then, each vague discourse is associated with a unique intended model—but it is a rubbery model, not a standard fuzzy model (i.e. a model involving sftv's).

On the second way of spelling out the ordinal approach, there are no rtv's, only sftv's. We represent the facts about the truth of statements by assigning sftv's to them, but there are many acceptable ways of doing so: for any acceptable way of mapping sentences to sftv's, any mapping obtained by composing it with an order-preserving and endpoint-fixing transformation of $[0, 1]$ is equally acceptable. Thus there is no unique intended model of a discourse: there is a whole class of acceptable models, differing from one another by such transformations of the sftv's assigned—but they are all standard fuzzy models.

On either way of spelling out the ordinal approach to the artificial precision problem, a serious difficulty looms. In trying to resolve the conflict between basic fuzzy theories and the facts about how meanings are determined, we have generated a new conflict: with facts about the nature of vagueness (which was an area where basic fuzzy theories had no problem). On the view that truth is measured on an ordinal scale, it *makes no sense* to say that two sentences $P$ and $Q$ are *very close* in respect of truth: it makes sense to say only that one sentence is *truer* than another. But the idea of two sentences being *very close* in respect of truth is at the heart of the Closeness definition—and so a view that makes no room for this notion lacks the resources to accommodate predicates that satisfy Closeness.[32]

The fourth and final response to the artificial precision problem that we shall consider here is *fuzzy plurivaluationism*, due to [106].[33] Fuzzy plurivaluationism is just like classical plurivaluationism except that its models are fuzzy, not classical. It stands to basic fuzzy theories of vagueness (on which a vague discourse is associated with a unique intended standard fuzzy model) in just the way that classical plurivaluationism stands to the classical semantic picture (on which a discourse is associated with a unique intended classical model). That is, everything about the original view is retained (so in the classical case, only standard classical models are countenanced, and in the

[32] Recall that it was already noted in Section 4.1 that if the only structure on the truth degrees was an ordering then we would not be able to accommodate Closeness.

[33] For a more detailed presentation and motivation of this view, see [106, §2.5, ch. 6]. See also [9].

fuzzy case, only standard fuzzy models are countenanced), *except* the idea that each discourse is associated with a unique intended model. The latter idea is replaced with the thought that each discourse is associated with multiple acceptable models. Which models? The fuzzy plurivaluationist answer to this question is different from the answer given by the second way of spelling out the ordinal approach discussed above. On the ordinal approach, the acceptable models are closed under order-preserving and endpoint-fixing transformations of the truth degrees. On the plurivaluationist approach, by contrast, the acceptable models are those models not ruled out as incorrect by the meaning-determining facts. (For example, suppose that speakers in a discourse are all disposed to count certain paradigm individuals as tall and certain others as definitely not tall; in that case, acceptable models of the discourse must assign 'tall' an extension that maps the former individuals to 1 and the latter to 0.) As we have seen, the meaning-determining facts do not suffice to single out a unique intended standard fuzzy model of vague discourse: this is precisely what generates the artificial precision problem for basic fuzzy theories. A fortiori, there are multiple acceptable models (where 'acceptable' means 'not ruled out as incorrect by the meaning-determining facts'). Note that fuzzy plurivaluationism avoids the problem faced by the ordinal view of not being able to say meaningfully that two sentences are very close in respect of truth. One plausible constraint on acceptable models is that vague predicates come out as satisfying Closeness relative to them. Hence if $x$ and $y$ are very similar in $F$-relevant respects, $Fx$ and $Fy$ must be very similar in respect of truth on every acceptable model (if $F$ is vague). In that case we will indeed be able to talk as if these two sentences are simply very close in respect of truth.

Fuzzy plurivaluationism successfully solves the artificial precision problem. The problem is that the meaning-determining facts do not suffice to pick out a unique standard fuzzy model of vague discourse as the intended model. The fuzzy plurivaluationist solution is to abandon the notion of a unique intended model in favour of the idea of multiple acceptable models—where an acceptable model is one that is not ruled out as incorrect by the meaning-determining facts. As the problem is precisely that there is not a unique acceptable fuzzy model of vague discourse—because too many models are compatible with the constraints imposed by the meaning-fixing facts—it follows *a fortiori* that fuzzy plurivaluationism—the view that there are multiple equally correct models—is correct. The upshot of fuzzy plurivaluationism is that 'Bob is tall' (say) does *not* have a uniquely correct sftv: it is assigned multiple different sftv's—one on each acceptable model—and none of these is more correct than any of the others. This was the desired result: that it was not the case on basic fuzzy theories was precisely the artificial precision problem. Furthermore, we have avoided assigning a unique sftv to each vague sentence *for the right reason*: because doing so is incompatible with the meaning-determining facts.

### 5.2 Truth functionality

The second objection to fuzzy theories is that they are incompatible with ordinary usage of compound propositions in the presence of borderline cases. This is a very common objection and there are many different versions of it to be found in the literature. Many of the objectors falsely assume that fuzzy logic is simply PFL—but some versions of the objection are more general: they are directed against any truth(-degree)-functional account. Here is a sample of the objections:

1. *Fine I and Osherson and Smith I.* Suppose that a certain blob is on the border of pink and red and let $P$ be the sentence 'the blob is pink' and $R$ the sentence 'the blob is red'—so $P$ and $R$ are neither clearly true nor clearly false. Fine thinks that $P \vee R$ is clearly true and that $P \wedge R$ is clearly false. This is not predicted by a fuzzy account based on PFL.[34] On a related note, Osherson and Smith think that where $Ax$ means that $x$ is an apple, $Aa \wedge \neg Aa$ should be true to degree 0 and $Aa \vee \neg Aa$ should be true to degree 1, whatever $a$ is. This conflicts with PFL.[35]

2. *Kamp.* Kamp thinks that $\alpha \wedge \neg\alpha$ is clearly true to degree 0, even when $[\alpha] = 0.5$; and that $\alpha \wedge \alpha$ is clearly true to a degree strictly greater than 0, when $[\alpha] = 0.5$. Assuming $[\neg\alpha] = 1 - [\alpha]$, no truth function for $\wedge$ can predict this. So this is an argument not just against PFL but against any degree-functional account that agrees with PFL about negation.[36]

3. *Fine II.* With $P$ and $R$ as in 1, Fine claims that $P \wedge P$ is equivalent to $P$ and hence is neither clearly true nor clearly false, while (as already discussed) $P \wedge R$ is clearly false. Given that $P$ and $R$ have the same degree of truth, this is an argument against any degree-functional account of conjunction.[37]

4. *Osherson and Smith II.* Consider an apple (a), illustrated thus:

(a)

Osherson and Smith claim that (a) is psychologically less prototypical of the concept 'apple' than of the concept 'striped apple'. If we equate prototypicality with degree of membership/truth and take 'striped apple' to be formed from the two components 'striped' and 'apple' by intersection/conjunction, then this is an objection to the minimum rule for conjunction—and more generally to any account according to which $[\alpha \wedge \beta]$ can never be strictly greater than $[\alpha]$ (or $[\beta]$).[38]

---

[34] See [30, pp. 269–70].

[35] See [79, pp. 45–6]. Osherson and Smith present their argument in terms of degrees of membership of objects in sets rather than degrees of truth of statements.

[36] See [53, p. 546].

[37] See [30, p. 269].

[38] See [79, pp. 43–5]. Essentially the same objection (using the example of *pussy willow* and *willow*) was made earlier by Kay [56, p. 153]. Osherson and Smith also present a dual objection concerning alleged disjunctions that are apparently less true than either disjunct [79, pp. 46–8].

The objections are often based solely on the intuitions of their authors—for example:

> we would have $[\phi \wedge \neg\phi] = \frac{1}{2}$, which seems absurd. For how could a logical contradiction be true to *any* degree? [53, p. 546][39]

> a given object $x$ may be a triangle (say) to degree 0.9; $f_\triangle(x) = 0.9$. If the complement of $f_\triangle$ represents 'is not a triangle' and union disjunction, then $\max(f_\triangle, 1 - f_\triangle)$ should represent 'is a triangle or isn't a triangle' *and should be the constant 1 function*; but it isn't. [119, p. 108; my emphasis][40]

> the same value must be given to (e) 'if Tek is tall then Tek is not tall' as to (f) 'if Tek is tall then Tek is tall' since the respective values of the antecedent and consequent of these two conditionals are the same. But (f) is intuitively true and (e) is not, so again no choice of value will capture our intuitions about both of these cases [58, p. 97].

So one perfectly legitimate response is to express alternative intuitions—for example:[41]

> [Fine] claims that 'red' and 'pink', even though vague and admitting of borderline cases of applicability, are nevertheless logically connected so that to say of some color shade that it is both red and pink is obviously to say something false. I must confess being completely insensitive to that intuition of a penumbral connection. [72, p. 77, n. 2]

> An aspect of the theory of fuzzy sets which Osherson and Smith find objectionable is that, in the theory, the union of A and its complement, A′, is not, in general, the whole universe of discourse. This relates, of course, to the long-standing controversy regarding the validity of the *principle of the excluded middle*... The principle of the excluded middle is not accepted as a valid axiom in the theory of fuzzy sets *because it does not apply to situations in which one deals with classes which do not have sharply defined boundaries*. [126, p. 292; second emphasis mine]

> the failure of contradiction and excluded-middle laws is typical of fuzzy logic as emphasized by many authors. This is natural with gradual properties like 'tall'. [21, p. 152]

Sometimes the objections are based on empirical data, but there are also considerable empirical data in support of various fuzzy theories. For example, studies in [5], [6], [92] and [97] show a significant willingness of subjects to agree with statements such as

[39] I have omitted a superscript and a subscript from Kamp's notation because they add complexity that is irrelevant in the present context. Cf. [30, p. 270] "Surely $P$ & $-P$ is false even though $P$ is indefinite"; [50, p. 199] "This consequence is absurd, because a self-contradiction surely merits a truth value of zero"; [123, p. 136] "How can an explicit contradiction be true to any degree other than 0?"; [55, p. 134] "According to the most familiar versions of fuzzy logic the degree to which $a$ satisfies the conjunctive concept *apple which is not an apple* is... greater than 0. Clearly this is not the right result"; and [43, pp. 366–7] "...an obviously counterintuitive conclusion".

[40] Cf. [55, pp. 146–7], [16, pp. 389–90] and [43, pp. 366–7].

[41] Cf. also [51, pp. 287–8], [65, p. 141], [33, pp. 323–4], [58, p. 164] (NB Keefe, unlike fellow supervaluationists Fine and Kamp & Partee, does not take $Fa \vee \neg Fa$ to be assertable when $a$ is a borderline case of $F$), [12, p. 578], [13, p. 31], [106, p. 86] and [93, p. 341].

'$X$ is tall and not tall' and 'The circle is near the square and it is not near the square'.[42] Furthermore, we must also be very careful to ensure that any data that are established are actually relevant to the assessment of fuzzy theories. For example, as we saw above, Osherson and Smith [79, pp. 43–5] take it as a datum that "There can be no doubt that [(a)] is psychologically less prototypical of an apple... than of an apple-with-stripes" and then take this to mean that (a)'s degree of membership in the set of striped apples is greater than its degree of membership in the set of apples. But one could accept the former datum about *prototypicality* and yet maintain that when it comes to degrees of *membership* (and *truth*), (a) is a degree 1 member of 'striped apple' and of 'apple' (and '(a) is a striped apple' and '(a) is an apple' are both quite simply true, to degree 1).[43]

At this point, then, it has not been established that there are consistent patterns of ordinary usage that pose even a prima facie threat to fuzzy theories of vagueness. But suppose for the sake of argument that we take the intuitions of the objectors as genuine data. There are in fact many ways in which fuzzy theories could accommodate such data. We can divide them broadly into semantic and pragmatic approaches; and we can further divide the semantic approaches into those that employ degree-functional truth rules and those that employ non-degree-functional truth rules. Let's consider these approaches in turn.

**Degree-functional semantics:** Some of the intuitions of the objectors can be accommodated straightforwardly using fuzzy resources that have already been mentioned in this chapter. For example:

1. *Fine I and Osherson and Smith I.* In Łukasiewicz t-norm logic, when $[P] = [R] = 0.5$, $[P \vee R] = 1$ and $[P \wedge R] = 0$.[44] This meets Fine's desiderata. Likewise, in Łukasiewicz logic, *whatever* the degree of truth of $\alpha$, $[\alpha \wedge \neg\alpha] = 0$ and $[\alpha \vee \neg\alpha] = 1$ [13, p. 31], [11, p. 138] and [81].[45] This meets Osherson and Smith's desiderata.

2. *Kamp.* We do not have to define $[\neg\alpha] = 1 - [\alpha]$. In Gödel t-norm logic, $\alpha \wedge \neg\alpha$ is 0 true and $\alpha \wedge \alpha$ is 0.5 true. In product t-norm logic, $\alpha \wedge \neg\alpha$ is 0 true and $\alpha \wedge \alpha$ is 0.25 true. This meets Kamp's desiderata.

3. *Fine II.* In Łukasiewicz t-norm logic, when $[P] = [R] = 0.5$, '$P$ and $P$' is true to degree 0.5 where 'and' is read as weak conjunction and '$P$ and $R$' is true to degree 0 where 'and' is read as strong conjunction [81]. This meets Fine's desiderata.

Furthermore, there are oceans of functions on $[0, 1]$ available to fuzzy theorists that we have not yet mentioned. For example, recall Osherson and Smith's idea that (a)'s degree of membership in 'striped apple' should be greater than its degree of membership in 'apple'. This sort of situation could be modelled by taking 'striped apple' to be formed from the sets 'striped' and 'apple' not by a conjunction/intersection operation

[42] See [102] for further discussion and references.

[43] The point that degrees of membership and truth on the one hand and degrees of typicality on the other hand need to be carefully distinguished has been made by numerous authors including [126, p. 293], [100, pp. 51–2], [52], [55, p. 131, p. 133], [80, p. 191], [12, p. 578] and [11, pp. 132–3].

[44] Where $\wedge$ is the Łukasiewicz t-norm and $\vee$ is its dual: $x \vee y = 1 - ((1 - x) \wedge (1 - y))$.

[45] Where $\wedge$ is the Łukasiewicz t-norm and $\vee$ is its dual.

but by an averaging operation [12, p. 578]. For another example, one can introduce a 'determinately' operator $\Delta$ [114] and [7]:

$$[\Delta\alpha] = \begin{cases} 1 & \text{if } [\alpha] = 1 \\ 0 & \text{if } [\alpha] \neq 1 \end{cases}$$

When $[\alpha] = 0.5$, then even using the conjunction and negation of PFL:

$$[\Delta\alpha \wedge \Delta\neg\alpha] = [\Delta\alpha \wedge \neg\Delta\alpha] = 0$$

**Pragmatics:** A moment ago we saw one way in which apparently non-degree-functional data could be accommodated within a degree-functional semantics: Fine's idea that '$P$ and $R$' should be clearly false while '$P$ and $P$' is neither clearly true nor clearly false—which seems to rule out a degree-functional treatment of 'and'—could be accommodated by reading 'and' as weak conjunction in '$P$ and $P$' and as strong conjunction in '$P$ and $R$'. It is not easy, however, to make this sort of approach—which posits an ambiguity in 'and'—work in detail,[46] so it is important to note that there is a second way in which apparently non-degree-functional data could be accommodated within a theory that employed a degree-functional semantics: via pragmatics. Suppose that a consistent pattern of usage of the sort Fine finds intuitive was in fact discovered. There is no reason why a fuzzy theorist should not accommodate this by saying that while 'and' is unambiguously interpreted as some particular truth function, nevertheless the *assertability* of conjunctions sometimes goes by a different rule, viz.: assert $\Phi$ iff $\Phi$ would certainly be true no matter how the language were precisified. (Compare the following well-known view [66], [47], [48] and [69]: the truth conditions of the indicative conditional 'If $A$ then $B$' are the same as those of the material conditional $A \supset B$; however *assertability* for indicative conditionals goes not by truth but by conditional probability (of $B$ given $A$).) In other words, if the data appear to support a supervaluationist picture, this does not in fact mean that we need a supervaluationist semantics: a supervaluationist approach could instead be located in the pragmatics.

**Non-degree-functional semantics:** Finally, even if it were for some reason important to treat some connective in a non-degree-functional way (i.e. in the semantics), this does not rule out fuzzy theories! As mentioned in Section 3.2, there is no reason why fuzzy theories cannot use non-degree-functional approaches at the 'truth rules' stage of developing fuzzy models.

## 6 Historical remarks and further reading

### Section 1

The characterisation of vagueness in terms of borderline cases can be traced at least as far as [83], while the blurred boundaries characterisation can be traced at least as far as Frege's statement that if we represent concepts in extension by areas on a plane, then vague concepts do not have sharp boundaries, but rather fade off into the background (*Grundgesetze* vol. II, §56) [8, p. 259]. A few authors still introduce vagueness in terms of just one characteristic (e.g. possession of borderline cases) but since at least [60] the three-feature characterisation has become standard.

[46] See [77, p. 13] for one suggestion in this area and [28, pp. 200–1] for discussion.

**Section 2**

Advocates of epistemic theories of vagueness include [18], [17], [112, ch. 6], [113], [122], [123, ch. 7–8] and [44]. For recursive tripartite approaches to vagueness see [116], [117] and [93]. For supervaluationist approaches see [75], [22], [30], [53], [84], [14] and [58, ch. 7–8]. For a subvaluationist approach to vagueness see [45]. For the degree-theoretic form of supervaluationism see [53], [54, pp. 234–5], [67, pp. 228–9] and [68, pp. 69–70]. Plurivaluationism was explicitly distinguished from supervaluationism in [106]; prior to that, the two views were conflated in the literature. [86] is a possible early example of a plurivaluationist approach. Contextualist treatments of vagueness include [54], [115], [89], [90], [111, ch. 7], [26] and [99]. Note that there are various contextualist theories in the literature which differ in more or less subtle ways (see e.g. [91, p. 245], [27, p. 329], [2], [3] and [1]) and that the necessarily brief Section 2.5 presents just one prominent kind of contextualist position; nevertheless a version of the objection that the contextualist solution to the sorites turns on speakers not believing that the contextualist story is the correct account of vague language applies also to other versions of contextualism (e.g. note that the contextualist solutions to the sorites considered in [3] turn on attributing *confusion* to ordinary speakers; see also [1, §3.1] and [2, §4]; and cf. also [59, §§4–5]).

**Section 3**

The study of infinite-valued logics begins with Łukasiewicz; see e.g. [70]. Fuzzy set theory was born in [125]; a related earlier idea is Post's notion of an $n$-valued set [73, p. 47, p. 98]. For overviews of the history of fuzzy logic and set theory see [39, ch. 10], [78, ch. 8], [62] and [41]. For technical introductions to fuzzy logic and set theory, see e.g. [63], [39], [78], [76] and of course this Handbook. Note that 'Zadeh logic' has other names including 'Kleene–Zadeh logic' and sometimes simply 'fuzzy logic'. For discussion of non-truth-functional fuzzy logics see e.g. [35] and [41, p. 298]; cf. also [23–25]. Key early sources for the fuzzy view of vagueness are [36], [64] and [72]; [15] is a relevant earlier work. [106] is a recent comprehensive elaboration and defence of a fuzzy theory of vagueness.

**Acknowledgements**

I am very grateful to Stewart Shapiro and Chris Fermüller for detailed comments on a draft version.

# BIBLIOGRAPHY

[1] Jonas Åkerman. Contextualist theories of vagueness. *Philosophy Compass*, 7(7):470–480, 2012.

[2] Jonas Åkerman and Patrick Greenough. Vagueness and non-indexical contextualism. In Sarah Sawyer, editor, *New Waves in Philosophy of Language*, pages 8–23. Palgrave Macmillan, 2009.

[3] Jonas Åkerman and Patrick Greenough. Hold the context fixed—vagueness still remains. In Richard Dietz and Sebastiano Moruzzi, editors, *Cuts and Clouds: Vagueness, its Nature, and its Logic*, pages 275–288. Oxford University Press, Oxford, 2010.

[4] Ken Akiba and Ali Abasnezhad, editors. *Vague Objects and Vague Identity: New Essays on Ontic Vagueness*. Springer, 2014.

[5] Sam Alxatib and Jeff Pelletier. On the psychology of truth-gaps. In Rick Nouwen, Robert Van Rooij, Uli Sauerland, and Hans-Christian Schmitz, editors, *Vagueness in Communication*, pages 13–36. Springer, 2011.

[6] Sam Alxatib and Jeff Pelletier. The psychology of vagueness: Borderline cases and contradictions. *Mind and Language*, 26:287–326, 2011.

[7] Matthias Baaz. Infinite-valued Gödel logic with 0-1-projections and relativisations. In Petr Hájek, editor, *Gödel'96: Logical Foundations of Mathematics, Computer Science, and Physics*, volume 6 of *Lecture Notes in Logic*, pages 23–33. Springer, Brno, 1996.

[8] Michael Beaney, editor. *The Frege Reader*. Blackwell, Oxford, 1997.

[9] Libor Běhounek. In which sense is fuzzy logic a logic for vagueness? In Thomas Lukasiewicz, Rafael Peñaloza, and Anni-Yasmin Turhan, editors, *Proceedings of the First Workshop on Logics for Reasoning about Preferences, Uncertainty, and Vagueness (PRUV 2014)*, pages 26–39, Vienna, 2014.

[10] Libor Běhounek, Petr Cintula, and Petr Hájek. Introduction to mathematical fuzzy logic. In Petr Cintula, Petr Hájek, and Carles Noguera, editors, *Handbook of Mathematical Fuzzy Logic - Volume 1*, volume 37 of *Studies in Logic, Mathematical Logic and Foundations*, pages 1–101. College Publications, London, 2011.

[11] Radim Bělohlávek and George J. Klir. Fallacious perceptions of fuzzy logic in the psychology of concepts. In Radim Bělohlávek and George J. Klir, editors, *Concepts and Fuzzy Logic*, pages 121–148. MIT Press, Cambridge, MA, 2011.

[12] Radim Bělohlávek, George J. Klir, Harold W. Lewis III, and Eileen C. Way. On the capability of fuzzy set theory to represent concepts. *International Journal of General Systems*, 31:569–585, 2002.

[13] Radim Bělohlávek, George J. Klir, Harold W. Lewis III, and Eileen C. Way. Concepts and fuzzy sets: Misunderstandings, misconceptions, and oversights. *International Journal of Approximate Reasoning*, 51:23–34, 2009.

[14] Brandon Bennett. Modal semantics for knowledge bases dealing with vague concepts. In Anthony G. Cohn, Leonard Schubert, and Stuart C. Shapiro, editors, *Principles of Knowledge Representation and Reasoning: Proceedings of the 6th International Conference (KR-98)*, pages 234–244. Morgan Kaufmann, 1998.

[15] Max Black. Vagueness: An exercise in logical analysis. *Philosophy of Science*, 4(4):427–455, 1937.

[16] Nicolao Bonini, Daniel Osherson, Riccardo Viale, and Timothy Williamson. On the psychology of vague predicates. *Mind and Language*, 14:377–393, 1999.

[17] Richmond Campbell. The sorites paradox. *Philosophical Studies*, 26:175–191, 1974.

[18] James Cargile. The sorites paradox. *British Journal for the Philosophy of Science*, 20:193–202, 1969. Reprinted in [61] pp. 89–98.

[19] B. Jack Copeland. Vague identity and fuzzy logic. *Journal of Philosophy*, 94:514–534, 1997.

[20] Steven Davis and Brendan S. Gillon, editors. *Semantics: A Reader*. Oxford University Press, Oxford, 2004.

[21] Didier Dubois and Henri Prade. Can we enforce full compositionality in uncertainty calculi? In *Proceedings of the Twelfth National Conference on Artificial Intelligence* (AAAI-94), pages 149–154. AAAI Press, 1994.

[22] Michael Dummett. Wang's paradox. *Synthese*, 30:301–324, 1975. Reprinted in [61] pp. 99–118.

[23] Dorothy Edgington. Validity, uncertainty and vagueness. *Analysis*, 52:193–204, 1992.

[24] Dorothy Edgington. Vagueness by degrees. In Rosanna Keefe and Peter Smith, editors, *Vagueness: A Reader*, pages 294–316. MIT Press, Cambridge, MA, 1997.

[25] Dorothy Edgington. The philosophical problem of vagueness. *Legal Theory*, 7(4):371–378, 2001.

[26] Delia Graff Fara. Shifting sands: An interest-relative theory of vagueness. *Philosophical Topics*, 28: 45–81, 2000. Published under 'Delia Graff'.

[27] Delia Graff Fara. Profiling interest relativity. *Analysis*, 68(4):326–335, 2008.

[28] Christian G. Fermüller. Comments on 'Vagueness in language: The case against fuzzy logic revisited' by Uli Sauerland. In Petr Cintula, Christian G. Fermüller, Lluís Godo, and Petr Hájek, editors, *Understanding Vagueness: Logical, Philosophical, and Linguistic Perspectives*, volume 36 of *Studies in Logic*, pages 199–202. College Publications, London, 2011.

[29] Hartry Field. Theory change and the indeterminacy of reference. *The Journal of Philosophy*, 70: 462–481, 1973.

[30] Kit Fine. Vagueness, truth and logic. *Synthese*, 30:265–300, 1975. Reprinted with corrections in [61] pp. 119–150.

[31] Graeme Forbes. Thisness and vagueness. *Synthese*, 54:235–259, 1983.

[32] Graeme Forbes. *The Metaphysics of Modality*. Clarendon Press, Oxford, 1985.

[33] Gy. Fuhrmann. "Prototypes" and "Fuzziness" in the logic of concepts. *Synthese*, 75:317–347, 1988.

[34] Nikolaos Galatos, Peter Jipsen, Tomasz Kowalski, and Hiroakira Ono. *Residuated Lattices: An Algebraic Glimpse at Substructural Logics*, volume 151 of *Studies in Logic and the Foundations of Mathematics*. Elsevier, Amsterdam, 2007.

[35] Giangiacomo Gerla. *Fuzzy Logic—Mathematical Tool for Approximate Reasoning*, volume 11 of *Trends in Logic*. Kluwer and Plenum Press, New York, 2001.

[36] Joseph Amadee Goguen. The logic of inexact concepts. *Synthese*, 19(3–4):325–373, 1969.

[37] Joseph Amadee Goguen. Fuzzy sets and the social nature of truth. In Madan M. Gupta, Rammohan K. Ragade, and Ronald R. Yager, editors, *Advances in Fuzzy Set Theory and Applications*, pages 49–67. North-Holland, Amsterdam, 1979.

[38] Susan Haack. Do we need "fuzzy logic"? *International Journal of Man–Machine Studies*, 11:437–445, 1979.

[39] Petr Hájek. *Metamathematics of Fuzzy Logic*, volume 4 of *Trends in Logic*. Kluwer, Dordrecht, 1998.

[40] Petr Hájek. Ten questions and one problem on fuzzy logic. *Annals of Pure and Applied Logic*, 96: 157–165, 1999.

[41] Petr Hájek. What is mathematical fuzzy logic. *Fuzzy Sets and Systems*, 157(5):597–603, 2006.

[42] Petr Hájek and Vilém Novák. The sorites paradox and fuzzy logic. *International Journal of General Systems*, 32(4):373–383, 2003.

[43] James A. Hampton. Typicality, graded membership, and vagueness. *Cognitive Science*, 31:355–384, 2007.

[44] Paul Horwich. *Truth*. Clarendon Press, Oxford, second revised edition, 1998.

[45] Dominic Hyde. From heaps and gaps to heaps of gluts. *Mind*, 108:641–660, 1997.

[46] Dominic Hyde. *Vagueness, Logic and Ontology*. Ashgate, Aldershot, 2008.

[47] Frank Jackson. On assertion and indicative conditionals. In Frank Jackson, editor, *Conditionals*, pages 111–135. Oxford University Press, Oxford, 1979.

[48] Frank Jackson. *Conditionals*. Basil Blackwell, Oxford, 1987.

[49] Frank Jackson and Graham Priest, editors. *Lewisian Themes: The Philosophy of David K. Lewis*. Oxford University Press, Oxford, 2004.

[50] Philip N. Johnson-Laird. *Mental Models*. Harvard University Press, Cambridge, MA, 1983.

[51] Gregory V. Jones. Stacks not fuzzy sets: An ordinal basis for prototype theory of concepts. *Cognition*, 12:281–290, 1982.

[52] Charles W. Kalish. Essentialism and graded membership in animal and artifact categories. *Memory & Cognition*, 23:335–353, 1995.

[53] Hans Kamp. Two theories about adjectives. In Edward L. Keenan, editor, *Formal Semantics of Natural Language*, pages 123–155. Cambridge University Press, Cambridge, 1975. Reprinted in [20] pp. 541–562.

[54] Hans Kamp. The paradox of the heap. In Uwe Mönnich, editor, *Aspects of Philosophical Logic*, pages 225–277. D. Reidel, Dordrecht, 1981.

[55] Hans Kamp and Barbara Partee. Prototype theory and compositionality. *Cognition*, 57:129–191, 1995.

[56] Paul Kay. A model-theoretic approach to folk taxonomy. *Social Science Information*, 14:151–166, 1975.

[57] Rosanna Keefe. Vagueness by numbers. *Mind*, 107:565–579, 1998.

[58] Rosanna Keefe. *Theories of Vagueness*. Cambridge University Press, 2000.

[59] Rosanna Keefe. Vagueness without context change. *Mind*, 116(462):275–292, 2007.

[60] Rosanna Keefe and Peter Smith. Introduction: theories of vagueness. In Rosanna Keefe and Peter Smith, editors, *Vagueness: A Reader*, pages 1–57. MIT Press, Cambridge, MA, 1997.

[61] Rosanna Keefe and Peter Smith, editors. *Vagueness: A Reader*. MIT Press, Cambridge, MA, 1997.

[62] George J. Klir. Foundations of fuzzy set theory and fuzzy logic: A historical overview. *International Journal of General Systems*, 30(2):91–132, 2001.

[63] George J. Klir and Bo Yuan. *Fuzzy Sets and Fuzzy Logic: Theory and Applications*. Prentice Hall, 1995.

[64] George Lakoff. Hedges: A study in meaning criteria and the logic of fuzzy concepts. *Journal of Philosophical Logic*, 2(4):458–508, 1973.

[65] George Lakoff. *Women, Fire, and Dangerous Things: What Categories Reveal about the Mind*. University of Chicago Press, Chicago, 1987.

[66] David Lewis. Probabilities of conditionals and conditional probabilities. In Frank Jackson, editor, *Conditionals*, pages 76–101. Oxford University Press, Oxford, 1976.

[67] David Lewis. General semantics. In *Philosophical Papers*, volume I, pages 189–232. Oxford University Press, New York, 1983.

[68] David Lewis. Survival and identity. In *Philosophical Papers*, volume I, pages 55–77. Oxford University Press, New York, 1983.

[69] David Lewis. Probabilities of conditionals and conditional probabilities II. In Frank Jackson, editor, *Conditionals*, pages 102–110. Oxford University Press, Oxford, 1986.

[70] Jan Łukasiewicz and Alfred Tarski. Untersuchungen über den Aussagenkalkül. *Comptes Rendus des Séances de la Société des Sciences et des Lettres de Varsovie, cl. III*, 23(iii):30–50, 1930.
[71] John MacFarlane. Fuzzy epistemicism. In Richard Dietz and Sebastiano Moruzzi, editors, *Cuts and Clouds: Vagueness, its Nature, and its Logic*, pages 438–463. Oxford University Press, Oxford, 2010.
[72] Kenton F. Machina. Truth, belief, and vagueness. *Journal of Philosophical Logic*, 5(1):47–78, 1976. Reprinted in [61] pp. 174–203.
[73] Grzegorz Malinowski. *Many-Valued Logics*. Clarendon Press, Oxford, 1993.
[74] Vincenzo Marra. The problem of artificial precision in theories of vagueness: A note on the rôle of maximal consistency. *Erkenntnis*, 79(5):1015–1026, 2013.
[75] Henryk Mehlberg. *The Reach of Science*. University of Toronto Press, Toronto, 1958. Extract from §29, pp. 256–259 reprinted under the title 'Truth and Vagueness' in [61] pp. 85–88.
[76] Hung T. Nguyen and Elbert A. Walker. *A First Course in Fuzzy Logic*. Chapman and Hall/CRC, Boca Raton, FL, second edition, 2000.
[77] Vilém Novák and Antonín Dvořák. Fuzzy logic: A powerful tool for modeling of vagueness. Technical Report 158, Institute for Research and Applications of Fuzzy Modeling, University of Ostrava, Ostrava, 2011.
[78] Vilém Novák, Irina Perfilieva, and Jiří Močkoř. *Mathematical Principles of Fuzzy Logic*. Kluwer, Dordrecht, 2000.
[79] Daniel N. Osherson and Edward E. Smith. On the adequacy of prototype theory as a theory of concepts. *Cognition*, 9:35–58, 1981.
[80] Daniel N. Osherson and Edward E. Smith. On typicality and vagueness. *Cognition*, 64:189–206, 1997.
[81] Francesco Paoli. Truth degrees, closeness, and the sorites. To appear in Otavio Bueno and Ali Abasnezhad, editors, *On the Sorites Paradox*. Springer.
[82] Francesco Paoli. A really fuzzy approach to the sorites paradox. *Synthese*, 134:363–387, 2003.
[83] Charles Sanders Peirce. Vague. In James Mark Baldwin, editor, *Dictionary of Philosophy and Psychology*, page 748. Macmillan, New York, 1902.
[84] Manfred Pinkal. *Logic and Lexicon: The Semantics of the Indefinite*, volume 56 of *Studies in Linguistics and Philosophy*. Kluwer, Dordrecht, 1995. Translated by Geoffrey Simmons.
[85] Graham Priest. Fuzzy identity and local validity. *Monist*, 81:331–342, 1998.
[86] Marian Przełęcki. Fuzziness as multiplicity. *Erkenntnis*, 10:371–380, 1976.
[87] Hilary Putnam. The meaning of 'meaning'. In *Mind, Language and Reality*, volume 2 of *Philosophical Papers*, pages 215–271. Cambridge University Press, Cambridge, 1975.
[88] Willard Van Orman Quine. *Word and Object*. MIT Press, Cambridge, MA, 1960.
[89] Diana Raffman. Vagueness without paradox. *Philosophical Review*, 103:41–74, 1994.
[90] Diana Raffman. Vagueness and context relativity. *Philosophical Studies*, 81:175–192, 1996.
[91] Diana Raffman. How to understand contextualism about vagueness: Reply to Stanley. *Analysis*, 65: 244–248, 2005.
[92] David Ripley. Contradictions at the borders. In Rick Nouwen, Robert Van Rooij, Uli Sauerland, and Hans-Christian Schmitz, editors, *Vagueness in Communication*, pages 169–188. Springer, 2011.
[93] David Ripley. Sorting out the sorites. In Koji Tanaka, Francesco Berto, Edwin Mares, and Francesco Paoli, editors, *Paraconsistency: Logic and Applications*, pages 329–348. Springer, 2013.
[94] Bertil Rolf. Sorites. *Synthese*, 58:219–250, 1984.
[95] Gideon Rosen and Nicholas J.J. Smith. Worldly indeterminacy: A rough guide. *Australasian Journal of Philosophy*, 82(1):185–198, 2004. Reprinted in [49] pp. 196–209.
[96] David H. Sanford. Borderline logic. *American Philosophical Quarterly*, 12:29–39, 1975.
[97] Uli Sauerland. Vagueness in language: The case against fuzzy logic revisited. In Petr Cintula, Christian G. Fermüller, Lluís Godo, and Petr Hájek, editors, *Understanding Vagueness: Logical, Philosophical and Linguistic Perspectives*, volume 36 of *Studies in Logic*, pages 185–198. College Publications, London, 2011.
[98] Stephen P. Schwartz. Intuitionism versus degrees of truth. *Analysis*, 50:43–47, 1990.
[99] Stewart Shapiro. *Vagueness in Context*. Oxford University Press, Oxford, 2006.
[100] Edward E. Smith and Daniel N. Osherson. Compositionality and typicality. In Stephen Schiffer and Susan Steele, editors, *Cognition and Representation*, pages 37–52. Westview Press, Boulder, CO, 1988.
[101] Nicholas J.J. Smith. Consonance and dissonance in solutions to the sorites. To appear in Otavio Bueno and Ali Abasnezhad, editors, *On the Sorites Paradox*. Springer.
[102] Nicholas J.J. Smith. Undead argument: The truth-functionality objection to fuzzy theories of vagueness. To appear in *Synthese*, doi: 10.1007/s11229-014-0651-7.
[103] Nicholas J.J. Smith. Vagueness and blurry sets. *Journal of Philosophical Logic*, 33:165–235, 2004.

[104] Nicholas J.J. Smith. A plea for things that are not quite all there: Or, Is there a problem about vague composition and vague existence? *Journal of Philosophy*, 102:381–421, 2005.
[105] Nicholas J.J. Smith. Vagueness as closeness. *Australasian Journal of Philosophy*, 83:157–183, 2005.
[106] Nicholas J.J. Smith. *Vagueness and Degrees of Truth*. Oxford University Press, Oxford, 2008. Paperback 2013.
[107] Nicholas J.J. Smith. Fuzzy logic and higher-order vagueness. In Petr Cintula, Christian G. Fermüller, Lluís Godo, and Petr Hájek, editors, *Understanding Vagueness: Logical, Philosophical, and Linguistic Perspectives*, volume 36 of *Studies in Logic*, pages 1–19. College Publications, London, 2011.
[108] Nicholas J.J. Smith. Many-valued logics. In Gillian Russell and Delia Graff Fara, editors, *The Routledge Companion to Philosophy of Language*, pages 636–651. Routledge, London, 2012.
[109] Nicholas J.J. Smith. Measuring and modeling truth. *American Philosophical Quarterly*, 49:345–356, 2012.
[110] Nicholas J.J. Smith. Truthier than thou: Truth, supertruth and probability of truth. To appear in *Noûs*, doi: 10.1111/nous.12108.
[111] Scott Soames. *Understanding Truth*. Oxford University Press, New York, 1999.
[112] Roy Sorensen. *Blindspots*. Clarendon Press, Oxford, 1988.
[113] Roy Sorensen. *Vagueness and Contradiction*. Oxford University Press, Oxford, 2001.
[114] Gaisi Takeuti and Satoko Titani. Globalization of intuitionistic set theory. *Annals of Pure and Applied Logic*, 33:195–211, 1987.
[115] Jamie Tappenden. The liar and sorites paradoxes: Toward a unified treatment. *Journal of Philosophy*, 90:551–577, 1993.
[116] Michael Tye. Vague objects. *Mind*, 99:535–557, 1990.
[117] Michael Tye. Sorites paradoxes and the semantics of vagueness. *Philosophical Perspectives*, 8: 189–206, 1994. Reprinted in [61] pp. 281–293.
[118] Michael Tye. Vagueness: Welcome to the quicksand. *Southern Journal of Philosophy*, 33(Suppl.):1–22, 1995.
[119] Alasdair Urquhart. Many-valued logic. In Dov M. Gabbay and Franz Guenther, editors, *Handbook of Philosophical Logic*, volume III, pages 71–116. D. Reidel, Dordrecht, 1986.
[120] Robert Van Rooij. Vagueness and linguistics. In Giuseppina Ronzitti, editor, *Vagueness: A Guide*, pages 123–170. Springer, 2011.
[121] Brian Weatherson. True, truer, truest. *Philosophical Studies*, 123:47–70, 2005.
[122] Timothy Williamson. Vagueness and ignorance. *Proceedings of the Aristotelian Society, Supplementary Volumes*, 66:145–162, 1992. Reprinted in [61] pp. 265–280.
[123] Timothy Williamson. *Vagueness*. Routledge, London, 1994.
[124] Crispin Wright. Further reflections on the sorites paradox. *Philosophical Topics*, 15:227–290, 1987.
[125] Lotfi A. Zadeh. Fuzzy sets. *Information and Control*, 8(3):338–353, 1965.
[126] Lotfi A. Zadeh. A note on prototype theory and fuzzy sets. *Cognition*, 12:291–297, 1982.

NICHOLAS J.J. SMITH
Department of Philosophy
Main Quadrangle A14
University of Sydney
NSW 2006 Australia
Email: njjsmith@sydney.edu.au

# INDEX

Lightning Source UK Ltd.
Milton Keynes UK
UKHW021518240419
341530UK00004B/330/P